Plant Biotechnology and
Molecular Markers

Plant Biotechnology and Molecular Markers

Edited by
P.S. Srivastava
Alka Narula
Centre for Biotechnology, Faculty of Science
Jamia Hamdard, New Delhi, India

Sheela Srivastava
Department of Genetics
University of Delhi South Campus
New Delhi, India

Springer-Science+Business Media, B.V.

A C.I.P. catalogue record for the book is available from the Library of Congress

ISBN 978-94-017-4044-9   ISBN 978-1-4020-3213-4 (eBook)
DOI 10.1007/978-1-4020-3213-4

Sant Saran Bhojwani was born to Mrs. Nam Adhari and Mr. Parmanand on 20th November, 1940 in the serene and tranquil environment of Dayalbagh, about 3 km from the hustle-bustle of the Agra city. He had his early education in Dayalbagh and graduated and postgraduated from Agra University. Soon after finishing M.Sc. (Botany), Dr Bhojwani served his alma-mater (R.E.I. Dayalbagh, Agra) as lecturer for one year before joining the University of Delhi as a doctoral student. His supervisor, late Professor B.M. Johri assigned him a challenging research problem, with the warning that his Ph.D. degree would depend on his demonstrating the cellular totipotency of endosperm, a completely unorganized, short-lived, triploid tissue. Earlier, many students of Professor Johri and scientists elsewhere in the world could establish tissue cultures of endosperm but failed to induce the organogenic differentiation. It was remarkable that within six months of his joining Delhi University, Dr Bhojwani achieved differentiation of normal shoot buds from the endosperm of *Exocarpus cupressiformis*; a parasitic flowering plant (*Nature*, 1965). At this stage a very renowned American plant physiologist, Prof. F.C. Steward, visited the University of Delhi who even after observing the cultures could not believe that the endosperm tissue could form shoots and remarked, "Young man, take a bet with me. All the shoots in the cultures are diploid. If that is the case remember me or else forget me". However, when the shoots of endosperm origin were cytologically analysed, all of them were found to be triploid, which is of considerable practical importance in plant genetics and improvement. Subsequently, Dr Bhojwani established the cellular totipotency of endosperm cells by reporting regeneration of triploid shoots and/or plants in *Scurrula pulverulenta, Acacia nilotica* (Garg *et al.*, 1996), *Morus alba*

(Thoma *et al.*, 2000) and *Azadirachta indica* (Chaturvedi *et al.*, 2003). In the meantime, many other scientists confirmed the observations of Bhojwani.

Dr. Bhojwani and his students worked on a range of basic and applied aspects of *in vitro* plant morphogenesis. During 1971-1972 he worked with Dr Norman Sunderland at the John Innes Institute, Norwich, U.K. under the British Council Fellowship Programme and reported quantitative changes in nucleic acid and protein contents of microspores during the induction of androgenesis in tobacco using histochemistry and cytophotometry (*J. Exp. Bot.* 1973). In 1972, he spent three months in the laboratory of Professor Edward C. Cocking, FRS, at the University of Nottingham, U.K. and reported for the first time isolation of microspore protoplasts using helicase enzyme. The report appeared in *Nature, New Biology* (1972).

Dr Bhojwani had another opportunity to work in the U.K. for a year during 1975-1976 under the Royal Society Commonwealth Bursary. This time he spent the whole year with Professor Cocking and worked on wheat tissue culture (*Z. Planzenphysiol.*, 1977) and protoplast isolation and culture in cotton (*Plant Sci. Lett.* 1977). At this point of time there was considerable interest in the application of biotechnological techniques to crop improvement. However, a major limitation in achieving this goal was the recalcitrance of legumes, cereals and other major crop plants for plant regeneration from cultured cells, an essential step in genetic engineering and somatic hybridization. This prompted Dr Bhojwani to critically review the literature on tissue culture of crop plants which was presented as an invited lecture in a meeting organised by the Agricultural Research Council, London and later published in *Euphytica* (1977). The review, discussing the progress and problems of tissue culture of major crop plants and emphasizing the need for extensive further research in the area, was a highly cited publication which paved the way for a fresh spurt of research to achieve high frequency regeneration in tissue cultures of these plants.

In 1978 Dr Bhojwani was awarded the prestigious Senior Fellowship of the National Research Advisory Council of New Zealand, and the family moved to Palmerston North to join the Plant Physiology Division of the D.S.I.R., New Zealand. Before the expiry of the term of the Fellowship, the D.S.I.R. offered Dr Bhojwani a position of Senior Scientist (Scientist 105) and the Government of New Zealand granted Permanent Residence to him and his family. In 1980 he was confirmed in the job.

The stay of Dr Bhojwani in New Zealand was very productive. He published numerous papers on the micropropagation of Willow (*N.Z.J. Bot.*, 1980), Garlic (*Sci. Hortic.*, 1980), Clover (*Physiol. Plant.* 1981), Japanese Pear (*Sci. Hortic*, 1984), and Feijoa (*Acta Hortic.* 1987). He also worked on *Trifolium* spp and reported, for the first time, regeneration of full plants from mesophyll protoplasts of white clover (*Plant Sci.* 1982, *Euphytica,* 1984). Virus-free garlic plants of a Japanese variety imported into New Zealand were produced by shoot tip culture to facilitate its release through quarantine (*Sci. Hortic* 1982/83). Impressed by the work and publications of Dr Bhojwani the D.S.I.R. decided to promote him to Scientist 106, an opportunity which was pre-empted by his decision to return to India in 1981. However, his post in the D.S.I.R. was not filled for at least two years expecting that Dr Bhojwani might decide to return to New Zealand. He did return to New Zealand in 1983 but only as a Visiting Scientist for three months to finish some experiments which remained incomplete in 1981 and process the data for publication.

After his sojourn in New Zealand, Dr Bhojwani made a modest beginning as a Research Associate at the University of Delhi and started guiding Ph.D. students in 1981. Fortunately, a major research project on "Micropropagation of Important Horticultural and Silvicultural species of India" was sanctioned to him by the UGC, under which he and his students developed an efficient protocol for clonal propagation of the leguminous tree species, *Leucaena leucocephala,* and *in vitro* nodulation of micropropagated plants by *Rhizobium* to enhance their field survival. He also demonstrated that sugar cubes, produced by Daurala Sugar Mills, was a fair substitute of 'Analar' Grade Sugar used in Plant tissue culture media. The sugar cubes were more than 10 times cheaper than the 'Analar' Grade sucrose. In 1985 the Department of Environment and Forests, Government of India, awarded another major research project to Dr Bhojwani to work on "In Vitro Conservation of Endangered Plants". It led to the development of protocols for micropropagation and cold storage of Himalayan Species of three medicinally important plants, viz., *Picrorhiza kurroa, Podophyllum hexandrum* and *Saussurea lappa.* In collaboration with the scientists at the Biochemical Engineering and Biotechnology Department of IIT Delhi, Dr Bhojwani studied the kinetics of cell growth in suspension cultures of *Podophyllum hexandrum* and *in vitro* production of Podophyllotoxin, an anticancerous drug (*Biotechnol. Lett.* 2001, *J. Biosci. Bioengg.*, 2002).

Dr Bhojwani guided six Ph.D.'s on plant regeneration alone from somatic and gametic cells of *Brassica* spp and published several papers (*Plant Cell Tissue Organ Cult.* 1985, 1991; *Biol. Plant.*, 1989; *Plant Sci.* 1990a,b; *Euphytica* 1993). A detailed investigation on direct shoot regeneration from excised cotyledons of *B. juncea* proved a viable system for genetic transformation of this important oleiferous crop of India. His group also achieved high frequency androgenesis and selection of agronomically useful androclones in *B. juncea.* This work was supported by funds from MOMBUSHO, Japan and European Commission, Brussels.

Dr Bhojwani undertook two major projects on mulberry biotechnology and investigated micropropagation of some elite clones and production of gynogenic haploids (*Euphytica*, 1999) and endosperm derived triploids (*Pant Cell Rep.* 2000) of this invaluable tree for silk industry, the sole source of feed for silkworms. Recently, he has reported the production of gynogenic haploids (*Plant Cell Rep.* 2003) and triploids (*J. Plant Physiol.*, 2003) of Neem.

Dr Bhojwani has published 75 research papers in journals of international repute, 10 critical reviews and 19 invited chapters in books published from India and abroad. In addition, he has authored and edited several books. His first book "The Embryology of Angiosperms" (Vikas Publishers, New Delhi) has been a popular text book for graduate and post-graduate students in India and many other countries. Running into its 5th edition, the book has been translated into Japanese (1995) and Korean (2001). In 1983, Dr Bhojwani brought out another book titled "Plant Tissue Culture : Theory and Practice", published by Elsevier, The Netherlands. This has been regarded as the first standard text book on the subject and became so popular worldwide that the publishers brought out its paperback edition in 1986. It was translated into Korean in 1986. Under a project funded by the Department of Biotechnology, Dr Bhojwani completed a mammoth task of compiling 'A Classified Bibliography of Plant Tissue Culture', covering the entire literature on the subject up to 1984. He spent two weeks in the U.K. under the INSA-Royal Society Exchange Programme to complete the volume. It soon became a popular reference book. A supplement to this volume, covering the literature of the next five years, was brought out in 1989. Both the volumes were published by Elsevier, The Netherlands. Dr Bhojwani has edited four volumes, viz., "Plant Tissue Culture : Applications and Limitations" (1990; Elsevier), "Morphogenesis in Plant Tissue Cultures" (1999; Kluwer Academic Publishers, The Netherlands),

"Current Trends in the Embryology of Angiosperms" (2002; Kluwer Academic Publishers) and "Agrobiotechnology and Plant Tissue Culture" (2003; Science Publishers, U.S.A.).

Dr Bhojwani has been in great demand by the organisers of conferences, seminars, workshops, training courses and refresher courses because of his contributions and in-depth knowledge in the subject of Plant Tissue Culture. He is a voracious speaker and scientists and students look forward to his informative and thought-provoking lectures. In one of the meetings of the Indian Association of Plant Tissue Culture held at NBRI, Lucknow, in 1976-77, the late Professor P.N. Mehra, Padamshri, who chaired the lecture of Dr Bhojwani, was so impressed by his lecture that he asked the audience to give standing ovation to the young scientist. Dr Bhojwani has participated in several National and International Conferences in India and overseas. He was invited to deliver a lecture at the conference on "Problems Related to Mass Propagation of Horticultural Species", Belgium (1985). The organisers of the Conference on Tissue Culture of Tropical Plants in Bagota, Colombia invited him to deliver a Plenary Lecture and Chair a session. Dr Bhojwani was a member of the International Advisory Committee of the VIII Conference of the International Association of Plant Tissue Culture held in Florence, Italy, where he organised a workshop. He was also a member of the International Advisory Committees of the 1st, 2nd and 3rd Asia-Pacific Conferences in Taejon, South Korea (1993), Shanghai, China (1997) and Singapore (2000). He delivered plenary lectures in the Conferences held in South Korea and Singapore. In 1987 an International Symposium on Gene Manipulation for Plant Improvement was organised in Kuala Lumpur, Malaysia and Dr Bhojwani was invited to deliver a plenary lecture. Dr Bhojwani also attended the VI Conference of the International Association of Plant Tissue Culture held in Minnesota, U.S.A. (1986) and the International Botanical Congress in Yokohama, Japan (1993). He also delivered an invited talk in the latter. He was the only Indian invited as a Resource Person to a workshop on "Production and Utilization of Double-Haploid Lines in Rice Breeding" organised by the International Agency for Atomic Energy in Suwon, South Korea, in 1999. Dr Bhojwani delivered plenary lectures and chaired sessions in International Conferences in Dhaka, Bangladesh. Recently, he was invited to participate in the 15th Biennial Conference of the New Zealand Chapter of International Association of Plant Tissue Culture and Biotechnology" at Leigh, New Zealand and presented a paper on "Pollen Embryogenesis in *Brassica* ssp.

Dr Bhojwani has been a recipient of many honours. He was elected Full Member of the New Zealand Institute of Agricultural Sciences (1981). In 1990 he became Invited Member of the Technology Transfer Association of Japan. He was awarded the Nawashina Memorial Medal. Dr Bhojwani was elected as a Fellow of the National Academy, Allahabad in 1994 and has been awarded many National and International Fellowships to visit laboratories in other countries. Besides the British Council Fellowship and Royal Society Bursary to work in U.K., Dr Bhojwani was awarded the Senior Fellowship of the NRAC, New Zealand; Fellowship of the Japanese Society for the Promotion of Science; Biotechnology Overseas Associateship, Government of India; CIDA/NSERC Research Associateship, Canada; INSA-KOSEF Fellowship of South Korea, and Fellowship of the Kernforschungsanlage, Germany.

Dr Bhojwani has been on the Editorial Board of many journals. To mention a few, *Scientia Horticulture,* Holland; *Journal of Biochemistry and Biotechnology*, New Delhi; *Phytomorphology*, Delhi, *Plant Tissue Culture*, Dhaka and *Chromosome*, Calcutta.

Dr Bhojwani has been a member of the Academic Council's of the TERI School of Advanced Study and C.C. Singh University, Meerut. He is the Chairman of the Research Advisory Committee of the Central Tassar Research & Training Institute, Ranchi. He was a Visiting Senior Fellow of the Tata Energy Research Institute, New Delhi and made a major contribution to the designing and production and planning of the DBT-Sponsored Plant Tissue Culture Pilot Plant. He was also a consultant to the Commercial Plant Tissue Culture Laboratories such as A.V. Thomas, Cochin and Aranaya Micropropagation, New Delhi.

After serving the University of Delhi for 35 years, Professor Bhojwani took voluntary retirement to serve the Dayalbagh Educational Institute (Deemed University), Agra as its Honorary Director. Married to Shaku, Dr Bhojwani discharged his family obligations well and timely with both the children married and settled happily with their families. His daughter, Anjli Sarup, married to Mr Gursewak Maneesh, is living in Allahabad whereas his son Nova, with his dentist wife, Kokila has recently moved to the U.S.A. as a Software Engineer. His wife Shaku in the true Indian tradition extended her full support to the husband and deserves appreciation for her forbearance and active interest throughout his career, especially during the long periods when Professor Bhojwani was away completing academic assignments.

EDITORS

Preface

The genesis of the volume, Plant Biotechnology and Molecular Markers, has been the occasion of the retirement of Professor Sant Saran Bhojwani from the Department of Botany, University of Delhi. For Professor Bhojwani, retirement only means relinquishing the chair as being a researcher and a teacher which has always been a way of life to him. Professor Bhojwani has been an ardent practitioner of modern plant biology and areas like Plant Biotechnology and Molecular Breeding have been close to his heart. The book contains original as well as review articles contributed by his admirers and associates who are experts in their area of research.

While planning this contributory book our endeavour has been to incorporate articles that cover the entire gamut of Plant Biotechnology, and also applications of Molecular Markers. Besides articles on *in vitro* fertilization and micropropagation, there are articles on forest tree improvement through genetic engineering. Considering the importance of conservation of our precious natural wealth, one article deals with cryopreservation of plant material. Chapter on molecular marker considers DNA indexing as markers of clonal fidelity of *in vitro* regenerated plants and prevention against bio-piracy. A couple of write-ups also cover stage-specific gene markers, DNA polymorphism and genetic engineering, including raising of stress tolerant plants to sustain productivity and help in reclamation of degraded land.

The readiness with which the colleagues acceded to our request and the quality of articles reflect the esteem in which they hold Professor Bhojwani. It is hoped that in honouring Professor Bhojwani, this volume will further the frontiers of knowledge in Biotechnology as a whole and Plant Biotechnology in particular.

While finalizing this volume we have received unsolicited support from all our colleagues and friends which we acknowledge gratefully. Mr M.S. Sejwal and Mr Manish Sejwal, Anamaya Publishers, New Delhi, have been forthcoming with suggestions and have shown utmost patience while preparing the proof and final publication. They deserve the appreciation of all contributors as well as of the editors.

Editors

Contents

Professor Sant Saran Bhojwani   *v*
Preface   *xi*

1. *In Vitro* Androgenesis: Events Preceding Its Cytological Manifestation   1
   Shashi B. Babbar, Nishi Kumari and Jitendera K. Mishra

2. Doubled Haploids: A Powerful Biotechnological Tool for Genetic
   Enhancement in Oilseed Brassicas   18
   Deepak Prem, Kadambari Gupta and Abha Agnihotri

3. Double Fertilisation *in vitro* and Transgene Technology   31
   Erhard Kranz, Yoichiro Hoshino, Takashi Okamoto and Stefan Scholten

4. Polymorphism of Sexual and Somatic Embryos as Manifestation of Their
   Developmental Parallelism Under Natural Conditions and in Tissue Culture   43
   Tatyana B. Batygina

5. Molecular Biology and Genetic Engineering of Polyamines in Plants   60
   M.V. Rajam, R. Kumria and S. Singh

6. Biotechnological Approaches Towards Improvement of Medicinal Plants   78
   Alka Narula, Sanjeev Kumar, K.C. Bansal and P.S. Srivastava

7. Production of Phytochemicals in Plant Cell Bioreactors   117
   Saurabh Chattopadhyay, A.K. Srivastava and V.S. Bisaria

8. Development of Biotechnology for *Commiphora wightii*; A Potent Source
   of Natural Hypolipidemic and Hypocholesterolemic Drug   129
   Sandeep Kumar, S.S. Suri, K.C. Sonie and K.G. Ramawat

9. Biotechnology in Quality Improvement of Oilseed Brassicas   144
   Abha Agnihotri, Deepak Prem and Kadambari Gupta

10. Role of Biotechnology for Incorporating White Rust Resistance in
    Brassica Species   156
    Kadambari Gupta, Deepak Prem and Abha Agnihotri

11. Current Trends in Forest Tree Biotechnology   169
    E.M. Muralidharan and Jose Kallarackal

12. Cloning Forestry Species   183
    Vibha Dhawan and Sanjay Saxena

13. Micropropagation of Woody Plants   195
    J.S. Rathore, Vinod Rathore, N.S. Shekhawat, R.P. Singh,
    G. Liler, Mahendra Phulwaria and H.R. Dagla

14. Biotechnology in Mulberry (*Morus* spp.) Crop Improvement: Research
    Directions and Priorities   206
    S.B. Dandin and V. Girish Naik

15. Development of High Efficiency Micropropagation Protocol of an
    Adult Tree—*Wrightia tomentosa*   217
    S.D. Purohit, P. Joshi, K. Tak and R. Nagori

16. *In Vitro* Regeneration and Improvement in Tropical Fruit Trees: An
    Assessment   228
    Madhulika Singh, Uma Jaiswal and V.S. Jaiswal

17. Tissue Culture of Cashewnut   244
    Sumita Jha and Sudripta Das

18. Changing Scenarios in Indian Horticulture   261
    Sanjay Saxena and Vibha Dhawan

19. Cryopreservation: A Potential Tool for Long-term Conservation of
    Medicinal Plants   278
    Sonali Dixit, Sangeeta Ahuja, Alka Narula and P.S. Srivastava

20. Molecular Mapping and Marker Assisted Selection of Traits for Crop
    Improvement   289
    Anushri Varshney, T. Mohapatra and R.P. Sharma

21. Studies on Male Meiosis in Cultivated and Wild *Vigna* Species   331
    S. Rama Rao and S.N. Raina

22. Transgenic Crops for Abiotic Stress Tolerance   346
    Deepti Tayal, P.S. Srivastava and K.C. Bansal

23. Cell Differentiation in Shoot Meristem: A Molecular Perspective   366
    Jitendra P. Khurana, Lokeshpati Tripathi, Dibyendu Kumar, Jitendra
    K. Thakur and Meghna R. Malik

INDEX   387

Plant Biotechnology and Molecular Markers

Plant Biotechnology and Molecular Markers
P.S. Srivastava, Alka Narula and Sheela Srivastava (Editors)
Copyright © 2004 Anamaya Publishers, New Delhi, India

1. *In Vitro* Androgenesis: Events Preceding Its Cytological Manifestation

Shashi B. Babbar, Nishi Kumari and Jitendera K. Mishra

Department of Botany, University of Delhi, Delhi 110 007, India

Abstract: *In vitro* sporophytic development from the microspores of angiosperms through a process referred to as androgenesis is an important field of research because of its fundamental as well as applied importance. Besides providing possibility of developing haploid plants in large numbers having application in plant breeding, the phenomenon offers an experimental system for studying the events associated with transition from gametophytic to sporophytic phase. Of particular interest are the changes occurring in the pollen prior to the onset of androgenic divisions, as these are likely to throw light on the very process of androgenic induction. The present article reviews the ultrastructural, biochemical and molecular changes that take place prior to cytological manifestation of androgenesis.

1. Introduction

In angiosperms, the gametophytic phase is short lived and is completely dependent on the sporophyte. Male gametophytes in these plants are referred to as pollen grains or microspores. Pollen grains being haploid possess each gene in a single copy. Though destined to function as male gametophytes, under suitable conditions, pollen grains are capable of developing into sporophytes through a process called androgenesis. A new field of *in vitro* androgenesis was initiated due to the landmark discovery of Guha and Maheshwari [1, 2], who reported development of embryos from microspores in cultured anthers of *Datura innoxia*. Since this path-breaking discovery, investigations in this field have aimed to: (i) extend the technique to more and more taxa, (ii) identify the intrinsic and extrinsic requirements for successful development of plants from microspores and (iii) understand the mechanism of induction. It is the last aspect, which has remained enigmatic. The most baffling aspect of *in vitro* androgenesis is the transition from gametophytic to sporophytic development.

The exact mechanism underlying this event, conditioning the male gametophyte to embark upon an altogether different mode of development, is yet unknown. Studies using cytochemical, biochemical, electron microscopic and molecular techniques have indicated that some structural and biochemical changes do take place in microspores when they switch over to sporophytic pathway. However, the critical turning point, at which the entity pre-ordained to become a gametophyte switches over to an altogether different development pathway is yet not identified. Therefore, the changes that take place at ultrastructural, biochemical and molecular levels, preceding initiation of divisions leading to sporophytic organization of microspore, may either be the cause or result of onset of androgenesis, depending on whether they are taking place before or after this unidentified transition point [3]. Nevertheless, a brief review of such events which precede the expression of androgenesis is necessary, to reflect upon the induction mechanism initiating sporophytic development of a microspore.

2.   Trigger for Androgenic Induction

As early as 1975, Vasil and Nitsch [4] stated that angiosperm pollen is a versatile entity and its normal course of development is precisely controlled by certain factors present within the anther. Further, they opined that severing the contact with the plant and culturing of anthers on a nutrient medium may be causing an imbalance in this precisely controlled influence, resulting in the sporophytic development in pollen. However, the information gathered subsequently indicates that developments preceding sporophytic divisions in microspore under *in vitro* conditions can take place even in the absence of culture medium [5]. Thus, it seems that it is not the prerogative of culture medium to suppress gametophytic development of pollen as the intrinsic control can also be broken down by other factors.

2.1   Stress as a Major Trigger

It is known for a long time that stress treatment or sub-optimal conditions can alter development programs. Stress is an important component of androgenic induction. The role of stress in androgenic induction is explained by the following two hypotheses: (i) stress may cause developmental defects by increasing the expression or stabilization of a critical target gene and (ii) repression of critical genes below a threshold level [6].

In case of *Brassica napus*, the type of stress treatment given to cultured microspores for the induction of androgenesis can vary from heat to gamma irradiation and colchicine [7–10]. Heat treatment is the most commonly used pretreatment to initiate androgenesis in *B. napus* [11]. The temperature must be around 32°C for 8 h in order to induce androgenesis sufficiently. This is time and dose dependent and any interruption in between, leads to the failure of androgenesis [12, 13]. This temperature of 32°C is near the temperature above which most microspores and pollen grains of *B. napus* die [14]. It appears that stress alone, rather than in combination with tissue culture conditions, is needed to initiate androgenesis. Thus, 32°C temperature treatment itself can initiate the redirection process *in situ* before initiating *in vitro* culture [13]. Gamma irradiation especially in combination with the temperature treatment has a stimulatory effect on the induction of androgenesis [8].

In tobacco, androgenic process is usually induced in bi-cellular pollen grains, which are initially cultured under glutamine and sugar starvation conditions before being transferred to a high glucose medium. This is a very efficient method for androgenic induction in a large percentage of pollen grains [15–18].

The elicitation of a general stress response in young microspores is associated with the appearance of small heat-shock protein (smHSP) transcripts that precedes induction of androgenesis [14, 19]. Generally, larger the temperature difference between the donor plant growth conditions and *in vitro* culture conditions, stronger is smHSP signal. Similar results have been obtained when colchicine and gamma irradiation are used as stress stimuli [8, 10]. No smHSP was, however, produced below 25°C, a temperature too low to elicit a stress response. Thus, appearance of smHSPs may be used as a molecular marker to indicate whether or not the pollen grains have responded to the stress elicitors and therefore, are capable of initiating androgenesis [13, 14].

The changes in the pollen differentiation process leading to androgenesis can be initiated only in microspores of specific developmental stages. Therefore, stress response has to be considered together with the stage of microspore development [13].

3. Impact of Microspore Developmental Stage

For most of the species, a suitable stage for the induction of androgenesis lies between just before or just after first pollen mitosis. During this phase of development, the microspores are non-committal in their developmental potential, as most of the sporophyte-specific gene products are eliminated from the cytoplasm before meiosis [20] and the gametophyte-specific genes are generally transcribed only after first pollen mitosis [21]. After the first mitosis, the cytoplasm gets populated with gametophytic information and it gradually becomes irreversibly programmed to form the male gametophyte [22]. A variety of external stimuli are applied during the microspore development/culture in order to mask the gametophytic program and induce the expression of sporophyte-specific genes, thereby making them to switch over to sporophytic mode of development.

Based on two model systems, i.e. *Brassica napus* and *Nicotiana tabacum* and also on the findings on wheat (*Triticum aestivum*), Touraev et al. [23] suggested that microspores are competent to change this developmental program within a relatively wide developmental window. In cereals, the uninucleate stage up to first pollen mitosis, with species-specific variations, has frequently been recommended [24]. A stage around first pollen mitosis has been considered optimal for the induction of androgenesis in *Secale cereale* [25, 26].

It is found that the developmental stage of microspore affects the induction in a major way. In callus/embryoid induction, a positive curvilinear trend was observed in each case and the highest induction was obtained when *B. napus* microspores had undergone mitosis [10].

4. Pathways of Androgenesis

Based on the studies in different plants, the five routes of androgenesis that have been identified are: (i) by repeated divisions of the vegetative cell, (ii) by repeated divisions of the generative cell, (iii) by repeated divisions of both, (iv) through symmetrical divisions in uninucleate microspore giving rise to two identical cells rather than unequal generative and vegetative cells (B-pathway) and (v) origin from fusion product of generative and vegetative. A species can exhibit predominance of one or the other pathway [27].

Thus, division of otherwise quiescent vegetative cell, more than one division in generative cell or formation of two-celled unit with identical cells from a uninucleate microspore can be taken as the first sign of deviation from gametophytic development.

Of the abovementioned pathways, the fourth pathway is the most widely studied and is considered to be the major pathway of androgenesis. This pathway involves symmetrical division of microspore. Stress serves as a major signal for this symmetrical division. Since colchicine treatment has been shown to increase the number of symmetrically dividing embryogenic pollen grains in cultures of *Brassica napus* [9, 10, 28], it has been suggested that symmetry during cytokinesis is an important factor in deflecting the gametophytic program of the pollen grain towards the androgenic one.

The significance of microspore division symmetry for vegetative cell-specific transcription and generative cell differentiation has been addressed in microspores of transgenic tobacco plants transformed with promotor of vegetative cell-specific tomato lat52 gene fused to reporter gus gene [29]. *In vitro* maturation, in the presence of high concentrations of colchicine, blocks the first pollen mitosis effectively, resulting in the formation of uninucleate pollen grains expressing both the abovementioned genes which are capable of germination and a pollen tube growth, despite the absence of a generative cell. Lower amounts of colchicine induced symmetric division producing two similar daughter cells, both expressing the *gus* gene. These results demonstrate that division asymmetry, at the first pollen mitosis, is essential for the correct generative cell differentiation. Moreover, the activation of vegetative cell-specific transcription and functional maturation may be uncoupled from cytokinesis [29].

Touraev et al. [17] on the other hand, have shown that cultivation of pollen grains containing even two equal sized cells under the maturation conditions, lead to the development of mature pollen grains. This indicates that rather than the symmetry of first pollen mitosis irreversible commitment to embryogenesis is essential. In an interesting study, Zonia and Tupy [30] have shown that lithium disrupts the partitioning of membrane-associated calcium, blocks polar nuclear migration and subsequently, induces a symmetrical mitosis in microspores of tobacco.

Eady et al. [29] proposed two models to explain the significance of first pollen mitosis for pollen determination and differentiation. According to the first model of passive repression, low levels of gametophytic expressed factors (which are the result of asymmetric division), are present in the generative cell. On the contrary, the symmetrically dividing cells, or in the case of colchicine-blocked uninucleate microspores, no such repression of the vegetative cell-specific genes occurs. According to the second model, there exists an active repressor, blocking transcription in the generative cell, which upon asymmetric division, is again selectively retained in the generative cell.

5.   Sub-Cellular Changes Associated with Androgenic Induction

As the microspore switches from its normal gametophytic to sporophytic pathway, numerous structural, biochemical and molecular changes take place at the cell level.

5.1   Structural Changes

The first ever report, mentioning any cytological change, preceding the sporophytic cell division, was that of Sunderland and Wicks [31]. They observed that in tobacco anthers, cultured either at uninucleate microspore stage or during first haploid mitosis, the grains undergo either normal mitosis resulting in typical vegetative and generative cells or rarely modified first mitosis resulting into two identical cells. Subsequently, after a lag phase of some days, two types of grains can be distinguished. These types are characterized by their differential staining; one stains lighter than the other. The former develops into an embryoid, whereas in the latter, starch is deposited and some of these may even germinate [31, 32]. These observations imply that before cell divisions, a dedifferentiation process takes place in such grains, which are destined to develop into embryoids. This dedifferentiation process mainly involves degradation of gametophytic information in embryogenic grains. This is substantiated to some extent by electron microscopic investigations conducted on cultured anthers of *Nicotiana tabacum* [33]. During the first two days of culture, microspores underwent normal gametophytic differentiation at the same rate as under *in vivo* conditions. In anthers cultured for 8-12 days, two types of grains could be distinguished. In the first type, presumed to be embryogenic, vegetative cell was occupied by multi-vesiculate structures resembling lysosomes. In these grains, cytoplasm was scarce in organelles and by twelfth day these grains were virtually devoid of organelles, except plastids. In contrast, vegetative cell of other grains was with full complements of cytoplasmic organelles, however, lysigenous cavities, encountered in embryogenic type, were conspicuously absent [34]. With the first division of vegetative cell, the embryogenic microspores were re-populated with various cell organelles, thus, once again becoming rich in cytoplasm [35]. Based on these observations, authors suggested that prior to the first division of vegetative cell, leading to sporophytic development, there is a controlled degradation of organelles in embryogenic microspores. In contrast, similar ultrastructural studies conducted on *Datura innoxia* showed that there was no degradation and re-synthesis of cytoplasm in its vegetative cell prior to androgenic induction [36]. The incongruity in observations was ascribed to the difference in post-mitotic development in these taxa. The post first haploid mitotic cytoplasm synthesis in *N. tabacum* is fast and microspores pass on to the stage 6 [37] from 5 quickly. Whereas, stage 5 in *D. innoxia* is extended. Dunwell and Sunderland [36] believed that in *N. tabacum* degradation of gametophytic cytoplasm was necessitated because the signal for sporophytic development is perceived only after its synthesis is over. On the other hand, slow development of *D. innoxia* microspores results in triggering of androgenesis before gametophytic information is at all or completely synthesized.

The microspores of maize after the culture possess cytoplasm scarce in organelles and low density of ribosomes. The first structural changes occurring at the sub-cellular level, include occurrence of nuclear chromatin at de-condensed stage, nucleolus comprising exclusively of the fibrillar component, cytoplasm with scarce organelles, a low ribosomal density and 2-3 fold increase in the number of nuclear pores, before the onset of the first pollen mitosis [38].

In *Brassica napus*, the first division of cultured microspores destined to become embryogenic is generally symmetrical. The first stage of differentiation in culture is the dispersion of the central vacuole. The centre of the cell then becomes occupied with a highly pleomorphic single nucleus. The large central nucleolus is gradually lost, being replaced by 3-6 smaller nucleoli. The plastids, which were previously dispersed throughout the cytoplasm subsequently become aggregated around the nucleus and lay down large quantities of starch granules. Another cytoplasmic feature is the appearance of large numbers of aggregated globules, which are not bound by membrane but may at times have small aggregates of ribosomes on their surface [39].

The central nucleus of the microspore undergoes a symmetrical mitotic division to form two cells within the exine. A normal middle lamella, followed by fibrillar wall, is then laid down in each of these cells. The cytoplasm of each of these cells contains two principal domains. The vacuolar domain contains dispersed or aggregated material while the second domain contains evenly staining globules. Both the vacuoles and the globules are distributed evenly in the cytoplasm. Over the course of this first division, the ER cisternae increase in number and the plastids continue to accumulate starch and become more irregular in their outline [39].

In *Hyoscyamus niger*, which exhibits sporophytic development predominantly through the generative cell, potentially embryogenic uninucleate microspore could be identified within 6 h of culture. Such microspores had an increased ratio of volume densities of the nucleolar granular zone to the fibrillar zone and dispersed to condensed chromatin [40]. The de-condensation of chromatin is associated with the early and rapid synthesis of DNA during development. The dispersed distribution of chromatin in potentially embryogenic grains may thus reflect increased DNA synthesis early in culture. These investigations have shown that continued DNA synthesis in generative cell in cultured microspores of *H. niger*, followed by mitosis and cytokinesis, result in the development of pollen embryos [41]. After first pollen mitosis, the generative cell maintains its large granular nucleolus. The volume fraction of the cytoplasm occupied by mitochondria and plastids and the area fraction occupied by RER and golgi cisternae, differ in the generative cells of potentially embryogenic and non-embryogenic pollen [42].

5.1.1   Cause of Symmetrical Division

Symmetrical division, rarely found in normal gametophytic development, is essential for the sporophytic embryogenesis in some taxa [28, 43, 44]. Simmonds [45] reported that during heat treatment a pre-prophase band (PPB) of microtubules develops in *Brassica* microspores, which was thought to determine the division symmetry.

During gametophytic development of microspore, the first pollen mitosis lacks a PPB [46, 47] and results in a non-consolidated cell plate. Such a cell plate is considered to be an important feature in this division because the generative cell is destined to be mobile within the vegetative cell. Normally microtubules are involved in anchoring the nucleus at the cell edge. Electron microscopic studies have revealed the presence of microtubules connecting the nuclear and plasma membranes [48]. Heat or colchicine treatments induce de-polymerization of microtubules [10, 45, 49], which results in displacement of the nucleus from its peripheral position, indicating thereby microtubular disruption in the nuclear-cortical zone [49].

Pre-prophase band (PPB) of microtubules is a good indicator of the embryogenic potential in microspores of *Brassica*. PPB is cortical in location and constitute the attachment site of the future cell plate. Moreover, wall maturation does not occur if the cell plate is attached at a site not previously occupied by PPBs [49]. The appearance of the PPB in heat-treated microspores of *Brassica napus* predicts a cytokinesis leading to a stable cell wall, a critical event in the initiation of multicellular organism comprising stationary cells separated by stable cell wall. Due to cell wall consolidation, deviation from the normal pollen mitosis occurs and symmetric division takes place [28, 39, 44, 50, 51, 52]. These observations suggest that symmetric division blocks the normal microspore development, which in turn, results in a default developmental pathway leading to androgenic induction [28, 39, 44, 45, 52]. Thus, at least in *B. napus*, the PPB is both a marker of embryogenic development and its integrity is critical to the development of first consolidated wall, which marks the beginning of a multicellular structure leading to embryogenesis.

During normal microspore ontogeny, there is a polar distribution of cytoplasmic organelles away from the microspore nucleus, prior to the first haploid mitosis, resulting in a generative cell that is deficient in organelles following division [46]. Ultrastructural examination of the early stages of androgenesis have revealed that potentially embryogenic microspores could be identified by the loss of polar distribution of organelles, resulting in generative cells possessing a full complement of organelles required for the continuous growth and division of embryos [42]. Since components of the cytoskeleton are also known to mediate the movement and position of cytoplasmic organelles, these results could be interpreted as an indication that the early processes of androgenic induction involve alteration in the structure and function of the cytoskeleton even in those microspores in which the first division is asymmetric [28].

5.2   Biochemical Changes

Biochemical changes mainly include changes in protein synthesis [19, 53-56], phosphorylation [57–59], and changes in secondary metabolite metabolism [38, 60, 61].

5.2.1 Change in Protein Synthesis

The induction of microspore embryogenesis must be accompanied by the activation of specific transcriptional factors which result in altered patterns of gene expression [55]. Change in the protein expression patterns during induction of microspore embryogenesis has been investigated by 2-D gel electrophoresis in *Brassica* and tobacco [19, 62].

Since culture of *Brassica* microspores at 32°C for 8 h leads to irreversible commitment to the sporophytic pathway [19], changes in protein synthesis during this period was examined by using in *situ* [^{35}S] methionine labeling followed by 2-D gel electrophoretic analysis [55]. The qualitative and quantitative analysis of 2-D [^{35}S] methionine protein patterns revealed that six polypeptides are specifically labeled under embryogenic culture conditions. Eighteen polypeptides incorporated [^{35}S] methionine at higher rate under embryogenic culture conditions (32°C) than in controls (18°C). These results indicated that only a limited number of proteins detectable in the 2-D gels of microspore extracts were associated with the induction of androgenesis [55].

The microspores of *Brassica* are irreversibly induced towards androgenic pathway during 4-8 h of the 8-h high temperature pretreatment. Analysis of *in vitro* translated total mRNA indicates that proteins of molecular weight 84, 67 and 66 kDa and to some extent, 27 kDa are synthesized during this period. Interestingly, these proteins were absent or present in low amounts in freshly isolated (0 h), potentially embryogenic microspores. Microspores, unable to undergo embryogenesis, contained very low amount of these proteins [19].

In a hybrid cultivar of *Zea mays*, during the induction of androgenesis, a 32 kDa protein (MAR 32) is induced which accumulates in the anthers during the cold pre-treatment. Different responsive and non-responsive genotypes have been evaluated and accumulation of MAR 32-like proteins observed only in certain responsive genotypes [54].

5.2.1.1 Heat-shock and Protein Synthesis

Heat-shock induces a program of gene expression in which synthesis of a family of proteins so-called heat-shock proteins (HSPs) takes place [63, 64]. In microspore culture, elevation of temperature to 32°C for 8 h is accompanied by *de novo* synthesis of a number of heat-shock proteins of 70 kDa class [56]. The HSPs act as molecular chaperones in the folding, refolding, assembly and transport of cellular proteins and are as such essential for cell survival [64]. Detailed analysis has shown that out of eight isoforms of HSP68, only one shows a three-fold increase. An immuno-cytochemistry study has revealed a co-distribution of HSP68 with DNA-containing organelles, presumably mitochondria. Of the six HSP70 isoforms detected, one increased to six-fold in the embryogenic culture condition. During normal pollen development, HSP70 is localized in the nucleoplasm during the S-phase of cell cycle and later in the cytoplasm. In early bi-cellular pollen of *Brassica*, the nucleus of the vegetative cell, which normally does not divide and never expresses HSP70, shows intense labeling of nucleoplasm with anti-HSP70 after 8 h of culture under embryogenic condition [56]. Such studies demonstrate a strong correlation between the phase of cell cycle and the nuclear localization of HSP70 with the induction of embryogenesis. On the basis of these studies, it is speculated that HSPs might be involved in the altered pattern of cell division, which leads to the induction of androgenesis [56].

In *Brassica* microspore culture, six proteins were identified; which were exclusively synthesized under embryogenic (32°C) conditions [55]. Of these, four that were specifically synthesized during the first 8 h at 32°C were not synthesized further after two days of culture. In a study on tobacco, a dramatic increase in the level of a low molecular weight HSP transcript has been detected in embryogenic microspores, following the inductive starvation treatment [65]. Based on these studies, it is speculated that these proteins represent ideal markers for the induction phase of microspore embryogenesis [66].

5.2.1.2   *Phosphorylation of Proteins*

Phosphorylation of proteins plays an important role in the reception of signals from exogenous factors by cells and in the expression of cellular functions [67, 68]. The embryogenic microspores of *Nicotiana rustica* as well as *N. tabacum* exhibited a specific pattern of protein phosphorylation. The characteristic pattern of phosphorylation was neither observed in microspores following gametophytic development nor in non-embryogenic microspores [57, 58]. Kyo and Harada [58] speculated that these phosphoproteins might be the essential factors for the onset of microspore embryogenesis. Moreover, as these were not detected after the beginning of sporophytic cell divisions, their function in the process of pollen embryogenesis appears to be transient [58]. Using density gradient centrifugation, it has been shown that these phosphoproteins were localized in the plasma membrane [59], thus, indicating their role in signal perception.

Comparison of the 2-D patterns of phosphorylated proteins, in 2-day-old embryogenic and non-embryogenic microspore cultures of *Brassica napus,* revealed a much higher phosphorylation state of HSP70 under embryogenic conditions [55]. It was reported that in cultured microspores of *Brassica napus*, HSP70 immuno-reacted with the monoclonal antibody MPM-2, which recognizes a mitosis-specific phosphorylated epitope, in embryogenic microspores [69]. In another study, Cordewener et al. [56] showed the difference in the rate of synthesis and the intracellular translocation of HSP70 in the embryogenic and non-embryogenic cultures. In a recent study, Cordewener et al. [66] have reported that change in synthesis of HSP70, translocation and protein phosphorylation are associated with the switch in the developmental pathway of *Brassica* microspores from gametophytic to sporophytic development.

5.2.1.3   *Ubiquitin-mediated Degradation Pathway*

A recent study based on immuno-cytochemistry using polyclonal antibody to ubiquitin revealed a developmentally regulated loss of free ubiquitin and ubiquitinated proteins in embryogenic microspores of maize. After immuno-localization experiments, a steady low level of UBQ/UBQ-Ps was revealed in most cell types of anthers excluding degenerated and non-induced microspores [38]. The localization of ubiquitinated compounds correlated particularly with those MCMs (multi-cellular microspores) which were considered potentially androgenic on the basis of their ultrastructural characteristics, than with cells displaying symptoms of elevated proteolytic activity and degradation [70]. These results confirmed that the ubiquitin-mediated pathway is involved in gene expression and regulation of cellular processes [71]. Similarly, Callis and Bedinger [72] had also reported a positive correlation between loss of free ubiquitin and ubiquitinated proteins, and pollen development and maturation. The results obtained by Alche et al. [38] confirmed that the return to the sporophytic pathway is once again accompanied by an increase in the levels of UBQ and UBQ-Ps species. The increase in the level of these proteins was opined to represent not only a consequence of the deviation of the microspore to the sporophytic pathway, but even a direct factor responsible for the androgenic induction [38].

5.2.2 Development of Phenolic Compounds

Delalonde et al. [60] reported that prior to the induction of androgenesis in maize, cold pretreatment is required and this leads to the accumulation of phenolic compounds. A possible role of phenolic compounds has been shown in the modulation of IAA-oxidase activity [73, 74]. IAA protection or degradation effects may be partially linked to the individual phenolic capacity existing in different varieties. Some diphenols could protect auxin by inhibiting IAA-oxidase [74] and some monophenols on the contrary, could increase this activity, thus increasing the degradation of IAA [75]. This accumulation of phenolic compounds can be attributed to the protection of IAA *in vitro* from IAA-oxidase. Consequently, the genotype with maximum *in vitro* protection for IAA is regarded as the best genotype for androgenesis [60]. This is interesting as one of the reasons for observed enhancement due to cold pretreatment was earlier considered to be the delayed browning of anthers. In fact, in *Datura metel* androgenic response was considerably enhanced if anthers were cultured on medium incorporated with cysteine (an anti-oxidant, presumably inhibiting the activity of phenol oxidases) and polyvinylpyrrolidone (an adsorbent of phenols). However, as the polyphenols are known to inhibit IAA oxidase activity, the observed enhancement due to decreased levels of IAA because of the increased activity of IAA oxidase was not ruled out [76].

5.3 Molecular Changes

The molecular basis of developmental switch from pollen maturation to embryogenesis is still not well understood. A large amount of data is available on gene expression during pollen development *in vivo* [22, 77]. Microspores isolated from tobacco anthers at different stages of development show an increase in the transcriptional and translational activities, accompanying a new program of gene expression in the period immediately following the first pollen mitosis [77-79]. In maize, this change in gene expression has been correlated with the cytological stage at which the microspores become incompetent for embryogenesis [80]. However, in comparison, information available on the molecular events during the developmental switch to embryogenesis is fragmentary.

Bhojwani et al. [53], for the first time, reported that in an embryogenic microspore, there is a change in the nucleic acid and protein content prior to embryogenesis. They reported a decrease in the RNA content of vegetative cell of the embryogenic pollen grains of *Nicotiana tabacum* prior to division, suggesting thereby that suppression of the gametophytic program already acquired, is the first step in the induction process. Later, Garrido et al. [62] reported a decrease in the overall synthesis of RNA and protein in tobacco pollen during 7-day starvation treatment. In another study, Kyo and Harada [15] reported an increase in the rate of protein synthesis in tobacco pollen cultured under maturation conditions. Kyo and Harada [15] proposed that the degradation of proteins and/or the suppression of synthesis of proteins were necessary to switch from normal pollen development to embryogenesis.

In contrast, studies conducted on *Hyoscyamus niger* highlighted the need of *de novo* RNA synthesis in embryogenic grains. The metabolic changes in embryogenic grains become discernible within 1-2 h of culture [40]. This study revealed that embryogenic grains are the only one to be labeled during 24 h of culture of anther segments on $5\text{-}^3\text{H}$ uridine incorporated medium for 1-2 h, thus indicating *de novo* RNA synthesis in embryogenic grains. The extended exposure time to the labeled uridine for 6 h increased the number of labeled grains, suggesting that in all the embryogenic grains, RNA synthesis is not started simultaneously. Further experiments utilizing actinomycin D, incorporated to the basal medium, revealed that even one hour of culture followed by transfer to inhibitor incorporated medium is sufficient for 15 per cent of the embryogenic grains to escape inhibition. The segments cultured on the basal medium for 24 h before being transferred to inhibitor adjuvated medium, developed embryos at the same frequency as observed in fragments continuously cultured on the basal medium. This observation implied that RNA required for initiation of androgenesis is synthesized during first 24 h of culture, whereas for subsequent development (up to at least heart-shaped stage of embryos) either RNA synthesis is not required or its synthesis remains unaffected by actinomycin D. Raghavan [81] further reported that within one hour of culture, in some of the microspores, poly (A) containing RNA (presumably mRNA triggering embryogenic development) content, monitored by (^{3}H) polyuridylic acid binding, is increased. Contrary to Raghavan's observations, Sopory [82] did not find any effect on the frequency of embryo production, if anthers of dihaploid *Solanum tuberosum* were cultured on actinomycin D incorporated medium for initial four days. However, continuous culture of anthers in the presence of this transcriptional inhibitor reduced the response, thus, indicating that fresh RNA synthesis is required only after four days of culture. In contrast, puromycin (a translational inhibitor) totally inhibited the response. Based on these observations it was proposed that for induction of androgenesis in *Solanum tuberosum*, conserved messenger exists and for initiation of sporophytic development only translation is required [82].

The studies on *Hyoscyamus niger* also revealed that in this plant the transcription in the generative nucleus is a more important prerequisite for the subsequent embryogenic pathway, in comparison to that in the vegetative one. Thus, embryogenic divisions are initiated only in those microspores in which the generative nucleus alone or along with the vegetative nucleus synthesizes RNA and pollen grains in which RNA synthesis occurs almost exclusively in the vegetative nucleus become starch filled and non-embryogenic [40, 81]. These observations are consistent with data obtained from previous studies by Raghavan [41, 83], according to which in *H. niger* organogenetic part of embryoid is formed by repeated divisions of generative nuclei.

In tobacco microspores, starvation treatment as an induction stimulus, results in the dedifferentiation of male gametophyte, followed by a redifferentiation process including the acquisition of embryogenic competence and de-repression of the cell division. During normal pollen development, the generative nucleus passes through S phase to G_2 phase soon after first pollen mitosis, while the vegetative nucleus remains arrested in the G_1-phase [84]. During starvation, a large fraction of the pollen shows DNA replication in the vegetative cell. However, inhibition of DNA replication in the vegetative cell, caused by the addition of hydroxy-urea to the starvation medium, did not affect the formation of embryos after transfer to a hydroxy-urea-free medium with sucrose [84]. Thus, Zarsky et al. [84] concluded that DNA replication during starvation was not essential for embryogenic induction, but emphasized that an event preceding S-phase is important. In addition, the induction of changes in development is characterized by the activation of specific transcription factors, which in turn, cause altered pattern of gene expression. RNA and protein synthesis cultured in tobacco pollen showed a gradual decrease during the starvation treatment [15, 62, 65].

Two major approaches have been utilized to identify the molecular markers for pollen embryogenesis [85]. In one, gene products expressed during zygotic embryogenesis have been used as probes for differentiating pollen embryos. This method led to the characterization of one of the first markers, the ^{12}S storage glycoprotein, found in microspore-derived embryos of *Brassica napus* [86]. It is a useful marker since it demonstrates that non-zygotic embryos do accumulate proteins characteristic of zygotic embryos. Boutilier et al. [87], applying the above approach prepared a cDNA library of *Brassica* microspore embryoids and found that the expression of the napin seed storage proteins coincides with the induction of microspore embryogenesis and could therefore be used as a molecular marker for earliest stage of induction. Napin genes were highly expressed in the embryogenic microspores, but not in the microspores undergoing pollen development or in the somatic tissues. Three members of the distinct Bnm NAP sub-family of napin seed storage protein genes (*Bnm NAP2, Bnm NAP3* and *Bnm NAP4*) were responsible for the majority of napin gene expression in embryogenic microspores. The elevated temperatures induced expression of napin gene only in embryos and microspores that were competent for embryogenesis [87].

The second approach for the identification of developmental markers is based on a comparison of gene expression during micro-gametogenesis and induced embryogenesis in microspores. Reynolds and Kitto [88] prepared a cDNA library of young pollen embryoids of wheat and screened it with cDNA probes prepared from pollen at different stages of development [88]. Two clones, *pEMB4* and 94 were expressed very early during culture, suggesting that these genes are associated with morphogenesis and are not simply expressed as a consequence of differentiation. The accumulation patterns of clones may indicate the activation of specific genes associated with the major morphological and physiological activities connected with the formation and differentiation of pollen embryoids *in vitro*. These genes are spatially and temporally specific, and were not expressed in microspore culture. *pEMB4* may be an example of a "transition" gene, which is normally expressed only at the time of first haploid mitosis. However, as a consequence of embryogenic induction, this gene remains turned on in the developing embryos [88].

A cysteine-labelled metallothionein (*EcMt*) gene, isolated from a wheat pollen embryoid, was transcribed only in embryogenic microspores, pollen embryoids and developing zygote embryos of wheat. Increase in the transcript was directly correlated to the synthesis of abscisic acid [89]. Treatment of the cultures with fluridone, the inhibitor of ABA biosynthesis, suppressed not only ABA accumulation, but also the *EcMt* gene transcripts and the ability of microspores to become embryogenic [89]. To demonstrate the direct involvement of ABA in this process, exogenous ABA was added to fluridone treated cultures. The inhibitory effect on ABA on both gene expression and androgenesis was negated. Kawashima et al. [90] speculated that based on sequence similarities, the wheat *EcMt* is a functional analogue of the animal Mt and that it may play a role in zinc homeostasis during androgenesis in which zinc Mt sequester or disperse zinc dependant DNA and RNA polymerases as well as in transacting zinc fingers proteins during differentiation. Since the *EcMt* gene transcript is not expressed during normal pollen development, but only in the microspore-derived embryoids or developing zygotic embryos, this may be an example of a new expressed sporophytic gene [89]. Also, since the *EcMt* transcripts appear only in the embryogenic microspores after 6 h *in vitro*, this may serve as a marker for the early events of microspore embryogenesis, as the first structural change associated with embryogenic induction is observed within 12 h of culture [88–90].

Zarsky et al. [65] identified a cDNA clone for a low molecular weight heat shock protein from a library prepared from RNA isolated from *Nicotiana tabacum* binucleate pollen grains. This study revealed that although the gene was expressed normally during the later stages of pollen ontogeny, it was transcriptionally activated by starvation-induced pollen embryogenesis and is therefore, an example of a developmentally regulated gene associated with androgenic induction.

In another study, Vrinten et al. [91] prepared cDNA clones corresponding to genes differentially expressed during the early stages in barley microspore culture. These genes were isolated and characterized. Three cDNAs representing genes, not previously identified in barley, were isolated. The first gene *ECA*1 (early culture abundant) was expressed only during the early stages of culture and with a reduced expression in low-density culture. This lacked significant homology with any other known gene or protein. The second one, *ECGST* (early culture glutathione S-transferase), having homology with members of group glutathione S-transferase genes was thought to be important in protecting cells from oxidative stress during culture process. The third one *ECLTP* (early culture lipid transfer proteins) had homology with lipid transfer proteins (LTP's) and an expression pattern similar to that of an LTP known to be a marker of early stages of embryogenesis in the carrot somatic embryogenesis system.

Significant genotypic effects suggest that genetic factors are also important in determining the androgenic potential of microspores. There has been considerable work to isolate genes that control these events. In maize, the products of a number of crosses between highly embryogenic and non-embryogenic genotypes were analyzed using RFLP markers [92, 93]. Cowen et al. [92] used 98s families to map genes which were associated with the anther culture response. These families were derived from the cross of a highly embryogenic line (139/39-05) and a non-responsive line (B73). The analysis showed that the anther culture response is associated with two major recessive genes on chromosome 3 and 9, which are epistatic, two minor genes on chromosomes 1 and 10. Thus, tightly linked RFLP markers may serve as starting points for the characterization of genes conferring high androgenic capacity. Since only a small percentage of the microspores within an anther form embryos, it has been suggested that the anther culture response is limited to those microspores bearing certain favorable genetic factors.

6. Conclusions and Prospects

Despite many years of efforts by a number of groups, the mechanism of induction of *in vitro* androgenesis is still poorly understood. Though, the information on cellular, biochemical and molecular changes preceding its cytological expression is meager and limited only to a few taxa, certain generalizations can be made that may provide the framework for future investigations.

Induction of androgenesis is developmentally regulated. Once the pollen maturation gene products begin to accumulate after first pollen mitosis, the developmental pathway is determined and androgenesis cannot take place. Thus, only microspores at certain stages of development can be redirected to undergo androgenesis. The optimal developmental stage for induction of androgenesis is around the first pollen mitosis. In most of the cases, if cultured at bi-celled stage it is the vegetative nucleus that contributes to the embryo formation. However, if cultured at uninucleate stage, besides development through this pathway, B-pathway also becomes operative.

The androgenic induction requires some external stress to competent microspores. The stress may be in form of heat, cold, chemicals, starvation or water stress. The stress response during induction involves some disruption and organization of the cytoskeleton, altering the polarity of the cytoplasmic components of competent microspores. Imposition of stress is correlated with the appearance of smHSP transcripts in the affected cells. Without elicitation of the stress response, as indicated by the appearance of smHSP transcripts, androgenesis cannot proceed.

Induction results in an altered pattern of synthesis and accumulation of RNA and proteins in potentially embryogenic microspores, leading to the first sporophytic divisions. Although the identity of most of these genes is unknown, in several cases, they are stress-related or are associated with zygotic embryogenesis. The stress-related genes may be concerned with a general reprogramming of the cell or provide some type of protection from that stress. Perhaps stress is the single most important factor determining whether androgenesis will be initiated or not.

Some of the potential areas for future research in the field of androgenic induction could be:

(a) The emphasis so far has been on gene products, which are up regulated during the androgenic induction period. Future studies should concentrate on gene products, which are down regulated.

(b) Up-regulation of many genes may also be due to certain other factors and this might not be a potential molecular marker for induction. This can be verified by using molecular techniques such as antisense technology against gene in question so as to determine whether or not the same is really involved in the induction process.

(c) There is no information as to how division in the vegetative cell becomes self-sustaining. Studies in this area could be rewarding.

(d) It has not been unequivocally demonstrated whether there are any cytological and histochemical differences between the two similar looking cells or nuclei, after the equal division of microspore. This requires further investigation.

(e) Information about the signal transduction pathway operating during androgenesis is practically lacking. The physiological, biochemical and molecular changes that occur within the competent microspores and/or pollen in response to this information are poorly understood.

The investigation, on the abovementioned aspects of androgenesis, will require an integrated approach using a combination of the physiological, biochemical, molecular and genetic dissections of the regulatory relationship between gene expression and androgenic induction.

References

1. S. Guha, S.C. Maheshwari, *In vitro* production of embryo from anthers of *Datura,* Nature 204 (1964) 497.
2. S. Guha, S.C. Maheshwari, Cell division and differentiation of embryos in the pollen grains of *Datura in vitro*, Nature 212 (1966) 97–98.
3. S.B. Babbar, Studies on *in vitro* sporophytic development of pollen in some angiosperms, Ph.D. Thesis, Univ. of Delhi, Delhi (1982).
4. I.K. Vasil, C. Nitsch, Experimental production of pollen haploids and their uses, Z. Pflanzenphysiol. 76 (1975) 191–212.
5. S.B. Babbar, S.C. Gupta, Chilling induced androgenesis in anthers of P*etunia hybrida* without any culture medium, Z. Pflanzenphysiol. 100 (1980) 279–283.
6. P. Smykal, Pollen embryogenesis: the stress mediated switch from gametophytic to sporophytic development, current status and future prospects, Biol. Plant. 43 (2000) 481–489.
7. P.M. Pechan, W.A. Keller, Identification of potentially embryogenic microspores in *Brassica napus*, Physiol. Plant. 64 (1988) 377–384.
8. P.M. Pechan, W.A. Keller, Induction of microspore embryogenesis in *Brassica napus* by gamma irradiation and ethanol stress, *In Vitro* Cell Dev. Biol. 25 (1989) 1073–1075.

9. M.A. M. Zaki, H.G. Dickinson, Modification of cell development *in vitro*: the effect of colchicine on anther and isolated microspore culture in *Brassica napus,* Plant Cell Tiss. Org. Cult. 40 (1995) 255–270.

10. J.P. Zhao, D.H. Simmonds, W. Newcomb, Induction of embryogenesis with colchicine instead of heat in microspores of *Brassica napus* L. cv. Topas. Planta 198 (1996) 433–439.

11. P.M. Pechan, D. Bartels, D.C.W. Brown, J. Schell, Messenger-RNA and protein changes associated with induction of *Brassica* microspore embryogenesis, Planta 184 (1991) 161–165.

12. J.B.M. Custers, J.H.G. Cordewener, Y. Nollen, H.J.M. Dons, M.M.L. van Campagne, Temperature controls both gametophytic and sporophytic development in microspore cultures of *Brassica napus.* Plant Cell Rep. 13 (1994) 267–271.

13. P.M. Pechan, P. Smykal, Androgenesis: affecting the fate of the male gametophyte, Physiol. Plant., 111 (2001) 1–8.

14. P. Smykal, P.M. Pechan, Stress as assessed by the appearance of smHSPs transcripts, is required but not sufficient to initiate androgenesis, Physiol. Plant, 110 (2000) 135–143.

15. M. Kyo, H. Harada, Control of the developmental pathway of tobacco pollen *in vitro*, Planta 168 (1986) 427–432.

16. E. Heberle-Bors, Isolated pollen culture in tobacco: plant reproductive development in a nutshell, Sex. Plant Reprod. 2 (1989)1–10.

17. A. Touraev, A. Ilham, O. Vicente, E. Herberle-Bors, Stress induced microspore embryogenesis in tobacco: an optimized system for molecular studies, Plant Cell Rep. 15 (1996a) 561–565.

18. A. Touraev, M. Pfosser, O. Vicente, E. Herberle-Bors, Stress as a major signal controlling the developmental fate of tobacco microspores: towards a unified model of induction of microspore embryogenesis, Planta 200 (1996b) 144–152.

19. P.M. Pechan, Heat shock proteins and cell proliferation, FEBS Lett. 280 (1991) 1-4.

20. E.C. Porter, D. Parry, J. Bird, H.G. Dickinson, Nucleic acid metabolism in the nucleus and cytoplasm of angiosperm meiocytes, in: C. Evans, H.G. Dickinson (Eds), Controlling Events in Meiosis, Company of Biologists, Cambridge,1984, pp. 363–369.

21. R. Scott, E. Dagless, R. Hodge, P. Wyatt, I. Soutlemi, J. Draper, Patterns of gene expression in developing anthers of *Brassica napus,* Plant Mol. Biol. 17 (1991)195–207.

22. J.P. Mascarenhas, RNA and protein synthesis during pollen development and tube growth, in: J.Heslop-Harrison (Ed), Pollen-Development and Physiology, Butterworths, London,1971, pp. 201–222.

23. A. Touraev, O. Vicente, E. Heberle-Bors, Initiation of microspore embryogenesis by stress, Trends Plant. Sci. 8 (1997) 297–302.

24. S.K. Sopory, M. Munshi, Anther culture, in: S.M. Jain, S.K. Sopory, R.E.Veilleux (Eds), *In Vitro* Haploid Production in Higher Plants Vol. I, Kluwer Academic Publishers, Dordrecht,1996, pp. 145–176.

25. G. Wenzel, E. Thomas, Observations on growth in culture of anthers of *Secale cereale,* Z. Pflanzenzücht. 72 (1974) 89–94.

26. F.T. Roux, S. Deimling, H.H. Geiger, Anther culture ability in *Secale cereale* L., Plant Breed. 114 (1995) 259–261.

27. S.B. Babbar , N. Walia, S.N. Raina, *In vitro* androgenesis in the improvement of horticultural crops, in: K.L. Chadha ,P.N. Ravindran, L. Sahijram (Eds), Biotechnology in Horticultural and Plantation Crops, Malhotra Publ. House, New Delhi, 2000, pp. 79–119.

28. M.A.M. Zaki, H.G. Dickinson, Microspore-derived embryos in *Brassica*: the significance of division symmetry in pollen mitosis I to embryogenic development, Sex. Plant. Reprod. 4 (1991) 48–55.

29. C. Eady, K. Lindsey, D. Twell, The significance of microspore division and division symmetry for vegetative cell-specific transcription and generative cell differentiation. Plant Cell 7 (1995) 65–75.

30. L.E. Zonia, J. Tupy, Lithium treatment of *Nicotiana tabacum* microspores blocks polar nuclear migration, disrupts the partitioning of membrane associated Ca and induces symmetrical mitosis, Sex. Plant Reprod. 8 (1995) 152–160.

31. N. Sunderland, F.M. Wicks, Cultivation of haploid plants from tobacco pollen, Nature 224 (1969) 1227–1229.

32. N. Sunderland, F.M. Wicks, Embryoid formation in pollen grains of *Nicotiana tabacum*, J. Exp. Bot. 22 (1971) 215–226.
33. J.M. Dunwell, N. Sunderland, Pollen ultrastructure in anther cultures of *Nicotiana tabacum*, I. Early stages of culture, J. Exp. Bot. 25 (1974a) 352–361.
34. J.M. Dunwell, N. Sunderland, Pollen ultrastructure in anther cultures of *Nicotiana tabacum*, II. Changes associated with embryogenesis, J. Exp. Bot. 25 (1974b) 363–373.
35. J.M. Dunwell, N. Sunderland, Pollen ultrastrucutre in anther cultures of *Nicotiana tabacum*, III. The first sporophytic division, J. Exp. Bot. 26 (1975) 240–252.
36. J.M. Dunwell, N. Sunderland, Pollen ultrastructure in anther cultures of *Datura innoxia*, I. Division of the presumptive vegetative cell, J. Cell. Sci. 22 (1976) 469–480.
37. N. Sunderland, Anther culture as a means of haploid induction, in: K.J. Kasha (Ed), Haploids in Higher Plants: Application and Potential, Guelph Univ., Guelph,1974, pp. 91–122.
38. J. D. Alche, A.J. Castro, M. Solymoss, I. Timar, B. Barnabas, M.I. Rodriguez-Garcia, Cellular approach to the study of androgenesis in maize anthers: immunocytochemical evidence for the involvement of the ubiquitin degradative pathway in androgenic induction, J. Plant Physiol. 156 (2000), 146–155.
39. M.A.M. Zaki, H.G. Dickinson, Structural changes during the first division of embryos resulting from anther and free microspore culture in *Brassica napus*, Protoplasma 156 (1990) 149–162.
40. V. Raghavan, Embryogenic determination and ribonucleic acid synthesis in pollen grains of *Hyoscyamus niger* (henbane), Am. J. Bot. 66 (1979) 784–795.
41. V. Raghavan, Patterns of DNA synthesis during pollen embryogenesis in henbane, J. Cell Biol. 73 (1977) 521–526.
42. T.L. Reynolds, An ultrastructural and stereological analysis of pollen grains of *Hyoscyamus niger* during normal ontogeny and induced embryogenic development, Am. J. Bot. 71(1984) 490–504.
43. Y. Hamoaka, Y. Fujita, S. Iwai, Effects of temperature on the mode of pollen development in anther culture of *Brassica campestris*, Physiol. Plant. 82 (1991) 67–72.
44. C.A. Telmer, W. Newcomb, D.H. Simmonds, Microspore development in *Brassica napus* and the effect of high temperature on division *in vivo* and *in vitro*, Protoplasma 172 (1993) 154–165.
45. D.H. Simmonds, Mechanism of induction of microspore embryogenesis in *Brassica napus* : significance of the preprophase band of microtubules in the first sporophytic division, in: N. Akkag (Ed), Biomechanics of Active Movement and Division of Cells NATO ASI series, Vol. H84, Springer-Verlag, Berlin,1994, pp. 569–574.
46. A.A.M. van Lammeren, C.J. Keijzer, M.T.M. Willemse, H. Kieft, Structure and function microtubular cytoskeleton during pollen development in *Gasteria verrucosa* (mill) H. Duval, Planta 165 (1985) 1–11.
47. D.H. Simmonds, C. Gervais, W.A. Keller, Embryogenesis from microspores of embryogenic and non-embryogenic lines of *Brassica napus*, in: D.I. McGregor (Ed), Rapeseed in a Changing World, Proc. VIIIth Int. GCIRC Rapeseed Cong., Saskatoon, GCIRC+Canola Council of Canada, 1991, pp. 306–311.
48. G. Hause, B. Hause, van A.A.M. van Lammeren, Microtubular and actin filament configurations during microspore and pollen development in *Brassica napus* cv. Topas, Can. J. Bot. 70 (1992) 1369–1376.
49. D.H. Simmonds, W.A. Keller, Significance of preprophase bands of microtubules in the induction of microspore embryogenesis of *Brassica napus*, Planta 208 (1999) 283–391.
50. R.S. Sangwan, B.S. Sangwan-Norreel, Ultrastructual cytology of plastids in pollen grains of certain androgenic and non-androgenic plants, Protoplasma 138 (1987) 11–22.
51. N. Sunderland, J.M. Dunwell, Anther and pollen culture, in: H.E. Street, Plant tissue and cell culture, Blackwell Sci. Publ., Oxford, 1977, pp. 223–264.
52. C.A. Telmer, D.H. Simmonds, Cellular changes during heat shock induction and embryo development of cultured microspores of *Brassica napus* cv. Topas, Protoplasma 185 (1995) 106–112.
53. S.S. Bhojwani, J.M. Dunwell, N. Sunderland, Nucleic acid and protein contents of embryogenic tobacco pollen, J. Exp. Bot. 24 (1973) 863–871.
54. P. Virgne, F. Riccardi, M. Beckert, C. Dumas, Identification of a 32-kDa anther marker protein for androgenic response in maize (*Zea mays* L.), Theor. Appl. Genet. 8 (1993) 843–850.

55. J.H.G. Cordewener, R. Busink , J.A. Traas , J.B.M. Custers, M.M.L. van Campagne, Induction of microspore embryogenesis in *Brassica napus* L. is accompanied by specific changes in protein synthesis, Planta 195 (1994) 50–56.

56. J.H.G. Cordewener, G. Hause, E. Gorgen, R. Busink, B. Hause, H.J.M. Dons, A.A.M. van Lammeren, M.M.L. van Campagne, P.M. Pechan, Changes in synthesis and localization of members of the 70-kDa class of heat-shock proteins accompany the induction of embryogenesis in *Brassica napus* L. microspores, Planta 196 (1995) 747–755.

57. M. Kyo, H. Harada, Phosphorylation of proteins associated with embryogenic dedifferentiation of immature pollen grains of *N. rustica*. J. Plant Physiol. 136 (1990a) 716–722.

58. M. Kyo, H. Harada, Specific phosphoproteins in the initial period of tocacco pollen embryogenesis, Planta 182 (1990b) 58–63.

59. M. Kyo, T. Ohkawa, Investigation of sub-cellular localization of several phosphoproteins in embryogenic pollen grains of tobacco. J. Plant Physiol.137 (1991) 525–529.

60. M. Delalonde, Y. Barret, M.P. Coumans, Development of phenolic compounds in maize anthers during cold pretreatment prior to androgenesis, J. Plant Physiol. 149 (1996) 612–616.

61. D. Schulze, K.P. Pauls, Flow cytometric analysis of cellulose tracks development of embryogenic *Brassica* cells in microspore cultures, New Phytologist 154 (2002) 249–254.

62. D. Garrido, N. Eller N, E. Heberle-Bors, O. Vicente, *de novo* transcription of specific mRNAs during the induction of tobacco pollen embryogenesis, Sex. Plant Reprod. 6 (1993) 40–45.

63. R.I .Morimoto, A. Tissiens, C. Georgoprulos, The stress response, function of the proteins and perspectives, in: R.I. Morimoto, A. Tissieres, C. Georgopoulos (Eds), Stress Proteins in Biology and Medicine, Cold Spring Harbor Press, New York,1990, pp. 1–36.

64. J.P. Hendrick, F.U. Hartl, Molecular chaperone function of heat-shock protein, Ann. Rev. Biochem. 62 (1993) 349–384.

65. V. Zarsky, D. Garrido, N. Eller, J. Tupy, O. Vicente, The expression of a small heat shock gene is activated during induction of tobacco pollen embryogenesis by starvation, Plant Cell Environ. 18 (1995) 139–147.

66. J.H.G. Cordewener, J. Bergervoet, C.M. Liu, Changes in protein synthesis and phosphorylation during microspore embryogenesis in *Brassica napus,* J. Plant Physiol. 156 (2000) 156–163.

67. P. Choen, The role of protein phosphorylation in neural and hormonal control of cellular activity, Nature, 296 (1982) 613–619.

68. C.W. Ronson, B.T. Nixon, F.M. Ausubel, Conservative domain in bacterial regulatory proteins that respond to environmental stimuli. Cell 49 (1987) 579–581.

69. B. Hause, G. Hause, P. Pechan, A.A.M. van Lammeren, Cytolskeletal changes and induction of embryogenesis in microspore and pollen cultures of *Brassica napus* L. Cell Biol. Int. 17 (1993) 153–163.

70. Y.Q. Li, Southworth, H.F. Linskens, D.L. Mulcahy, M. Crest, Localization of ubiquitin in anthers and pistils of *Nicotiana,* Sex. Plant Reprod. 8 (1995) 123–128.

71. D. Finley, V. Chau, Ubiquitination, Ann. Rev. Cell Biol. 7 (1991) 25–69.

72. J. Callis, P. Bedinger, Developmentally regulated loss of ubiquitin and ubiquitinated proteins during pollen maturation in maize, Proc. Natl. Acad. Sci., USA 91 (1994) 6074–6077.

73. H.J. Grambow, S.B. Langenbeck, The relationship between oxidase activity, peroxidase activity, hydrogen peroxide and phenolic compounds in the degradation of IAA *in vitro*, Planta 157 (1983) 131–137.

74. R. Pressey, Anions activate the oxidation of IAA by peroxidase from tomato and other sources. Plant Physiol. 93 (1990) 798–804.

75. G.G. Gross, C. Janse, E.F. Elstner, Involvement of malate, monophenols, and the superoxide radical in hydrogen peroxide formation by isolated cell walls from horseradish (*Armoraciala pathifolia* Gilib), Planta 136 (1977) 271–276.

76. S.B. Babbar, S.C. Gupta, Promotory effect of polyvinylpyrrolidone and l-cysteine HCl on pollen plantlet production in anther cultures of *Datura metel*, Z. Pflanzenphysiol. 106 (1982) 459–464.

77. P. Bedinger, M.D. Edgerton, Developmental staging of maize microspores reveal a transition in developing microspore proteins, Plant Physiol. 92 (1990) 474–479.

78. J.A.M .Schrauwen, P.F.M. de Groot, M.M.A. van Herpen, van T. van der Lee, W.H. Reynen, K.A.P. Weterings, G.J. Wullems, Stage-related expression of mRNA during pollen development in lily and tobacco, Planta 182 (1990) 298–304.

79. J.J. Tupy, E. Suss, Hrabetova. L Rihova, Developmental changes in gene expression during pollen differentiation and maturation in *Nicotiana tabacum* L. Biol. Plant. 25 (1983) 231–237.

80. P. Mandaron, M.F. Niograet, R. Mache, F. Moneger, *In vitro* protein synthesis in isolated microspores of *Zea mays* at several stages of development. Theor. Appl. Genet. 80 (1990) 134–138.

81. V. Raghavan, Distribution of poly (A)-containing RNA during normal pollen development and induced pollen embryogenesis in *Hyoscyamus niger,* J. Cell Biol 89 (1981) 593–606.

82. S.K. Sopory, Effect of sucrose, hormones and metabolic inhibitors on the development of pollen embryoids in anther culture of dihaploid *Solanum tuberosum.* Can. J. Bot. 57 (1979) 2691–2694.

83. V. Raghavan, Role of generative cell in androgenesis in henbane, Science. 191 (1976) 388–389.

84. V. Zarsky, D. Garrido, L. Rihova, J. Tupy, O. Vicente, E. Herbrle-Bors, Derepression of the cell cycle by starvation is involved in the induction of tobacco pollen embryogenesis, Sex. Plant Reprod. 5 (1992) 89–194.

85. T.L. Reynolds, Pollen embryogenesis, Plant Mol. Biol. 33 (1997) 1–10.

86. M.L. Crouch, Non-zygotic embryos of *Brassica napus* L. contain embryo-specific storage, Planta 156 (1982) 520–524.

87. K.A. Boutilier, M.J. Gine, J.M. DeMoor, V.N. Iyer, B.L.Miki Expression of the BnmNAP sub-family of napin genes coincides with the induction of *Brassica* microspores embryogenesis, Plant Physiol. 26 (1994) 1711–1723.

88. T.L. Reynolds, S.L. Kitto, Identification of embryoid abundant genes that are temporally expressed during pollen embryogenesis in wheat anther cultures, Plant Physiol. 100 (1992) 1650.

89. T.L. Reynolds, R.C. Crowford, Changes in abundance of an abscisic acid-responsive early cysteine-labelled metallothionein transcript during pollen embryogenesis in bread wheat (*Triticum aestivum*), Plant. Mol. Biol. 32 (1996) 823–829.

90. I. Kawashima, T.D. Kennedy, M. Chino, B.G. Lane, Wheat Ecmetallothionein genes like mammalian Zn^{2+} metallothionein genes are conspicuously exepressed during embryogenesis, Eur. J. Biochem. 209 (1992) 971–976.

91. P.L. Vrinten, T. Nakamura, K.J. Kashu, Characterization of cDNAs expressed in the early stages of microspore embryogenesis in barley (*Hordeum vulgare*) L. Plant Mol. Biol. 41(1999) 455–463.

92. N.M. Cowen, C.D. Johnson, M. Miller, A. Woosley, Mapping genes conditioning *in vitro* androgenesis in maize using RFLP analysis, Theor. Appl. Genet. 84 (1992) 720–724.

93. Y. Wan, T. R. Rocheford, J. M. Wedham, RFLP analysis to identify putative chromosomal regions involved in the anther culture response and callus formation of maize, Theor. Appl. Genet. 85 (1992) 362–365.

Plant Biotechnology and Molecular Markers
P.S. Srivastava, Alka Narula and Sheela Srivastava (Editors)

2. Doubled Haploids: A Powerful Biotechnological Tool for Genetic Enhancement in Oilseed Brassicas

Deepak Prem[2], Kadambari Gupta[2] and Abha Agnihotri[1]

[1]Bioresources and Biotechnology Division, TERI, Habitat Place, Lodhi Road, New Delhi 110003, India
[2]Centre for Bioresources and Biotechnology, TERI-School of Advanced Studies, Habitat Place, Lodhi Road, New Delhi 110 003, India

Abstract: The review presents the detailed advantages of using doubled haploids for crop improvement programs. It also elaborates the use of the doubled haploid technique in terms of its conjugation with other biotechnological approaches. The hall mark of doubled haploid technique lies in versatile potential of its use in various breeding programs and its ability to compress the time taken for breeding a desired genotype in comparison to conventional breeding methods. The factors affecting the production of haploids/doubled haploids have been discussed in the light of relevant research undertaken towards genetic enhancement for value addition in oilseed brassicas.

1. Introduction

The oils of plant origin have important edible and non-edible uses in human life. Various oil yielding crop species have been domesticated to produce high oil yielding seeds for edible or industrial purposes. The oilseed bearing crops include perennial trees like coconut and palm and annuals like groundnut, soybean, sunflower and oleiferous brassicas. The oilseed brassicas include *B. carinata*, *B. nigra* and rapeseed-mustard (collective term for *B. napus*, *B. juncea* and *B. campestris*). Among various oilseed crops mentioned above oilseed brassicas command a substantial market proportion. Rapeseed mustard with an average world production of 3.6×10^7 million ton and an acreage of 24.7 million Ha, ranks third at global level preceded by soybean and cotton seed (FAO on line statistics, http://www.fao.org/).

A considerable proportion of the world production of oilseed brassicas has been contributed by developing countries, in particular China and India (FAO year book 1990-1999, http://www.fao.org/). In India, among the nine annual oilseed crops grown in the country, oilseed brassicas collectively rank second in terms of production and acreage next only to groundnut. Among the various *Brassica* species, *B. juncea* occupies the maximum acreage followed by *B. campestris* especially in the north gangetic plain [1]. *B. napus* is a relatively new introduction to India but is gaining rapid popularity in Punjab and Himachal Pradesh [2]. Thus, in India too, the oilseed brassicas are a major contributor to vegetable oil production and a considerable amount of research effort is directed towards their genetic enhancement.

On account of its economic importance, oilseed brassica attracted the early plant breeders and geneticists, and pioneering research was undertaken for elucidating its cytogenetic composition [3-6]. The cytogenetic relationship between various oilseed brassicas has been represented by U [7] and is popularly referred to as the U's triangle. Brassicas comprise three diploid species namely *B. nigra*, *B. oleracea* and *B. campestris* and three allotetraploids namely *B. carinata*, *B.*

juncea and *B. napus* which have arisen out of interspecific hybridization between the diploid species followed by spontaneous chromosomes doubling. Evidence in support of this relationship has been accumulated from cytological, biochemcial and molecular investigations [8, 9].

The oilseed brassicas show variable pollination behaviour in terms of existence of self-compatible and incompatible forms. *B. campestris* is cultivated in the form of three ecotypes namely brown sarson, yellow sarson and toria. These ecotypes comprise both—self-compatible and incompatible plant types. *B. napus*, *B. carinata* and *B. juncea* are predominantly self-pollinated whereas *B. nigra* and *B. oleracea* are often cross-pollinated species [10, 11]. Therefore, a variety of breeding methodologies ranging from inbred development through pure line selection to hybrid cultivar development have been utilized for genetic enhancement of oilseed brassicas. The choice of the breeding methodology, irrespective of the breeding goal, largely depends upon the predominance of self- or cross-pollination, and the availability of naturally occurring genetic variations [2]. Nevertheless, the recognition of breeding goals are essential before a strategy to attain the same is spelt out.

2. Need for Genetic Enhancement in Oilseed Brassicas

The major breeding objectives for genetic enhancement of oilseed brassica are aimed at enhancing the commercial value of the end product either quantitatively or qualitatively. On one hand breeding objectives like development of high yielding varieties and generation of varieties having resistance to abiotic stress (salt, drought and frost) and biotic stress (insects and diseases) target production in terms of quantitative enhancement. On the other hand, improvement in oil content, and quality of oil and meal target the qualitative improvement of the produce. In recent years greater emphasis has been laid upon nutritional quality enhancement of oilseed brassicas with the aim of providing better nutrition and value addition [12].

Various methods are being routinely used for generation of genetic variability and genetic enhancement. These range from conventional tools, such as selection from available germplasm and hybridization for transfer of desired genes (either *in vivo* or *in vitro* through embryo rescue and somatic hybridization) to the development of transgenics and use of molecular markers for selection. Among the biotechnological tools used for genetic improvement doubled haploids have emerged as an exciting tool for brassica breeders since this technology has a versatile ability to blend with and expedite the existing approaches [13].

3. Doubled Haploids: The Concept and Its Utility

Doubled haploids are plants produced by spontaneous or artificial doubling of the chromosomes of haploid plants. Such a plant is valuable because the chromosomes that are created by artificial/ spontaneous doubling are exact copies of the chromosomes that were present in the haploid plant-justifying the term doubled haploid. Doubled haploids offer a major advantage by attainment of homozygosity in a single step thus significantly reducing the breeding cycle along with its use in conjugation with other biotechnological methods for expediting the crop improvement programs [14]. The major advantages of the doubled haploids are briefly summarized as follows.

3.1 Attainment of Homozygosity

According to conventional breeding approaches, homozygosity may be achieved by repeated selfing and rigorous selection for several generations [15]. Normally this exercise requires about

10 to 12 years for varietal development programs, and is most efficient in self-pollinated crops that do not show inbreeding depression. Doubled haploids greatly reduce the time required for obtaining homozygous plants if an efficient haploid generation protocol is available. Moreover, attaining homozygosity for recessive and quantitatively controlled traits is an even greater mammoth task because of the involvement of many loci and masking of recessive allels in heterozygous state [16]. Doubled haploids may be generated in a single step thus fixing the genotypic combinations in a single generation [17, 18]. This technique also considerably reduces the time required for homozygous line development in cross-pollinated crops and may be especially useful in parent development for hybrid production [19]. Thus doubled haploids essentially compress the breeding cycle by accelerating the development of homozygous lines [14].

3.2 Utilization of Gametic Gene Combinations
Doubled haploids offer the unique advantage of utilization of the haploid phase for selection. This fact assumes greater importance for selection in case of induced mutations, which are generally recessive in nature or other quantitative traits, controlled by recessive allels [20]. Since haploids would express recessive genes, transgressive segregants for recessive traits can efficiently be recovered through diploidization of chromosomes. This concept also offers the advantage of significantly smaller population size required to find the least likely recombinant in case of quantitative traits, since in a doubled haploid population selection is actually effective on gametic gene combinations [17, 18]. Therefore, doubled haploid populations have been used to study the inheritance of important quantitative traits [21–23].

3.3 Versatile Compatibility with Other Approaches
In addition to the above stated advantages of the doubled haploids, they can also be profitably utilized for mutation breeding, disease resistance, biotechnological gene transfer etc. Some of the applications of doubled haploids in conjugation with other breeding approaches are reviewed below:

- *Mutation breeding*: Mutations are immediately expressed in haploid and doubled haploid plants, hence these are very lucrative targets for mutation research. In *B. napus* imidazoline herbicide resistance has been introduced using doubled haploid technique in conjugation with chemical mutagenesis. These resistant lines have been evaluated in field trials [24, 25]. Besides this, chemical as well as physical mutagens have been utilized to develop resistance to *Phoma lingam* [26] and *Alternaria brassicola* [27] in *B. napus* by treating cultured microspores. *In vitro* mutagenesis at haploid level also led to the development of high oleic acid, thinner seed coat, high oil and protein, and low fibre content in *B. napus* lines [22, 28] and modified erucic acid content in *B. carinata* [29].
- *Disease resistance*: Microspore cultures are one of the most excellent targets for *in vitro* selection for disease resistance, provided that the disease defence system is active at such an early stage of plant development [13]. Gametoclonal variation exhibited by haploids generated through anther/microspore culture, along with host specific and non-host specific toxins as medium supplements, could be used for *in vitro* selection of resistant genotypes.
- *Biotechnological gene transfer*: Microspores or anther form a good explant source for gene transfer systems such as PEG, electroporation, microinjection and biolistic methods. Microspore

derived embyros have been used as recipient cell system for *Agrobacterium*-mediated gene transfer in *B. napus* [30–34].

- *Molecular breeding*: DNA based procedures such as RFLP/AFLP analysis are being increasingly employed in plant breeding due to their enormous range of application. DNA markers provide unprecedented refinement in genetic analysis through the construction of nearly saturated genetic maps. This provides the breeder with a highly efficient marker aided selection tool. Double haploids being truly homozygous for all loci are now being routinely used for genetic mapping of brassicas since they reduce the time required for making RFLP maps and for generating polymorphic mapping populations [35–38]. Doubled haploid production have been utilized to study gene linkages and interactions [39].

- *Breeding for desired oil profile*: Doubled haploid technique has also proved useful in the development of mustard cultivars expressing specific oil profile, such as modified erucic acid, reduced linolenic acid, increased linoleic, palmitic and oleic acid content for specific purposes [28, 40, 41]. Hence double haploids provide a powerful breeding tool for obtaining designer mustard varieties having specific fatty acid profile.

4. Development of Haploids and Doubled Haploids

Availability of haploids is a prerequisite for doubled haploid production. Spontaneously occurring haploid plants having half the normal number of chromosomes were discovered in 1920s, but utilization of haploid plants was not a practical technique until methods for the controlled production of haploid plants were developed. Apart from spontaneous occurrence, which is rare and is confined to a few species [42], haploid plants may be produced by interspecific/intergeneric crossing followed by selective chromosome elimination, for example, as in barley and wheat × maize. Developments in tissue culture techniques in the early 1960's opened the doorway for haploid plant production by *in vitro* culturing of unfertilized ovules (gynogenesis) or from the mature/immature pollen grains (androgenesis). Although doubled haploids have been obtained via gynogenesis [43–45], the majority of published successes have resulted from anther and microspore culture. A breakthrough in this direction was achieved when Guha and Maheshwari [46] for the first time demonstrated that anthers of *Datura innoxia* cultured *in vitro* produce embryos that originate from immature pollen grains or microspores. However, the difficulties associated with anther culture are low frequencies of haploids, difficulty in distinguishing spontaneous doubled haploids from diploids which regenerate from somatic tissue, and the considerable time and labour which may be needed to generate the desired doubled haploid population required for successful utilization in breeding program [47].

Developments in isolated microspore culture for several crop species attracted the interest of brassica breeders for generation of doubled haploids. Oilseed brassicas being responsive to cell and tissue culture techniques have been extensively researched upon for both anther and isolated microspore culture [13, 48–50].

5. Anther Culture in Oilseed Brassicas

Successful anther culture in brassicas was first reported by Canadian scientists in early 1970s. Keller and Armstrong [51] reported development of embryoids from cultured anthers of *B. napus*. Following this several reports on *B. napus* anther culture were published [49, 52–58]. Similar to these, anther culture has also been reported in *B. nigra* [59], *B. campestris* [60–62] and in *B. juncea* [63–67]. These reports on anther culture of various oilseed brassicas primarily elucidate the possibility of doubled haploid production. However, the major bottleneck of this technique lies in low frequency of embryogenesis resulting in the realization of a very few embryos. Several factors may be responsible for this low embryo yield. It has been proposed that anther wall generated toxins may inhibit microspore embryo development inside cultured anthers [68, 69] and the congested conditions inside the anther may limit the nutrient supply to growing microspores [52]. Much of these drawbacks of anther culture have been effectively overcome by the development of isolated microspore culture technique.

6.   Microspore Culture in Oilseed Brassicas

As in the case of anther culture, the first report of successful isolated pollen culture was also reported in *B. napus* [49]. This, in fact, was the first report of isolated pollen grain or microspore embryogenesis in plant species other than those belonging to the family *Solanacae*. Following this many reports have elucidated the potential of isolated microspore culture in oilseed brassicas for developing haploid embryos.

Among the various oilseed brassicas, *B. napus* is primarily the most researched upon species for doubled haploid production using isolated microspore culture. Many laboratories have developed specialized protocols for isolated microspore culture of *B. napus* [56, 70–74]. Several reports on isolated microspore culture have also been published in *B. carinata* [18, 70, 75], *B. nigra* [76, 77], *B. campestris* [62, 76, 78–83] and *B. juncea* [23, 77, 83–87]. However, apart from *B. napus*, limited success has been achieved in other species towards cultivar development, due to the lack of efficient microspore embryogenesis.

7.   Factors Influencing Microspore Embryogenesis

The ability to induce totipotency in anther cultures/isolated microspore cultures is greatly influenced by several factors. These include genetic and exogenic factors that may have profound implications on microspore development *in vitro*. Various factors that influence microspore embryogenesis in oilseed brassicas are:

7.1   Genotype

The genotype of the donor plants have been reported to have a profound effect on the microspore embryogenic response. Genotypic variations in haploid embryo development have been observed in several brassica species [88]. Genotypic variability for microspore embryogenesis response in isolated microspore culture has been reported in *B. campestris* [18, 79, 80, 82], *B. juncea* [23, 77, 85] and in most reports sited for *B. napus* above. Recently microspore embryogenic ability has been studied as a stably inherited trait using the diallele mating system in *B. napus* by Zhang and Takahata [89]. The study elucidates that both additive and dominant effects are significant for microspore embryogenesis as a genetically controlled trait. Similar dominant gene action has been reported by Cloutier et al. [39] reiterating the strong genotype influence on microspore embryogenic ability. Ajisaka et al. [90] have further reported two putative chromomosome regions associated with microspore embryogenic ability in *B. campestris*.

7.2   Donor Plant's Growth Condition

The donor plant's growth condition has a marked effect on the physiological processes of the plant thus it invariably influences the microspore embryogenic ability. Proper light, temperature, humidity and nutrients are all necessary to develop healthy plants. Varying growth conditions of donor plants have been tested for their influence on microspore embryogenesis, ranging from plants grown under field conditions to plants grown under artificial or growth room conditions. In *B. campestris*, plants grown under cold temperature conditions, 10/5°C day/night cycle showed enhanced microspore embryogenic capability [80]. However, plants grown under relatively higher temperature regime 28/15°C day/night cycle upto bolting and then transferred to low temperature conditions as mentioned above, have also been reported to show enhanced microspore embryogenic response [82]. Low temperature regime for donor plants has been reported to show a positive correlation with increased microspore embryogenesis in *B. napus* [72, 74, 76, 91–93]. However, in this species too, donor plants grown till bolting at higher temperature regime (25–28°C day/ 12–15°C night) and then shifted to a low temperature (10/5°C day/night cycle) have been reported to show higher microspore embryogenic ability [74]. Limited reports are available on the influence of donor plant's growth condition on microspore embryogenesis in *B. juncea*. A relatively higher temperature regime (20–21°C day/15–18°C night) has been reported to be congenial for microspore embryogenic response in isolated microspore cultures of *B. juncea* [23, 77, 85]. In view of this it seems likely that the plants may be grown under normal *in vivo* temperature conditions for healthy and vigorous vegetative growth, however, a shift to lower temperature regime is essential for enhancing microspore embryogenic response.

7.3 Microspore Development Stage

The microspore development stage is a prime important factor that influences the microspore's ability to turn totipotent. This is primarily due to the fact that microspores would only respond to embryo formation at a developmental stage when they are not committed to develop into pollen grains [58, 64, 92]. Moreover, in oilseed brassicas the microspore development is asynchronous and microspores of different developmental stages may be observed in a developing anther. Therefore, selection of buds that have maximum proportion of embryogenic microspores is essential for efficient microspore embryo yield. The microspore development may be divided into three basic stages viz. the tetrad stage (when the microspore mother cell splits into four haploid cells), the uninucleate stage (when the uninucleate microspore prepares for the nuclear division to form the vegetative and generative nuclei) and the binucleate stage (when the microspore contains a generative and a vegetative nucleus). Each of the abovementioned microspore development stage has been extensively researched upon in *B. napus* to determine the exact stage at which the microspore is not under a differentiation pressure or is not committed towards pollen development.

It has been established that there is an optimum development stage (embryogenic window) that corresponds to the late uninucleate to early binucleate stage of development, during which large number of microspores could undergo embryogenesis [14, 94]. Further to this it has been proposed that non-embryogenic microspores produce inhibitory substances that suppress embryo development in the embryogenic microspores [68, 69]. This may be because of the rupturing of non-embryogenic binucleate microspores [71] thus reducing the embryogenic frequency and altering the morphology of embryos [69, 92]. Replacement of culture media after microspore isolation helps in reducing the autotoxins thus allowing normal embryo development. Similar influence of microspore developmental stage on microspore embryogenesis has been reported in

B. campestris [80, 82], *B. carinata* [75], and *B. juncea* [66, 77, 83, 85]. These studies have established that the late uninucleate stage is most responsive to embryogenesis and that selection of buds with majority of late uninucleate microspores increases the frequency of embryo formation in isolated microspore culture in oilseed brassicas.

7.4 Microspore Density

The density of isolated microspores in the culture media is another essential factor responsible for normal development of embryos. Microspore culture density ranging from 1 to 10,000 cells/ml has been reported for different species. Varying density of microspores have been studied by various scientists to get the optimum embryo yield such as 1×10^4 microspores/ml [83], 2×10^4/ml [92], $3-4 \times 10^4$/ml [95–98], 8×10^4/ml [74] and 10×10^4/ml [99]. However, the most favourable density has been found to range between 5 and 8×10^4 cells/ml for *B. napus* and 1 and 4×10^4 cells/ml for *B. juncea*.

7.5 Media Composition

The basic media composition for microspore culture protocols in various oilseed brassicas have mostly been the same over the years. Initial reports of anther culture in oilseed brassicas emphasized the role of increased sucrose concentration in culture media [51, 60]. Subsequent to this, it was established that high sucrose concentration (10 to 13%) is essential for microspore embryogenesis in anther as well as isolated microspore culture in *B. napus* [49, 52, 56]. Keller et al. [60] demonstrated that L-serine was an important constituent of the anther culture media in *B. napus*. Later, Lichter [52] demonstrated that basal Nitsch and Nitsch [100] medium supplemented with glutamine, glutathione and L-serine was most congenial for anther culture in the same species. Off late Nitsch and Nitsch medium [100] modified by Lichter [52], commonly known as NLN medium, has become the most frequently used media for isolated microspore culture in oilseed brassicas.

7.6 Growth Additives

Activated charcoal has been reported to be beneficial for embryo growth and normal development in both anther and microspore culture of *B. napus* [55, 72]. However, its role in triggering microspore embryogenesis has not been reported, rather it has been reported to be beneficial for normal growth of induced embryos [72, 83, 101]. Growth regulators have been used in both isolated microspore as well as anther culture of oilseed brassicas. Initial reports on microspore culture emphasised the role of 6-benzyl amino purine and 1-naphthalene acetic acid for induction of microspore embryogenesis [49, 52, 102]. However, off late it has been reported that growth regulators are not essential for inducing microspore embryogenesis [71, 77, 94]. Several scientists have also propounded the role of colchicine as an embryogenesis inducing agent [57, 91, 103-105]. However, owing to high toxicity, potential hazard and great care required in handling colchicine, this concept has not been utilized extensively. Till date the major role of colchicine has been limited for doubling chromosome number of haploid plants.

Apart from the abovementioned factors, post culture incubation conditions are also reported to have profound effect on induction and development of microspore derived embryos. For most reports cited above an initial heat shock of 30–32°C for 3–10 days has been reported to be essential for microspore embryogenesis. Further to induction of microspore embryos, determination of ploidy of the produced embryos and colchiploidy for chromosome doubling are essential part of any successful doubled haploids production protocol. Determination of ploidy is usually done either through root cytology or using flow cytometry. Colchiploidy may be carried out by axillary bud treatment or by dipping the plantlet's roots in colchicine [74]. Having overcome the successful induction of the embryogenesis, the doubling of the chromosomes and regeneration of doubled haploid plants is not a severe bottleneck for utilizing this technique for various crop improvement programs.

8. Conclusions

The oleiferous brassica species have been extensively utilized as an important source of edible oil. Oilseed brassicas stand third in world's oilseeds production and acreage, whereas in India, it ranks second next to groundnut. A considerable amount of work has been undertaken for genetic enhancement of oilseed brassicas to increase its commercial value through various conventional as well as modern approaches. Among the biotechnological techniques used, double haploids have emerged as a promising tool. In addition to compressing the breeding cycle by accelerating the development of homozygous plants, it has versatile compatibility with other approaches like mutation breeding, transgenics, molecular breeding, etc. However, for successful utilization of doubled haploids in crop improvement programs, generation of substantial doubled haploid population is essential, which in turn is possible with the availability of an efficient haploid production protocol. Haploids have been produced by selective chromosome elimination, gynogenesis or androgenesis. Among the three, androgenesis, i.e. anther/microspore culture has been the most successfully utilized approach. Both anther and microspore culture have been researched upon in oilseed brassicas, however, extensive work has been done elucidating the potential of the isolated microspore culture technique for developing haploid embryos. A number of genetic and exogenic factors influence the efficiency of microspore embryogenesis. This article has presented some of the important factors affecting microspore development and their implications on establishing an efficient microspore culture protocol. The doubled haploids are already being extensively utilized for the quality and agronomic improvement in *B. napus* but its practical utilization in other economically important oilseed brassicas is yet to be realized. The work in this direction is in progress at several institutions opening many a new vistas for utilization of this versatile technique for genetic enhancement and value addition in the oilseed brassicas.

References

1. Anonymous 2000. In: Annual Progress Report of the AICRP, on Rapeseed-Mustard, Sewar, Bharatpur (Rajasthan), ICAR, Govt. of India.
2. Banga, S.K. 1996. Breeding for oil and meal quality. In: Chopra, V.L. and Prakash, S. (Eds). Oilseed and Vegetable Brassicas: Indian Perspective, Chap. 11, pp. 234–249.
3. Takamine, N. 1916. Uber die rebenden and die prasynaptischen Phasen der Reduktionsteilung. Bot. Mog. Tokyo, 30: 293–230. Cited in: Chopra, V.L., Prakash, S. (Ed) 1996. Oilseed and Vegetable Brassicas: Indian Perspective, Chap. 2, pp. 6–34.
4. Karpechenko, G.D. 1922. The number of chromosomes and the genetic correlation of cultivated *Cruciferae*. Bull. Appl. Bot., Genet. Pl. Breed, 13: 3–14.

5. Karpechenko, G.D. 1929. A contribution towards the synthesis of a constant hybrid of three species. Proc. USSR Cong. Genet. Pl. Breed. 2: 277–294.

6. Morinaga, T. 1928. Preliminary notes on interspecific hybridization in *Brassica*. Proc. Imp. Acad. 4: 620–622.

7. UN 1935. Genome analysis in *Brassica* with special references to the experiment formation of *B. napus* and peculiar mode of fertilization. Japan J. Bot. 7: 389–452

8. Prakash, S., Hinata, K. 1980. Taxonomy, cytogenetics and origin of crop brassicas, A review. Opera Botanica, 55: 1–57.

9. Prakash, S., Chopra, V.L. 1990. Reconstruction of alloploid *Brassicas* through non-homologous recombination: Introgression of resistance to pod shattering in *Brassica napus*. Genet. Res. Comb. 56: 1–2.

10. Dhillon, S.S., Labana, K.S. 1988. Outcrossing in Indian mustard *Brassica juncea* (L.) Coss. Ann. Biol. 4: 100–102.

11. Olsson, G. 1955. Investigation of the degree of cross pollination in white mustard and rape. Sver Utsaedesfoeren Tidshr, 62: 311–322. Cited in: Chopra, V.L., Prakash, S. (Ed) 1996. Oilseed and Vegetable Brassicas: Indian Perspective, Chap. 2, pp. 6–34.

12. Agnihotri, A., Kaushik, N. 2000. Incorporation of superior nutritional quality traits in Indian *B. juncea*. Indian J. Pl. Genet. Res. 13(3): 352–358.

13. Downey, R.K., Rimmer, S.R. 1993. Agronomic improvements in oilseed *Brassica*s. Advances in Agronomy. 50: 1–66.

14. Kott, L.S. 1997. Application of doubled haploid technology in breeding of oilseed *Brassica napus*. Crop Science 51(1/2): 28–32.

15. Allard, R.W. 1960. Principles of plant breeding. John Wiley, New York.

16. Baily, T.R., Comstock, R.E. 1976. Linkage and synthesis of better genotypes in self-fertilizing species. Crop Sci. 16: 363–370.

17. Rajhathy, T. 1976. Haploid flax revisited. Z Pflanzenzuecht 76: 1–10.

18. Ferrie, A.M.R., Palmer, C.E. and Keller, W.A. 1995. Haploid embryogenesis, in: Thrope, T.A. (Ed) *In vitro* embryogenesis in plants. Kluwer Academic Publisher, Netherlands, pp. 309–344.

19. Ulrich, A., Furtan, W.H., Downey, R.K. 1984. Biotechnology and rapeseed breeding: some economic considerations. Sci. Counc. Can. Rep., 1–67.

20. Maluszynski, M., Ahloowalia, B.S., Sigurbjornsson, B. 1995. Application of *in vivo* and *in vitro* mutagenesis technique for crop improvement. Euphytica 85 (1–3): 303–315.

21. Siebel, K., Pauls, K.P. 1989. Inheritance patterns of erucic acid content in populations of *Brassica napus* microspore-derived spontaneous diploids. Theor. Appl. Genet. 77: 489–494.

22. Henderson, C.A.P., Pauls, K.P. 1992. The use of haploidy to develop plants that express several recessive traits using light seeded Canola (*Brassica napus*) as an example. Theor. Appl. Genet. 83: 476–479.

23. Thiagarajah, M.R., Stringam, G.R. 1993. A Comparison of genetic segregation in traditional and microspore-derived populations of *Brassica juncea* L. Czern and Coss. Plant Breeding 111(4): 330–334.

24. Swanson, E.B., Coumans, M.P., Brown, G.L., Petal, J.D., Beversdorf, W.D. 1988. The characterization of herbicide tolerant plants in *Brassica napus* L. after *in vitro* selection of microspores and protoplasts. Plant Cell Rep. 7: 83–87

25. Swanson, E.B., Herrgesell, M.J., Arnoldo, M., Sippell, D.W., Wong, R.S.C. 1989. Microspore mutagenesis and selection. Canola plants with field tolerance to the imidazolinones. Theor. Appl. Genet. 78: 525–530.

26. MacDonald, M.V., Ahmed, I., Ingram, D.S. 1989. Mutagenesis and haploid culture for disease resistance in *Brassica napus*. XII Eucarpia Cong. Sci. Plant Breed. 25: 22–27.

27. MacDonald, M.V., Ahmad, I., Menten, J.O.M., Ingram, D.S. 1990. Haploid culture and *in vitro* mutagenesis (UV light, X-rays, and gamma rays) of rapid cycling *Brassica napus* for improved resistance to disease. Proc. Int. Sym., IAEA and FAO, UN, Vienna, 18–22 June 2: 129–138.

28. Wong, R.S.C., Zee, S.Y., Swanson, E.B. 1996. Isolated microspore culture of Chinese flowering cabbage (*Brassica campestris* ssp. *parachinensis*). Plant Cell Rep. 15: 396–400.

29. Barro, F., Escobar, F., Vega, M., Martin, A. 2001. Doubled haploid lines of *Brassica carinata* with modified erucic acid content through mutagenesis by EMS treatment of isolated microspores. Plant Breeding 120(3): 262–264.

30. Swanson, E.B., Erickson, L.R. 1989. Haploid transformation in *Brassica napus* using an octapine producing strain of *Agrobacterium tumefaciens*. Theor. Appl. Genet. 78: 831–835.

31. Oleck, M.M., Phan, C.V., Eckes, P., Donn, G., Rakow, G., Keller, W.A. 1991. Field resistance of canola transformants (*Brassica napus* L.) to ignite (Phosphinotricin). Proc. 8th Internat. Rapeseed Conf. Saskatoon, Canada. pp 292–297.

32. Kazan, K., Curtis, M.D., Goulter, K.C., Manners, J.M. 1997. *Agrobacterium tumefaciens* mediated transformation of doubled haploid canola (*Brassica napus*) lines. Aust. J. Pl. Physiol. 24(1): 97–102.

33. Goll 2000. *Brassica napus* transformation protocol [www.usask.ca/agriculture/plant sci/classes/plsc 416 / projects_2002 / eftoda.htm].

34. Fukuoka, H., Ogawa, T., Matsuoka, M., Ohkawa, Y., Yano, H. 1998. Direct gene delivery into isolated microspores of rapeseed (*B. napus* L.) and the production of fertile transgenic plants. Plant Cell Rep. 17: 323–328.

35. Ecke, W., Uzunova, M., Weissleder, K. 1995 Mapping the genome of rapeseed (*Brassica napus*) 11. Localization of genes controlling erucic acid synthesis and seed oil content. Theor. Appl. Genet. 91(6/7): 972–977.

36. Chopra, P.K., Kirti, P.B. 1996. Biotechnology. In: Chopra, V.L., Prakash, S. (Eds). Oilseed and vegetable *Brassicas*: Indian Perspective, Chapter 12, pp. 250–269.

37. Cheng, W.Y., Landry, B.S., Raney, P., Rakow, G.F.W. 1998. Molecular mapping of seed quality traits in *Brassica juncea*. Acta. Hort. 459: 139–147.

38. Sakova, L., Curn, V. 1998. Identification and classification of selected cruciferous species and rape doubled haploid lines using RAPD markers. J. Genet. Breed. 34(2): 61–67.

39. Cloutier, S., Cappadocia, M., Landry, B.S. 1995. Study of microspore culture responsiveness in oilseed rape (*Brassica napus* L.) by comparative mapping of a F$_2$ population and two microspore-derived population. Theor. Appli. Genet. 91: 841–847.

40. Tiwari, A.S., Sequin-Swatrz, G., Downey, R.K. 1998. Zero erucic acid doubled haploids in *Brassica juncea*. Genome 30(1): 464.

41. Cegielska, T., Szaal, L., Krzymaski, J. 1999. An *in vitro* mutagens selection system for *Brassica napus*. Proc 10th GCIRC, International Rapeseed Cong. Sept. 26–29, Canberra, Australia.

42. Kimber, G., Riley, R. 1983. Haploid angiosperms. Botanical Review, 29: 480–531.

43. Bossoutrot, D., Hosemans, D. 1983. Gynogenesis in *Beta vulgaris* L.: from *in vitro* culture of unpollinated ovules to the production of doubled haploid plants in soil. Plant Cell Rep. 4: 300–303.

44. Keller, J. 1990. Haploids from unpollinated ovaries of *Allium cepa*-single plant screening, haploid determination and long term storage, in: Nijkamp, H.J.J. Van Der Plas, L.H.W. Van, Aartrijk, J. (Eds). Progress in Plant Cellular and Molecular Biology, Kluwer Academic Publishers, 275–279.

45. Zhu, Z., Wu, H. 1979. *In vitro* production of haploid plantets from the unpollinated ovaries of *Triticum aestivum* and *Nicotiana tabacum*. Acta Academia Sinica, 6: 181–183.

46. Guha, S. and Maheshwari, S.C. 1964. *In vitro* production of embryo from anthers of *Datura*. Nature 204, 497.

47. Smith, M.K., Drew, R.A. 1990. Current applications of tissue culture in plant propagation and improvement. Aust. J. Plant. Physiol. 17: 267–289.

48. Keller, W.A., Arnison, P.G., Cardy, B.J. 1987. Haploids from gametophytic cells – Recent developments and future prospects. Plant Tissue Cell Cult. Proc. 6th Int. Cong., 1986, 223–241.

49. Lichter, R. 1982. Induction of haploid plants from isolated pollen. Z. Pflanzenphysial. Bd. 105.5 427–434.

50. Beversdorf, W.D. 1990. Micropropagation in crop species, in: Nijkamp, H.J.J, Van Der Plas, L.H.W. Van Aartrijk, J. (Eds.). Progress in Plant Cellular and Molecular Biology, Kluwer Academic Publishers, Dordrecht, 3–12.

51. Keller, W.A., Armstrong, K.C. 1977. Embryogenesis and plant regeneration in *Brassica napus* cultures. Can J. Bot. 55: 1383–1388.

52. Lichter, R. 1981. Anther culture of *B. napus* in a liquid culture medium. Z. Pflanzenphysial. Bd. 103.5. 229–237.

53. Loh, Chiang, S., Ingram, D.S. 1982. Production of haploid plants from anther cultures and secondary embryoids of winter oilseed rape, *Brassica napus* ssp. *oleifera*. New Phytol. 91: 507–516.

54. Martha, L.C. 1982. Non zygotic embryos of *Brassica napus* contain embryo specific storage proteins. Planta 156: 520–524.

55. Johansson, L. 1983. Effects of activated charcoal in anther cultures. Physiol. Plant. 59: 397–403.

56. Dunwell, J.M., Cornish, M. and Decourcel, A.G.L. 1983. Induction and growth of microspore-derived embryos of *B. napus* ssp. *oleifera*. J. Exp. Bot. 34 (149): 1768–1778.

57. Mathias, R., Robbelen, G. 1991. Effective diplodization of microspore derived haploids of rapeseed (*B. napus* L.) by *in vitro* colchicine treatment. Plant Breeding 106: 82–84.

58. Zaki, A.M., Dickinson, H.G. 1990. Structural changes during the first divisions of embryos resulting from anther and free microspore culture in *B. napus*. Protoplasma, 156: 149–162.

59. Leelavathi, S., Reddy, V.S., Sen, S.K. 1984. Somatic cell genetic studies in *Brassica* species: High frequency production of haploid plants in *Brassica alba*. Plant Cell Rep. 3: 102–105.

60. Keller, W.A., Rajhathy, T., Lacapra, J. 1975. *In vitro* production of plants from pollen in *Brassica campestris*. Can. J. Genet. Cytol., 17: 655–66.

61. Keller, W.A., Armstrong, K.C. 1979. Simulation of embryogenesis and haploid production in *Brassica campestris* anther cultures by elevated temperature treatments. Theor. Appl. Genet. 55: 65–67.

62. Aslam, F.N., Mac Donald, M.V., Loudon, P., Ingrain, D.S. 1990. Rapid cycling *Brassica* species : in breeding and selection of *B. campestris* for anther culture. Annal. of Bot. 65 (5): 557–566.

63. George, L., Rao, P.S. 1983. *In vitro* induction of pollen embryos and plantlets in *Brassica juncea* through anther culture. Plant Sci. Lett. 26: 111–116.

64. Sharma, K.K., Bhojwani, S. 1985. Microspore embryogenesis in anther culture of two Indian cultivars of *Brassica juncea*. Pl. Cell Tis. Org. Cul. 4: 235–239.

65. Aggarwal, P.K., Bhojwani, S.S. 1993. Enhanced pollen grain embryogenesis and plant regeneration in anther culture of *Brassica juncea*. Euphytica 70: 191–196.

66. Prabhudesai, V., Bhaskaran, S. 1993. A continuous culture system of direct somatic embryogenesis in microspore-derived embryos of *Brassica juncea*. Plant Cell Rep. 12: 289–292.

67. Malik, M.R., Rangaswamy, N.S., Shivanna, K.R. 2001. Induction of microspore embryos in a CMS line of *B. juncea* and formation of the androgenic plantlets. Euphytica 120: 195–203.

68. Bors, H.E. 1984. Genotypic control of pollen plant formation in *Nicotiana tabacum* L. Theor. Appl. Genet. 67: 475–479.

69. Kott, L.S., Polson, L., Ellis, B., Beversdorf, W.D. 1988. Autotoxicity in isolated microspore cultures of *Brassica napus*. Can. J. Bot. 66: 1665–1670.

70. Chuong, P.V., Beversdorf, W.D. 1985. High frequency embryogenesis through isolated microspore cultures in *Brassica napus* L. and *B. carinata* Braun. Plant Sci. (Limerich, Irel.) 39: 219–226.

71. Swanson, E.B., Coumans, M.P., Wu, S.C., Barsby, T.L., Beversdorf, W.D. 1987. Efficient isolation of microspores and the production of microspore-derived embryos from *B. napus*. Plant Cell Rep. 6: 94-97.

72. Gland, A., Lichter, R., Schweiger, H.G. 1988. Genetic and exoneous factors affecting embryogenesis in isolated microspore cultures of *B. napus* L. J. Plant Physiol: 132: 613–617.

73. Kott, L.S., Polsoni, L., Beversdorf, W.I. 1988. Cytological aspects of islolated microspore culture of *Brassica napus*. Can. J. Bot. 66: 1658–1664.

74. Coventy, J., Kott, L.S., Beversdorf, W.D. 1988. Manual for microspore culture technique for *Brassica napus*. Dep. Crop Sci. Technol. Bull. OAC publication 0489, Univ of Guelph, Guelph, Ont. Canada.

75. Barro, F., Martin, A. 1999. Response of different genotypes of *Brassica carinata* to microspore culture. Plant Breeding 118: 79–81.

76. Lichter, R. 1989. Efficient yield of embryoids by culture of isolated microspore of different Brassicaceae species. Plant Breeding 103: 119–123.

77. Lionneton, E., Beuret, W., Delaitre, C., Ochatt, S., Rancillae, M. 2001. Improved microspore culture and doubled haploid plant regeneration in the brown condiment mustard (*B. juncea*). Plant Cell Rep. 20: 126–130.

78. Sato, T., Nishio, T., Hirai, M. 1989. Plant regeneration from isolated microspore culture of Chinese cabbage (*Brassica compestris* spp. *pekinensis*). Plant Cell Rep. 8: 486–488.

79. Burnett, L., Yarrow, S., Huang, B. 1992. Embryogenesis and plant regeneration from isolated microspores of *Brassica rapa* L. ssp. *oleifera*. Plant Cell Rep. 11: 215–217.

80. Baillie, A.M.R., Epp, D.J., Hutcheson, D., Keller, W.A. 1992. *In vitro* culture of isolated microspores and regeneration of plants in *Brassica campestris*. Plant Cell Rep. 11: 234–237.

81. Wong, S.C., Swanson, E. 1991. Genetic modification of canola oil: High oleic acid canola, in: Fat and Cholesterol Reduced Food, C. Haberstroh, C.E. Morris (Eds), pp 154–164. Gulf, Houstan, Texas.

82. Guo, D.Y., Pulli, S. 1996. High frequency embryogenesis in *B. campestris* microspore culture. Pl. Cell Tis. Org. Cul. 46: 219–225.

83. Parihar, D.S., Aradhye, S. 1999. Microspore culture for induction of doubled haploidy in *B. juncea* and *B. rapa*. Proc. 10th GCIRC International Rapeseed Congress, Sept. 26–29, Canberra, Australia.

84. Thiagaraj, M.R., Stringam, G.R. 1990. High frequency embryo induction from microspore culture of *Brassica juncea* L. Czern and Coss. Abst. 7th Int. Cong. On Plant Tiss. Cell Cult. Amsterdam pp. 190.

85. Hiramatsu, M., Odahara, K., Matsue, Y. 1995. A survey of microspore embryogenesis in leaf mustard (*Brassica juncea*). Acta Horticulturae 392: 139–145.

86. Agnihotri, A., Ginette-Seguin-Swartz, Downey, R.K. 1996. Microspore embryogenesis in *B. juncea*, in: Pareek, L.K. (ed). Trends in Plant Tissue Culture and Biotechnology. pp 218–221.

87. Purnina and Rawat, S. 1997. Regenerating *Brassica juncea* plants from *in vitro* microspore culture. Cruciferae News 19: 47–48.

88. Palmer, C.E., Keller, W.A., Arnison, P.G. 1996. Experimental haploidy in brassica species, in: Jain, S.M., Sapory, S.K., Veilleux, R.E. (Ed). *In vitro* Haploid Production in Higher Plants Volz. Kluwer, Dordrecht, pp. 143–172.

89. Zhang, E.L., Takahata, Y. 2001. Inheritance of microspores embryogenic ability in *Brassica* crops. Theor. Appl. Genet. 103: 254–258.

90. Ajisska, H., Kuginuki, Y., Hidak Siratori, M., Ishigurok Enamotos, S., Hirai, M. 1999. Mapping of loci affecting the culture efficiency of microspore culture of *Brassica rapa* L. Syn *campestris* L. using DNA polymorphism. Breed Sci. 49: 187–192.

91. Chen, Z.Z., Synder, S., Fan, Z.G., Loh, W.H. 1994. Efficient production of doubled haploid plants through chromosome doubling of isolated microspores in *B. napus*. Plant Breeding, 113: 217–221.

92. Telmer, C.A., Simmonds, D.H., Newcomb, W. 1992. Determination of development stage to obtain high frequencies of embryogenic microspores in *B. napus*. Physiologia Plantarum. 84: 417–424.

93. Chuong, P.V., Pauls, K.P., Beversdorf, W.D. 1988. High frequency embryogenesis in male sterile plants of *Brassica napus* through microspore culture. Can. J. Bot. 66: 1676–1680.

94. Pechan, P.M., Keller, W.A. 1988. Identification of potentially embryogenic microspores in *B. napus*. Physiologia Plantarum. 74: 377–384.

95. Haung, B., Keller, W.A. 1989. Microspore culture technology. J. Tissue. Cul. Meth. 12: 171–178

96. Takahata, Y., Keller, W. 1991. High frequency embryogenesis and plant regeneration in isolated microspore culture of *Brassica oleracea*. Plant Science 74: 235–242.

97. Duijs, J.G., Voorrips, R.E., Visser, D.L., Custers, J.B.M. 1992. Microspore culture is successful in most crop types of *Brassica oleracea* L. Euphytica 60: 45–55.

98. Custers, J.B.M., Cordewener, J.G.H., Nollen, Y., Dons, H.J.M. 1994. Temperature controls both gametophytic and sporophytic development in microspore cultures of *Brassica napus*. Plant Cell Rep. 13: 267–271.

99. Polsoni, L., Kott, L.S., Beversdorf, W.D. 1988. Large scale microspore culture technique for mutation selection in *Brassica napus*. Can. J. Bot. 66: 1681–1685.

100. Nitsch, C., Nitsch, J.P. 1967. The induction of flowering *in vitro* in stem segments of *Plumbago indica* in the production of vegetative buds. Planta 72: 335–370.

101. Dias, J.C.S. 1999. Effect of activated charcoal on *Brassica oleracea* microspore culture embryogenesis. Euphytica, 108: 65–69.
102. Charne, D.G., Beversdorf, W.D. 1988. Improving microspore culture as a rapeseed breeding tool: the use of auxins and cytokinins in an induction medium. Can. J. Bot. 66: 1671–1675.
103. Mollers, C., Iqbal, M.C.M., Robbelen, G. 1994. Efficient production of doubled haploid *B. napus* plants by colchicine treatment of microspores. Euphytica, 75: 95–104.
104. Iqbal, M.C.M., Molers, C., Robbelem, G. 1994. Increased embryogenesis after colchicine treatment of microspore cultures of *B. napus* L. J. Plant Physiol. 143: 222–226.
105. Zhao, J., Simmonds, D.H., Newcomb, W. 1996. High frequency production of doubled haploid plants of *B. napus* cv. Topas derived from colchicine induced microspore embryogenesis without heat shock. Plant Cell Rep. 15: 668–671.

Plant Biotechnology and Molecular Markers
P.S. Srivastava, Alka Narula and Sheela Srivastava (Editors)

3. Double Fertilisation *in vitro* and Transgene Technology

Erhard Kranz,[1] **Yoichiro Hoshino**[2]**, Takashi Okamoto**[3] **and Stefan Scholten**[1]

[1]Centre for Applied Plant Molecular Biology, AMP II, Institute for General Botany,
University of Hamburg, Ohnhorststr. 18, D-22609 Hamburg, Germany

[2]Field Science Centre for Northern Biosphere, Hokkaido University, Kita 11, Nishi 10,
Kita-Ku, Sapporo 060-0811, Japan

[3]Department of Biological Science, Tokyo Metropolitan University, Minami-Osawa 1-1,
Hachioji, Tokyo 192–0397, Japan

Abstract: The procedure of *in vitro* fertilisation with single isolated maize gametes is the well characterised model system to study fertilisation and early zygotic embryogenesis of higher plants. It allows individual development of zygotes and primary endosperm cells. Both *in vitro* produced zygotes and primary endosperm cells are able to develop into embryos, fertile plants and endosperm in culture. These zygotes and primary endosperm cells are able to self-organise independently from maternal tissue. Many developmental steps of both the *in vitro*-produced embryo and endosperm are comparable to the situation in planta. Application of molecular techniques to the *in vitro* fertilisation system can dissect specific expression patterns of known genes, for example, cell cycle regulators and to isolate unknown genes and their products. Expression of foreign genes is possible in gametes and zygotes. This allows to unravel the roles of genes during fertilisation and early development. The ability of gametes and zygotes to express transgenes enable us to follow the expression of GFP based reporter genes for the visualisation of subcellular components in these living cells.

1. Introduction

Two fertilisation events occur in angiosperm species [1, 2]. During these processes one sperm fuses with the egg and the resulting zygote subsequently develops into an embryo. The other sperm fuses with the secondary nucleus in the central cell forming a primary endosperm cell which develops into endosperm. Now these two fertilisation events can be accomplished *in vitro*. As has been possible for a long time with animal and lower plant gametes, *in vitro* fertilisation (IVF) can be performed with single higher plant gametes. It has been performed mainly with maize (for reviews, see for example [3-7]). The application of single cell culture techniques allows single zygotes to develop into embryos and fertile plants, as well as single *in vitro* fertilised central cells to develop into endosperm. In culture, the zygote without an endosperm and the primary endosperm cell without an embryo are able to develop in a manner similar to that *in vivo*. They are able to self-organise without mother tissue. Thus, an *in vitro* model system for investigations of zygotic embryogenesis and endosperm development is now available to dissect more precisely the early processes which are developmentally important.

To date comprehensive cytological and ultrastructural *in vivo* data on double fertilisation in maize are available [8-13], but there is a lack of molecular information on gamete interactions and only few data exist on fertilisation-induced molecular events after zygote formation. Therefore, cDNA libraries of egg cells and zygotes were generated to explore gene expression after fertilisation [14, 15]. Analyses of these libraries showed that expression of several genes is up- or downregulated after *in vitro* gamete fusion [16]. Expression of some cell cycle genes was investigated in single gametes and zygotes of maize to follow the re-entering of the gametes into the cell cycle between *in vitro* fertilisation and first cell division [17]. This is possible, because the gene expression status of single cells can be investigated by the use of reverse transcriptase polymerase chain reaction (RT-PCR) methods [17, 18].

This article is focused on advances in zygote and primary endosperm cell development *in vitro* and describes the application of transgene technology to study early developmental processes.

2.   *In vitro* Fertilisation

Whereas animal and lower plant IVF-systems can easily make use of naturally free-living gametes, sperm, egg and central cells of angiosperms presuppose their isolation, because the embryo sac is generally deeply embedded in the ovule, and the sperm cells are enclosed in pollen grains or tubes. Micromanipulation techniques and skills are prerequisites for the isolation, fusion and culture of single cells. These methods were developed originally for experiments with somatic cells [19-22]. They were adapted and improved for investigations with gametic cells [23-25]. By use of these methods, experimental access to single gametes, fertilisation and post-fertilisation events under continuous microscopic observation with defined conditions are possible, for example, isolation, selection and fusion of pairs of gametic protoplasts. Also, this allows to design detailed experiments to follow precisely timed early events of zygote, embryo and endosperm formation after gamete fusion.

Isolated gametes (Fig. 1 a and b) are protoplasts and therefore can be fused by techniques that have proved to be successful in the fusion of somatic protoplasts. These are electrofusion and the fusion methods using polyethylene glycol or calcium to induce cell fusion. Sperm and egg cell fusion are electrically induced in maize [17, 23, 25-27] and in wheat [28]. Using maize, the same method was applied to central cell fertilisation [29, 30]. Calcium mediated cell fusion of sperm and egg cells [31-33] and of sperm and central cells [29] were also performed in maize. Possibly attributed to the large differences between the cell sizes isolated sperm cells fuse fast, generally in less than one second with egg and central cells [25].

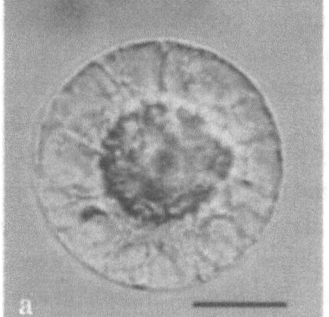 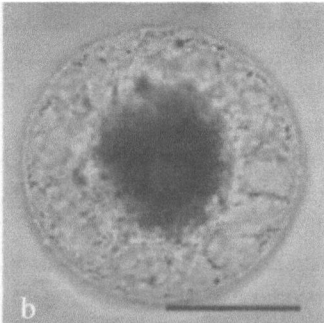 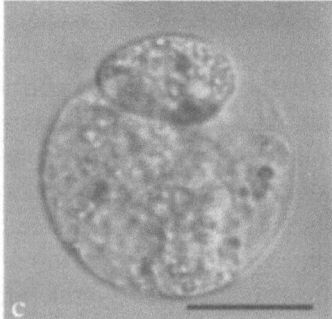

Fig. 1.   Isolated female gametic cells and two-celled embryo derived from *in vitro* fertilised egg cell from maize. (a) egg cell. Bar = 33 μm. (b) central cell. Bar = 105 μm. (c) two-celled embryo, 43 h after *in vitro* fertilisation. Bar = 38 μm.

3. Embryo and Endosperm Development

Development of a single isolated egg cell to the zygote, embryo and finally to a fertile hybrid plant or from an isolated central cell to endosperm after IVF have exclusively been reported in maize [27, 29].

Embryo and endosperm development occurs in culture independently from each other and without female tissue. Sustained growth of *in vitro* fertilised egg cells or isolated zygotes has been achieved by co-cultivation of zygotes and feeder cells. The feeder effect depends on the medium composition which must fulfil the demands of both, the zygote and the feeder cells for optimal growth. *In vitro* development of maize zygotes turned out to be genotype-independent. The generally high developmental capacity may be attributed to the natural predestination of the zygote to form an embryo. Originating from a single *in vitro* zygote, transition stage embryos, consisting of a meristematic region and a suspensor form a scutellum-like compact white tissue, and subsequently, a coleoptile and a plantlet. Clearly, the plant formation occurs without a maturation period, as seed formation is circumvented. Seeds which are obtained from regenerated plants are of the F_2 generation [27].

The maize *in vitro* zygote is metabolically highly active. Newly formed cell wall material can be detected as early as 30 sec after *in vitro* gamete fusion [26]. After IVF, karyogamy was observed as early as 35 min [27] to 45 min [34] in egg cells and 1 h in central cells [29]. It is completed both in the egg and in the central cell within 2 h after fertilisation *in vitro* (HAF). The time course of karyogamy was determined by using isolated, DAPI-stained nuclei of fertilised egg and central cells [29]. Two types of karyogamy were observed in *in vitro* fertilised central cells. The sperm nucleus fuses either with one of the two polar nuclei or with the secondary nucleus which can be formed prior to pollination and fertilisation. This was also found in maize [35]. *In vitro* produced maize zygotes [17, 23, 26, 27, 32, 33] and *in vitro* produced wheat zygotes [28] divide in culture. Depending on culture conditions, in maize it occurs as early as 29 HAF (E. Kranz, unpublished data), but generally 42-46 HAF (Fig. 1c) [26, 27]. Maize zygotes divide in plants about 16 h after karyogamy [35].

Zygotic polarity is mostly of maternal origin. The distinct polarity [23, 26, 27, 36] of *in vitro* and *in vivo* maize zygotes may mainly derive from the uneven distribution of cytoplasm within the egg. Comparable to the situation in the embryo sac in plants, where the cell wall generally surrounds the egg only at the micropylar region, the isolated and cultured egg restores its polarity by forming new cell wall material in a polar fashion [26]. Maize eggs fused either with barley, *Coix* or *Sorghum* sperm cells divide asymmetrically, just as the maize egg divides after homologous fusion [26]. The maize egg also divides asymmetrically after fusion with a wheat sperm. However, when a wheat egg is fused with a maize sperm, the plane of the dividing zygote is rather symmetrical and characteristic to the situation in the wheat egg fused previously with a wheat sperm [6]. Thus, the underlying processes performing the plane of the first cell division asymmetrically are also of maternal origin.

In higher plants, the function of cell cycle regulatory genes during the first zygotic cell cycle remains to be investigated. In maize, cyclin genes are differentially expressed during the first embryonic cell division cycle which is regulated zygotically rather than maternally as in many animal zygotes. Maize sperm cells express the cell division cycle-specific genes *cdc2ZmA/B* and the mitotic cyclin Zeama; *CycA1; 1*. However, the other mitotic cyclins Zeama; *CycB1; 2* and Zeama; *CycB2; 1* are not expressed in the male gametes [17]. What is generally the contribution of the sperm cell in egg cell division? Isolated egg cells of maize and fusion products of two maize egg cells do not divide [23, 26]. However, as in somatic cell culture [37], a short treatment of high amounts of 2,4-D can trigger cell division in cultured isolated egg and central cells [26, 38]. Also, in mutants of *Arabidopsis*, unfertilised central cells can develop into endosperm [39, 40]. In animal and lower plant systems, egg activation and fertilisation-induced signalling events have been widely studied. Investigations like these are now also feasible in angiosperms by using single gametes. In maize egg cells and *in vitro* zygotes, membrane Ca^{2+} and the calcium receptor protein calmodulin are mainly localised in the vicinity of their nuclei [27]. It is well known that calcium ions play a central role in the regulation of metabolic processes and signal transduction [41]. A localised elevation to micromolar Ca^{2+} levels from the increased Ca^{2+} influx across the plasma membrane is needed for early fertilisation events, for example, the generation of the fertilisation potential and cell wall secretion in the brown alga *Fucus serratus* [42]. In maize, a transient elevation of free cytosolic Ca^{2+} in egg cells after fertilisation was reported [33]. An influx of extracellular Ca^{2+} induced by gamete fusion was measured by the use of an extracellular Ca^{2+} selective vibrating probe. The Ca^{2+} influx spread subsequently through the whole egg cell plasma membrane as a wave front, starting in the vicinity of the sperm cell fusion side [43, 44].

In maize, central cell fertilisation can also be performed [29]. The isolated maize central cell does not divide without fertilisation, as the egg cell generally does not divide in culture. However, single fertilised central cells develop into a characteristic tissue, comparable to the *in vivo* situation. The transition from the syncytium to the stage of cellularisation of *in vitro* endosperm occurs within 3-5 days after fertilisation. As found in plants, cell divisions are highly frequent and synchronised after cellularisation. In maize endosperm develops initially more rapidly in the micropylar than in the antipodal area of the fertilised embryo sac [45]. It is characterised by densely cytoplasmic cells predominately located at the base of the suspensor and larger vacuolated cells in other regions near the embryo [46]. *In vitro* produced endosperm consists of one globular part containing small cells with dense cytoplasm and one oblong part with more large cells. Compared to the oblong part, the globular part develops more rapidly in culture. The similarity in morphological polarisation both of the embryo and the endosperm might indicate underlying similar developmental processes and might have a common origin. The central cell might be regarded as a modified egg cell and early endosperm, evolved from a second embryo, develops as a special kind of embryogenesis [5, 47, 48, 49].

In plants, endosperm development is terminated. Plant regeneration from *in vitro* produced endosperm has not been observed. However, shoot bud development from isolated and cultured endosperm of several species was reported [50]. Also, plant regeneration was achieved from callus cultures which originated from excised endosperm (for review see [51]). In maize, plant regeneration from excised immature endosperm derived callus and suspension cultures [52-54], has not been reported.

4.   Transgene Expression in Gametes and Early Development

Transgenic technology provides a way to gain insights of gene function by altering the expression level of a given gene, for example, by overexpression or expression of antisense RNA. A lot of studies have shown that these techniques are well suited to unravel the role of genes important for development, as for example transcription factors [55, 56]. Moreover, the novel marker, green fluorescent protein (GFP), isolated from *Aequorea victoria* extends the possibilities of transgenic technology. Due to its non-toxic nature and the non-invasive visualisation by fluorescence microscopy, GFP permits real-time observations of dynamic changes in living cells. GFP fusion proteins can be used to study subcellular localisation, movements of proteins and organelles *in vivo* [57, 58]. Fusions of GFP with entire proteins of known or unknown function have shown where these proteins are located and whether they move from one compartment to another [59]. The GFP based cameleon calcium indicator, developed by Miyawaki et al. [60] may be used to characterise the spatial and temporal distribution of calcium ions during fertilisation and early embryonic development *in vivo*. Recently the function of this indicator was shown in guard cells of *Arabidopsis* [61]. Other GFP based approaches, being of special interest for the application to the *in vitro* fertilisation system, enable visualisation of cytoskeleton components. A microtubule reporter gene (*gfp-mbd*) was constructed by fusing a GFP gene to the microtubule binding domain of the mammalian microtubule-associated protein 4 (MAP4) gene. GFP-MBD labels cortical microtubules after transient expression of the reporter gene in living epidermal cells of faba bean [62]. Granger and Cyr [63] showed that constitutive expression of the microtubule reporter gene in stable transformed tobacco BY-2 cells allows spatial and temporal resolution of microtubule arrays as they reorganise throughout the cell cycle. Labelling of microtubular structures in intact *Arabidopsis* plants was recently shown by Camilleri et al. [64]. By using GFP fusion proteins, which bind to actin [65] the visualisation of dynamic changes of this component of the cytoskeleton might be achieved.

In addition to the cytoskeleton, GFP that possesses specific intracellular sorting signals for defined cell compartments can be used to tag, for example, endoplasmic reticulum, golgi apparatus and vacuoles. Dynamic changes or reorganisation of these cellular components during zygote and endosperm development can be observed by using transgenic gametes for *in vitro* fertilisation. These examples show that expression of transgenes in isolated gametes and *in vitro* produced zygotes will become a valuable tool for cytological and functional analyses of these developmental stages. So far mainly two strategies are followed to study expression of foreign genes in gametes and zygotes: direct delivery of DNA into these untransformed cells via microinjection and the use of transgenic gametes and zygotes derived from stable transformed plant lines.

4.1 Microinjection

Transient expression of transgenes after microinjection of plasmid DNA in zygotes was reported by Leduc et al. [36]. In this study the *gus* gene under control of the maize histone H3C4 promoter followed by an actin intron and two anthocyanin regulatory genes under control of the 35S promoter were used as reporter genes. They were injected in zygotes of maize and isolated 24 h after pollination. Transient expression, with a frequency of 3.5% on an average was reported in zygotes 4 days after injection. Pónya et al. [66] demonstrated transient expression of reporter genes after microinjection of plasmid DNA into egg cells and isolated zygotes of wheat. A *gfp* gene under control of the ubiquitin promoter was injected into egg cells, whereas the *gus* gene driven by the 35S promoter was injected into zygotes. Transient expression frequencies of 46 and 52% on an average for egg cells and zygotes, respectively were reported. High-frequency AC fields, applied to immobilise the cells on an electrode were suggested by the authors to be a possible reason for such high expression frequencies. However, this remains to be determined.

In general, immobilisation of cells for microinjection is performed with a holding capillary or by embedding them in low melting point agarose. After injection of embedded isolated maize zygotes we obtained transient expression frequencies up to 30% (E. Kranz, unpublished results). In these experiments the GFP gene under control of an enhanced 35S promoter followed by the first intron of the *hsp70* gene [67] was used. GFP fluorescence was monitored about 18 h after injection and culture. The described studies focus on the transient expression of transgenes after microinjection of plasmids into egg cells or zygotes. The advantage of this method is that results can be obtained immediately after injection of DNA into a cell of interest. It might be a suitable method for evaluation of promoter activities in the target cells.

Holm et al. obtained stable transformed plants via microinjection of DNA into isolated zygotes [68]. Basis of these experiments was an efficient regeneration system for isolated barley zygotes. This co-culture system with barley microspores undergoing embryogenesis allows isolated zygotes to develop into embryo-like structures with a frequency of 75%. Fertile plants were regenerated from approximately 50% of these embryo-like structures [69]. After microinjection of the *gus* gene under control of the rice actin promoter into isolated barley zygotes, presence of the construct was confirmed by PCR with a mean frequency of 21% of the derived structures. GUS expression was found in few cases. Two lines of green plants were shown to be transgenic, one of them for an intact copy of the expression cassette beside fragments of the construct. However, the *gus* gene was not expressed. Degradation of the introduced DNA was discussed to be a possible reason for the rarely found expression of the transgene after microinjection [68]. After circumvention of these problems stable transformation via microinjection of zygotes would be of great advantage for applied purposes, since the use of selectable marker genes is not required. The regenerants can be screened directly for the presence of the transgene. Efficient regeneration systems for isolated zygotes which are the basis for this transformation method, were established for wheat [70, 71] and maize [36]. *In vitro* produced maize zygotes can also be efficiently regenerated into plants [27]. *In vitro* fertilisation provides the possibility to inject DNA into egg cells before fertilisation. This option might have an impact on the integration event.

4.2 Transgenic Plant Lines

For investigation of stable integrated transgene expression in maize, gametes and zygotes transgenic plant lines can be generated by microprojectile bombardment of immature embryos. An advantage of stable transformation over transient expression assays is the fact that transgenic lines can be used for various experiments without the need for new time-consuming microinjection experiments. Nevertheless, due to the long generation time of maize, the time needed to establish and characterise transgenic maize lines has to be considered.

There is little information on transcription and translation activity in maize gametes and zygotes. The competence of maize gametes and zygotes to express stable integrated transgenes was shown in our laboratory with plants transgenic for the *gfp* gene. The same *gfp* vector (35S: *gfp*, [67]), as used for microinjection experiments, was introduced. It was optimised for high expression levels of GFP in monocotyledonous plants and codes for a plant codon usage optimised S65T version of the *gfp* gene. In female gametophytes the 35S: *gfp* construct was expressed. Egg cells, synergids, and central cells showed GFP fluorescence, whereas no fluorescence was detected in transgenic male gametes [72]. After fertilisation of non-transgenic egg cells and central cells with transgenic sperm cells expression of the transgene was induced early after fertilisation, and GFP was detected in zygotes and early endosperm (S. Scholten, unpublished results). The 35S promoter construct was active in egg cells, central cells, zygotes, embryos and early endosperm. This opens the possibility to design new experiments and to express various transgenes under control of this promoter construct during fertilisation and early development.

Therefore, we constructed expression vectors adapted to the requirements of maize to label microtubules and actin filaments in living cells according to the constructs described by Marc et al. [62] and Kost et al. [65]. Double labelling experiments with spectral GFP variants [73] or the recently isolated red fluorescent protein [74] might enable analyses of dynamic changes and the interactions of actin filaments and microtubules *in vivo*. We chose GFP to label microtubules and the red fluorescent protein to label actin filaments. Transient expression analyses of these two constructs revealed that double labelling of both cytoskeletal components is possible in living cells (Fig. 2). The next step will be the generation and characterisation of transgenic plant lines expressing both constructs in gametes and zygotes. Once established, these lines could be used to study dynamics and interactions of the main cytoskeletal components during early development *in vivo*.

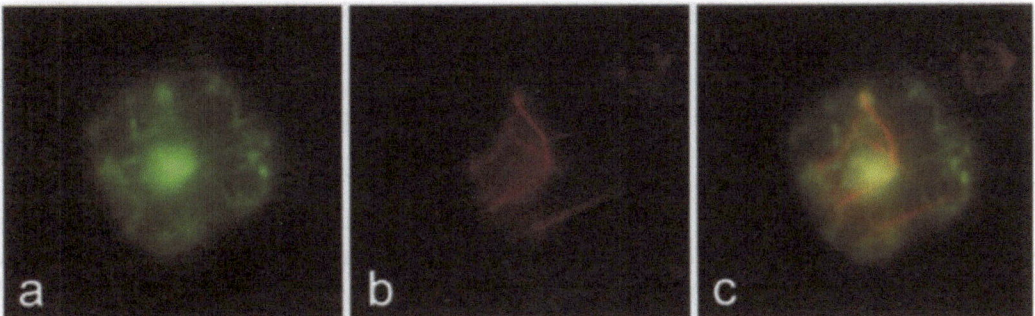

Fig. 2. **Transient expression of microtubule and actin tagged fluorescent proteins. Scutellar tissue of immature maize embryos were bombarded with constructs (see text for details) to tag both major cytoskeletal components with fluorescent proteins. (a-c) Cell with tagged microtubules and actin filaments. (a) Microtubules tagged with green fluorescent protein. (b) Actin filaments tagged with red fluorescent protein. (c) Overlay of (a) and (b) showing both major cytoskeletal components tagged with fluorescent proteins within the same cell.**

5. Prospects

IVF with single gametes can now be used for wide hybridisation approaches to create new hybrid and cybrid plants. Possibly due to zygotic and postzygotic incompatibility mechanisms, resulting hybrid plants might be restricted to hybridisation between more closely related species. This has been demonstrated in egg activation studies: Cell divisions were triggered in isolated maize eggs by sperm cells of several cereal species. However, zygotic incompatibility was observed after *in vitro* fusion of maize eggs with *Brassica* sperm cells [26].

Also, IVF techniques are valuable experimental tools for the elucidation of various processes of double fertilisation and early development of embryo and endosperm under defined conditions. Clearly, important progress towards a better understanding of these processes will continue to come from analyses of mutants. However, experimental access to single higher plant gametes and zygotes will facilitate studies on fertilisation and early developmental processes which are difficult to investigate in plants. These studies together with gene cloning, protein isolation and characterisation will certainly allow a comparison with fertilisation-induced processes occurring in lower plants and animals [75, 76].

Fertilisation-induced signal transduction events, changes in the endoplasmic reticulum, cytoskeleton, and nuclear movement are now possible to be studied under defined conditions, e.g., an exact time point after gamete fusion. Such studies can be performed both in the zygote and in the primary endosperm cell allowing comparative studies. Thus, they will provide a more precise picture of co-ordinated processes during early developmental stages of the embryo and the endosperm [5, 77, 78].

Molecular analyses are possible with few cells. In maize, cDNA libraries from egg cells and *in vitro* zygotes were constructed by using RT/PCR techniques to isolate and to study the function of the cloned egg and fertilisation induced genes [6, 14, 15]. PCR protocols were adapted for expression studies of known genes by use of single cells [18], for example, to follow gene expression of cell cycle regulatory genes in a time course during zygote development [17]. Also, fertilised central cells and primary endosperm cells are promising target cells for the isolation of unknown genes and expression studies by using especially endosperm specific genes [77, 79-81]. For functional analyses of gene products, existing protocols such as immunocytochemical techniques for protein detection and methods for protein isolation are being currently adapted to single cells and to small cell aggregates in our laboratory. These tools will provide a valuable contribution to the elucidation of common features and differences in zygotic and somatic developmental processes.

Many processes involved in early endosperm development might well be studied during development of *in vitro* produced or isolated primary endosperm cells by using defined culture conditions. These are, for example, the suppression of phragmoplast formation between nuclei, the mitotic hiatus, the synchronised re-initiation of mitosis, the periclinal phragmoplast formation, the initiation of cellularisation via formation of nucleocytoplasmic domains (NCD) of a radial microtubular array, alveolation, the programming of nuclear location and division planes during cell wall formation in the syncytium [82].

In this respect the expression of GFP based marker genes might be a valuable tool. Stable transgenic lines showed that central cells and early endosperm as well as egg cells and early embryos are competent to express transgenes. *In vitro* fertilisation and culture systems enable direct observation and monitoring of the development of individual cells. Combination of this option with the expression of GFP based marker genes for subcellular structures will facilitate new strategies to analyse cytological characteristics during fertilisation, early zygotic and endosperm development. The use of fluorescent protein based markers for cytoskeletal components is of high interest, since the plant cytoskeleton has crucial functions in cellular processes that are essential for cell morphogenesis and development [83]. Once established through transgenic maize plants, cytological markers might be of value to correlate expression data with specific developmental stages, e.g., cell cycle phases through visualisation of microtubular structures. Also, the GFP based cameleon calcium indicator [60] might be a possibility to characterise the spatial and temporal distribution of calcium ions during fertilisation and early embryonic development *in vivo*. The function of this indicator was demonstrated in guard cells of *Arabidopsis* [61].

Additionally, transgenic approaches provide the opportunity for functional analyses during fertilisation and very early zygotic and endosperm development. Transcription factors being expressed in egg cells, and cell cycle regulators, both might have a critical role during fertilisation, further development or morphogenesis and thus are interesting candidates for antisense and overexpression studies. A more comprehensive view on the fertilisation processes and early development will certainly be the result of linking *in vitro* fertilisation with transgenic technology.

Acknowledgements

This article is dedicated to Sant S. Bhojwani, an outstanding scientist and a person we highly regard, on the occasion and in honour of his 62nd birthday.

References

1. S. Nawaschin, Resultate einer Revision der Befruchtungsvorgänge bei *Lilium martagon* und *Fritillaria tenella*, Bull. Acad. Imp. Sci. St. Petersburg 9 (1898) 377–382.
2. M.L. Guignard, Sur les anthérozoides et la double copulation sexuelle chez les végétaux angiospermes, Rev. Gen. Bot. 11 (1899) 129–135.
3. E. Kranz, T. Dresselhaus, *In vitro* fertilization with isolated higher plant gametes, Trends Plant Sci. 1 (1996) 82–89.
4. E. Kranz, J. Kumlehn, Angiosperm fertilization, embryo and endosperm development *in vitro*, Plant Sci. 142 (1999) 183–197.
5. J.-E. Faure, C. Dumas, Fertilization flowering plants. New approaches for an old story, Plant Physiol. 125 (2001) 102–104.
6. E. Kranz, J. Kumlehn, T. Dresselhaus, Fertilization and zygotic embryo development *in vitro*, in: M. Cresti, G. Cai, A. Moscatelli (Eds), Fertilization in Higher Plants, Springer, Heidelberg, 1998, pp. 337–349.
7. E. Kranz, *In vitro* fertilization, in: S. S. Bhojwani, W.Y. Soh (Eds), Current Trends in the Embryology of Angiosperms, Kluwer Academic Publishers, 2001, pp. 143-166.
8. V.H. Rhoades, A study of fertilization in *Zea mays*. Thesis for the degree of Master of Science. Cornell University, 1934.
9. A.G. Diboll, D.A. Larson, An electron microscopic study of the mature megagametophyte in *Zea mays*, Am. J. Bot. 53 (1966) 391–402.
10. A.G. Diboll, Fine structural development of the megagametophyte of *Zea mays* following fertilization, Am. J. Bot. 55 (1968) 787–806.
11. A.A.M. Van Lammeren, A comparative ultrastructural study of the megagametophytes in two strains of *Zea mays* L. before and after fertilization, Agric. Univ. Wageningen Papers 86–1, 1986, 1–37.
12. A.A.M. Van Lammeren, H. Kieft, Cell differentiation in the pericarp and endosperm of developing maize kernels (*Zea mays* L.) with special reference to the microtubular cytoskeleton, in: Embryogenesis in *Zea mays* L. A Structural Approach to Maize Caryopsis Development *in vivo* and *in vitro*. Proefschrift A. A. M. van Lammeren (Wageningen University), 1987, pp. 115–148.
13. J.H.N. Schel, H. Kieft, An ultrastructural study of embryo and endosperm development during *in vitro* culture of maize ovaries (*Zea mays*), Can. J. Bot. 64 (1986) 2227–2238.
14. T. Dresselhaus, H. Lörz, E. Kranz, Representative cDNA libraries from few plant cells, Plant J. 5 (1994) 605–610.
15. T. Dresselhaus, C. Hagel, H. Lörz, E. Kranz, Isolation of a full-size cDNA encoding calreticulin from a PCR-library of *in vitro* zygotes of maize, Plant Mol. Biol. 31 (1996) 23–34.
16. T. Dresselhaus, S. Cordts, S. Heuer, M. Sauter, H. Lörz, E. Kranz, Novel ribosomal genes from maize are differentially expressed in the zygotic and somatic cell cycles, Mol. Gen. Genet. 261 (1999) 416–427.
17. M. Sauter, P. von Wiegen, H. Lörz, E. Kranz, Cell cycle regulatory genes from maize are differentially controlled during fertilization and first embryonic cell division, Sex. Plant Reprod. 11 (1998) 41–48.

18. J. Richert, E. Kranz, H. Lörz, T. Dresselhaus, A reverse transcriptase polymerase chain reaction assay for gene expression studies at the single cell level, Plant Sci. 114 (1996) 93–99.

19. H.U. Koop, H.G. Schweiger, Regeneration of plants from individually cultivated protoplasts using an improved microculture system, J. Plant Physiol. 121 (1985) 245–257.

20. H.U. Koop, H.G. Schweiger, Regeneration of plants after electrofusion of selected pairs of protoplasts, Eur. J. Cell Biol. 39 (1985) 46–49.

21. H.G. Schweiger, J. Dirk, H.U. Koop, E. Kranz, G. Neuhaus, G. Spangenberg, D. Wolff, Individual selection, culture and manipulation of higher plant cells, Theor. Appl. Genet. 73 (1987) 769–783.

22. G. Spangenberg, H.U. Koop, Low density cultures: microdroplets and single nurse culture, in: K. Linsey (Ed), Plant Tissue Culture Manual A10, Kluwer Academic Publishers, Dordrecht, 1992, pp. 1–28.

23. E. Kranz, J. Bautor, H. Lörz, In vitro fertilization of single, isolated gametes of maize mediated by electrofusion, Sex Plant Reprod. 4 (1991) 12–16.

24. E. Kranz, In vitro fertilization of maize mediated by electrofusion of single gametes, in: K. Lindsey (Ed), Plant Tissue Culture Manual E1, Kluwer Academic Publishers, Dordrecht, 1992, 1–12.

25. E. Kranz, In vitro fertilization with isolated single gametes, in: R. Hall (Ed), Methods in Molecular Biology, 111, Plant Cell Culture Protocols, Humana Press Inc., Totowa, NJ, 1999, pp. 259–267.

26. E. Kranz, P. von Wiegen, H. Lörz, Early cytological events after induction of cell division in egg cells and zygote development following in vitro fertilization with angiosperm gametes, Plant J. 8 (1995) 9–23.

27. E. Kranz, H. Lörz, In vitro fertilization with isolated, single gametes results in zygotic embryogenesis and fertile maize plants, Plant Cell 5 (1993) 739–746.

28. M. Kovács, B. Barnabás, E. Kranz, Electro-fused isolated wheat (Triticum aestivum L.) gametes develop into multicellular structures, Plant Cell Rep. 15 (1995) 178–180.

29. E. Kranz, P. von Wiegen, H. Quader, H. Lörz, Endosperm development after fusion of isolated, single maize sperm and central cells in vitro, Plant Cell 10 (1998) 511–524.

30. E. Kranz, J. Bautor, H. Lörz, Electrofusion-mediated transmission of cytoplasmic organelles through the in vitro fertilization process, fusion of sperm cells with synergids and central cells, and cell reconstitution in maize, Sex Plant Reprod. 4 (1991) 17–21.

31. J.E. Faure, C. Digonnet, C. Dumas, An in vitro system for adhesion and fusion of maize gametes, Science 263 (1994) 1598–1600.

32. E. Kranz, H. Lörz, In vitro fertilisation of maize by single egg and sperm cell protoplast fusion mediated by high calcium and high pH, Zygote 2 (1994) 125–128.

33. C. Digonnet, D. Aldon, N. Leduc, C. Dumas, M. Rougier, First evidence of a calcium transient in flowering plants at fertilization, Development 124 (1997) 2867–2874.

34. J.E. Faure, H.L. Mogensen, C. Dumas, H. Lörz, E. Kranz, Karyogamy after electrofusion of single egg and sperm cell protoplasts from maize: Cytological evidence and time course, Plant Cell 5 (1993) 747–755.

35. R. Mól, E. Matthys-Rochon, C. Dumas. The kinetics of cytological events during double fertilization in Zea mays L., Plant J. 5 (1994) 197–206.

36. N. Leduc, E. Matthys-Rochon, M. Rougier, L. Mogensen, P. Holm, J.-L. Magnard, C. Dumas, Isolated maize zygotes mimic in vivo embryonic development and express microinjected genes when cultured in vitro, Dev. Biol. 177 (1996) 190–203.

37. D. Dudits, J. Györgyey, L. Bögre, L. Bakó, Molecular biology of somatic embryogenesis, in: T.A. Thorpe (Ed), In Vitro Embryogenesis in Plants, Kluwer Academic Publishers, Dordrecht, 1995, pp. 267–308.

38. Y. Fu, M.X. Sun, H.Y. Yang, C. Zhou, In vitro divisions of unfertilized central cells and other embryo sac cells in Nicotiana tabacum var. Macrophylla, Acta Bot. Sin. 39 (1997) 778–781.

39. N. Ohad, L. Margossian, Y.-C. Hsu, C. Williams, P. Repetti, R.L. Fischer, A mutation that allows endosperm development without fertilization, Proc. Natl. Acad. Sci. USA 93 (1996) 5319–5324.

40. A.M. Chaudhury, L. Ming, C. Miller, S. Craig, E.S. Dennis, W.J. Peacock, Fertilization-independent seed development in Arabidopsis thaliana, Proc. Natl. Acad. Sci. U.S.A. 94 (1997) 4223–4228.

41. G. Zhang, D.D. Cass, Calcium signalling in sexual reproduction of flowering plants, Recent Res. Dev. Plant Physiol. 1 (1997) 75–83.

42. S.K. Roberts, I. Gillot, C. Brownlee, Cytoplasmic calcium and *Fucus* egg activation, Development, 120 (1994) 155–163.

43. A.F. Antoine, J. Faure, S. Cordeiro, C. Dumas , M. Rougier, J-A. Feijo, A calcium influx is triggered and propagates in the zygote as a wavefront during *in vitro* fertilization of flowering plants, Proc. Natl. Acad. Sci. USA 12 (2000) 10643–10648.

44. A.F. Antoine, J. Faure, C. Dumas, J-A. Feijo, Differential contribution of cytoplasmic Ca^{2+} and Ca^{2+} influx to gamete fusion and egg activation in maize, Nat. Cell Biol. 3 (2001) 1120–1123.

45. L.F. Randolph, Developmental morphology of the caryopsis in maize, J. Agric. Res., 53 (1936) 881–916.

46. J.H.N. Schell, H. Kieft, A.A.M. Van Lammeren, Interactions between embryo and endosperm during early developmental stages of maize caryopses (*Zea mays*), Can. J. Bot. 62 (1984) 2842–2853.

47. M. Favre-Duchartre, Homologies and phylogeny, in: B.M. Johri (Ed), Embryology of Angiosperms, Springer, Berlin, 1984, pp. 697–734.

48. E. Sargant, Recent work on the results of fertilization in angiosperms, Annals of Bot. 14 (1900) 689–712.

49. W.E. Friedman, Organismal duplication, inclusive fitness theory, and altruism: Understanding the evolution of endosperm and the angiosperm reproductive syndrome, Proc. Natl. Acad. Sci. USA 92 (1995) 3913–3917.

50. B.M. Johri, S.S. Bhojwani, Growth responses of mature endosperm in cultures, Nature, 208 (1965) 1345–1347.

51. S.S. Bhojwani, Culture of endosperm, in: I.K. Vasil, (Ed), Cell Culture and Somatic Cell Genetics of Plants, 1, Academic Press, Inc., Orlando, 1984, pp. 258–268.

52. J. Straus, Maize endosperm tissue grown *in vitro*. II. Morphology and cytology, Am. J. Bot. 41 (1954) 833–839.

53. M. Tabata, F. Motoyoshi, Hereditary control of callus formation in maize endosperm cultures *in vitro*, Japan. J. Genet. 40 (1965) 343–355.

54. J.C. Shannon, J.W. Lui, A simplified medium for the growth of maize (*Zea mays*) endosperm tissue in suspension culture, Physiol. Plant. 40 (1977) 285–291.

55. S. Ramachandran, K. Hiratsuka, N.-H. Chua, Transcription factors in plant growth and development, Current Opin. Gene Dev. 4 (1994) 642–646.

56. L. Meisel, E. Lam, Switching on gene expression: Analysis of the factors that spatially and temporally regulate plant gene expression, in: Setlow J.K. (Ed) Genetic Engineering, Plenum Press, New York, 19 (1997) 183–198.

57. R.J. Grebenok, E. Pierson, G.M. Lambert, F.-C. Gong, C.L. Afonso, R. Halderman-Cahill, J.C. Carrington, D.W. Galbraith, Green-fluorescent protein fusions for efficient characterization of nuclear targeting, Plant J. 11 (1997) 573–586.

58. R.H. Köhler, GFP for *in vivo* imaging of subcellular structures in plant cells, Trends Plant Sci. 3 (1998) 317–320.

59. M.R. Hanson, R.H. Köhler, GFP imaging: methodology and application to investigate cellular compartmentation in plants, J. Exp. Bot. 356 (2001) 529–539.

60. A. Miyawaki, J. Llopis, R. Heim, J.M. McCaffery, J.A. Adams, M. Ikura, R.Y. Tsien, Fluorescent indicators for Ca^{2+} based on green fluorescent proteins and calmodulin, Nature 388 (1997) 882–887.

61. G.J. Allen, J.M. Kwak, S.P. Chu, J. Llopis, R.Y. Tsien, H.F. Harper, J.I. Schroeder, Cameleon calcium indicator reports cytoplasmic calcium dynamics in *Arabidopsis* guard cells, Plant J. 19 (1999) 735–747.

62. J. Marc, C.L. Granger, J. Brincat, D.D. Fisher, T.-H. Kao, A.G. McCubbin, R.J. Cyr, A *gfp-map4* reporter gene for visualizing cortical microtubule rearrangements in living epidermal cells, Plant Cell 10 (1998) 1927–1939.

63. C.L. Granger, R.J. Cyr, Microtubule reorganization in tobacco BY-2 cells stably expressing GFP-MBD, Planta 210 (2000) 502–509.

64. C. Camilleri, J. Azimzahdeh, M. Pastugila, C. Bellini, O. Grandjean, D. Bouchez, The *Arabidopsis TONNEAU2* gene encodes a putative novel protein phosphatase 2A regulatory subunit essential for the control of the cortical cytosteleton, Plant Cell 14 (2002) 833–845.

65. B. Kost, P. Spielhofer, N.-H. Chua, A GFP-mouse talin fusion protein labels plant actin filaments *in vivo* and visualizes the actin cytoskeleton in growing pollen tubes, Plant J. 16 (1998) 393–401.

66. Z. Pónya, P. Finy, A. Fehér, J. Mitykó, D. Dudits, B. Barnabás, Optimisation of introducing foreign genes into egg cells and zygotes of wheat (*Triticum aestivum* L.) via micoinjection, Protoplasma 208 (1999) 163–172.

67. S.-Z. Pang, D.L. DeBoer, Y. Wan, G. Ye, J.G. Layton, M.K. Neher, C.L. Armstrong, J.E. Fry, M.A.W. Hinchee, M.E. Fromm, An improved green fluorescent protein gene as a vital marker in plants, Plant Physiol. 112 (1996) 893–900.

68. P.B. Holm, O. Olsen, M. Schnorf, H. Brinch-Pedersen, S. Knudsen, Transformation of barley by microinjection into isolated zygote protoplasts, Transgenic Res. 9 (2000) 21–32.

69. P.B. Holm, S. Knudsen, P. Mouritzen, D. Negri, F.L. Olsen, C. Roué, Regeneration of fertile barley plants from mechanically isolated protoplasts of the fertilized egg cell, Plant Cell 6 (1994) 531–543.

70. J. Kumlehn, R. Brettschneider, H. Lörz, E. Kranz, Zygote implantation to cultured ovules leads to direct embryogenesis and plant regeneration of wheat, Plant J. 12 (1997) 1473–1479.

71. J. Kumlehn, H. Lörz, E. Kranz, Differentiation of isolated wheat zygotes into embryos and normal plants, Planta 205 (1998) 327–333.

72. S. Scholten, E. Kranz, *In vitro* fertilization and expression of transgenes in gametes and zygotes, Sex. Plant Reprod. 14 (2001) 35–40.

73. T.-T. Yang, P. Sinai, G. Green, P.A. Kitts, Y.-T. Chen, L. Lybarger, R. Chervenak, G.H. Patterson, D.W. Piston, S.R. Kain, Improved fluorescence and dual color detection with enhanced blue and green variants of the green fluorescent protein, J. Biol. Chem. 273 (1998) 8212–8216.

74. M. Matz, A.F. Fradkov, Y.A. Labas, A.P. Savitsky, A.G. Zaraisky, M.L. Markelov, S.A. Lukyanov, Fluorescent proteins from nonbioluminescent anthozoa species, Nature Biotechnol. 17 (1999) 969–973.

75. D. Epel, The initiation of development at fertilization. Cell Diff. Devel. 29 (1990) 1–12.

76. V.D. Vacquier, Evolution of gamete recognition proteins. Science 281 (1998) 1995–1998.

77. O.A. Olsen, R.H. Potter, R. Kalla, Histo-differentiation and molecular biology of developing cereal endosperm, Seed Sci. Res., 2 (1992) 117–131.

78. M.A. Lopes, B.A. Larkins, Endosperm origin, development and function. Plant Cell 5 (1993) 1383–1399.

79. G. Hueros, S. Varotto, F. Salamini, R.D. Thompson, Molecular characterization of *BET1*, a gene expressed in the endosperm transfer cells of maize, Plant Cell 7 (1995) 747–757.

80. D.N.P. Doan, C. Linnestad, O.A. Olsen, Isolation of molecular markers from the barley endosperm coenocyte and the surrounding nucellus cell layers, Plant Mol. Biol. 31 (1996) 877–886.

81. H.G. Opsahl-Ferstad, E. Le Deunff, C. Dumas, P.M. Rogowsky, *ZmEsr*, a novel endosperm-specific gene expressed in a restricted region around the maize embryo, Plant J. 12 (1997) 235–246.

82. O.A. Olsen, Endosperm developments, Plant Cell, 10 (1998) 485–488.

83. B. Kost, J. Mathur, N.-H. Chua, Cytoskeleton in plant development, Current Opin. Plant Biol. 2 (1999) 426–470.

Plant Biotechnology and Molecular Markers
P.S. Srivastava, Alka Narula and Sheela Srivastava (Editors)

4. Polymorphism of Sexual and Somatic Embryos as Manifestation of Their Developmental Parallelism Under Natural Conditions and in Tissue Culture

Tatyana B. Batygina

Department of Embryology and Reproductive Biology, Komarov Botanical Institute of the
Russian Academy of Science, Prof. Popov str., 2, 197376 St. Petersburg, Russia

Abstract: The new approach and strategy of investigation have permitted to reveal the existence of somatic embryo not only *in vitro*, but also in natural conditions.

Plant organism is able to form somatic embryos at all stages of its development and on different organs (vegetative and generative) along with sexual reproduction. This ability enlarges the plasticity and tolerance of reproductive system. The system approach also permitted to introduce the new notion—"embryoidogeny", as a special category of asexual reproduction (vegetative) *in situ*, *in vivo* and *in vitro* and consider its role in the reproductive system of flowering plants [1, 2]. The embryoidogeny includes the following forms: ovular (nucellar and integumental), embryonic (monozygotic-cleavage) and homophasic vivipary (foliar, cauligenic and rhizogenic). It is somatic embryo that is the elementary structural unit of all these reproductive forms.

The development of sexual and somatic embryos proceeds in parallels manifesting their great polymorphism. The only difference between them is the origin: zygotic embryo of heterophasic reproduction, embryoid of homophasic one. Various transitional forms from embryo, embryoid to bud can be revealed.

There are sexual embryos, embryoids and buds that appear as three elementary structural units of reproduction and propagation.

The correlation of different modes of reproduction in plants, i.e. the reproductive strategy of species [3] is predominantly determined by the process of adaptive evolution. This prolonged process has led to the emergence of a great diversity of reproductive structures. These are zygotic embryos, adventive buds, propagules, somatic embryos, etc. All the diversity of embryo-like structures appear in plant tissue culture *in vitro* and on plants *in situ* and *in vivo* that was generalised by the term "embryoid" [4]. Later the following definition was created to expand the term: embryoid (somatic embryo)—the originative structure arising asexually *in situ*, *in vivo* and *in vitro*. Embryoid is bipolar at all stages of development as also the sexual embryo. Unlike these two structures, a bud develops as a pole itself and can give rise to an individual only after regeneration of roots [3, 5].

Similarity of sexual and somatic embryo development is considered to be one of the main points under the discussion of terms of the morphogenetic pathways such as embryogenesis, embryoidogenesis and gemmorhizogenesis. The point of discussion is: to what extent the morphogenesis of somatic embryo corresponds to that of sexual embryo, i.e. to what extent it obeys Errera's, Sachs's and Hertwig's laws of cell division and the laws of embryogeny [6, 7].

The attention paid to the role of embryoidogeny in general system of plant reproduction is insufficient. Terminology connected with embryo-like structures and their classification also remains debatable.

Only separate aspects of somatic embryo morphogenesis *in vitro* have been studied in terms of comparative embryology. The comparison of somatic embryos under natural conditions (cleavage, nucellar, integumentary, cauligenous, rhizogenous, foliar) with sexual embryos and with somatic embryos from *in vitro* culture is practically absent.

Well known Indian embryologists Swamy and Krishnamurthy [8] came to a conclusion that the development of somatic embryo *in vitro* would never stand comparison with that of zygotic one. They have stated that embryoid is devoid of the main features of sexual embryo and from the morphological standpoint more closely related to a shoot bud (Fig. 1). Their arguments are the following: embryoid is devoid of the internal differentiation because its initial cell lacks polarisation; embryoid lacks an orderly pattern of cell divisions; the laws of cell divisions of Errera and Sachs are not obeyed; suspensor does not form; the protoderm formation is belated and also incomplete; embryoid lacks typical centres of polar organization—the hypophysis and epiphysis; it often lacks the initials of hypocotyls as well; the organization of main root is suppressed, plerome and periblem do not develop, and so on.

The latest data on the development of somatic embryo *in situ* and *in vitro* give evidence with regard to the abovementioned signs proposed by Swamy and Krishnamurthy to distinguish sexual embryos from somatic embryos, as optional. They can not be used in all cases of definition of morphological status of sexual embryo, as well as of embryoid.

The system approach [9, 10] allowed to compare the modes of sexual and somatic embryo formation on the model objects *in situ*, *in vivo* and *in vitro* for the first time. It has been revealed that the developmet of a somatic embryo *in situ* and *in vitro* recapitulates a sexual embryo with the origin being the only difference (sexual embryo results from heterophasic reproduction, somatic embryo—homophasic reproduction). A new type of asexual reproduction in flowering plants—embryoidogeny—has been established. Somatic embryo occurs to be an elementary structural unit of embryoidogeny [4, 11-13].

In connection with this it seems necessary to carry out a comparative analysis of sexual and somatic embryo development (*in situ*, *in vivo* and *in vitro*). The following structures have been chosen for comparison: (1) the initial cells of sexual and somatic embryos [14]; (2) sexual embryos of plants with different types of embryogenesis, and of seedling development with respect to their reproductive biology and ecology; (3) somatic embryos developing on different parts of plant (seed, leaf, stem, root); (4) somatic embryos, obtained from different explants (plants from diverse taxa) with various modes of initiation and of development in *in vitro* culture.

The present article examines sexual and somatic embryos in terms of certain characteristics, for example: polarity, symmetry, the formation of epiphysis and hypophysis, development of main organs, shoot and root apices, etc.

1. Sexual Embryos Under Natural Conditions

A high totipotency peculiar to the mature zygote, preserved in *ca* and *cb* cells has been recognized (Fig. 2). Souéges took the degree of participation of these cells in embryo formation as a basis for the classification of embryogenic types. He established six basic types (megarchetypes) [15].

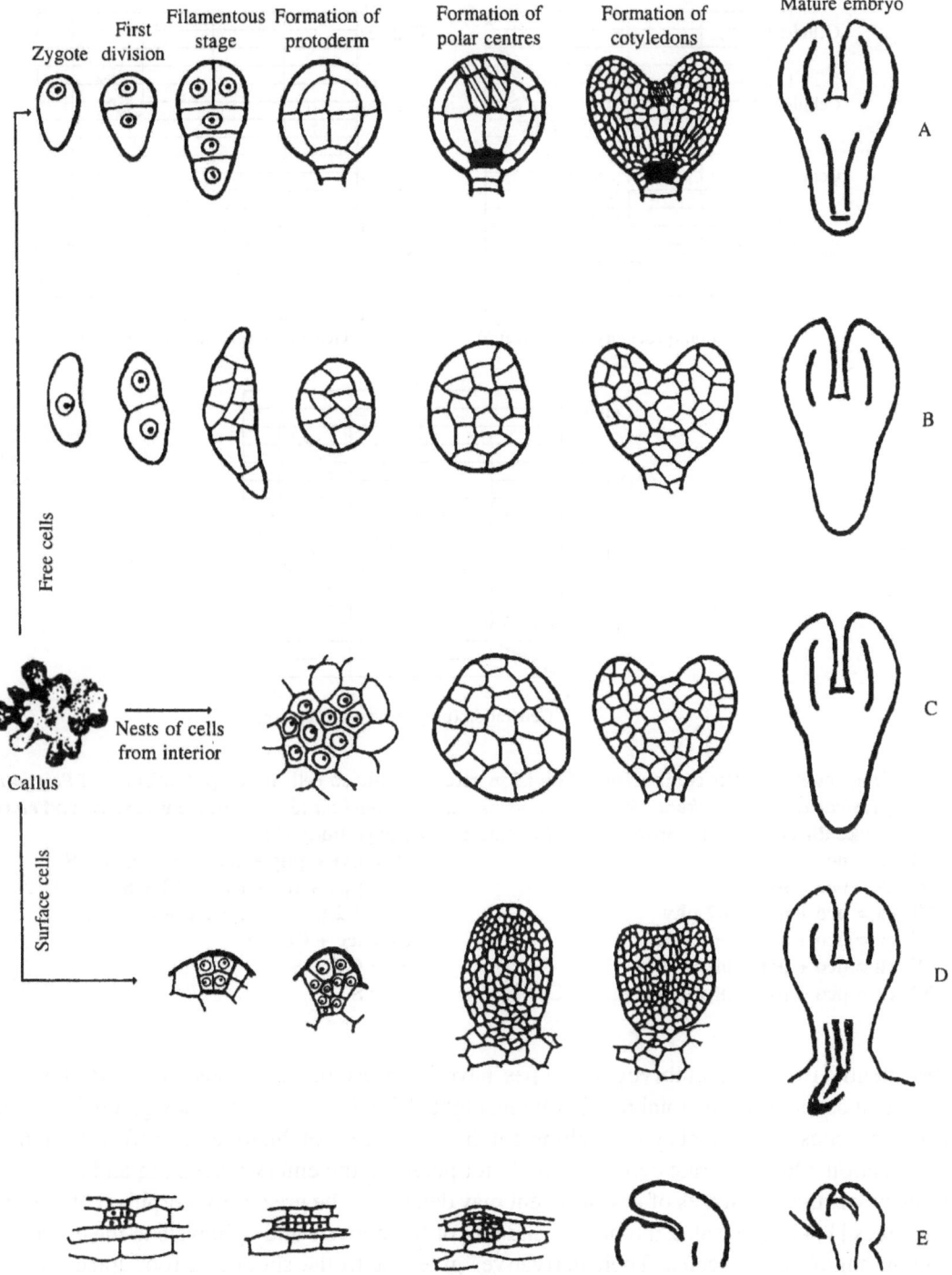

Fig. 1. Diagrammatic representation of the development of a sexually produced typical dicotyledon embryo (A), of embryoids (B, C and D) and of an adventitious shoot bud (E) [8].

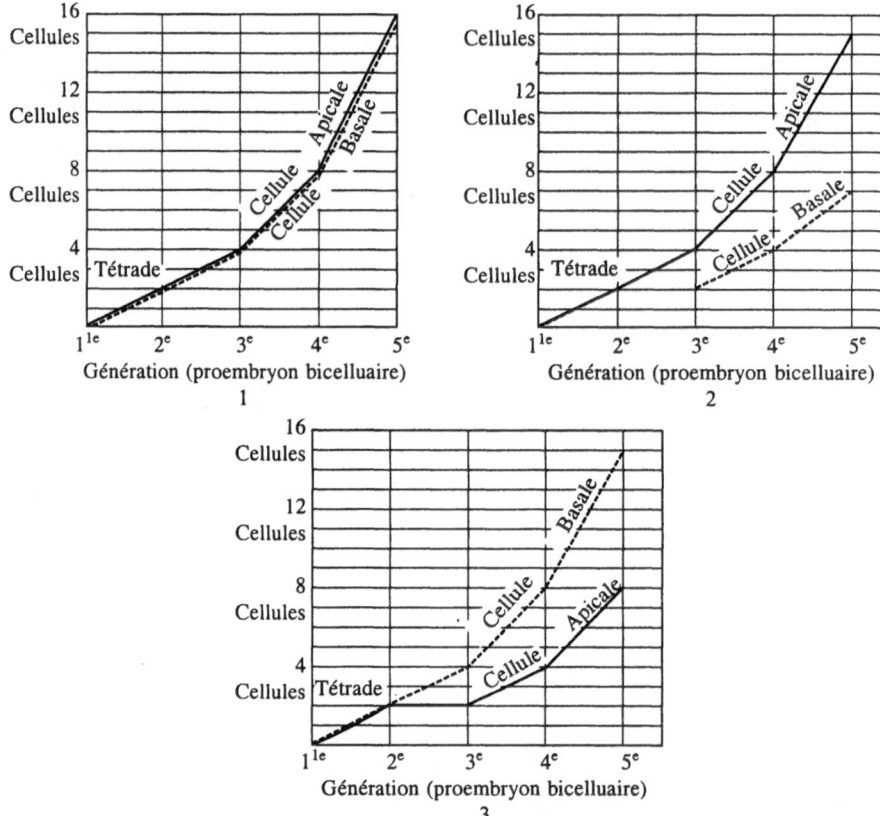

Fig. 2. Diagrams show the comparison of division rate of *ca* and *cb* cells in the proembryos of *Polygonum persicaria* (1), *Oenothera biennis* (2) and *Erodium cicutarium* (3). In different taxa *ca* and *cb* cells make different contribution to the formation of embryo body [15].

I	ca = pco	cb = pvt + phy + icc + icc + CO + S
II	ca = pco + pvt	cb = phy + icc + iec + CO + S
III	ca = pco + pvt + 1.2 phy	cb = 1/2 phy + icc + icc + CO ± S
IV	ca = pco + pvt + phy + ice	cb = icc + Co + S
V	ca = pco + pvt + phy + pee + iec	cb = CO + S
VI	ca = pco + pvt + phy + icc + iec + Co	cb = S

Subsequently two more embryogenic types have been established (Paeonad- and Graminad-types), that account for the total number being eight [16] (Fig. 3). Totipotency gradually reduces during the subsequent embryo development in the course of histogenic differentiation and specialization. Only separate cell loci remain totipotent in the embryo, seedling and plant. High totipotency at the first stages of development may determine the great polymorphism of embryos.

Souéges [17] has revealed the initials and loci of hypophysis and epiphysis in the embryos of different angiosperm species. Their derivatives give rise to the shoot and root apices (Fig. 4). The formation of hypophysis and epiphysis initials and loci takes place at the early stages of embryogenesis, but the point of their differentiation is taxon-specific. Ontophylogenetic approach allows to reveal, at least, following five groups of zygotic embryos (Fig. 5):

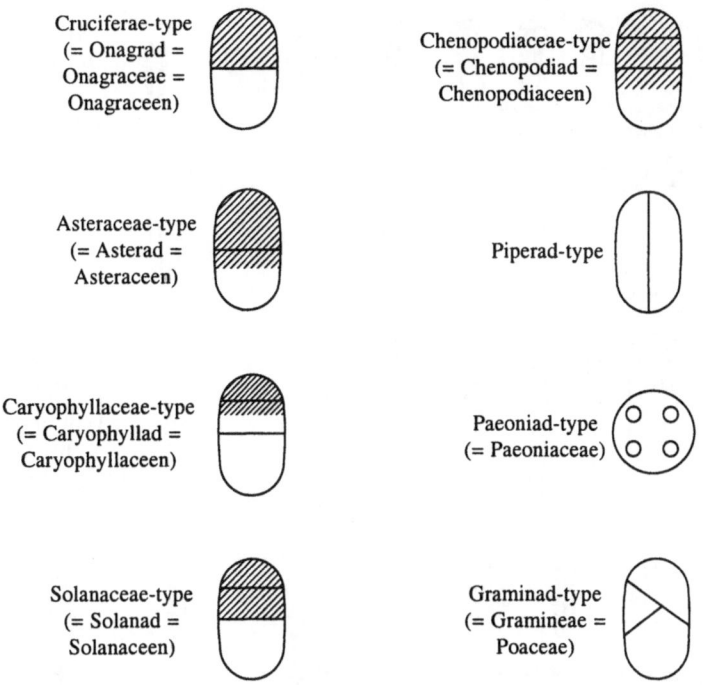

Fig. 3. The main types of embryogenesis in angiosperms [16].

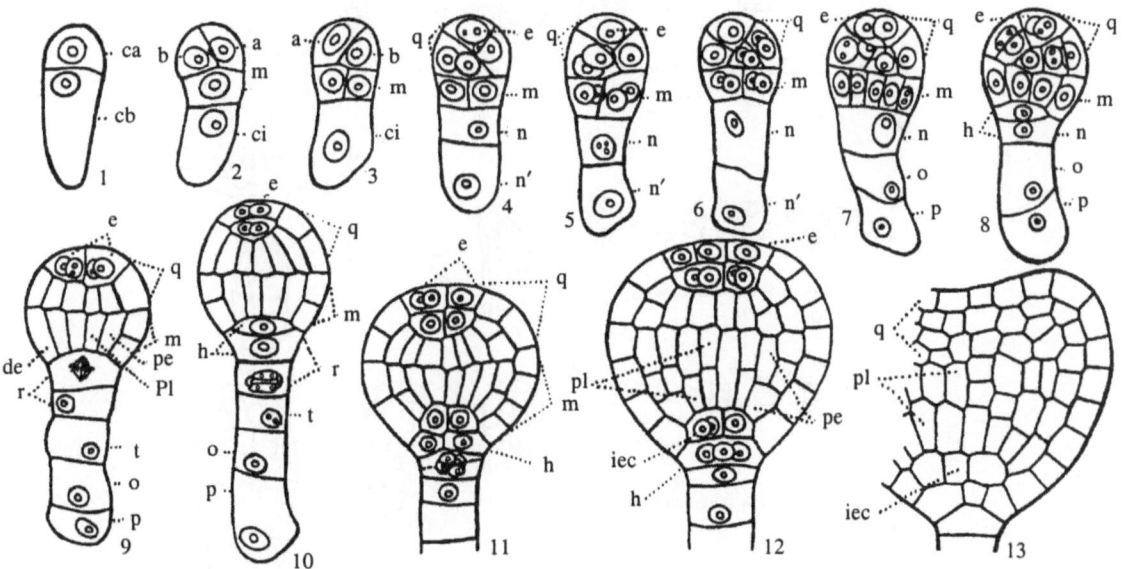

Fig. 4. Early and middle stages of *Geum urbanum* L. embryo development [18].

1. The embryo exhibits typical initials and loci of hypophysis and epiphysis (for example, *Geum urbanum*—Asterad-type of embryogenesis [18]);
2. The embryo exhibits only a typical initial of hypophysis and its locus (for example,

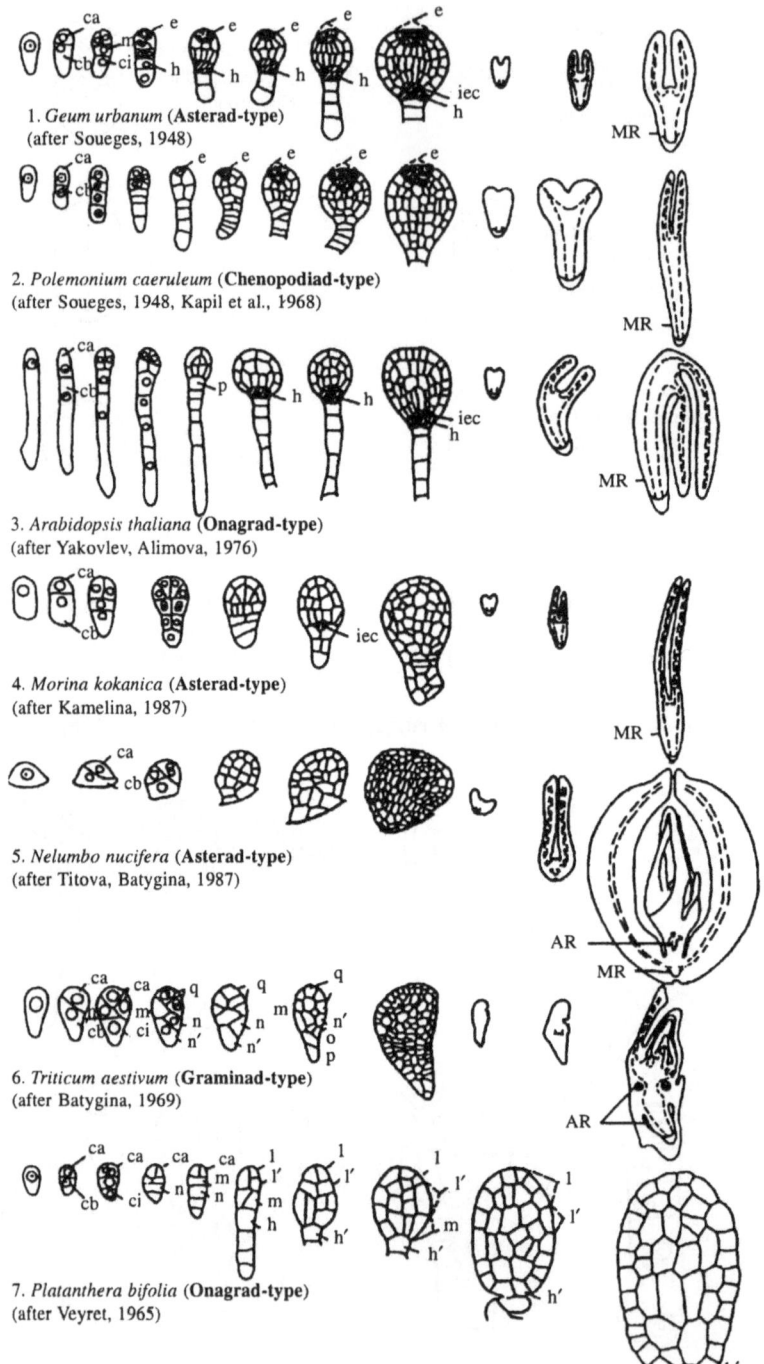

1. *Geum urbanum* (**Asterad-type**)
(after Soueges, 1948)

2. *Polemonium caeruleum* (**Chenopodiad-type**)
(after Soueges, 1948, Kapil et al., 1968)

3. *Arabidopsis thaliana* (**Onagrad-type**)
(after Yakovlev, Alimova, 1976)

4. *Morina kokanica* (**Asterad-type**)
(after Kamelina, 1987)

5. *Nelumbo nucifera* (**Asterad-type**)
(after Titova, Batygina, 1987)

6. *Triticum aestivum* (**Graminad-type**)
(after Batygina, 1969)

7. *Platanthera bifolia* (**Onagrad-type**)
(after Veyret, 1965)

**Fig. 5.   Polymorphism of sexual embryos. AR: adventive root, MR: main root
[18, 19, 21-23, 39, 50].**

Arabidopsis thaliana and *Capsella bursa-pastoris*—Onagrad-type of embryogenesis [19, 20]);

3. The embryo exhibits only a typical epiphysis initial and its locus (for example, *Polemonium caeruleumn*—Chenopodiad-type of embryogenesis [17, 21]);

4. The embryo lacks typical initials and loci of hypophysis and epiphysis (for example, *Triticum aestivum*—Graminad-type of embryogenesis [22, 11, 5]);

5. The initials of hypophysis and (or) epiphysis arise in the embryo, but they do not develop the typical loci of hypophysis and (or) epiphysis (for example, *Platantera bifolia* and *Gymnadenia conopsea*—Onagrad-type of embryogenesis [23, 24].

Polarity and morphological axis are normally established in the embryos of first four groups, and shoot and root apices develop orderly in the first three groups—the apex of the main root, and in the fourth—the apices of the adventive roots. The mature embryo of the fifth group exhibits morphological polarity, but shoot and root apices are not distinguished exomorphically. The embryos of all groups can be classified as normal types, not irregular, though they lack traits of the classical embryo according to Swamy and Krishnamurthy.

The development of epicotyl varies greatly from taxon to taxon (for example, from epicotyl locus to a well developed plumule). The same situation applies to the root (a developed main root, or just its initials, or the adventive roots, or the full absence of roots can be observed in the embryo).

Comparative analysis of structure and genesis of the mature embryo in different angiosperm taxa gives the evidence that almost all signs proposed by Swamy and Krishnamurthy (see before) and many other scientists to characterise sexual embryo vary greatly.

Somatic embryos under natural conditions are shown in Fig. 6. 'Asexual (homophasic) reproduction in flowering plants is accomplished by two elementary structural units of different morphological essence: a somatic embryo (individual as the whole) and an adventive bud (only a part of an organism). The majority of botanists when speak about vegetative reproduction bear in mind only gemmorhizogenesis. Both structures seem to develop in parallel during evolution, and realised side-by-side with sexual embryo (heterophasic reproduction) to a diverse extent in different taxa. Embryoids in natural conditions could be provisionally divided into two groups: embryoids developing in the flower and those developed on vegetative organs. Morphogenesis of somatic embryo, and subsequently of a seedling *in situ* occurs in the whole system of parental organism under its influence. The situation is similar to sexual embryo. This peculiarity of the development probably is the reason for the orderly embryogenesis and a high per cent of seedlings. On the contrary, the development of somatic embryos *in vitro* never produces high quantity of normal plant-regenerants.

Unique form of embryoidogeny (monozygotic-cleavage) can be observed in *Paeonia* (Fig. 7) [25, 26]. The zygotic embryo phase in the seed of all *Paeonia* species finishes at the stage of protoderm formation in coenocyte-cellular structure (this being the heterophasic reproduction). Then sexual embryo is cloned, when its epidermal cells give rise to somatic embryos. The somatic embryo does not differ from sexual one of a typical dicot (Asterad-type of embryogenesis) in morphogenesis. The establishment of polar axis takes place since the stage of initial cell. The orderly development of root and shoot apices subsequently occurs. The genesis of *Paeonia* somatic embryo and its structure seems to be determined by the conditions within embryo sac

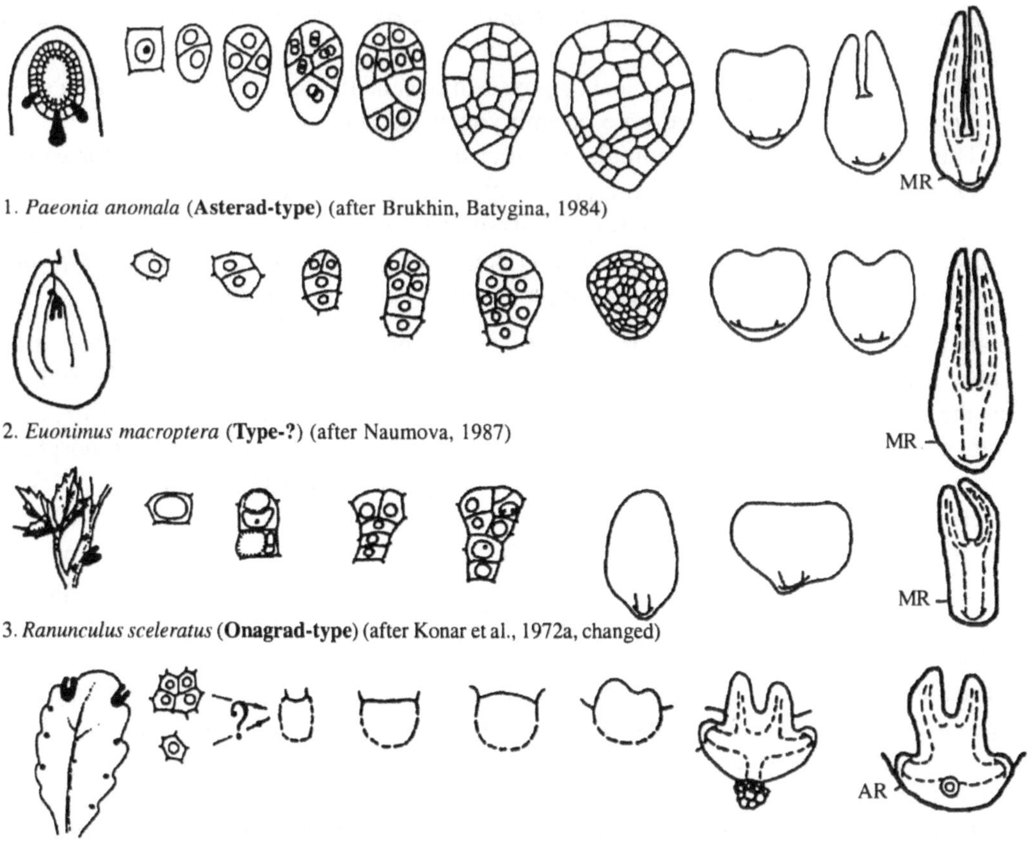

1. *Paeonia anomala* (**Asterad-type**) (after Brukhin, Batygina, 1984)
MR
2. *Euonimus macroptera* (**Type-?**) (after Naumova, 1987)
MR
3. *Ranunculus sceleratus* (**Onagrad-type**) (after Konar et al., 1972a, changed)
MR
AR

Fig. 6. Polymorphism of somatic embryos in natural conditions. AR: adventive root, MR: main root [27, 28, 35, 51].

and ovule, where it takes place. Probably this also guides the development of nucellar and integumentary embryoids. However, unlike the *Paeonia* embryo, these develop during cloning of a parental sporophyte. The first divisions in them are irregular, the initials and loci of hypophysis and epiphysis are not observed, but the development of protoderm, shoot and root apices and the differentiation of the main root with all its elements correspond to those of sexual embryos [27].

Sexual embryo
Somatic embryos
and degenerating sexual embryo
2n
n + n
2n
2n
2n
Heterophasic reproduction
Homophasic reproduction

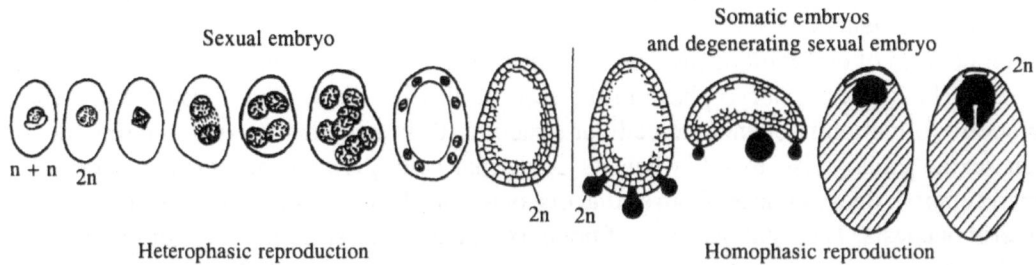

Fig. 7. Switching over the programme from heterophasic to homophasic reproduction in *Paeonia* seed [26].

The external shape and internal structure of *Paeonia* somatic embryo and of nucellar embryos (ovular embryoidogeny) are similar to that of zygotic embryo.

The development of somatic embryo on the stem of *Ranunculus sceleratus in situ* takes place in similar way [28]. It occurs according to onagrad-type. The authors have mentioned the absence of suspensor differentiation, unlike sexual embryos of this species. The formation of somatic embryos from epidermal leaf cell derivatives was observed in *Crassula multicava* [29].

The inner structure in these embryos, particularly of shoot and root apices may be compared with that of the majority of zygotic embryos in spite of absence of typical initials and loci of hypophysis and epiphysis. The development of sexual and somatic embryos of the same species in natural conditions occurs according to the same type of embryogenesis.

Thus, the conducted analysis of somatic embryos in different taxa, developing in natural conditions shows a large variety. The structure and genesis of somatic embryos *in situ* are taxon-specific and to a considerable extent determined by the place of their formation, and also by the environment. For example, somatic embryos arisen in the seed more often develop the main root. Those formed on vegetative organs usually develop adventive roots.

2. Somatic Embryos in *in vitro* Culture

As is known the *in vitro* culture provides to the researchers two different model systems (callus and suspension) to obtain somatic embryos. These systems are distinguished by many parameters that result in different structure and behaviour of initiating cells and determine the whole genesis of somatic embryos. The development of sexual and somatic embryos of the same species in natural conditions occurs according to the same type of embryogenesis. The development *in vitro* of somatic embryos may be traced by the example of several species, contrasted by a number of features [30].

In the callus culture of *Triticum aestivum* the endogenous initiation of meristematic zone has been revealed (Fig. 8). This zone occurs to be a special tissue which consists of cell rows strictly oriented to the callus surface [31, 32, 33]. Later, the cells of this zone become the initials of somatic embryos. It is noticeable that in wheat microspore culture the embryoids arise exogenously (from callus epidermal cells). At the early stages of development the sequence and place of divisions in sexual embryo and embryoids are relatively similar (Graminad-type) [22, 5], though the embryoids have some variability, preserved at the subsequent stages of their development. The initials are differently oriented inside the cluster, the polar axes of young embryoids are situated in different planes in reference to the callus surface. The delay of divisions of *ca* derivatives has also been observed as compared with *cb* derivatives. Subsequently, it causes the disturbance in their division sequence and of vacuolisation character and as a result, the appearance of embryoids with "linear" structure of the apical pole and with abnormal histogenesis. In zygotic embryogenesis the apical pole gives rise to the most of scutellum (cotyledon) and to the plumule, while in the abnormal embryoids apical cells are destroyed. Finally the majority of abnormal embryoids degenerate. In certain cases the normal plumule formation has been observed, though the development of adventive roots disturbed (these embryoids may be possibly considered as a transitional forms). Usually in the studied cultivars of *Triticum* the regeneration of plants by means of embryoidogenesis does not occur.

The comprative analysis of the development of somatic embryo in callus culture and of sexual embryo *in situ* in *Aconitum heterophyllum* (Ranunculaceae) has shown relative similarity.

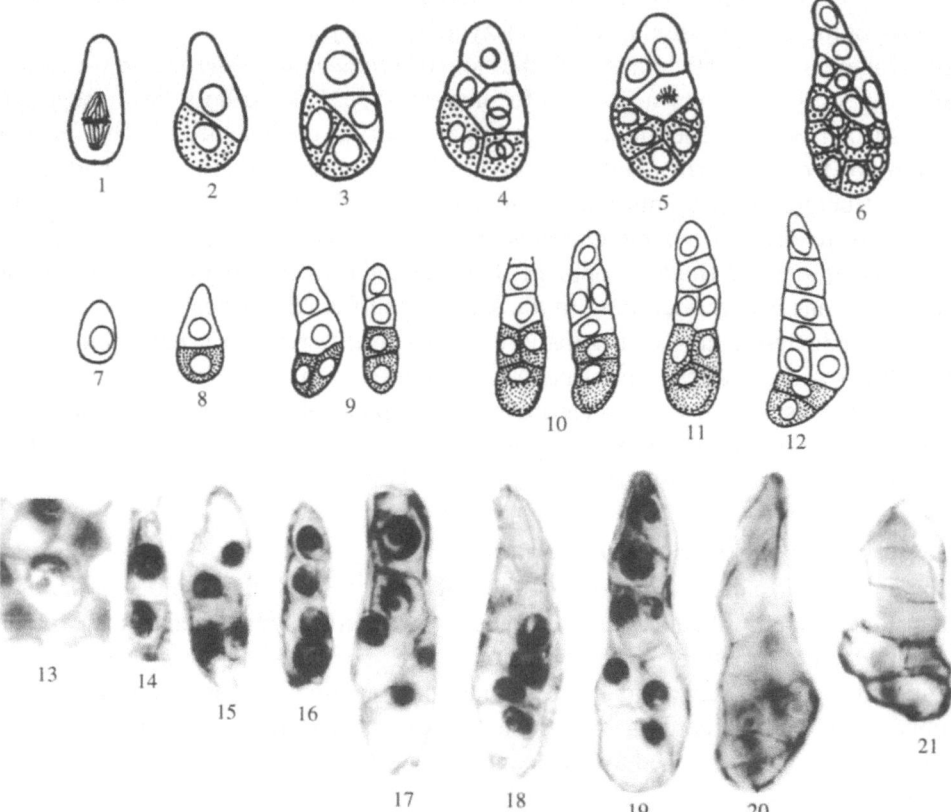

Fig. 8. Early stages of wheat zygotic embryo development in natural conditions (1–6), and somatic embryos in tissue culture at different stages of development (1–12 drawings and 13–21 LM [49]).

Single initial cells of callus give rise to embryoids which subsequently pass all main stages of normal embryogenesis—tetrads (T-shaped), quadrants, octants, globule, heart- and torpedo-shaped. Embryoidogenesis in *A. heterophyllum* occurs as a whole according to Onagrad-type of embryogenesis, which is peculiar for Ranunculaceae [35]. It is noticeable that embryoids exhibit well differentiated suspensors. In the course of the experiment different abnormalities have been observed in the embryoid development, the most frequent has been the precocious differentiation of xylem in the plerome area at the heart-shaped stage. This is evidently connected with the disbalance of carbohydrates and hormones in the culture medium and callus tissue. As a result the number of regenerants essentially decreases.

The comparative analysis of data on structure and development of embryoids in *Daucus carota* (produced in callus and suspension culture) has revealed its relative similarity to that of zygotic embryo [35, 36].

Thus a considerable resemblance of sexual and somatic embryo development has been revealed in the main morphogenetic regularities: polarity (bipolarity), symmetry (radial, bilateral, dorsoventral), the pattern of cell divisions and histogen differentiation (formation of morphogenetic fields), morphogenetical and morphophysiological correlations and allometry. Same critical stages are also exhibited in the development of sexual and somatic embryos: the laying down of the first cell wall, protoderm formation, differentiation of organs, autonomy, etc.

The embryoidogenesis does not recapitulate zygotic embryogenesis only in the cases of abnormal development (first divisions are irregular in the embryo, and the laws of embryogeny are disobeyed, the protoderm formation is abnormal, the formation of root and shoot apices is distorted, and so on).

3. Transitional Forms from Embryo to Bud (Fig. 9)
The analysis of literature and the original data on the development of sexual and somatic embryos in natural conditions and *in vitro* has led us to conclude that there are structures which differ from typical sexual and somatic embryos and from bud by their morphology. In connection with this we introduce now a term "transitional form" (in the terms of evolution) which means a structure exhibiting traits of an embryo (for example, globular, heart- and torpedo-shaped stages of development) and of a bud (formation of adventitious roots during regeneration) [5, 12, 37]. The embryos of Nelumbonaceae (*Nelumbo nucifera*), Ceratophyllaceae (*Ceratophyllum demersum*), Poaceae, Orchidaceae and Orobanchaceae illustrate the possible ways of such transition. A similar phenomenon can be observed among somatic embryos.

The embryos of *Nelumbo* (hydrophyte) and *Ceratophyllum* (hydatophyte) lack epiphysis and hypophysis. In the embryo of *Nelumbo* the main root is substituted by the adventive roots during germination. The latter originate at the base of plumule leaves at the later stages of embryogenesis. However, the embryo of *Nelumbo* is bipolar from the very beginning of its development [38, 39].

The embryo of *Ceratophyllum* seems to be bipolar at the first stages of its genesis, as it exhibits a group of cells which can be taken for initials of a main root. However, in the mature embryo only a well developed plumule can be found, and no main root. As for the adventive roots, they do not arise neither in the seed, nor in the seedling. The seedling of *Ceratophyllum*, thus, lacks typical bipolarity [40, 41].

In the majority of Poaceae (xerophytes) the mature embryo exhibits well formed plumule and developed adventive roots (their number depends on species, where and when the parental plant grows). The main root had been transformed into coleorhiza in the course of evolution [42–45]. However, a number of investigators are of the opinion that the main root exists in the grasses embryo. The embryo is bipolar and dorsoventral from the first stages of development, although the question remains, whether the grasses embryo preserve the primary polarity or it is replaced by a secondary polarity in the course of embryogenesis.

In Orchidaceae the shoot and root apices are not exhibited morphologically in the mature embryo. Later during protocorm formation the bud and the adventive root arise and the secondary polarity establishes [9, 46, 47].

The embryos of parasitic plants can serve as best model for investigation on reduction of typical initials of epiphysis and hypophysis of shoot and root apices (Fig. 10). A considerable peculiarity of its genesis is great variability of the first developmental stages. It is displayed in the diverse contribution of *ca* and *cb* derivatives into the formation of embryo body. For example, in *Aeginetia indica* even the first few divisions are irregular, so that the type of embryogenesis can not be elucidated.

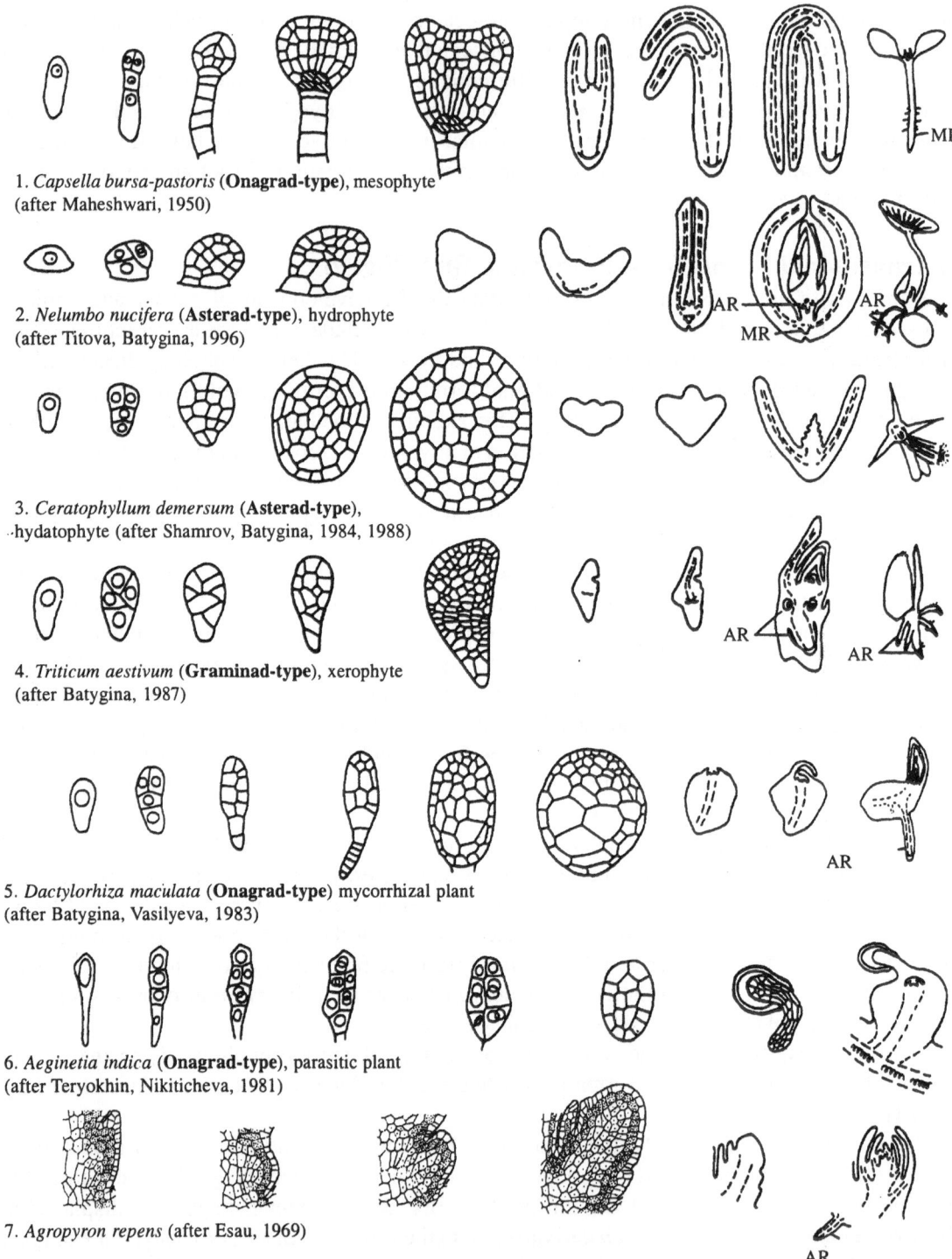

1. *Capsella bursa-pastoris* (**Onagrad-type**), mesophyte
(after Maheshwari, 1950)

2. *Nelumbo nucifera* (**Asterad-type**), hydrophyte
(after Titova, Batygina, 1996)

3. *Ceratophyllum demersum* (**Asterad-type**),
hydatophyte (after Shamrov, Batygina, 1984, 1988)

4. *Triticum aestivum* (**Graminad-type**), xerophyte
(after Batygina, 1987)

5. *Dactylorhiza maculata* (**Onagrad-type**) mycorrhizal plant
(after Batygina, Vasilyeva, 1983)

6. *Aeginetia indica* (**Onagrad-type**), parasitic plant
(after Teryokhin, Nikiticheva, 1981)

7. *Agropyron repens* (after Esau, 1969)

**Fig. 9. Parallellism of the first stages of morphogenesis in sexual and vegetative reproduction.
AR: adventive root, MR: main root [9, 11, 20, 39-41, 46, 52].**

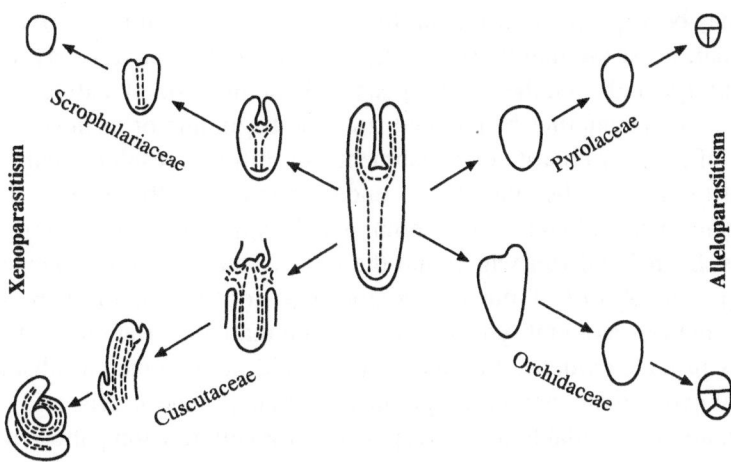

Fig. 10. **Four main types of embryo reduction corresponding to four forms of parasitism in different angiosperm families [48].**

However, the structure of mature embryo provides evidence that embryogenesis in *A. indica* as well as in other Orobanchaceae corresponds to onagrad-type. The differentiation of hypocotyl and radicle initials is distorted to different extents in Orobanchaceae. Besides that, in holoparasitic plants the protoderm formation is irregular in some parts of the embryo [46–48].

The analysis of sexual embryos in parasitic plants with diverse degree of reduction proves that bipolarity can be observed from the very first stages of embryogenesis. But it vanishes in the course of development and secondary polarity is established only during sprouting (in contact with host-plant).

The comparison of morphology and embryology of the abovementioned sexual embryos provides evidence to their transitional forms (because they exhibit morphological features of sexual embryo and of bud). The whole complex of enumerated peculiarities can be interpreted as an evidence of evolutionary tendencies towards a transition from normal sexual embryo to a bud (or *vice versa*).

Such reproductive structures as propagules that arise on vegetative organs of plants (homophasic reproduction) bear similarity to the abovementioned transitional forms (heterophasic reproduction). In *Bryophyllum*, bipolar propagules arise on the leaf. They lack the initials and loci of hypophysis and epiphysis. Globular-, heart- and torpedo-shaped stages are observed in their development (and this brings them closer to somatic embryos). However, the absence of main root and the development of adventive roots show that propagules are similar to the adventive bud. Morphogenesis of these structures may be compared with that of sexual embryo in Poaceae [5] which also may be considered as a transitional form.

The investigation of somatic embryogenesis of *Triticum in vitro* has revealed that some of the obtained embryoids are the transitional forms. They have a well developed plumule, but the morphogenesis of their adventive roots is distorted, if compared with sexual embryos [32].

Polymorphism of sexual and somatic embryos (*in situ, in vitro*) shows that morphogenetic reorganisations may occur at the level of cell, organ and organism and they can affect different stages of embryo and seedling development. At the most early stages of embryoidogenesis these reorganisations may be displayed as irregular divisions, with contribution of *ca* and *cb* derivatives into embryo formation; at the middle stages, hypophysis and epiphysis may vanish and at the later stages the embryo and even the seedling can lack the main or adventive roots. The embryo differentiation as a whole can thus be reduced from the beginning of its development. In sexual embryo the point of primary polarity establishment can shift as a result of adaptive evolution, or the primary polarity can be substituted by secondary one during the seedling formation [37].

The development of sexual and somatic embryos (*in situ* and *in vitro*) obeys the laws of cell division (Errera's, Sachs' and Hertwig's) and the laws of embryogeny (Souèges', Johansen's). However, with the variability of the first zygote division and of the subsequent stages of embryogenesis (eight types of embryogenesis and more than 50 variations [7]), the existence of transitional forms provide evidence that these laws could not be taken as absolute.

In every intitial cell, zygote, embryo, regardless of embryogenesis type and plant species, all the genetic information is available that is responsible for cell division pattern in the proembryo and for subsequent morphogenetic events. In the course of evolution a certain type of embryogenesis had been determined for each taxon, but the initial morphogenetic potential of both sexual and somatic embryos is usually realised under stress conditions (hybridisation, mutations, tissue culture, etc.).

4. Conclusions

1. A sexual embryo, an embryoid and a bud are three elementary structural units of seed and vegetative reproduction. A plant is able to form somatic embryos at all stages of its development and on different organs (vegetative and generative), side by side with the sexual reproduction. This ability enlarges the plasticity and tolerance of the reproductive system.

A high heterogeneity of seeds increases the adaptivity of a plant and of the whole population.

2. Sexual and somatic embryos formed *in situ* and *in vitro* reveal the great polymorphism. However, the obvious uniformity of morphogenesis and of the main regularities of shoot and root apex development appear in sexual and somatic embryos. During its development a plant individual exhibits also formation of shoot and root apices. polarity, symmetry and so on, whether it is the development of sexual embryo, embryoid, or regeneration of a bud.

3. The development of sexual and somatic embryos exhibits certain parallels, manifested in the great polymorphism of these structures, which depends on their high adaptive abilities. The only difference between them concerns the origin: zygotic embryo as a result of heterophasic reproduction and embryoid of homophasic one. Transitional forms from one structure to another can be revealed.

Somatic embryo does not stand comparison with zygotic embryo only in the case of developmental abnormalities.

Sexual and somatic reproduction appear not to be strictly divided.

Acknowledgements

I would like to express my gratitude to Dr. valentina E. Vasilyaeva for valuable advises and to Miss Elena Bragina for the help in illustrating and technical help in manuscript preparation.

The research was carried out with the support of Russian Foundation for Fundamental Research, grants number 02–04–49807.

References

1. T.B. Batygina, A new approach to the system of flowering plants, Phytomorphology 39 (1989) 311–325.
2. T.B. Batygina, Embryoidogeny is a new type of vegetative propagation, in: T.B. Batygina (Ed), Embryology of Flowering Plants. Terminology and Concepts. Vol. 3. Reproductive Systems, World and Family, 2000, pp. 334–350 (in Russian).
3. T.B. Batygina, Embryological bases of plasticity and the adaptive abilities of reproductive systems in flowering plants, in: B.A. Ursev (Ed), Biological Diversity: Approaches to the Study and Conservation, 1992, pp. 201–212 (in Russian).
4. I.K. Vasil, A.C. Hildebrandt, Variations of morphogenetic behaviour in plant tissue cultures, II *Petroselinum hortense*. Am. J. Bot. 25 (1966) 7–11.
5. T.B. Batygina, Parallel development of somatic and sexual embryos, in: Abstr. 14th Int. Congr. on Sexual Plant Reproduction, Lorne, Australia, 1996, pp. 4.
6. R. Souéges, Exposes d'embryologie et de morphologie végétales, VII. Les lois du développement, Act. Sci. Industr. 521 (1937) 1–94.
7. D.A. Johansen, Plant Embryology, Waltham Mass., USA 1950, pp. 1–305.
8. B.G.L. Swamy, K.V. Krishnamurthy, On embryos and embryoids. Proc. Indian Acad. Sci. 96 (1981) 401–414.
9. T.B. Batygina, V.E. Vasilyeva, System of reproduction of Orchidaceae (on the example of *Dactylorhiza maculata* (L.) Soo), in: Proc. VII Int. Symp. Fertilization and Embryogenesis in Ovulated Plants, Bratislava, Czechoslovakia, 1983, pp. 27–33.
10. T.B. Batygina, V.E. Vasilyeva, The expediency of system approach to the problem of embryo differentiation in the Angiosperms. Ontogenez. 14 (1983) 304–311 (in Russian).
11. T.B. Batygina, The Grain of Cereals, nauka, Leningrad, 1987, pp. 1–103 (in Russian).
12. T.B. Batygina, Embryoidogeny—a new category of reproduction in flowering plants, in: E.S. Terekhin, (Ed), Problems of the Reproductive Biology of the Seminal Plants, Sankt-Petersburg, 1993, pp. 15–25 (in Russian).
13. T.B. Batygina, Apomixis, agamospermy and vivipary and their role in the system of reproduction in Angiosperms, in: Proc. Int. Symp. Apomixis in Plants: the Condition of Problem and Perspectives of Investigations, Saratov, Russia, 1994, pp. 16–18.
14. T.B. Batygina, V.E. Vasilyeva, Sexual reproduction of flowering plants: periodisation of egg cell and zygote development and possible types of caryogamy, in: B. Bhatia (Ed), Plant Form and Function, Angkor Publisher, 1997, pp. 170–198.
15. R. Souéges, Exposes d'embryologie et de morphologie végétales. X. embryogenie et classification. Deuxiéme fascicule: Essai d'un systeme embryogenique (Partie générale), Act. Sci. Industr (1939) 1–85.
16. T.B. Batygina, Embryology of wheat, Kolos, Leningrad, 1974, pp. 1–206 (in Russian).
17. R. Souéges, L'hypophyse et l'epiphyse: les problems d'histogenése qui leur sont lies. Bull. Soc. Bot. France 81 (1934) 737–748, 769–778.
18. R. Souéges, Exposes d'embryologie et de morphologie végétales. Embryogenie et classification. XI. Essai d'un systéme embryogénique (Partie speciale: Premiere period du systeme), Act. Sci. Industr 1060 (1948) 1–101.
19. M.S. Yakovlev, G.K. Alimova, Embryogenesis in *Arabidopsis thaliana* (L.) Heynh. (Cruciferae), Bot. Zhurn 61 (1976) 12–24 (in Russian).
20. P. Maheshwari, An Introduction to the Embryology of Angiosperms, New York, London, Sydney, 1950, pp. 1–453.
21. R.N. Kapil, P.N. Rustagi, V. Rukmani, A contribution to the embryology of Polemoniaceae, Phytomorphology 18 (1968) 403–412.
22. T.B. Batygina, On the possibility of separation of a new type of embryogenesis in Angiospermae, Rev. Cytol. et Biol. Vég 32 (1969) 335–341.

23. Y. Veyret, Embryogénie comparée et blastogénie chez les Orchidaceae-Monandrae, O.P.S.T.O.M. Paris, 1965. pp. 106.

24. I.I. Shamrov, Z.I. Nikiticheva, Morphogenesis of ovule and seed in *Gynmnadenia conopsea* (Orchidaceae). Structural and histochemical research, Bot. Zhurn. 77 (1992) 45–60 (in Rassian).

25. T.B. Batygina, Morphogenesis of somatic embryos developing in natural conditions, Biologija 3 (1998) 61–64.

26. T.B. Batygina, New type of embryogenesis—Paeoniad, in: T.B. Batygina (Ed), Embryology of Flowering Plants. Terminology and Concepts. Vol. II. Seed, World and Family, 1997, pp. 526–528 (in Russian).

27. T.N. Naumova, The family Celastraceae, in: T.B. Batygina, M.S. Yakovlev (Eds), Comparative embryology of Flowering Plants. Davidiaceae-Asteraseae, Nauka, Leningrad, 1987, pp. 49–54 (in Russian).

28. R.N. Konar, E. Thormas, H.E. Street, Origin and structures of embryoids arising from epidermal cells of the stem of *Ranunculus sceleratus* L., J. Cell Sci. 11 (1972) 77–93.

29. J. McVeigh, Regeneration in *Crassula multicava*, Amer. J. Bot. 25 (1938) 7–11.

30. T.A. Thorpe, C. Stasolla, Somatic embryogenesis, in: S.S. Bhojwani, W.Y. Soh (Eds), Current Trends in the Embryology of Angiosperms, Kluwer Academic Publishers, 2001, pp. 279–336.

31. I.I. Shamrov. T.I. Kjachuk, T.B. Batygina, P.A. Djachuk, Embryoidogenous type of asexual reproduction and classification of anomalies in the anther culture on the example of wheat, in: Proc. Int. Conf. Biology of Cultivated Cells and Biotechnology, Novosibirsk, Russia, 1988, pp. 210–211 (in Russain).

32. T.B. Batygina, T.B. Mametjeva, V.E. Vasilyeva, Problems of *in vivo* and *in vitro* morphogenesis, Bot. Zhurn. 68 (1978) 87–111 (in Russian).

33. G.E. Titiova, N.N. Kruglova, O.A. Seldimirova, T.B. Batygina, Morphogenesis of sexual embryo and embryoid forming from microspores *in vitro* in wheat, *Triticum aestivum* L. (resemblance and distinctions), Phiziologia rastenii. (2003) (in press).

34. T.B. Sokolovskaya, The family Ranunculaceae, in: M.S. Yakovlev (Ed), Comparative Embryology of Flowering Plants. Winteraceae-Juglandaceae, Nauka, Leningrad, 1981, pp. 130–138 (in Russian).

35. T.B. Batygina, E.A. Bragina, G.E. Titova, Morphogenesis of propagules in viviparous species *Bryophyllum diagremontianum* and *B.* calycinum, Acta Soc. Bot. Pol. 65 (1996) 127–133.

36. A. Komami, R. Kawahara, M. Matsumoto, S. Sunabori, T. Toya, A. Fujiwara, M. Tsukahara, J. Smith, M.Ito, M. Fukuda, K. Nomura, T. Fujimura, Mechanisms of somatic embtyogenesis in cell cultures: physiology, biochemistry and molecular biology, *In Vitro* Cell Dev. Biol. Plant 28 (1992) 11–14.

37. T.B. Batygina, Some aspects of morphogenetic polarity in the ontogenesis of plants, in: Abstr. III Congr. RSPPh, Sankt-Petersburg, Russia, 1993, pp. 258 (in Russian).

38. N.S. Snegirevskaya, Materials to the morphology and systematic of genus *Nelumbo adans*. Transactions of the Komarov Botanical Institute of the USSR Academy of Sciences. 13 (1964) 104–172 (in Russian).

39. G. E. Titova, T.B. Batygina, Is the embryo of Nymphaealean plants (Nymphaeales S.L.) a dicotyledonous?, Phytomorphology 46 (1996) 171–190.

40. I.I. Shamrov, T.B. Batygina, The embryo and endosperm development in some Ceratophyllaceae species, Bot. Zhurn. 69 (1984) 1328–1335 (in Russian).

41. I.I. Shamrov, T.B. Batygina, The family Ceratophyllaceae, in: A.L. Takhtajan, (Ed.), Comparative Anatomy of Seeds. Magnoliidae, Ranunculidae, Nauka, Leningrad, 1988, pp. 153–156 (in Russian).

42. G.D. Pashkov, On the morphological essence of root base in grasses, Bot. Zhurn. 36 (1951) 597–606 (in Russian).

43. H. Pankow, H.V. Guttenberg, Vergleichenge Studien uber die Entwicklung Monokotyler Embryonen und Keimpflanzen, V. Gustav Fischer Verlag. Hf F, 1957, pp. 1–39.

44. H. von Guttenberg, Grundzuge der Histogenese honerer Pflanzen. I. Die Angiospermen, Handbuch Pflanzenanatomie, Gebrüder Bornträder, Berlin, 1960, pp. 1–315.

45. J.L. Guignard, Recherches sur l'embryogenie des Graminées; rapports des Graminées avec les autres Monocotyledones, Ann. Sci. Nat. Bot., 12 ser. 2 (1961) 491–610.

46. E.S. Terykhin, S.I. Nikiticheva, The Family Orobanchaceae. Ontogenesis and Phylogenesis, Nauka, Leningrad, 1981, pp. 1–228 (in Russian).

47. E.S. Teryokhin, Z.I. Nikiticheva, The postseminal development of parasitic angiospermae. The metamorphosis, Bot. Zhurn. 53 (1987) 39–57 (in Russian).

48. E.S. Teryokhin, Parasitic Flowering Plants. The Evolution of Ontogenesis and the Mode of Life, Nauka, Leningrad, 1977, pp. 1–219 (in Russian).
49. T.B. Batygina, I.I. Shamrov, G.E. Titova, Somatic embryogenesis in cereals (comparative embryological approach), in: Abstr. XV Int. Bot. Congr. Yokohama, Japan, 1993, pp. 564.
50. O.P. Kamelina, The family morinaceae, in: T.B. Batygina, M.S. Yakovlev, (Eds), Comparative Embryology of Flowering Plants. Davidiaceae-Asteraceae, Nauka, 1987, pp. 177–184 (in Russian).
51. V.B. Brukhin, T.B. Batygina, Embryo culture and somatic embryogenesis in culture of *Paeonia anomala*, Phytomorphology, 44 (1994) 151–157.
52. K. Esau, Anatomy of Seed Plants, John Wiley & Sons, New York, 1977, pp. 1–550.

Plant Biotechnology and Molecular Markers
P.S. Srivastava, Alka Narula and Sheela Srivastava (Editors)

5. Molecular Biology and Genetic Engineering of Polyamines in Plants

M.V. Rajam, R. Kumria and S. Singh

Plant Polyamine and Transgenic Research Laboratory, Department of Genetics,
University of Delhi-South Campus, Benito Juarez Road, New Delhi 110021, India

Abstract: The involvement of polyamines in various cellular and metabolic processes has been well established, but their mechanism of action and extent of involvement and regulation in various responses is not clearly understood. The use of specific biosynthetic inhibitors as well as mutants has been employed to study the intricacies of their regulation but many queries still remain unanswered. The cloning of the genes of polyamine metabolism allowed for the generation of transgenic plants with over-expression or down-regulation of a particular gene. These transgenic plants could be used to study the effects on plant development, metabolic shifts as well as stress responses. Similarly, the up- or down-regulation of the entire polymine metabolism is also possible by introduction of two or more genes into plants which would provide greater insight into the mechanisms of polyamine functions.

1. Introduction

Polyamines (PAs) are naturally occurring polycationic aliphatic amines, which due to their ubiquity and versatility are involved in the regulation of various cellular and molecular processes. They are positively charged compounds with their charge distributed along the molecule. The common PAs, spermidine (SPD) and spermine (SPM) and their diamine precursor putrescine (PUT) play a critical role in the normal functioning of all cells [1]. They are involved in the cellular functioning both at the molecular and physiological levels due to their association with various macromolecules (DNA, RNA and proteins) and membranes as well as their high concentration in the cytosol thus behaving as osmolytes [2]. The role of PAs is much better studied in animal systems than plants, though they have been suggested to have a role as new plant growth regulators either by mediating the plant hormone effects or independently signalling other responses [3–5].

PAs exist in three forms in the cell, viz. as free cations, covalently bound to low molecular weight phenolic compounds like hydroxycinnamic acids (conjugated form of PAs) and bound to marcomolecules or membranes (bound form of PAs). Though the major form is the free cationic form of PAs, there are instances when the amounts of conjugated form exceed the free form and these are known to be critical in certain physiological processes including seed germination, flower development, defence responses and stress reactions [2, 6–8]. Besides PUT, SPD and SPM, there are certain unusual PAs found in nature, e.g. thermo-SPM which have been detected in bacteria residing in hot springs and they seem to be important in protecting the enzymes from heat denaturation [9–10] and aminobutylhomo-SPD found in fast growing cells of root nodule bacteria *Rhizobium* [11]. NorSPD and norSPM are found in thermophilic red algae, brown algae, and *Chlamydomonas*, *Nitella* and *Chlorella* [12, 13]. Similarly, some unusual PAs have been reported in plants, homo-SPD was first detected in sandalwood [14] and also in mosses and ferns [12]. In leguminous plants, other unusual PAs like canavalmine, homoagmatine, aminopropylcanavalmine and aminobutylcanavalmine have been detected [15, 16]. NorSPD and NorSPM have been detected in alfalfa grown under drought conditions and have been postulated to play a protective role under stress conditions [17]. As a matter of fact it has been suggested that PA distribution, especially of SPM, may serve as a phylogenetic marker [12].

The study of plant PAs has come a long way since the first report by Bagni [18] regarding the stimulatory effect of a PA (PUT) on growth of *Helianthus tuberosus* explants. Since then, PAs have been demonstrated to be associated with regulation of somatic embryogenesis [19–22], root and shoot formation [23–25], flower and fruit development [26], stress responses [2, 4, 5, 27] and senescence [28–29]. In fact, PAs may serve as 'biomarkers' for *in vitro* morphogenetic potential including plant regeneration via somatic embryogenesis [5, 29]. The multifaceted functions of PAs as well as the variations in their levels in response to changes in the physiological state, point towards their role as possible second messengers, though their high titres do not support the view. Various studies have been conducted to investigate the involvement of PAs in cell functioning, using mutants of PA biosynthetic genes and specific substrate-based inhibitors of PAs. Though much information could be generated regarding the involvement and possible mechanisms of action, no clear picture of their functioning emerged. Hence, transgenic plants expressing PA biosynthetic genes in constitutive and regulated manner were generated, with an aim to answer some of the queries regarding the functioning and role of PAs.

This article deals with the molecular biology of PAs in plants, with special reference to transgenic plants expressing PA biosynthesis genes.

2. Polyamine Metabolism

The diamine PUT is universally derived from ornithine by the rate limiting enzyme ornithine decarboxylase (ODC). Plants, bacteria and some fungi have an alternate pathway for PUT synthesis from arginine by arginine decarboxylase (ADC). In plants, ADC forms the major pathway for PA biosynthesis. The PUT, hence derived from either of the two pathways is converted into higher PAs, the triamine SPD and the tetraamine SPM by the addition of aminopropyl groups obtained from decarboxylated S-adenosylmethionine (dcSAM). The dcSAM is formed by the decarboxylation of SAM by SAMDC, SAM is in turn synthesized from methionine by SAM synthase [2]. SAM is a major methyl donor in the cellular metabolism and also forms a part of ethylene biosynthesis. SAM is converted to 1-aminocyclopropane-1-carboxylic acid (ACC) by ACC synthase, which is converted to ethylene by ACC oxidase (Fig. 1).

The catabolism of PAs involves the enzymes diamine oxidase (DAO) and polyamine oxidase (PAO). DAO preferentially acts on diamines (PUT, cadaverine) to form pyrroline, ammonia (NH_3) and hydrogen peroxide (H_2O_2), though it can break down SPD to aminopropyl pyrroline, NH_3 and H_2O_2 [3]. PAO, on the other hand, breaks down SPD and SPM to pyrroline and aminopyrroline respectively, and diamine propane (DAP) and H_2O_2 [3]. Pyrroline formed due to the activity of DAO and PAO is further converted to γ-aminobutyric acid (GABA) by a nicotinic acid diamine (NAD) dependent dehydrogenase [30]. Besides oxidases there might exist alternate pathways for PA catabolism as there are many plants in which oxidases have not been detected [31]. Alternative diversion of the PAs into other metabolic pathways also plays a role in the regulation and the dynamics of the PA metabolism.

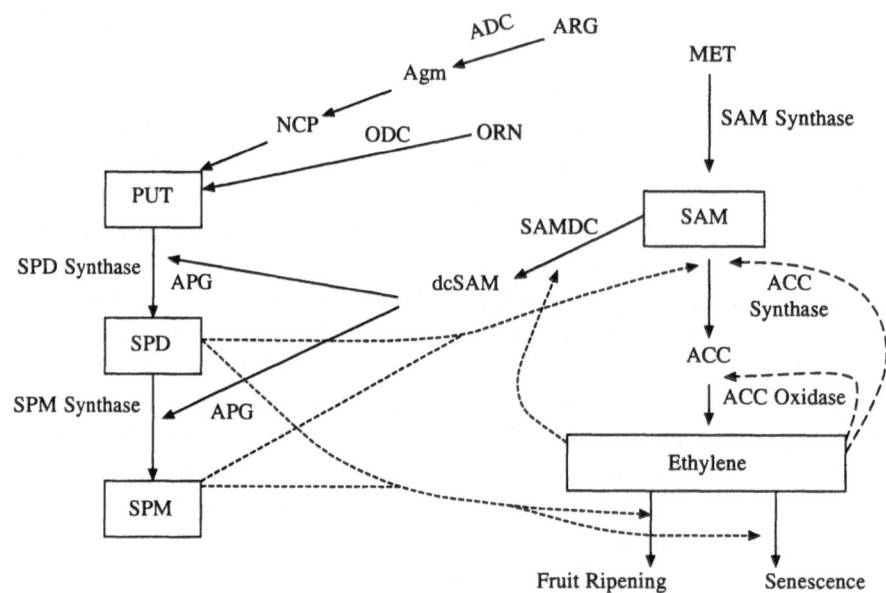

Fig. 1. **Metabolic and functional inter-relationships between polyamine and ethylene metabolism. Dotted lines show inhibitory effecs of respective metabolism on the other, whereas dashed lines depict stimulatory effects. ARG, arginine; ORN, Ornithine; MET, Methionine; PUT, Putrescine; SPD, Spermidine; SPM, Sperimine; ODC, Ornithine decarboxylase; ADC, Arginine decarboxylase; SAM, S-Adenosyl methionine; SAMDC, SAM decarboxylase; APG, Aminopropyl group, ACC, Aminocyclopropylcarboxylic acid; Agm, Agmatine; NCP, N-Carbomyl putrescine.**

PUT forms the precursor for pyrrolidine ring of the nicotine and tropane alkaloids. PUT is converted to N-methyl PUT by PUT methyl transferase (PMT) which then forms the pyrrolidine ring of nicotine and other tropane alkaloids [32].

3. Role of Polyamines in Biological Processes

Most of the PA functions can be attributed to their polycationic nature and the distribution of the charge along the molecule, which allows them to bind to a variety of molecules, including nucleic acids, protein and cell membranes in the cell and regulate their functions. A very crucial binding of PAs is with membrane phospholipids thus stabilizing them and reducing chlorophyll loss when bound to the thylakoid membrane [29, 33], as well as preventing the lowering of membrane potential and the Ca^{2+}, PO_4^{-3} fluxes of mitochondrial membrane under saline stress [34, 35]. Similarly, PAs have been found associated with cell wall components like lignin and pectins [36] and have been implicated in maintaining cell wall characteristics by strengthening the links between cell wall components [37]. PAs also play a role in cell wall expansion and are part of modulators involved in host-pathogen interactions [38, 39]. The binding of SPD to the plasma membrane proteins in zucchini hypocotyls has been characterized. SPD was found to have a specific binding to a 44 and a 66 kDa protein [40].

The binding of PAs with DNA is also known to be responsible for playing some part in the regulation of synthesis and function of DNA, including gene expression. It had been proposed that PAs affect growth by interacting with DNA [41-43] and SPM plays a part in B to Z DNA transitions [43, 44]. SPM has been reported to stabilize the triplex DNA formation and aggregation [45]. PAs have also been reported to stimulate DNA, RNA and protein synthesis [46]. They are involved in joining of okazaki fragments and the depletion of PAs leads to accumulation of short DNA pieces [47]. As a matter of fact *odc* has been suggested to be a proto-oncogene and its over expression leads to cell transformation [46].

PAs are also known to bind with RNA molecules, and protect them from RNases [48]. SPD and SPM are known to stimulate the reading of amber mutations and play an active role in the expression of specific genes [49–50]. The antisenescence effects of PAs are in part due to the inhibition of ribonuclease synthesis and activity by PAs [2, 51]. PAs also inhibit the protease activity, thereby delaying the degradative processes initiated during stress or senescence [52].

PAs regulate their own biosynthesis by inducing a ribosomal frame shifting in the translation of ODC antizyme [53]. Besides, being involved at the DNA and RNA levels, PAs are also known to regulate protein synthesis and activity. SPM has been reported to have a specific role in activity of cyclic AMP-independent caesin kinase [54, 55]. A branched quarternary PA, tetrakis (3-aminopropyl) ammonium along with SPM has been reported to support protein synthesis at high temperature in a thermophilic bacteria [56]. SPD stimulates protein synthesis in chloroplast especially in light [57]. PUT, SPD, SPM and cadaverine enhanced the phosphorylation of several plasma membrane proteins in tobacco, cucumber and *Arabidopsis* [58]. PUT has been reported to increase the phosphorylation of many soluble proteins, unlike SPD and SPM, which decreased the phosphorylation of soluble proteins [58]. PAs are known to regulate cell division and also prolong the cell division phase by inhibiting the synthesis of phenylpropanoids. The conjugation of PAs with phenolics regulated the free PA levels and therefore led to the cessation of cell division [59].

The critical role played by PAs in growth and development along with their role in protein phosphorylation/dephosphorylation and the binding of PAs (SPD) to specific protein in the thin-layer of tobacco together point towards the possibility of these being considered as signal transduction molecules [60, 61]. Though the high PA titres in the cell are quite unlike secondary messengers, which increase rapidly, and transiently in response to stimuli. Therefore, this aspect of PAs warrants more attention to be able to pinpoint their role and mechanism of involvement, if any, in signal transduction [58, 61, 62].

The study of physiology, biochemistry and genetics of the PA metabolism was initiated with the generation of mutants of PA metabolism in *E. coli*, yeast and plants [63]. The *E. coli* mutants defective in their PUT synthesis were the first to be isolated, these preferentially used the ODC pathway in the absence of any PA supplements and the ADC pathway in the presence of arginine [64, 65]. The substitution of PUT by its analogs which could not be converted into the higher amines could not restore the growth of the mutants, whereas the inclusion of SPD analogs was helpful, these results demonstrated the critical requirement of SPD for growth [66]. The *E. coli* mutants defective in the SAMDC function were also isolated, but their growth rate was almost unaffected. Further, an *E. coli* strain deficient in ADC, ODC and SAMDC function was isolated which was able to grow at a reduced growth rate [67]. This strain was utilized in elucidating the role of PAs in ribosomal complex formation and hence protein synthesis [68]. Mutant studies in *Saccharomyces cerevisiae* revealed the presence of a single biosynthetic pathway (ODC) for the synthesis of PUT as well as the absolute requirement of SPD and SPM for growth and sporulation in yeast [69].

PA mutants have also been raised in the model plants, *Arabidopsis* and tobacco [63]. The *Arabidopsis* mutants with lower levels of ADC and ODC activities had abnormal root, shoot and floral morphology [63]. Malmberg and McIndoo [70] have isolated tobacco mutants with abnormal floral morphologies, including flowers with large non-functional stigma, anthers with non-viable pollen and ovary with most ovules turned into anthers. These mutants were deficient in the function of SAMDC and brought out the role of SPD and SPM in flower development. The ODC mutants failed to flower, demonstrating the involvement of ODC in floral initiation [71].

Tobacco mutants have been raised by activation T-DNA tagging, such that the regions adjacent to the insertional position over-expressed the gene, and these were selected on selective concentration of MGBG [72]. The mutated plants showed altered phenotypes, abnormal floral morphology, male sterility and parthenocarpy, with higher SAMDC activity and SPD levels.

The study of PAs pertains much to the availability of specific, irreversible, substrate- or product-based inhibitors of its biosynthetic enzymes [73]. The substrate analogue of ODC, α-difluoromethylornithine (DFMO) was the first inhibitor to be synthesized [74]. Similarly α-difluoromethylarginine (DFMA) has been used as a potent inhibitor of ADC [75]. Another set of similar substrate-based analogous for ODC and ADC were monofluoromethylornithine (MFMO) and monofluoromethylarginine (MFMA), respectively. These have been reported to be much more potent than DFMO and DFMA [73]. The substrate-based inhibitor for lysine decarboxylase, α-difluoromethyllysine (DFML) is also available [73, 76]. A very potent inhibitor of the enzyme SAMDC is methylgloxyl *bis* (guanylhydrazone) (MGBG), though it has been put to limited use due to its non-specific effects on the respiratory enzymes [77]. The inhibitors have provided insight into the dynamics of inter-conversion and regulation of the levels of PAs in the cell. Yet no clear picture of their mechanism of action or extent, period or stage of involvement emerges. The polycationic nature of PAs allows them to interact with various molecules whose functioning is modulated and regulated by them. PAs also have a role in free radical scavenging due to their polycationic nature [78]. Although there are many other suggested possible mechanisms regarding the functions of PAs, the exact role, the extent of involvement and the mechanism of action of PAs is as yet not very clearly understood.

However, the development of transgenics expressing PA biosynthetic genes driven by constitutive promoters offered a good opportunity for the study of the PA functions. Since PAs are fundamental to the process of morphogenesis, the generation of such transgenics presented a problem. Also the transgenics recovered had abnormal phenotypes [61, 79–81]. Inducible promoters were applied to overcome these problems but ambiguous effect of the inducer on the PA metabolism and the problem of sustained induction remained as hurdles. The transgenics raised have been used to study the role of PAs in plant development [79–80, 82] and also the dynamics of their metabolism [83], but detailed studies were not conducted on the plant development and stress responses of the transgenics over-expressing PA biosynthesis genes as well as their response during *in vitro* morphogenesis. Therefore, much work is needed on the genetic manipulation of PAs in plants to clearly demonstrate the role of PAs in a variety of cellular and molecular processes.

4.   Cloning of Polyamine Metabolic Genes

The PA biosynthetic genes were first isolated from animal systems, yeast and bacteria. The *odc* gene has been cloned from human, rat, mouse, holstein, *Trypanosoma*, *Leishmania*, yeast,

Neurospora and *E. coli* [84–88]. Recently PA biosynthetic genes have been cloned from plants too. The *odc* gene has been cloned from *Datura*, tobacco and tomato [89–91].

The *adc* gene has been cloned from oat, tomato, pea, *Arabidopsis* and soybean [92–96]. The *samdc* gene has been cloned from *Arabidopsis*, *Datura*, potato [97], spinach [98], *Catharanthus roseus* [99], *Tritordeum* [100], *Pharbitis nil* [101], tomato, tobacco [63], rice [102] and also from human genome [103]. The *spd syn* gene has been cloned from *N. sylvestris*, *Hyocyamus niger* and *Arabidopsis* [104]. Recently *spm syn* gene has been cloned from human genome [105].

The genes coding for the enzymes involved in formation of conjugated PAs have also been cloned. The gene for the enzyme homo-SPD synthase has been cloned from bacteria (*Acetobacter*) [106], *Senecio vernalis* [107] and *Eupatorium cannavulgaris* [108]. Besides, the biosynthetic enzyme, the gene for the catabolic enzyme PAO has been cloned from maize [109] and DAO from lentil [110] and pea [111]. The cloned PA metabolic genes are summarized in Table 1.

Table 1.   The polyamine metabolic genes cloned from different organisms

PA metabolic gene	Source
odc	*E. coli*, *Trypanosoma*, *Leishmania*, *Datura*, tomato, mouse, yeast and human
adc	Oat, pea, soybean and tomato
samdc	Potato, *Arabidopsis*, Spinach, *Catharanthus*, *Tritordeum* and *Pharbitis nil*
spd syn	*Nicotiana sylvestris*, *Hyoscyamus niger* and *Arabidopsis*
spm syn	Human
homo-spd syn	*Acetobacter, Senecia vernalis* and *Eupatorium cannavulgaris*
dao	Lentil and pea
pao	Maize

The plant decarboxylases (both the ADC and ODC) belong to the group IV of decarboxylases and have the 23 amino acids which have been shown to be critical for the activity, conserved in them [90]. There exists a strong homology between the oat, tomato and soybean *adc* genes, further their catalytic motifs are more than 75% identical [93, 96]. The genes for related pathways like, homo-SPD synthase and PUT-N-methyltransferase seem to be evolved from the genes of the basic PA metabolism [108]. There are evidences for the presence of more than one copy of *adc* or *odc* gene in the plant [90, 96].

5.   Transgenic Plants Expressing Polyamine Metabolic Genes

Transgenic plants expressing PA biosynthetic genes were generated to gain better understanding of the PA metabolism, reconfirm the effects of the modulation of PA titres caused by the inhibitors at the molecular level and also to overcome the limitations of the inhibitor-based experiments. The transgenic approach was highly specific to the target gene and moreover it provided a tool for manipulating the metabolic flux with the persistant shift in the PA metabolism. Even though some of the plant PA biosynthesis genes have been isolated and characterised, most transgenics have been raised using genes from heterologous source as these were the first to be isolated. In most of the transgenics generated, CaMV35S promoter has been used to drive the transgene, though tetracycline (tet)-inducible promoter has also been used in cases where extreme deleterious effects of the transgene were expected [2, 5]. Some of the transgenic plants expressing PA biosynthesis genes are listed in Table 2.

Table 2.   Transgenic plants expressing polyamine metabolic genes

Gene	Gene source	Transgenic plant
odc	Yeast	Tobacco root cultures
	Mouse	Tobacco
odc	Mouse	Carrot
	Mouse	Rice
	Mouse (antisense)	Rice*, Tobacco*
adc	Oat	Tobacco
	Oat	Rice
samdc	Human	Tobacco
	Potato (sense)	Potato
	Potato (antisense)	Potato
samdc-odc	Human-Mouse	Tobacco*
dao	Pea (sense)	Pea
	Pea (antisense)	Pea

*Unpublished data from our laboratory.

The first report of the introduction of yeast *odc* gene was in root cultures of tobacco using *Agrobacterium rhizogenes* [112]. The study was aimed to increase the nicotine content of the culture as PUT is a precursor for nicotine. Hence, over-production of PUT was attempted by using a double enhancer sequence containing promoter but only a 3-fold increase in ODC activity and a 2-fold increase in nicotine was observed. This was suggested to have been the result of a tight regulation of nicotine biosynthesis or the activity of the enzyme PUT methyl transferase becoming limiting. PUT might be incorporated into many other secondary metabolic pathways, like alkaloids, which are present in significant amounts in Solanaceous plants and also conjugation of PUT could have been another pathway for the PUT synthesised due to the over-expression of the *odc* gene. No significant increase in SPD and SPM levels was observed as their biogenesis is also precisely regulated. No plants were regenerated from the transformed root cultures [112]. The tobacco transgenic plants expressing the mouse *odc* gene were the first PA transgenic plants raised by DeScenzo and Minocha [79]. Two constructs were used for the transformation of tobacco, one having complete coding sequence of *odc* and the other in which 350 bp of 3' end were removed. The truncated gene produced a functional peptide 37 amino acids less at the C terminus and an increased half-life. The enzyme activity when checked at the pH optimum for mouse ODC was much higher in transgenics compared to endogenous plant ODC activity in the controls, whereas at the pH optimum for the plant ODC, no significant change was observed. PUT was found to be 2–3 fold higher in leaves and 4–12 fold higher in callus, though no significant increase was observed in SPD and SPM content as the amounts of SAM were suggested to be limiting. The transgenics having high PUT titres were stunted with wrinkled leaves and reduced stamens.

Carrot cell lines are known to have no detectable ODC activity, only the ADC pathway is functional in them. Carrot cell lines were transformed with mouse *odc* gene driven by CaMV35S promoter [83] and the effect of the high PUT titres on somatic embryogenesis was studied. The transformed cell lines showed improved somatic embryogenesis which could be correlated to higher PUT amounts. The somatic embryos were formed even in the presence of DFMA which inhibited the carrot ADC, therefore all the PA requirements of the embryos were fulfilled by the introduced mouse *odc* gene. Exogenous addition of PUT was not found to be helpful thus suggesting that a fast turnover of PUT is also essential besides the high concentration for somatic embryogenesis. These transformed carrot cell lines were used to study the shift in metabolic flux as compared to the control cell lines [113]. ^{14}C labelled arginine, ornithine, methionine or PUT was fed to the cell cultures and amount of label incorporated in different PAs and their fractions was analyzed. ^{14}C labelled PUT was much higher in trasgenic cell lines when ^{14}C ornithine was given as substrate and there was no difference in labelled PAs when ^{14}C arginine was fed to the cultures. In correlation the conversion of ^{14}C-methionine to ethylene was much lower in transgenics due to a shift in the dynamics towards PA metabolism as more of PUT was available to be converted to SPD and SPM [113].

Alterations in the PA levels during *Agrobacterium*-mediated genetic transformation with a reporter gene (*gus*) and mouse *odc* gene were found to affect the regeneration potential of indica rice [114]. It has been suggested that the modulation of PA metabolism may be used to improve the regeneration from transformed calli in rice and other crops [114].

In a recent study, it was demonstrated that over-expression of human *odc* gene in transgenic rice plants alters PA pools in a tissue-specific manner [115]. It was suggested that ODC rather than ADC is responsible for the regulation of PUT synthesis in plants. In these transgenics, significant changes in the levels of all three major PAs were observed in seeds and also in vegetative tissues (leaves and roots) as compared to oat *odc* transgenics, wherein PUT and SPM levels were higher in seeds only.

Transgenic tobacco plants over-expressing human *samdc* gene driven by CaMV35S promoter were generated by Noh and Minocha [80]. These plants were found to have 2–6 fold higher SPD than the untransformed controls, there was an increase in SPM too though PUT decreased. Since high amounts of SPD is cytotoxic, the regenerants obtained might have been moderate accumulators of SPD as the increase in SPD and SPM was not comparable to the dramatic decrease in PUT. The cytotoxicity of the high amounts of SPD did not allow the regeneration of any plants over-expressing potato *samdc* in potato, hence a tet-inducible promoter was used to drive *samdc*. The sense *samdc* plants showed 7-fold increase in SPD, 3-fold in SPM and a decrease in PUT on tet-induction. Similarly potato plants expressing antisense *samdc* gene driven by both 35S promoter and tet-inducible promoter were raised [81]. The antisense *samdc* plants were stunted, branched, necrotic with few small tubers; this was attributed to increase in ethylene levels caused by the channelling of SAM for the formation of ethylene due to down-regulation of SAMDC. A decrease in PUT levels was observed due to down-regulation of ODC and ADC by the elevated levels of ethylene. Phenotypic abnormality, i.e. delay in flowering was also observed in case of tobacco plants transformed with *Agrobacterium rhizogenes* and was attributed to a delay in the appearance of conjugated PAs due to the decrease in ODC and ADC activities [116].

Bhatnagar et al. [117] studied the genetic manipulation of PA metabolism in poplar cells by introducing mouse *odc* gene. It was observed that over-expression of the heterologous gene resulted in high levels of PUT and increase in PUT degradation in the transgenic cells. In continuation of the above study [118], they showed that there was an increased turnover of PUT as well as its conversion to SPD as compared to the untransformed cells, the increase being proportionate to the cellular content of this diamine. Furthermore, the increase in PUT catabolism in the transgenic cells did not result in any major changes in the activity of DAO or the half-life of PUT [118].

The fact that elevated levels of PAs were detrimental for plant regeneration and were cytotoxic, led to the use of tet-inducible promoter for driving oat *adc* gene introduced into tobacco [119]. The PUT levels were increased by 16-46% on tet-induction and more significant increase was seen in the conjugated and bound fractions of PAs. A prolonged induction of the transgene at an early stage of development led to plant growth inhibition, necrotic, wrinkled leaves, but no such effects were seen in case of older plants on tet-induction of the transgene. These results clearly brought forward the differential role of PAs at various developmental stages [119].

Rice transgenics over-expressing the oat *adc* gene have also been raised [82]. These transgenics showed a 4-7 fold increase in the activity of the ADC enzyme, along with a 4-fold increase in the PUT titres. The high PUT titres were found to be inhibitory to plant regeneration from the transformed calli. The effect of the strength of the promotor driving the *adc* gene on the PA metabolism as well as the morphogenic capacity of the transformed calli has also been analyzed [120]. In this study, oat *adc* gene under the strong maize ubiquitin promoter 1 (ubi-1) was introduced into rice but even then no significant change in PA levels was observed in seeds or in the vegetative tissues. However, Noury et al. [121] reported that only one specific transgenic line showed a significant increase in PUT and SPM levels in vegetative tissues and seeds. R_1 generation rice transgenics expressing the oat *adc* gene driven by ABA responsive promoter were tested for their response to various environmental stresses [122].

Since PAs are known to play a role in stress responses, particularly the activity of the enzyme ADC is known to increase under stress along with increase in PUT levels, rice transgenics with *adc* gene were tested for their tolerance to abiotic stress (drought) and it was reported that no chlorophyll loss was observed after 8-days of drought as compared to the untransformed control plants [82].

Tobacco transgenics over-expressing mouse *odc* gene affects cellular PAs and *in vitro* morphogenesis, and confers salt stress tolerance [123]. Further, it was seen that favourable changes in PAs titres and the optimum PUT: SPD ratio in transgenic lines showed better regeneration, and previously similar results were reported in indica rice genotypes [124, 125].

Transgenic pea plants with PA catabolic gene *dao* in sense and antisense orientation have also been generated in order to study the role of DAO in nodulation. The sense plants showed an increased DAO activity and reduced PUT levels. It was observed that DAO activity was not involved in nodule formation, but may have a role in regulation of PA levels in host cells [126].

Though the SPD synthase and the catabolic enzyme PAO genes have been isolated, no transgenic plants have been raised with them, neither are there any transgenics reported expressing the *adc* and *odc* genes in antisense orientation for studying the effects of long-term down-regulation of these genes on plant development.

Transgenic plants of tobacco and/or rice were also generated in our laboratory with oat *adc*, human *samdc* and *spd syn*, and the transgenics have been analysed for the effects on cellular PA concentrations, PA biosynthetic enzyme activities, plant development and abiotic stress responses [unpublished data]. The PA titres in *adc* and *samdc* tobacco transgenics were also analysed. In case of *adc* transgenics a significant increase in the PUT and SPD levels with no apparent changes in SPM was observed. The increase in the PA levels was comparable to the concurrent increase in the activity of ADC as well as SAMDC. The ODC activity was found to be decreased in these transgenics. The activity of the PUT catabolic enzyme DAO was measured, and a higher activity was found in all the transgenic lines tested, suggesting that the increase in DAO activity may be important to maintain optimum PA levels in the cell [127]. In case of SAMDC transgenics, in addition to an increase in SPD and SPM levels, there was a significant increase in PUT levels, which might be due to the re-conversion of SPD to PUT via acetyl-SPD or γ-amino butyraldehyde. These transgenics have exhibited very high SAMDC activity, which was accompanied by higher DAO activity. They also showed marginal increase in ODC activity. Tobacco transgenics with *adc* and *samdc* genes were also tested for stress responses. They showed increased tolerance to salinity (250 mM NaCl) and PEG (10%) mediated drought. Interestingly, these transgenics also showed enhanced resistance against fungal (caused by *Verticillium dahliae* and *Fusarium oxysporum*) and bacterial (caused by *Ralstonia solanecearum*) wilts [127].

Some of the above single PA transgenics (e.g. *odc* transgenics) exhibited morphological aberrations like stunted plants with wrinkled leaves, which might be due to altered PA levels and PUT: SPD ratios. This problem may be overcome by up-gradation of the entire PA pathway in transgenic plants by the simultaneous introduction of the PUT synthesis gene (i.e. *odc*) and SPD synthesis gene (i.e. *samdc* or *spd syn*) by co-transformation. Indeed, such an attempt has been made in our laboratory, and double transgenic tobacco plants were produced with mouse *odc* and human *samdc* genes. It was observed that the double transgenic tobacco plants were normal and did not show any morphological abnormalities. Further, regeneration was also better from the transformed leaf explants with both *odc* and *samdc* genes as compared to explants from single transformants. These results further substantiate the role of PUT : SPD ratio in *in vitro* plant regeneration [124, 125]. Further, stress assays with double tobacco transgenics revealed increased tolerance to salinity and bacterial wilt [127].

Since most of the phytopathogenic fungi have only the ODC pathway for the synthesis of PAs, a novel method for the control of fungal plant infections by the selective inhibition of the fungal ODC by using its specific inhibitor (DFMO) has been reported [1, 2, 128]. The selective inhibition of the fungal ODC might also be achieved at the molecular level by the use of the antisense RNA technology [2, 128]. The endoparasitic fungi which infect the plant might take up the antisense *odc* transcripts produced in the transgenic plants along with the nutrients from the plant cells, thereby leading to the inhibition of the fungal ODC and growth [2, 128]. The above hypothesis was tested for the antisense *odc* transformed tobacco plants for the control of fungal wilt caused by *Verticillium dahliae*. The transgenic lines tested showed increased resistance to fungal infection as compared to the untransformed control plant. However, some more studies would be needed to prove this hypothesis [129].

There seemed to be distinctive roles played by either of the PA biosynthetic enzymes, with the ODC being critical in the early development of roots as the root development was affected during the regeneration of the antisense *odc* transformed rice and tobacco plants. The high ODC activity or perhaps the high PUT titres were on the other hand inhibitory for the early growth and development probably because they might lead to an increase in DAO activity as PUT forms the substrate for the enzyme, which results in the production of H_2O_2, higher concentrations of which are cytotoxic. Even though there existed distinctive roles for either of the two enzymes, it was observed in case of sense *odc* plants that the high titres of PUT could compensate for the decrease in the activity of the enzyme ADC, as was observed in the response of sense *odc* plants to abiotic stress.

6.   Conclusions and Future Prospects

The PA transgenics have provided a lot of information regarding the long-term effects of the shifts in the metabolism on plant growth and development, but they can be further used for gathering more information regarding the effects of the variations in the PA metabolism on other fundamental metabolic pathways in the cell. Such studies may provide an insight into the extent of involvement of each metabolic pathway in a specific response or phenotype. Also, there seems to be distinctive roles played by either of the PA biosynthetic enzymes, as can be deduced from the variable phenotypes observed in the various transgenics; this suggests a possible role for the improvement of modulation of the specific responses.

The transgenics also have the advantage of being distinctive from each other due to the random integration of the transgene, which influences the expression of the transgene and hence the shifts in the physiology in each plant were different. Therefore, an array of different transgenic lines each showing variability in the expression of transgene that were generated, could provide information regarding the physiological effects of variable gene expression. A clearer picture would emerge for these transgenics with the analysis of the fate of the PAs accumulated in these plants, as the PA catabolic enzymes as well as their channelling into the other pathways like ethylene and tropane alkaloids etc. also play a role in the metabolic flux of the cell. Besides the PA metabolism, the variations in the dynamics of other related metabolic pathways and their cumulative effect on the phenotype and physiological responses of the transgenics would give a better view of the cross-talk amongst pathways and its role in the functioning of the cell.

Also, the PA transgenics raised need to be tested for their tolerance to various stresses to elucidate the role played by the transgenics in stress responses and also the degree of tolerance imparted by them. These transgenics may also be used for studying various other biological processes, including senescence and fruit ripening. The use of the PA biosynthesis genes to generate stress tolerant plants has a major hurdle of the plants having an abnormal phenotype. Therefore a better approach to the problem would be to up-grade the entire metabolism (by the introduction of the *odc/adc* gene in conjunction with SPD and SPM synthesis gene *samdc/spd syn*) instead of a singular step, so that the plants would possess a normal phenotype and yet be tolerant to stresses. In fact, this approach was examined in our laboratory and proved to be correct. Further, it has been observed that the optimum PUT : SPD ratio is very important for normal plant growth and development [124, 125]. Therefore, the maintenance of a balance between PUT and SPD ratio appears to be very important for obtaining normal transgenics. This may be achieved by transforming the plants with the genes for both PUT and SPD synthesis. We were able to produce a large number of normal tobacco transgenic plants with mouse *odc* and human *samdc* genes by co-transformation as well as step-wise transformation. Such an approach would be quite useful in producing normal transgenic plants with PA biosynthesis genes and they would also impart abiotic stress tolerance.

Acknowledgements

The research work in our laboratory has been generously supported by the grants from the Department of Biotechnology (Grant No. BT/R & D/08/40/96), Department of Science and

Technology (Grant No. SP/SO/AO6/96), and the Indian Council of Agricultural Research [Grant No. F-1 (21)/96-FFC], New Delhi to MVR. Award of Senior Research Fellowship from the University Grants Commission, New Delhi to RK and Junior Research Fellowship from Council of Scientific and Industrial Research, New Delhi to SS is gratefully acknowledged. We are also thankful to Mr. Amit Arora for his excellent secretarial assistance.

References

1. M.V. Rajam, L.H. Weinstein, A.W. Galston, Prevention of a plant disease by specific inhibition of fungal polyamine biosynthesis, Proc. Nat. Acad. Sci. USA, 82 (1985) 6874–6878.

2. M.V. Rajam, In: Plant Ecophysiology, M.N.V. Prasad (Ed), John Wiley & Sons, New York, 1997, pp. 343–374.

3. T.A. Smith, P.J. Davies, J.B. Reid, Role of polyamines in giberrellin-induced internode growth in peas, Plant Physiol. 78 (1985) 92–99.

4. P.T. Evans, R.L. Malmberg, Do polyamines have role in plant development? Ann. Rev. Plant Physiol. Plant Mol. Biol. 40 (1989) 235–269.

5. M.V. Rajam, S. Dagar, B. Waie, J.S. Yadav, P.A. Kumar, F. Shoeb, R. Kumria, Genetic engineering of polyamine and carbohydrate metabolism for osmotic stress tolerance in higher plants, J. Biosci. 23 (1998) 473–482.

6. J. Martin-Tanguy, The occurrence and possible function of hydroxycinnamoyl acid amides in plants, Plant Growth Regul. 3 (1985) 381–399.

7. R. Kaur-Sawhney, A.F. Tiburcio, A.W. Galston, Spermidine and flower bud differentiation in thin layer explants of tobacco, Planta 173 (1988) 282–284.

8. H. Felix, J. Harr, Influence of inhibitors of polyamine biosynthesis on polyamine levels and growth of plants, Z. Naturforsch 44 (1989) 55–59.

9. T. Oshima, Unusual polyamines in an extreme thermophile, *Thermus thermophillus,* in: Advances in Polyamine Research (Eds U. Bachrach, A. Keye, R. Chayan), Raven Press, New York, 4 (1983) pp 479–487.

10. T. Oshima, N. Senshu, In: Polyamines: basic and clinical aspects, K. Imanoh, F. Suzuki, O. Suzuki, U. Bachrach (Eds) VNU Science Press, Netherlands, 1985, pp 113–117.

11. A.W. Galston, R. Kaur-Sawhney, Polyamines in plant physiology, Plant Physiol. 94 (1990) 406–410.

12. K. Hamana, S. Matsuzaki, Distribution of polyamines in prokaryotes, algae, plants and fungi, in: Polyamines: basic and clinical aspects (Eds K. Imanoh, F. Suzuki, O. Suzuki, U. Bachrach), VNU Science Press, Netherlands, 1985a, pp 105–112.

13. S.B. Agarwal, M. Agarwal, E.H. Lee, G.F. Kramer, P. Pillai, Changes in polyamine and glutathione content of green alga, *Chlorogonium elongatum* (Dang) France exposed to mercury, Envr. Exp. Bot. 32 (1992) 145–151

14. R. Kuttan, A.N. Radhakrishnan, T. Spande, B. Witkop, Sym-homo-spermidine, a naturally occurring polyamine, Biochem. 10(1971) 361–365.

15. K. Hamana, S. Matsuzaki, Natural occurrence of guanidinooxypropyl-amine in *Wistaria floribunch* and the swordbean *Canavalia gladiata*, Biochem. Biophys. Res. Commun. 129 (1985) 46–51.

16. S. Matsuzaki, K. Hamana, K. Isobe, M. Niitsu, K. Samejima, Novel polyamines and guanidinoamines in higher plants, in: The Biology and Chemistry of Polyamines (S.H. Goldberg, I.d. Algranati Eds), 12 (1989) 159–167.

17. B. Rodriguez-Garay, G.G. Phillips, G.D. Kuehn, Detection of norspermidine and norspermine in *Medicago sativa* L. (alfalfa), Plant Physiol. 89 (1989) 525–529.

18. N. Bagni, Aliphatic amines as a growth factor of coconut milk stimulating cellular proliferation of *Helianthus tuberosus* (Jerusalem artichoke), Experimentia 22 (1966) 732–733.

19. S.C. Minocha, R.K. Minocha, Role of polyamines in somatic embryogenesis, in: Biotechnology in

Agriculture and Forestry, Somatic Embryogenesis and Synthetic Seeds I (Ed Y.P.S. Bajaj), Springer-Verlag, Berlin Heidelberg, 30 (1995) pp 53–70.

20.   P. Sharma, M.V. Rajam, Genotype, explant and position effects on organogenesis and somatic embryogenesis in eggplant (*Solanum melongena* L.), J. Exp. Bot. 46 (1995) 135–141.

21.   J.S. Yadav, M.V. Rajam, Spatial distribution and temporal changes in free and bound polyamines in leaves of *Solanum melongena* L. Associated with differential morphogenetic capacity: efficient somatic embryogenesis with putrescine, J. Exp. Bot. 48 (1997) 1537–1545.

22.   J.S. Yadav, M.V. Rajam, Temporal regulation of somatic embryogenesis by adjusting cellular polyamine content in eggplant, Plant Physiol. 116 (1998) 617–625.

23.   A.W. Galston, H.W. Flores, Polyamines and plant morphogenesis, in: Biochemistry and Physiology of Polyamines in Plants (Eds R.D. Slocum, H.E. Flores), CRC Press, Boca Raton, London, 1991, pp 175–186.

24.   P. Sharma, J.S. Yadav, M.V. Rajam, Induction of laterals in root cultures of eggplant (*Solanum melongena* L.) in hormone free liquid medium: A novel system to study the role of polyamines, Plant Sci. 125 (1997) 103–111.

25.   M.B. Watson, K.K. Emory, R.M. Piatak, R.L. Malmberg, Arginine decarboxylase (polyamine synthesis) mutants of *Arabidopsis thaliana* exhibit altered root growth, Plant J. 13 (1998) 231–239.

26.   R.K. Kakkar, V.K. Rai, Plant polyamines in flowering and fruit ripening, Phytochem. 33 (1993) 1281–1288.

27.   A.W. Galston, In: The Physiology of Polyamines (Eds U. Bachrach, Y.M. Heimer), CRC Press, Boca Raton, Florida, 2 (1989) pp 99–105.

28.   A.F. Tiburcio, R.R. Besford, T. Capell, A. Borell, P.S. Testillano, M.C. Risueno, Mechanism of Polyamine action during senescence responses induced by osmotic stress, J. Exp. Bot. 45 (1994) 1789–1800.

29.   F. Shoeb, J.S. Yadav, S. Bajaj, M.V. Rajam, Polyamines as biomarkers for plant regeneration capacity: improvement of regenration by modulation of polyamine metabolism in different genotypes of indica rice, Plant Sci. 160 (2001) 1229–1235.

30.   N. Bagni, A. Tassoni, Biosynthesis, oxidation and conjugation of aliphatic polyamines in higher plants, Amino Acids 20 (2001) 301–317.

31.   A. Santanen, L.K. Simola, Catabolism of putrescine and spermidine in embryogenic and non embryogenic callus lines in *Picea abies*, Physiol. Plant. 90 (1994) 125–129.

32.   R.D. Slocum, R. Kaur-Sawhney, A.W. Galston, The physiology and biochemistry of polyamines in plants, Arch. Biochem. Biophys. 235 (1984) 283–303.

33.   R.T. Besford, C.M. Richardson, J.L. Campos, A.F. Tiburcio, Effect of polyamines on stabilization of molecular complexes in thyalkoid membranes of osmotically stressed leaves, Planta 189 (1993) 201–206.

34.   J. Martin-Tanguy, F. Cabanne, E. Perdrizet, C. Martin, The distribution of hydroxycinnamic amides in flowering plants, Phytochem. 17 (1978) 1927–1928.

35.   M. Bueno, D. Garrido, A. Matilla, Gene expression induced by spermine in isolated embryonic axes of chickpea seeds, Plant Physiol. 87 (1993) 381–388.

36.   J. Messiaen, P. Cambier, P. Van Cutsem, Polyamines and pectin I. Ion exchange and selectivity, Plant Physiol. 113 (1997) 387–395.

37.   G. Berta, M.M. Altamura, A. Fusconi, F. Cerruti, F. Capitiani, N. Bagni, The plant cell wall is altered by inhibition of polyamine biosynthesis, New Phytol. 137 (1997) 569–577.

38.   Bharti, M.V. Rajam, Involvement of polyamines in resistance of wheat to *Puccinia recondita*, Phytochemistry, 71 (1996) 1009–1013.

39.   A.M. Moustacas, J. Nari, M. Borel, G. Noat. J. Ricard, Pectin methyl esterase, metal ions and plant cell wall expansion, Biochem. J. 279 (1991) 351–354.

40.   A Tassoni, F. Antognoni, M.L. Battistini, O. Sanvido, N. Bagni, Characterization of spermidine binding to solubilized plasma membrane proteins from Zucchini hypocotyls, Plant Physiol. 117 (1998) 971–977.

41.   B.G. Feuerstein, N. Pattabiraman, L.J. Marton, Spermine-DNA interactions: A theoretical study, Proc. Natl. Acad. Sci. USA, 83 (1986) 5948–5952.

42. H. Ohishi, I. Nakanishi, K. Inubshi, G. Van Der Marel, J.H. Van Boom, A. Rich, A.H.J. Wang, T. Hakoshima, K.Tomita, Interactions between the left-handed Z-DNA and polyamine: The crystal structure of the d(CG)$_3$ spermidine complex, FEBS Lett. 391 (1996) 153–156.

43. D. Balasundarum, A.K. Tyagi, Polyamine-DNA nexus: Structural ramifications and biological implications, Mol. Cell. Biochem. 100 (1991) 129–140.

44. M.L. Howell, G.P. Schroth, P.S. Ho, Sequence dependent effects of spermine on thermodynamics of B-DNA to Z-DNA transition, Biochem. 35 (1996) 15373–15382.

45. M. Musso, T. Thomas. A. Shirahata, L.H. Sigal, M.W. Van Dyke, T.J. Thomas, Effects of Chain length modification and bis (ethyl) substitution of spermine analogs on purine-purine-pyrimidine triplex DNA stabilization, aggregation and conformation transition, Biochem. 36 (1997) 1441–1449.

46. M. Auvinen, A. Paasinen, L.C. Anderson, E. Holta, Ornithine decarboxylase activity is critical for cell transformation, Nature 360 (1992) 355–358.

47. P. Pohjanpelto, E. Hölttä, Phosphorylation of Okazaki-like DNA fragments in mammalian cells and role of polyamines in the processing of this DNA, EMBO J. 15 (1996) 1193–1200.

48. D. Serafini-Fracassini, P. Torrigiani, C. Branca, Polyamines bound to nucleic acids during dormancy and activation of tuber cells of *Helianthus tuberosus*, Physiol. Plant. 60 (1984) 351–357.

49. M.D. Morch, C. Bencourt, Polyamines stimulate expression of amber termination codons *in vitro* by normal tRNAs, Eur. J. Biochem. 105 (1980) 445–451.

50. R. Kaur-Sawhney, A.W. Galston, Histological and biochemical studies on the anti-senescence properties of polyamines in plants, in: Biochemistry and Physiology of Polyamines in Plants (Eds R.D. Slocum, H.E. Flores), CRC Press, Boca Raton, 1991, pp 201–211.

51. H. Tabor, C.W. Tabor, Polyamine requirement for efficient translation of amber codons *in vivo*, Proc. Natl. Acad. Sci. USA, 79 (1982) 7087–7091.

52. C.M. Isola, L. Franzoni, Inhibition of net synthesis of ribonuclease by polyamines in potato tuber slices, Plant Sci. 63 (1989) 39–45.

53. E. Rom, C. Kahana, Polyamine regulate the expression of ornithine decarboxylase antizyme *in vitro* by inducing ribosomal frame-shifting, Proc. Natl. Acad. Sci. USA, 91 (1994) 3959–3963.

54. N. Datta, M.B. Schell, S.J. Roux, Spermine stimulation of a nuclear NII kinase from pea plumule and its role in the phosphorylation of a nuclear polypeptide, Plant Physiol. 84 (1987) 1397–1401.

55. S.J. Roux, Casein kinase-2 type protein kinases in plants: Possible targets of polyamine action during growth regulation? Plant Grow. Regul. 12 (1993) 189–193.

56. T. Uzawa, N. Hamasaki, T. Oshima, Effects of novel polyamines on cell-free polypeptide synthesis catalyzed by *Thermus thermophilus* HB8 extract, J. Biochem. 114 (1993) 478–486.

57. R. Blatter, A. Ochsenvein, A. Boschetti, Polyamine and cytoplasmic preparations enhances light-driven protein synthesis in isolated chloroplasts of *Chlamydomonas reinhardtii*, Plant Physiol. 30 (1992) 743–752.

58. X.S. Ye, S.A. Avdishko, J. Kuc, Effects of polyamines on *in vitro* phosphorylation of soluble and plasma membrane proteins in tobacco, cucumber and *Arabidopsis thaliana*, Plant Sci. 97 (1994) 109–118.

59. J.C. Mader, D.E. Hanke, Polyamine sparing may be involved in the prolongation of cell division due to inhibition of phenlypropanoid synthesis in cytokinin-starved soyabean oils, Plant Growth Regul. 16 (1997) 89–93.

60. A. Apelbaum, Z.N. Canellakis, P.B. Applewhite, R. Kaur-Sawhney, A.W. Galston, Binding of spermidine to a unique protein in thin-layer tobacco tissue culture, Plant Physiol. 88 (1988) 996–998.

61. A.F. Tiburcio, J.L. Campos, X. Figueras, R.T. Besford, Recent advances in the understanding of polyamine function during plant development, Plant Grow. Regul. 12 (1993) 331–340.

62. T. Mustelin, H. Poso, S.P. Lapijoki, J. Gynther, S.E. Andersson, Growth signal transduction: rapid activation of covalently bound ornithine decarboxylase during phosphatidyl inositol breakdown, Cell 49 (1987) 171–176.

63. A. Kumar, T. Altabella, M.A. Taylor, A.F. Tiburcio, Recent advances in polyamine research, Trends Plant Sci. 2 (1997) 124–130.

64. D.R. Morris, C.M. Jorstad, Isolation of conditionally putrescine deficient mutant of *Escherichia coli*, J. Bacteriol. 101 (1970) 731–737.

65. D.R. Morris, C.M. Jorstad, Growth and macromolecular composition of a mutant of *Escherichia coli* during polyamine limitation, J. Bacteriol. 113 (1973) 271–277.

66. I.N. Hirshfield, H.J. Rosenfeld, Z. Leifer, W.K. Maas, Isolation and characterization of a mutant of *Escherichia coli* blocked in the synthesis of putrescine, J. Bacteriol. 101 (1970) 725–730.

67. C.W. Tabor, H. Tabor, *Escherichia coli* mutants completely deficient in adenosylmethionine decarboxylase and in spermidine biosynthesis, J. Bacteriol. Chem. 253 (1978) 3671–3676.

68. H. Tabor, C.W. Tabor, Biochemical and genetic studies of polyamine in *Saccharomyces cerevisiae*, Adv. Polyamine Res. 4 (1983) 455–465.

69. M.S. Cohn, C.W. Tabor, H. Tabor, Isolation and characterization of *Saccharomyces cerevisiae* mutants deficient in S-adenosylmethionine decarboxylase, Spermidine and Spermine, J. Bacteriol. 134 (1978) 208–213.

70. M.L. Malmberg, J. McIndoo, Abnormal floral development of a tobacco mutant with elevated polyamine levels, Nature 305 (1983) 623–625.

71. R.L. Malmberg, J. McIndoo, A.C. Hiatt, B.A. Lowe, Genetics of polyamine synthesis in tobacco: Developmental switches in flower, Cold Spring Harbour Symp. 50 (1985) 475–482.

72. K. Fritz, I. Czaja, R. Walden, T-DNA tagging of genes influencing polyamine metabolism: Isolation of mutant plants lines and rescue of DNA promoting growth in the presence of polyamine biosynthetic inhibitor, Plant J. 7 (1995) 261–271.

73. P. Bey, C. Danzin, M. Jung, in: Inhibition of Polyamine Metabolism, Biological Significance and Basis for New Therapies (Eds P.P. McCann, A.E. Pegg, A Sjoerdsma,), Academic Press, San Diego, 1987, pp 1–32.

74. B.W. Metcalf, P. Bey, C. Danzin, M. Jung, P. Casara, J.P. Vevèrt, Catalytic irreversible inhibition of mammalian ornithine decarboxylase by substrate and product analogues, J Amer. Chem. Soc. 100 (1978) 2251–2253.

75. A. Kallio, P.P. McCann, P. Bey, DL-difluoromethylarginine: A potent enzyme activated inhibitor of bacterial arginine decarboxylase, Biochem. 20 (1981) 3163–3166.

76. A.J. Bitonti, P.J. Casara, P.P. McCann, P.P. Bey, Catalytic irreversible inhibition of bacterial and plant. arginine decarboxylase activities by novel substrate and product analogs, Biochem. J. 242 (1987) 69–74.

77. A.E. Pegg, H.G. William-Ashman, Pharmacologic interference with enzymes of polyamine biosynthesis and of 5-methylthioadenosine metabolism, in: Inhibition of Polyamine Metabolism. Biological Significance and the Basis for New Therapies (Eds. P.P. McCann, A.E. Pegg, A. Sjoerdsma), Academic Press, San Diego, 1987, pp 33–48.

78. G. Drolet, E.B. Dumbroff, R.L. Legge, J.E. Thompson, Radical scavenging properties of polyamines, Phytochem. 25 (1986) 367–371.

79. R.A. Descenzo, S.C. Minocha, Modulation of cellular polyamines in tobacco by transfer and expression of mouse ornithine decarboxylase cDNA, Plant Mol. Biol. 22 (1993) 113–117.

80. W.N. Noh, S.C. Minocha, Expression of a human S-adenosylmethionine decarboxylase cDNA in transgenic tobacco and its effects on polyamine biosynthesis, Transgenic Res. 3 (1994) 26–35.

81. A. Kumar, M.A. Taylor, S.A. Madarif, H.V. Davies, Potato plants expressing antisense and sense S-adenosylmethionine decarboxylase (SAMDC) transgene shows altered levels of polyamines and ethylene: antisense plants display abnormal phenotypes, Plant J. 9 (1996) 147–158.

82. T. Capell, C. Escobar, H. Liu, D. Burtin, O. Lepri, P. Christou, P., Over-expression of oat arginine decarboxylase cDNA in transgenic rice (*Oryza sativa* L.) affects normal development pattern *in vitro* and results in putrescine accumulation in transgenic plants, Theor. Appl. Genet. 97 (1998) 246–254.

83. D.R. Bastola, S.C. Minocha, Increase putrescine biosynthesis through transfer of mouse ornithine decarboxylase cDNA in carrot promotes somatic embryogenesis, Plant Physiol. 109 (1995) 63–71.

84. H.J. Van Kranen, L. Van De Zen, C.F. Van Kreijl, A. Bisschop, B. Wieringa, Cloning and nucleotide sequence of rat ornithine decarboxylase cDNA, Gene 60 (1987) 145–155.

85. A. Katz, C. Kahana, Isolation and characterization of the mouse ornithine decarboxylase gene, J. Biol. Chem. 263 (1988) 7604–7609.

86. N.J. Hickok, J. Wahlfors, A. Crozat, M. Halmekyto, L. Alhonen, J. Jahne, O.A. Jahne, Human ornithine decarboxylase-encoding loci: nucleotide sequence of the expressed gene and characterisation of a pseudogene, Gene 93 (1990) 257–263.

87. L.J. Williams, G.R. Barnett, J.L. Ristow, J. Pitkin, M. Perriere, R.H. Davis, Ornithine decarboxylase gene of *Neurospora crassa*: Isolation, sequence and polyamine mediated regulation of its mRNA, Mol. Cell. Biol. 12 (1992) 347–359.

88. J. Yao, D. Zadworny, U. Kuhnlein, J.F. Hayes, Molecular cloning of a bovine ornithine decarboxylase cDNA and its use in the detection of restriction fragment length polymorphism in Holsteins, Genome 38 (1995) 325-331.

89. V. Malik, M.B. Watson, R.L. Malmberg, A tobacco ornithine decarboxylase partial cDNA clone, J. Plant Biochem. Biotch. 5 (1996) 109–112.

90. A.J. Michael, J.M. Furze, M.J.C. Rhodes, D. Burtin, Molecular cloning and functional identification of a plant ornithine decarboxylase cDNA, Biochem. J. 314 (1996) 241–248.

91. D. Alabadi, J. Carbonell, Expression of ornithine cecarboxylase is transiently increased by pollination, 2,4–Dichlorophenoxyacetic acid, and gibberllic acid in tomato ovaries, Plant Physiol. 118 (1998) 323–328.

92. E. Bell, R.L. Malmberg, Analysis of a cDNA encoding arginine decarboxylase from oat reveals similarity to the *Escherichia coli* arginine decarboxylase and evidence of protein processing, Mol. Gen. Genet. 224 (1990) 431–436.

93. R. Rastogi, J. Dulson, S.J. Rothstein, Cloning of tomato (*Lycopersicon esculentum* Mill.) arginine decarboxylase gene and its expression during fruit ripening, Plant Physiol. 103 (1993) 829–834.

94. M.A. Perez Amador, J. Carbonell, A. Granell, Expression of arginine decarboxylase is induced during early fruit development and in young tissues of *Pisum sativum* (L.), Plant Mol. Bio. 28 (1995) 997–1009.

95. M.B. Watson, R.L. Malmberg, Regulation of *Arabidopsis thaliana* L. heynh arginine decarboxylase by potassium deficiency stress, Plant Physiol. 111 (1996) 1077–1983.

96. K.H. Nam, S.H. Lee, J.H. Lee, Differential expression of ADC mRNA during development and upon acid stress in soyabean (*Glycine max*) hypocotyls, Plant Cell Physiol. 38 (1997) 1156–1166.

97. M.A. Taylor, S.A. MadArif, A. Kumar, H.V. Davis, L.A. Scobie, S.R. Pearce, F.J. Flavell, Expression and sequence analysis of cDNAs induced during the early stages of tuberization in different organs of potato plant (*Solanum tuberosum* L.), Plant Mol. Biol. 20 (1992) 641–651.

98. C. Bolle, R.G. Herrmann, R. Oelmuller, A spinach cDNA with homology to S-adenosylmethionine decarboxylase, Plant Physiol. 107 (1995) 1461–1462.

99. G. Schroder, J. Schroder, cDNA for S-adenosyl-L-methionine decarboxylase from *Catharanthus roseus*, heterologous expression, identification of the proenzyme processing site, evidence for the presence of both subunits in the active enzyme and a conserved region in the 5′ messenger RNA leader, Eur. J. Biochem. 228 (1995) 74-78.

100. T. Dresselhaus, P. Barcela, C. Hagel, H. Lorez, K. Humbeek, Isolation and characterization of a *Tritordeum* cDNA encoding S-adenosylmethionine decarboxylase that is circadian clock-regulated, Plant Mol. Biol. 30 (1996) 1021–1033.

101. I. Yoshida, H. Yamagata, E. Hirasawa, Light regulated gene expression of S-adenosylmethionine decarboxylase in *Pharbitis nil*, J. Exp. Bot. 49 (1998) 617–620.

102. Z.Y. Li, S.Y. Chen, Differential accumulation of the SAMDC transcript in rice seedlings in response to salt and drought stresses, Theor. App. Gen. 100 (200) 782–788.

103. A. Pajunen, A. Gozat, O.A. Janne, R. Ihalainen, P.H. Laitinen, B. Stanley, R. Madhubala, A.E. Pegg, Structure and regulation of mammalian S-adenosylmethionine decarboxylase, J. Biol. Chem. 263 (1988) 17040–17049.

104. T. Hashimoto, K. Tamaki, K. Suzuki, Y. Yamada, Molecular cloning of plant spermidine synthases, Plant Cell Physiol. 39 (1998) 73–79.

105. V.P. Korhonen, M. Halmekytö, L. Kauppinen, S. Myöhänen, J. Wahlfors, T. Keinänen, T. Hyvönen, L. Alhonen, T. Eloranta, J. Jänne, Molecular cloning of a cDNA encoding human spermine synthase, DNA Cell Biol. 14 (1995) 841–847.

106. S. Yamamoto, S. Nagata, K. Kusaba, Purification and characterization of homospermidine synthase in *Actinobacer tartarogenes* ATCC 31105, J. Biochem. 114 (1993) 45–49.

107. D. Ober, Strategien zur immunologischen und molekularbiologischen untersuchung der homospermidin synthase, dem eingangsenzym der pyrrolizidinalkaloidbiosynthase, Ph.D dissertation (1997) TU Braunschweig.

108. A. Kaiser, Cloning and expression of a cDNA encoding homospermidine synthase from *Senecio vulgaris* (Asteraceae) in *Escherichia coli*, Plant J. 19 (1999) 195–201.

109. P. Tavadoraki, M.E. Schinia, F. Cecconi, S. Di Agostino, F. Manera, G. Rea, P. Mariottini, R. Federico, R. Angelini, Maize polyamine oxidase: primary structure from protein and cDNA sequencing, FEBS Lett. 426 (1998) 62–66.

110. A. Rossi, R. Petruzzelli, A.F. Agro, cDNA derived amino acid sequence of lentil seedlings amine oxidase, FEBS 301 (1992) 253–257.

111. A.J. Tipping, M.J. McPherson, Cloning and molecular analysis of the pea seedling copper amine, J. Biol. Chem. (1995).

112. J.D. Hamill, R.J. Robins, A.J. Parr, D.M. Evans, J.M. Furze, M.J.C. Rhodes, Over-expressing a yeast ornithine decarboxylase gene in transgenic roots of *Nicotiana rustica* can lead to enhanced nicotine accumulation, Plant Mol. Biol. 15 (1990) 27–38.

113. S.E. Anderson, D.R. Bastola, S.C. Minocha, Metabolism of polyamines in transgenic cells of carrot expressing a mouse ornithine decarboxylase cDNA, Plant Physiol. 116 (1998) 299–307.

114. R. Kumria, M.V. Rajam, Alteration in polyamine titres during *Agrobacterium*-mediated transformation of indica rice with ornithine decarboxylase gene affects plant regeneration potential, Plant Sci. 162 (2002) 769–777.

115. O. Lepri, L. Bassie, G. Safwat, P. Thu-Hang, P. Trung-Nghia, E. Holtta, P. Christou, T. Cappel, Over-expression of a cDNA for human ornithine decarboxylase in transgenic rice plants alters the polyamine pool in a tissue-specific manner, Mol. Genet. Genomics, 266 (2001) 303–312.

116. J. Martin-Tanguy, D. Tepfer, M. Paynot, D. Burtin, L. Heisler, C. Martin, Inverse relationship between polyamine levels and the degree of phenotypic alteration induced by the root inducing, left hand transferred DNA from *Agrobacterium rhizogenes*, Plant Physiol. 92 (1990) 912–918.

117. P. Bhatnagar, B.M. Glasheen, S.K. Bains, S.L. Long, R. Minocha, C. Walter, S.C. Minocha, Transgenic manipulation of the metabolism of polyamines in poplar (*Populus nigra × Maximowiczii*) cells, Plant Phyiol. 125 (2001) 2139–2153.

118. P. Bhatnagar, R. Minocha, S.C. Minocha, Genetic manipulation of the metabolism of polyamines in poplar cells. The regulation of putrescine catabolism, Plant Phyiol. 128 (2002) 1455–1469.

119. C. Masgrau, T. Altabella, R. Fareas, D. Flores, A.J. Thompson, R.T. Bestford, A.F. Tiburcio, Inducible over-expression of oat arginine decarboxylase in transgenic tobacco plants, Plant J. 11 (1997) 465–473.

120. L. Bassie, M. Noury, O. Lepri, T. lahaye, P. Christou, T. Capell, Promoter strength influences polyamine metabolism and morphogenic capacity in transgenic rice tissues expressing oat *adc* cDNA constitutively, Transgenic Res. 9 (2000) 33–42.

121. M. Noury, L. Bassie, O. Lepri, I. Kurek, P. Christou, T. Capell, A transgenic rice cell lineage expressing oat arginine decarboxylase (*adc*) cDNA constitutively accumulates putrescine in callus and seeds but not in vegetative tissue, Plant Mol. Biol. 43 (2000) 537–544.

122. M. Roy, R. Wu, Arginine decarboxylase transgene expression and analysis of environmental stress tolerance in transgenic rice, Plant Sci. 160 (2001) 869–875.

123. R. Kumria, M. V. Rajam, Ornithine decarboxylase transgene in tobacco affects polyamines, *in vitro*—morphogenesis and response to salt stress, J. Plant Physiol. (in press).

124. F. Shoeb, Regulation of plant regeneration by modulating cellular polyamine levels in fresh and long-term cultures of indica rice (*Oryza sativa* L.), Ph.D thesis (1999) University of Delhi New Delhi.

125. S. Bajaj, M.V. Rajam, Polyamine accumulation and near loss of morphogenesis in long term callus cultures of rice: restoration of plant regeneration by manipulation of cellular polyamine levels, Plant Physiol. 112 (1996) 1343–1348.

126. J.P. Wisniewski, E.A. Rathbun, J.P. Knox, N.J. Brewin, Involvement of diamine oxidase and peroxides in insolubilization of the extarcellular matrix: implications for pea nodule initiation by *Rhizobium leguminosarum*, Mol. Plant-Microbe Interact. 13 (2000) 413–420.

127. B. Waie, Genetic engineering of polyamine metabolism for osmotic stress tolerance in rice and tobacco, Ph.D thesis (2001) University of Delhi.

128. M.V. Rajam, Polyamine biosynthetic pathway: a potential target for plant chemotherapy, Curr. Sci. 74 (1998) 729–731.

129. R. Kumria, Modulation of polyamine biosynthesis, plant regeneration and stress response in transgenic rice and tobacco by introduction of ornithine decarboxylase gene, Ph.D. thesis (2000), University of Delhi.

Plant Biotechnology and Molecular Markers
P.S. Srivastava, Alka Narula and Sheela Srivastava (Editors)
Copyright © 2004 Anamaya Publishers, New Delhi, India

6. Biotechnological Approaches Towards Improvement of Medicinal Plants

Alka Narula, Sanjeev Kumar*, K.C. Bansal** and P.S. Srivastava

Centre for Biotechnology, Faculty of Science, Hamdard University, New Delhi 110 062, India
*Amity Institute of Biotechnology, Amity IT University Campus, Sector 125, Noida 201 303, India
**NRC on Plant Biotechnology, IARI, New Delhi 110012, India

Abstract: Herbs are now in great demand in both developed and developing countries because of their proven efficacy and little or no reported side effects. Secondary metabolites, the active principles expensive to produce and accumulate are usually biosynthesized in smaller quantities. This has resulted in ruthless exploitation of medicinally important plants creating imbalance in supply and demand. An alternative technology could be the application of *in vitro* culture of desirable medicinal plants to increase the plantation propagules and enhance the yield of specific drug components. Successful micropropagation protocols for various medicinal plants have been developed and their conservation has also become feasible through synthetic seeds and cryopreservation technologies. Besides other techniques, genetic engineering of medicinal plants using *Agrobacterium*-mediated transformation has many advantages that include fast growth and high level of stable production of secondary metabolites making them commercially and economically feasible. Genetic fidelity of tissue culture raised plants can be ascertained by using molecular markers.

1. Introduction

Medicinal plants have been the subject of man's curiosity and purpose since time immemorial. The importance of medicinal plants in the treatment of chronic diseases needs no elaboration. In fact, even with the tremendous advancement in the field of synthetic chemistry, almost 50% of the commercial drugs available in the market remain of plant origin. The herbal system was, however, pushed to the background with the advent of allopathic system. It is now back with a venegence and the age-old system of herbal medicine is being revived due to its long lasting curative effect, easy availability, natural way of healing and rare or no reported side effects. Due to growing world population, increased anthropogenic activities, rapidly eroding natural ecosystem etc., the natural habitat for a great number of plants are dwindling and many of them are facing extinction [1]. The inevitable ruthless exploitation of herbs leading to their rapid depletion from the wild is a cause for concern. In fact, the pace of depletion has outpaced the pace of conservation. New strategies are being therefore formulated for rapid multiplication and conservation of medicinal plants. Besides the conventional methods, biotechnology has proved useful in the improvement of herbs that yield drugs. In this resurgent era of herbal drugs it is very difficult to make an accurate assessment of the volume and value of herbal trade in India. Consequently, it varies widely [2]. According to estimates by the Ayurvedic Drug Manufacturers Association (ADMA), the current value of trade in Indian System of Medicine (chiefly Ayurveda, Siddha and Unani) and Homeopathy is around Rs. 4205 crores, roughly close to US$ 1 billion [see 3]. Therefore, the value of medicinal plants is also reflected in the economics of global market which was estimated to be 60 billion US dollars in 2000 (Fig. 1). The Asian region rich in biological wealth and genetic diversity also has a substantial share in herb trade (Fig. 2).

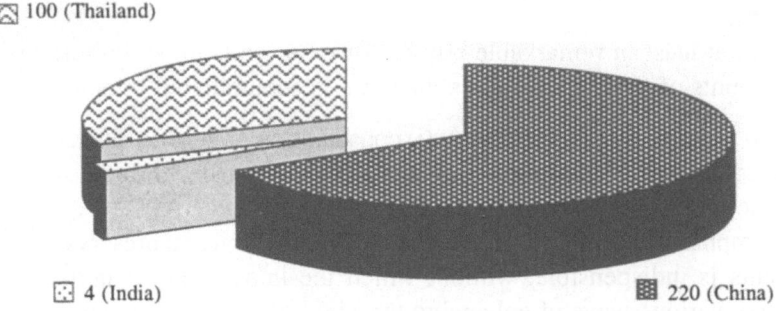

Fig. 1. **The market value of natural health products in 2000 was already worth US $ 500 billion of which 220 was projected for China, 100 for Thailand and only 4 billion for India.**

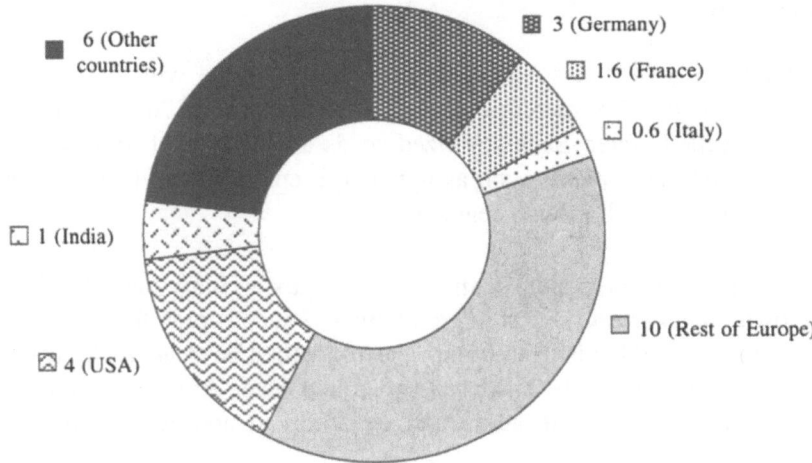

Fig. 2. **In 2000, country-wise share of herbal medicine was US $ 60 billions with India's share of only 1 billion. The predicted annual growth rate was 7% for this sector.**

It is surmountable that medicinal plant biotechnology has grown from cell technology, specifically plant tissue culture. Regeneration of plants has been achieved with cells and tissues excised from various medicinal herbs. The powerful techniques of plant cell and tissue culture, and recombinant DNA and bioprocessing technologies etc., coupled with sophisticated analytical tools such as NMR, HPLC, GC-MS, LC-MS etc., have offered mankind a great opportunity to exploit the totipotent, biosynthetic and biotransformation capabilities of plant cells under *in vitro* conditions. The scope for *in vitro* germplasm preservation and large-scale production of plant secondary metabolites has brightened.

Advantages of extracting secondary metabolites using plant tissue culture are:

1. The source of these metabolites, i.e., most of higher plants have specific agroclimatic

requirements. Hence specific metabolites can be produced in cultures all through the year even in places where these crops are not grown.

2. The already limited supply of these raw materials can not be exhausted considering the future needs.

3. If not in all, at least in remarkable number of cases cells under culture tend to produce greater amounts of these metabolites than that is accumulated in nature.

In addition, *in vitro* technology also facilitates: (i) conservation of genetic diversity and germplasm of medicinal plants through cryopreservation, and (ii) gene transfer through recombinant DNA technology and the molecular markers in the form of AFLP and RAPD.

In this article, emphasis has been laid on the fact that protection and preservation of germplasm of medicinal plants is indispensible, without which the knowledge of herbal medicines will remain futile. Also, various ways of enhancing the yield of active components are reviewed.

2. Materials and Methods

2.1 Micropropagation

Juvenile Explants: Seeds of the desired plant species are washed with 0.5–2.0% cetrimide followed by treatment with 0.1% mercuric chloride and dipped in 70% alcohol, thereafter washed with sterile distilled water. Such sterilized seeds are implanted on basal medium for germination. Various seedling explants such as hypocotyl, epicotyl, cotyledon and radicle are incoulated on suitable media with growth regulators.

Mature Explants: Explants such as stem segment, shoot apex, axillary buds, leaf, root, anther, etc., from field grown plants are surface sterilized with 1–2% cetrimide followed by treatment with streptomycin sulphate and bavistin solution. They may then be treated with 0.1% mercuric chloride, 70% alcohol and finally washed with sterile distilled water. After sequential sterilization, explants are implanted on media with auxin and cytokinin in appropriate combination and concentration.

Most of the cultures are maintained in a culture room at $25 \pm 2°C$ with $55 \pm 5\%$ relative humidity and 10–14 hr light/dark period with irradiance of 60–100 $\mu mol\ m^{-2}s^{-1}$ provided by white cool flourescent tubes. The cultures are monitored at regular intervals and the regenerants are maintained on the best suited medium. The rooted plantlets are hardened and then transplanted to pots and finally transferred to field.

2.2 Secondary Metabolite Analysis

Cultures harvested during different stages of growth and differentiation are analysed for the presence of secondary metabolites (alkaloids, steroids, flavonoids, glycosides, furanocoumarins, etc.). Quantification of isolated compounds is made either through spectrophotometry, High Performance Liquid Chromatography (HPLC) or Gas Liquid Chromatography (GLC). Stage showing highest yield of the active principle is selected as the harvesting stage for that particular culture.

As they are present in low amounts in plants, attempts can be made to enhance the yield by supplementing the medium with elicitors, precursors or manipulating the hormonal combination of the medium or subjecting biotic/abiotic stress to the cultures.

2.3 Cryopreservation of Cultures

Various explants have been used for cryopreservation of medicinally important plants. The general protocol involves treating the cultures with appropriate cryoprotectant such as DMSO, glycerol, sucrose or proline, etc. and then plunging in liquid nitrogen ($-196°C$). After freeze-storage, the cultures are thawed at 35–40°C, washed and recultured. Complete plantlets can be regenerated from such frozen cultures.

2.4 Synthetic Seeds

This technology involves the encapsulation of propagules (somatic embryos/axillary buds/shoot apices, etc.) which functionally mimic seeds and can develop into plantlets under suitable conditions. For encapsulation, the propagules may be embedded in a matrix that serves as endosperm, containing carbon source, nutrients, growth regulators and antimicrobial agents. Sodium alginate is commonly used. However, there are several other agents including guargum, calcium alginate, gelrite, sodium alginate with gelatin, potassium alginate, sodium pectate, etc. In addition, polyethylene oxide homopolymers, synthetic sodium-magnesium-lithium silicate, sodium crylate, etc. have been tried as coating agents. After mixing in the encapsulation matrix, the propagules are picked up by pipette and then dropped into a solution of calcium chloride. Thereafter, they are kept undisturbed for surface complexing to obtain encapsulated beads. The beads are kept in a solution of 2.5% $CaCl_2$ for 40 min on a shaker. After the completion of incubation period, the beads are recovered by decanting the $CaCl_2$ solution and washed 3–4 times with basal medium. Such encapsulated propagules cultured on nutrient medium or different substrates like filter paper, soilrite, etc. can develop into plants.

2.5 Transformation

In recent years, *Agrobacterium*-mediated transformation has emerged as an efficient method for genetic manipulation of plants. Although direct DNA transfer methods, particularly particle bombardment, are also being employed, other gene transfer methods include electroporation and electrophoresis, laser microbeam technique, microinjection, liposome fusion and injection. *Agrobacterium*-mediated transformation, however, has major advantages over these systems. After establishing a reliable protocol for micropropagation the explants can be incubated with *Agrobacterium* suspension, blotted dry on whatman filter paper and transferred to regeneration medium for co-cultivation. The co-cultivated explants are then transferred to the selection medium. After selection, the explants are allowed to grow on regeneration medium + Cefotaxime. The putative transgenics can be rooted and after hardening transferred to field. The transgenic nature of regenerated plants can be confirmed by polymerase chain reaction (PCR) and southern blot analysis.

2.6 Molecular Markers for Ascertaining Clonal Fidelity

DNA-based markers provide an efficient tool for screening tissue culture raised plants because these markers are not affected by environmental factors and present more reliable results. PCR-based markers such as RAPD have been used for detecting off-types from micropropagated plants. AFLP, and SNPs are now preferred as it combines the reliability of RFLP with RAPD.

For AFLP analysis, total genomic DNA can be isolated from desired plant parts by using a suitable method. It can then be restricted with restriction enzymes followed by ligation with specific adapters. Pre-amplification of the adapter-ligated DNA can be done by using selective nucleotides. The samples are electrophoresed on acrylamide gel and autoradiographed. AFLP amplification products are scored for their presence and absence across the individuals tested. Genetic similarity between pairs is estimated by Jaccard's coefficient. The phenetic dendrogram can be constructed by UPGMA (unweighed pair group method of arithmetic averages) in order to group individuals into discrete clusters.

3.   Results and Discussion

3.1   Micropropagation of Medicinal Plants and Yield of Secondary Metabolites

In vitro cultured cells and tissues can be induced to differentiate into plants through organogenesis [4-8] (Figs. 3 to 8) or somatic embryogenesis [9]. The response of any tissue *in vitro* is attributed to the composition of the medium besides other factors including a balance between growth regulators [10]. MS medium originally developed for rapid growth of tobacco tissue culture is the most frequently used for majority of the species. Already there are credible reports of *in vitro* propagation of medicinal plants by using various explants, such as leaf [11], stem [12, 13], shoot buds [14, 15] anthers [16], roots [17], shoot tips, nodal segments [18, 19], and seedlings [4, 20, 21] (Table 1).

The earliest detailed reference of plant cell cultures as an industrial route to natural product synthesis dates back to 1956. Despite the success and the related surge in information, the expected progress during the following decades remained slow. After 1973, a turning point in cell culture technology demonstrated reasonable yield of desired secondary metabolites [16, 22]. Earlier it was believed that enhancement in the yield of secondary metabolites was dependent on prolonged tissue cultures or organogenesis [23]. Subsequently, it has been revealed that the ability of product biosynthesis continues throughout during the culture regime and can be detected at various stages of growth and differentiation [24].

Yield of secondary metabolites can be enhanced by modifying the chemical milieu and culture conditions (Table 2). Zenk et al. [25] observed that the composition of culture medium not only affects growth and production of metabolites but also plays a critical role in initiation of morphogenic events in the culture. Consequently, almost all the major components of the growth medium have been tested for their varied effects on different types of differentiated and undifferentiated cultures [26].

The most commonly used carbon source for tissue culture media is sucrose. The other carbon sources tested for supporting growth include glucose, galactose and also complex carbohydrates such as milkwhey and molasses. Increased sugar concentration favoured synthesis of shikonin in cell cultures of *Lithospermum erythrorhizon*, diosgenin production in *Dioscorea*, and anthraquinone in cell cultures of *Gallium mollugo*. On the contrary, lesser amounts of sucrose favoured the production of ubiquinone 10 in *Coleus blumei* [see 27]. Saccharose as sugar source has shown strongest effect for secondary metabolite content increase in cultures of *Catharanthus*,

Fig. 3.   *Artemisia annua*. Stem segments cultured on MS + (in mg l^{-1}) NAA (0.1) + BAP (3.0) + GA$_3$ (0.1) + Asp (50) + Arg (50) + Glu (100) + Cyst hyd (5.0). (A) Multiple shoots in 10-week-old culture; (B) Further growth of A, after 15 weeks; (C) Closer view of B and (D) Plants at preflowering stage.

Fig. 4. *Bacopa monniera*. Nodal segments cultured on MS + (in mg l⁻¹) NAA (0.1) + BAP (0.5) + CH (500). (A) Multiple shoots after 6-weeks; (B) Plantlets, 8-week-old; (C) Closer view of B and (D) 10-week-old plantlets.

Fig. 5. *Crocus sativus*. **(A) Callus and direct root differentiation from bulb scale and (B) emergence of multiple shoot buds from cultured scales.**

Nicotiana, Chenopodium, Thalictrum, Dioscorea, and *Rhamnus* in tandem with the concentration supplied, and with a parallel increase in dry weight [28].

Higher concentrations of phosphate results in an increase in the production of indole alkaloids in *Catharanthus roseus*, whereas in callus cultures of *Peganum*, low phosphate levels stimulate the secondary metabolism [see 29]. Transfer of suspension cultures of *Thuja occidentalis* from

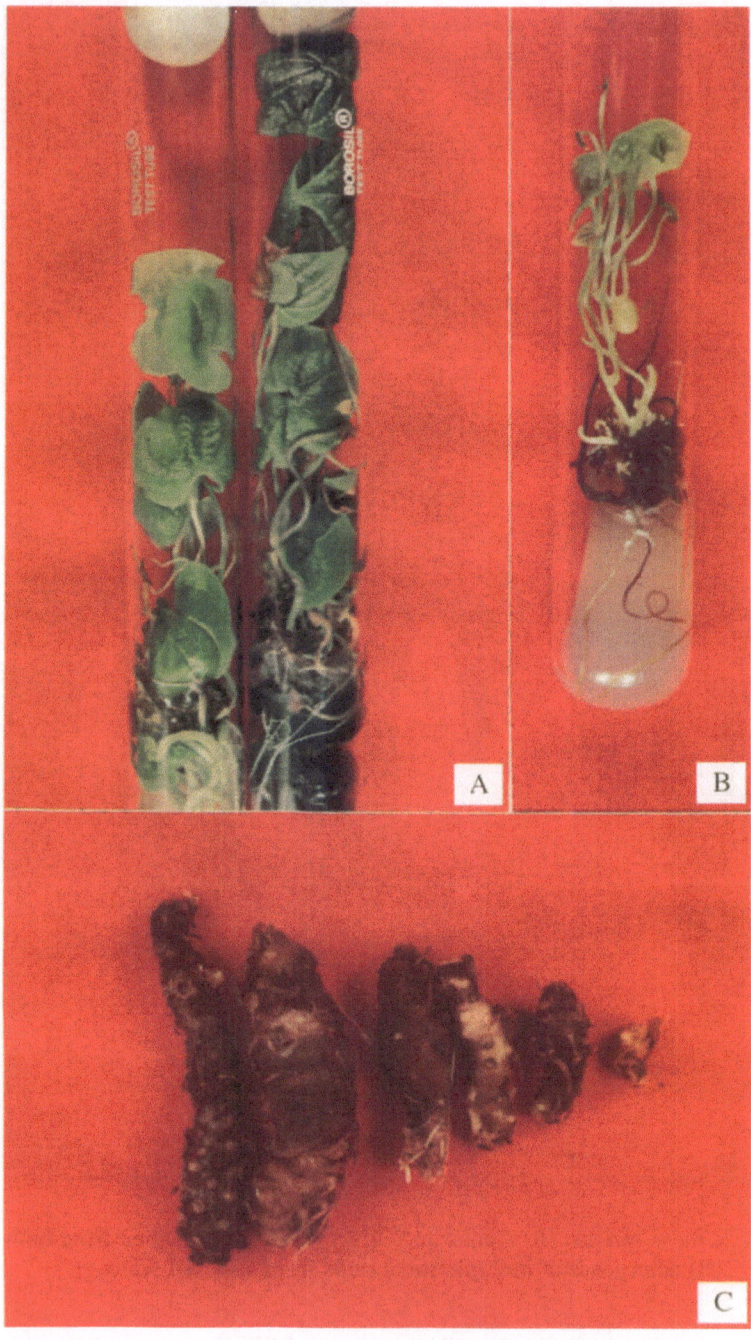

Fig. 6. *Dioscorea bulbifera.* Culture of nodal segments on MS + (in mg l^{-1}) IAA (0.1) + Kn (5.0) + CH (500). (A) 24-week-old regenerants; (B) Aerial bulbils on regenerants after 16 weeks and (C) *In vitro* formed tubers.

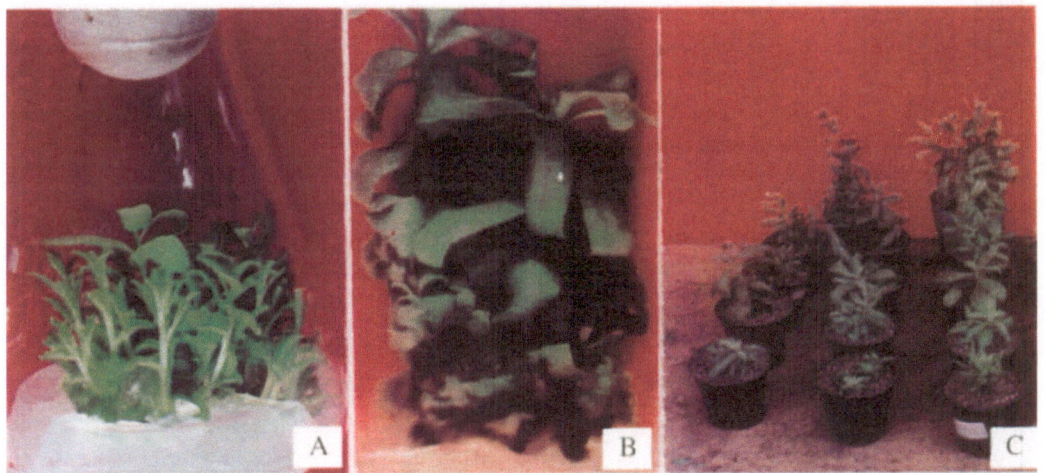

Fig. 7. *Pluchea lanceolata*. Culture of leaf explant on WB (2%) + Kn (5.0 mg l⁻¹). (A) Multiple shoots after 12 weeks; (B) Close up of a shoot and (C) Transplanted plantlets in soil: soilrite (1:1).

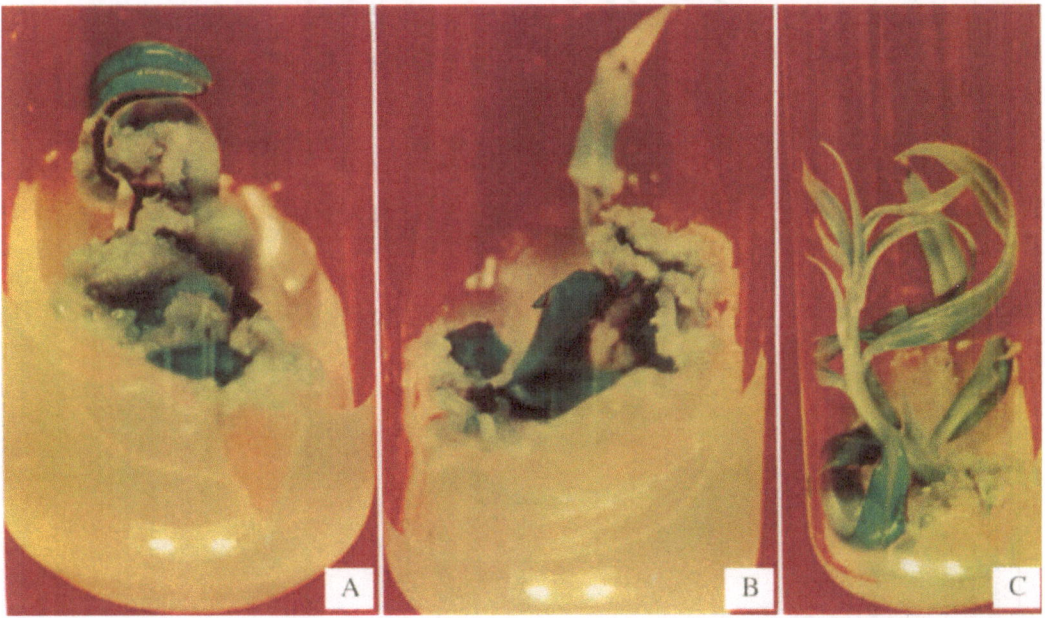

Fig. 8. *Thevetia neriifolia*. (A) Regenerating callus from juvenile leaves cultured on MS + (in mg l⁻¹) IAA (0.5) + BAP (1.0); (B) Regeneration of shoot from callus on MS + (in mg l⁻¹) IBA (0.1) + BAP (2.0) and (C) Growth of isolated shoot with callus at the base on MS + (in mg l⁻¹) IBA (0.1) + BAP (0.5).

MS to B5 medium induced the synthesis of terpenoids. Both, different NH_4^+ content and the stress due to transfer to specific media, seemed responsible for accelerated shikonin production in suspension cultures of *Lithospermum erythrorhizon* [30]. The type and amount of N source seems to affect the yield of secondary products. The ratio of nitrate and ammonia in the culture

Table 1. Examples of micropropagation of some medicinal plants

Plant	Therapeutic use	Cultured explant	Medium*	Response	Reference
Abrus precatorius	Abortifacient	Epicotyl segments	MS + NAA (0.1) + BAP (0.5) MS + NAA (0.1) + Kn (0.5)	Plantlets Plantlets	[20]
Ammi majus	Leucoderma	Nodal segments Isolated shoots	MS + IAA (0.5) + Kn (2.0) + CH (1000) MS + IAA (0.5) + Kn (5.0) + CH (500) + Ad (40) MS + IBA (0.2) + Glu (100)	Multiple shoots *In vitro* flowering and fruiting Plantlets	[136]
Artemisia annua	Antimalarial, anti-HIV	Cotyledonary leaves Stem segments Hypocotyl segments Immature inflorescence segments	MS + (in μM) NAA (0.5) + BAP (13.0) + GA_3 (0.3) + Glu(700) + Asp (300) + Arg (300) + Cyst HCl (30) MS + (in μM) NAA (1.0) + BAP (13) + CM (2 %) MS + (in μM) NAA (0.5) + BAP (13.0)	Multiple shoots Multiple shoots	[7]
Allium wallichii	Tuberculosis, nerve defects, cold, cough	Seedlings (without the root portion)	MS + Zt (20 μM)	Multiple shoots	[137]
Arnica montana	Anti-inflammatory	Nodal segments	MS + (in μM) NAA (5.3) + 2iP (5.0) + Phloroglucinol (0.6 mM) + Ad (0.2 mM)	Plantlets	[138]
Atropa acuminata	Antispasmodic, narcotic, analgesic, antiasthamatic	Shoot tips, nodal segments Isolated shoots	MS + BAP (1.0) + IBA (1.0) RT + IAA (1.0) RT + IBA (1.0)	Shoot buds Elongated shoots Plantlets	[139]
Bacopa monniera	Memory vitalizer	Nodal segments	MS + NAA (0.2) + BAP (5.0) + CH (500) MS + NAA (0.1) + BAP (0.5) + CH (500)	Callus, multiple shoots Plantlets	[6]
Catharanthus roseus	Antileukaemic	Juvenile explants and mature stem segments	MS + NAA (2.0) + BAP (5.0) + Asp (100) + CH (1000)	Callus	[13]

Plant	Medicinal use	Explant	Medium + growth regulators (mg/l)	Response	Reference
		Callus	MS + NAA (0.1) + BAP (5.0) + Zt (1.0) + Asp (100) + Glu (100)	Multiple shoots	
Clerodendrum inerme	Substitute for quinine	Leaf segments	MS + BAP (2.0) + NAA (0.5)	Callus, multiple shoots; Plantlets	[140]
Coleus forskohlii	Anti-inflammatory	Shoot tips	MS + IAA (2.0); MS + (in µM) IAA (0.57) + Kn (0.46)	Plantlets	[141]
Datura innoxia	Anticholinergic	Anthers	WB + NAA (1.0) + BAP (2.0) + CH (500); WB + IBA (0.1)	Multiple shoots; Plantlets	[16]
Dioscorea bulbifera	Antifertility	Nodal segments	MS + IAA (0.1) + Kn (5.0) + CH (500) + Charcoal (2000)	Plantlets	[142]
Holarrhena antidysentrica	Dysentry, colic pain	Node with axillary buds	MS + BAP (15 µM); MS + IBA (35 µM)	Multiple shoots; Plantlets	[143]
Isoplexis canariensis	Source of cardiac glycosides	Seedling explants; Isolated shoots	MS + BAP (5.0 µM); MS + (in µM) BAP (2.25) + IAA (0.17); MS + (in µM) BAP (0.22) + IAA (0.17)	Multiple shoots; Plantlets	[21]
Nothapodytes foetida	Antitumor	Leaf segments	MS +TDZ (1.36 µM); MS (1/2)) + (in µM) BAP (2.22) + IBA (0.49)	Multiple shoots; Plantlets	[144]
Peganum harmala	Abortifacient	Cotyledonary node with shoot tip	MS + (in µM) NAA (0.1) + BAP (5.0); MS + IBA (8.0 µM)	Multiple shoots; Plantlets	[145]
Phyllanthus caroliniensis	Antidiabetic and used against hepatitis B virus	Nodal segments; Isolated shoots	MS + (in µM) BAP (5.0) + Kn (1.25-5.0); MS	Multiple shoots; Plantlets	[146]
Plantago ovata	Laxative	Shoot buds; Callus	MS + (in µM) 2,4-D (4.5) + Kn (2.3); MS + (in µM) BAP (4.4) + NAA (2.7) + CM (10 %)	Callus; Somatic embryogenesis	[147]

(Contd)

Table 1. *(Contd)*

Plant	Therapeutic use	Cultured explant	Medium*	Response	Reference
Plumbago rosea	Anticancerous	Nodal segments	MS + IAA or IBA (0.1) + BAP (1.5) + Ad (50) MS (1/2) + IBA (0.25)	Multiple shoots Plantlets	[148]
P. zeylanica	Leprosy, skin diseases	Nodal segments	MS + (in µM) Ad (27.2) + IBA (2.46) MS + IBA (4.92 µM)	Multiple shoots Plantlets	[149]
Psoralea corylifolia	Psoriasis	Nodal segments	MS + BAP (0.5) MS + IAA (5.7)	Multiple shoots Plantlets	[150]
Rauwolfia serpentina	Cardiovascular diseases, sedative or tranquillizer	Shoot apices Nodal segments	MS + BAP (2.0) MS + BAP (1.0) + NAA (0.1) MS + NAA (2.0) + BAP (1.5)	Multiple shoots Multiple shoots Plantlets	[151]
Sterculia foetida	Abortifacient, diuretic	Cotyledonary nodes	MS + BAP (4.0) MS + IAA (2.0)	Multiple shoots Plantlets	[152]
Stevia rebaudiana	Antidiabetic	Nodal segments	MS + (in µM) BAP (8.87) + IAA (5.71) MS (1/2) + IBA (4.90 µM)	Multiple shoots Plantlets	[153]
Silybum marianum	Antihepatotoxic	Nodal segments	MS + IAA (0.1) + Kn (5.0) MS + NAA (0.1) + Zt (0.5)	Multiple shoots Plantlets	[154]
Solanum khasianum	Steroidal drugs	Leaf segments Callus Shoots	MS + 2,4-D (3.0) + Kn (1.0) MS + BAP (3.0) MS + NAA (2.0)	Callus Multiple shoots Plantlets	[155]
Tylophora indica	Asthma, bronchitis anti-tumorous	Leaf segments Callus Somatic embryos	MS + (in µM) 2,4–D (9.04) + Kn (0.05) MS + 2iP (9.84 µM) MS	Callus Somatic embryos Plantlets	[156]
Typhonium flagelliforme	Anticancerous	Shoot buds (from rhizome)	MS + (in µM) IBA (2.46) + BAP (1.33)	Plantlets	[8]
Valeriana jatamansi	Tranquillizer	Shoots buds	MS + BAP (4.44 µM) MS + (in µM) NAA (4.03) + BAP (4.44)	Multiple shoots Plantlets	[14]

Vitex negundo	Anticancerous	Nodal segments	MS + NAA (0.1) + BAP (2.0)	Multiple shoots and flowering	[157]
		Shoots with immature flowers	MS + NAA (0.5) + BAP (0.1)	Flower maturation, plantlets	
Withania somnifera	Antistress, antitumor, anti-inflammatory	Leaf segments	MS + BAP (1.0)	Multiple shoots	[158]
			MS + IBA (10.0)	Plantlets	

MS = Murashige and Skoog medium, WB = Wood and Braun medium, RT = Revised Tobacco medium (Khanna and Staba, 1968)
Ad = Adenine, Arg = Arginine, Asp = Asparagine, BAP = Benzylamino purine, CH = Casein hydrolysate, CM = Coconut milk;
2,4-D = 2,4-dichlorophenoxyacetic acid, Glu = Glutamine, IAA = Indole-3-acetic acid, IBA = Indole-3-butyric acid, 2iP = 2 iso-pentenyladenine,
Kn = Kinetin, NAA = α–naphthaleneacetic acid, TDZ = Thidiazuron, Zt = Zeatin.
*Concentration of growth hormones are in mgl^{-1} unless mentioned otherwise.

Table 2.   Effect of stage of culture and conditions on the yield of some secondary metabolites

Plant	Active constituent	Culture conditions and yield of secondary metabolite	Reference
Agaveamaniensis	Sapogenin steroid	Absence of calcium ions in media increased the sapogenin steriod content, while relatively high concentration of Mg, Co and Cu showed inhibitory effects	[35]
Ammi majus	Xanthotoxin	Xanthotoxin content monitored during different stage of growth and differentiation revealed highest content at plantlet differentiation stage bearing immature green fruits (*in vitro*)	[136]
Artemisia annua	Artemisinin	Enhanced artemisinin content was found *in vitro*	[7]
Beta vulgaris	Betalains	B5 medium supplemented with Co^{2+} (5 μM) increased the betalains production	[36]
Catharanthus roseus	Catharanthine and vindoline	Multiple shoot cultures raised directly from sterile seedlings inoculated on MS medium containing BA (4 μM) produced mainly catharanthine and vindoline in amounts higher than in the parent plant	[159]
	Indole alkaloids (ajmalicine, serpentine)	Zt or BA were more active than Kn in increasing the content of alkaloids. But all the three cytokinins enhanced the production of indole alkaloids	[160]
	Vinblastine	High degree of differentiation was correlative to the increased vinblastine production	[13]
Cinchona ledgeriana	Quinine	Shoot cultures contained much higher levels of quinine and related alkaloids than the cell suspensions	[161]
Datura stramonium	Hyoscyamine and scopolamine	Maximum contents were found in the stem and leaves of young plants; hyoscyamine being always the predominant component	[162]
Daucus carota	Anthocyanins	Increase of 63.41% in production of anthocyanins by addition of 1.0 nM Co^{2+}	[163]
Digitalis lanata	Digitoxin	Addition of Mn^{2+} (10 mM) at day zero of culture increased the digitoxin content	[164]
Dioscorea deltoidea	Diosgenin	Hypocotyl callus on RT+2,4-D (0.1)+ Cholesterol (10–100)+YE (0.5%) yielded higher diosgenin content	[see 142]
Ephedra andina *E. distachya* *E. equisitina* *E. fragilis* *E. gerardiana* *E. intermedia* *E. major* *E. minima* *E. saxatilis*	Alkaloids (1-ephedrine and d-pseudoephedrine) Lepidine	Trace quantities of alkaloids were present in cultures. The ability to produce alkaloids diminished to zero with successive subcultures	[165]
Lepidium sativum		Lepidine content was much higher in 8-month-old regenerants grown on $ZnSO_4$ (900 μM) or $CuSO_4$ (100 μM)	[37]

Papaver bracteatum	de-sanguinarine	Increase in concentration of Cu alone brought an increase in the content	[166]
Rauwolfia sellowii	Alkaloid	Increased alkaloid content in leaf callus	[167]
R. serpentina	Alkaloid	Total alkaloid content in the plantlets was higher as compared to field grown plants	[168]
		A group of new alkaloids, the raumaclines and some related alkaloids were isolated from cell suspensions fed with high level of ajmaline	[169]
Silybum marianum	Silybin	Higher silybin content in regenerants grown on $ZnSO_4$ (200 µM) or $CuSO_4$ (75 µM)	[170]
Solanum aviculare	Solasodine	Addition of cholesterol to the medium improved the yield	[171]
S. laciniatum	Solasodine	Decreasing the sucrose concentration increased the solasodine content significantly in shoot cultures	[172]
S. nigrum, and *S. nigrum* var. *judaicum*	Glycoalkaloids	Total glycoalkaloids were higher in plantlets	[173]
Stizolobium hassjoo	L-DOPA	Supplementation with 2.5 µM Co^{2+} stimulated 25 times the synthesis of L–DOPA	[174]
Withania somnifera	Withanolides	Maximum accumulation of withaferin A was noted in shoot tips proliferating on B5 medium; the withanolide D content was low. In MS medium withaferin A accumulation was low than in B5 medium, while withanolide D accumulation was higher Among the BA and Kn, BA favoured both shoot multiplication and withanolide synthesis. In the absence of any carbon source withanolide accumulation was very low in shoot tips (0.002%) Withaferin A accumulated maximum in the presence of 10% sucrose. The maximum accumulation of withanolide D was at 4% sucrose	[175]

media also influences growth and secondary metabolite production. Fujita et al. [30] report increase in the yield of shikonin with increase in the concentration of sole nitrogen source, nitrate till 6.7 mM, but the production decreased with above 10 mM nitrate level. Decreased levels of N are reported to stimulate the production of secondary metabolites such as, certain polyphenols, anthocyanins [13], etc.

Manipulation of concentrations of microelements in the nutrient media offers a strategy to increase the production of secondary metabolites in plant cell cultures [32]. Trace elements have indeed been considered as abiotic elictors or as inducing factors [33] that trigger the biosynthesis of secondary metabolites. There are results showing the effect of divalent ions; Co^{2+} and Cu^{2+} seem to have received more attention because of their positive effects on the production of secondary metabolites [30, 34, 35]. Increase in Co^{2+} from 1 to 5 µM resulted in the enhanced production of betalains in *Beta vulgaris* [36]. Enhanced shikonin production in the cultures of

Lithospermum erythrorhizon have been attributed to the increased concentration of both copper and sulphate [30].

Phytoxicity by heavy metals due to industrial pollution has caused degradation of cultivable land. Plants allowed to grow on such soil receiving sludge high in heavy metals show reduction in the quality and productivity. Efforts thus are required to raise metal tolerant plants. Tissue culture techniques have helped not only in raising metal tolerant plants but have also demonstrated that subjecting the cultures of medicinal plants to abiotic stress can be crucial in increasing the yield of secondary metabolites [37]. Several investigations have indeed demonstrated the possibility to raise metal tolerant plants *in vitro* [see 38]. Heavy metals have different role in metabolic functions. Some of them including Cu and Zn are required as micronutrients in biological systems to act as cofactor and/or as part of prosthetic groups of enzymes in a wide variety of developmental pathways [39], Cu a constituent of the medium is an essential microelement for plant growth [40, 41]. It is required for several biochemical and physiological pathways. Cu at higher concentrations exhibits strong toxicity and hamper plant growth as do some other heavy metals, such as Cd, Pb or Hg which have no function in plant metabolism. Copper is released as particulates in stack effluents primarily from Cu smelters. Greater concern of Cu comes from prolonged applications in fungicidal treatments [42]. Addition of Cu in the medium is reported to promote somatic embryogenesis as well as its subsequent development in *Citrus* [43]. Cu stimulated regeneration in wheat, *Nicotiana tabacum* and *Bacopa monniera* [44]. This was also so at lower concentrations with *Dioscorea bulbifera* (Narula, unpublished). Cu has proved to be more effective than Zn in enhancing the yield of xanthotoxin in *Ammi majus* and lepidine in *in vitro* cultures of *Lepidium sativum* [4, 5, 37]. Heavy metals and others have also induced a positive effect on alkaloid production in *Catharanthus roseus* [45]. Fe^{2+} [46] and Cu^{2+} [30] induced positive effects on the synthesis of shikonin. Endress [47] and Obrenovic [48] have already demonstrated profitable role of Cu^{2+} on the accumulation of betacyanins in callus cultures of *Portulacca grandiflora* and *Amaranthus caudatus* seedlings. Higher concentrations of Cu^{2+} in the media supported increased accumulation of sapogenin steroid in the *in vitro* cultures of *Agave amaniensis* [49, 50].

Plant growth regulators (auxins and cytokinins) are also effective triggers of secondary metabolites. An optimum concentration of 2,4-D (25 mgl^{-1}) favoured the production of L-Dopa in cell cultures of *Mucuna*. Low concentration of 2, 4-D (0.1 ppm) proved favourable for alkaloid production in cell cultures of *Cinchona ledgeriana*. In addition to concentration, the type of auxin used also exerts a strong influence on secondary product formation. 2,4-D in general proved less suitable for protein synthesis than IAA [51]. Zenk et al. [52] have reported that in *Morinda citrifolia* presence of 2,4-D reduced the production of anthraquinones but NAA enhanced the accumulation of anthraquinones. The alkaloid synthesis and biomass accumulation increased on nutrient media containing NAA as compared to 2,4-D in *Nothapodytes foetida* [53].

Like auxins, cytokinins also influence secondary metabolite production. BAP enhanced shikonin production in *L. erythrorhizon* and kinetin promoted L-Dopa synthesis in callus cultures of *Stizolobium hassjoo*. Likewise, in the presence of BAP maximum accumulation of withanolide occurred. There was a decline in withanolide in the cultures of *Withania somnifera* with an increase (2.0–5.0 mg l^{-1}) in the concentration of BAP [54].

In experiments conducted by Decendit et al. [55] Zt or BAP proved more effective than Kinetin in *Catharanthus* cell cultures. At 1 μM Zt or BAP production of alkaloids was doubled.

Higher concentrations resulted in the decrease of alkaloids. According to Bhatt et al. [56] besides growth regulators, a combination of IAA and sucrose in the medium can also stimulate the production of solasodine in the tissue cultures of *Solanum nigrum.*

Higher concentrations of NAA + Kn or IAA + Kn promoted the yield of diosgenin in *D. bulbifera.* Among the two auxins tried, NAA + Kn induced much higher content (Narula, unpublished). Corroborative results were obtained in *D. deltoidea* tissue cultures grown in the presence of 2,4-D, IBA, BA and GA singly and in combinations. The medium with 2,4-D favoured diosgenin production most consistently. GA and high BAP concentrations proved toxic [57]. GA or kinetin are otherwise reported to increase the steroid content in *Phaseolus aureus* and *Corylus avellana* and doubled production of diosgenin in *Solanum xanthocarpum* tissue cultures [57]. Zhao et al. [58] observed that an increase in jaceosidin production was accomplished by increased concentration of NAA. This concurs with the results of Matsumoto et al. [59] who used cell suspension cultures of *Populus.*

Addition of precursors of desired compounds to the culture medium also enhances the yield of secondary products. Ajmalicine production in *Catharanthus roseus* could be stimulated to approximately 10-fold by supplying secologanin [60]. Quinine in *Cinchona* cultures, rosemarinic acid in *Coleus blumei* and capsaicin production by cell cultures of *Capsicum frutescens* [61], and addition of loganin (precursor of secologanin) into the medium for enhanced yield of secologanin [62] are some examples where precursor addition caused an increase in the yield of related metabolites. Addition of various precursors (L-ornithine, L-arginine, L-phenylalanine, DL-β-phenyllactic acid and tropinone) alone was ineffective in stimulating hyoscyamine production in *Datura innoxia.* But, a combination of these precursors alongwith DL-β-phenyllactic acid and Tween 20 increased the yield [63].

The recognition that certain specific secondary metabolite products, such as phytoalexins are produced by plants which are active against microorganisms has led to the concept of using such stimulators for *in vitro* cultures also. These compounds have been described as 'elicitors' by Keen et al. [64]. Elicitors can be of biotic or abiotic origin [65]. Biotic elicitors are prepared from fungal, yeast or bacterial cultures, fungal mycelial extracts, culture filtrates, and fractions or compounds obtained from microbial cell walls. Autoclaved fungal mycelia induced the accumulation of diosgenin in *Dioscorea deltoidea* cultures [66]. The production of berberine and shikonin enhanced in the cultured cells treated with fungal extracts [67, 68]. A beta-glucan elicitor prepared by ethanol precipitation of yeast, *Saccharomyces cerevisiae* elicited alkaloid production in cultured cells of *Eschscholtzia.* In *Tabernaemontana divaricata* cultures, reserpine accumulation increased by treating the cells with an elicitor prepared from *Candida albicans* [69]. Purified fractions from bacteria also elicited diosgenin and capsaicin production [66]. Elicitation of capsaicin in *Capsicum frutescens* cultures could be achieved by supplementing the culture medium with chitosan, curdlan and xanthan.

The abiotic elicitors include physical and chemical stresses such as UV radiation, exposure to heat or cold, ethylene, fungicides, antibiotics, salts of heavy metals, salinity, etc. [70]. It has also been recorded that the synthesis of alkaloids can be similarly elicited with jasmonic acid and its esters playing a key role in regulating the response [71]. In fact, it is reported that fungal cell wall elicitors and methyl jasmonate (MeJa) can activate inducible secondary metabolism in soybean cell cultures by different mechanisms. Treatments with exogenous MeJa can elicit the accumulation of several classes of alkaloids in a wide range of plant species [72]. Hairy root cultures of *Datura stramonium* showed maximum alkaloid in the presence of MeJa followed by fungal elicitors and oligogalacturonide [73]. Jasmonate can elicit natural product formation not only in plants but also in cell cultures [74, 75]. Methyl jasmonate therefore could be an useful tool for the enhancement of lignan production in biotechnological processes. Feeding experiments with the precursor coniferyl alcohol resulted in fast increase in the pinoresinol content [76]. Some of the examples where elicitors caused enhancement in the yield of medicinal compounds are cited in Table 3.

Table 3.   Some examples of *in vitro* production of medicinal compounds when elicitors were used in cell suspensions

Plant	Elicitor Used	Active Principle	Reference
Catharanthus roseus	*Botrytis* species homogenate	Catharanthene	[176, 177]
	Fungal homogenate	Terpenoid, indole alkaloid	[178]
Eschscholtzia californica	Yeast	Sanguinarine	[179]
Hyoscyamus albus	*Phytophthora cinnamomi*	Lubimin	[180]
Lithospermum erythrorhizon	Oligogalacturonides	Dihydroechinofurane	[181]
Morinda citrifolia	Polysaccharides	Anthraquinones	[182]
Papaver bracteatum	Fungal	Sanguinarine	[183]
	Verticillum	Sanguinarine	[184]
P. somniferum	Fungal homogenate	Sanguinarine	[185]
	Botrytis species homogenate, *Pythium aphanidermatum*	Sanguinarine	[186]
Sanguinaria canadensis	*Verticillum*	Sanguinarine	[184]
Thalictrum rugosum	Yeast carbohydrate	Berberine	[67]
Tripterygium wilfordii	*Botrytis* species *Trichoderma virideae* *Rhodotorula rubra* *Sclerotinia sclerotiorum*	Oleanane triterpenes	[187]

It is not only the chemical milieu but also the physical factors which play a significant role in secondary metabolite production. Light as physical source, for example, has an effect on growth and development of plants as well as in stimulation of secondary metabolite production [77, 78]. In fact, quality, intensity and duration of light play a decisive role in the accumulation of secondary compounds [79, 58]. Production of diosgenin and related compounds seem to be controlled by different media ingredients as well as by light [80].

In some cases, direct effect of hydrogen ion concentration on secondary compound production has been demonstrated. For example, alkaloid synthesis in *Lupinus polyphyllus* cultures rose with a decrease in pH from 5.5 to more acidic, 3.5. Even physical conditions of the medium have proved crucial for the production of secondary metabolites in cultures. Cell suspension cultures have been favourites for the production of valuable secondary metabolites in cultures. These cultures initiated by transfer of most friable sector of an established callus tissue into an agitated liquid medium received more homogenous stimuli. A close correlation between the growth of cultures and yield of products has been envisaged. Since the product accumulates through the growth cycle, the product and biomass show a close correlation.

The first commercial production of a natural plant product by cell suspension cultures was developed in Japan for the production of naphthoquinone, shikonin. In suspension cultures of *Rauwolfia sellowii* alkaloid content was maximum at the end of the exponential growth phase. It has been argued that for industrial scale production of plant secondary metabolites, the cells should be suspended in liquid so that the entire operation of harvesting, inoculation and other treatments could be accomplished by pumping the suspended cells. Immobilized plant cells used in the same way as immobilized enzymes have also played an important role in the secondary product formation [81]. Although the enzymatic activity of immobilized cells is about half that of suspending cells, these have the advantage of being reusable as a biocatalyst over a considerable period [82].

3.2 Synthetic Seed

The concept of 'synthetic seed' was first introduced by Toshio Murashige in 1977 and later the use of synthetic seeds or artificial seeds was realized by Redenbaugh and coworkers [83] and others. Synthetic seeds help in reducing the cost of transport and in maintaining the uniformity. Besides, of much importance is the ability to provide large-scale delivery of elite genotypes selected from hand pollinated hybrids or genetically engineered plants. The first successful examples of synthetic seed technology have been in alfalfa [84] and celery [85]. Various vegetative propagules like axillary buds, shoot tips, bulbs, protocorms have been used [86]. The production of 'Syn' seeds has been reported in several medicinal plants like, *Atropa belladonna, Hyoscyamus muticus, Mentha arvensis, Picrorhiza kurroa* [87], *Dioscorea alata, D. floribunda,* [88], *Clitoria ternatea* [89] and *Guazuma crinita* [90].

3.3 Cryopreservation

Cryopreservation offers long-term conservation of germplasms. In addition to germplasm conservation, it also ensures genetic stability and retention of biosynthetic potential [91]. Cryopreservation has been achieved by using various explants (Table 4). The period over which the cultures retain viability vary widely with the species and a maximum of 3 years has been recorded in *Digitalis* [92].

Meristems have been preferred over cell and callus cultures because they are genetically more stable. Shoot tips of medicinal plants such as *Cichorium* sp. [93], *Dioscorea deltoidea, D. floribunda* [91], *Holostemma annulare* [94] and *Mentha* sp. [95] have been cryopreserved successfully. Genetic erosion due to periodic subculture and storage can be overcome by freeze preservation of callus and cell suspensions in liquid nitrogen. Cell suspensions of medicinal plants, e.g., *Atropa belladonna, Datura innoxia, Nicotiana tabacum, Panax ginseng,* etc. retain their biosynthetic potential after freezing. Cryopreservation of somatic embryos helps in storage at appropriate stage that can be used whenever required. The somatic embryos of carrot, orange and asparagus frozen in liquid nitrogen yielded high viability and regenerated complete plants. The potential of zygotic embryos is manifold in plants with recalcitrant seed, in fruit and timber trees and plantation crops. In wide hybridization programs, especially dealing with intergeneric crosses which are incompatible due to degeneration of embryos, can be possibly dissected out at immature stages and cryopreserved. Zygotic embryos of rice, wheat, barley, mustard and

Table 4. Examples of some cryopreserved medicinal plants

Plant	Explant/Culture	Method used	Reference
Anisodus acutangulus	Cell suspensions	Liquid nitrogen (−196°C)	[188]
Atropa belladonna	Pollen embryos Protoplasts	Liquid nitrogen (−196°C)	[163]
Catharanthus roseus	Cell suspensions	Low temperature (0-30°C), −196°C	[189]
Datura innoxia	Protoplasts	Exposed to vapors, immersed in liquid nitrogen	[163]
Dioscorea alata	Shoot tips	Encapsulation Dehydration	[88] [190]
D. balanica	Callus	Direct immersion in liquid nitrogen	[191]
D. bulbifera	Shoot tips	Encapsulation Dehydration	[88] [190]
D. floribunda	Shoot tips	Encapsulation Dehydration	[88]
Eucalyptus sp.	Shoot tips	Encapsulation Dehydration	[192]
Ipomea batatas	Shoot tips	Vitrification	[193]
Medicago sativa	Somatic embryos	Encapsulation	[83, 194]
Mentha sp.	Shoot tips	Encapsulation Vitrification	[195]
Nicotiana tabacum	Protoplasts	Liquid nitrogen (−196°C)	[163]
Panax ginseng	Cell suspensions	Hardening, −30, −70, then −196°C	[196]
Trifolium repens	Shoot tips	Vitrification	[197]

coconut cryopreserved by quick freezing, followed by thawing at 35-40°C produced viable plants but viability varied considerably [96].

The storage of pollen has been of prime interest to plant breeders. Cryopreservation of pollen enables en masse production of haploid plants, maintenance of stability of haploids and conservation of genetic resources. Segments of anthers and pollen embryos of *Atropa belladonna*, *Brassica campestris*, *Nicotiana tabacum* and *Primula obconica* have been successfully frozen and entire plants have been regenerated after one year of storage [97]. Freshly isolated protoplasts of *Atropa belladonna*, and *Nicotiana tabacum*, and *Datura innoxia* [98], and *Glycine max* [99] subjected to freezing in liquid nitrogen for various time periods have survived and retained their morphogenetic potential.

3.4 Molecular Markers

As micropropagation developed from a laboratory curiosity to commercial industry, different considerations became important concerning the feasibility of approaches for long-term economic benefits. The foremost concern has been the maintenance of the genetic integrity of micropropagated plants with regard to the explant source so that the advantages (high yield, uniform quality, shorter rotation period, etc.) in the use of elite genotypes over natural seedlings is maintained

[100,101]. Rani and Raina [101] have emphasized that micropropagation cannot be rewarding unless complete genetic fidelity is maintained. Thus for obtaining true-to-type plants, axillary branching or somatic embryogenesis have mostly been adopted. These two methods have generally been considered to be immune to genetic changes that may arise during cell division or differentiation under *in vitro* conditions [102]. Rani and Raina [101] showed that the field-transferred enhanced axillary branching derived plants of *Eucalyptus camaldulensis* were genetically stable in terms of genome size, RFLPs of nuclear and organellar genomes and RAPD and oligonucleotide fingerprinting patterns. The concept of uniformity among micropropagated plants, however, received a jolt when somaclonal variations were reported. Somaclonal variations can pose a threat to the genomic integrity of regenerated plants. Several strategies were therefore adopted to detect variants based on morphological traits, cytogenetical analysis for the determination of numerical and structural variation in the chromosomes and isozymes. But, these met with severe limitations. Molecular markers have thus been used to study genetic diversity, phylogeny and fingerprinting as well as to construct physical genetic maps in medicinal plants. The range of marker system includes RAPD, AFLP, microsatellites and RFLP.

RFLP was introduced as hybridization based marker for single copy loci. Since RFLP is capable of detecting multiple alleles, it reveals greater level of heterozygosity and has a higher information content. The major drawback of RFLP is that it screens very few loci per assay. It is expensive, labour intensive and technically complex as it involves the use of radioactive probes. It also requires larger amounts of genomic DNA making its application impractical for efficiently cataloguing of genetic resources [103]. RAPD technique is quite simple, inexpensive but less reliabile. AFLP has many advantages that make it applicable in assessment of genetic diversity, genetic mapping and tagging studies. AFLP has now become a preferred technique as it combines RFLP and RAPD [104]. AFLP markers offer best method for detecting mutations by randomly surveying the genome. This technique does not require prior sequence information. Besides, it has wide genome coverage as compared to other DNA-based markers [105] which makes it an ideal tool for detecting genetic variation. This technique has been used for analyzing genetic variation in somatic embryoids of pecan [106]. Singh et al. [107] reported application of AFLP markers for ascertaining clonal fidelity in tissue culture raised progenies of a medicinally important plant, *Azadirachta indica*. AFLP markers are now being routinely employed for assessment of genetic variation in economically important plant species including chichory [108], *Withania* sp. [109], etc. (Table 5).

3.5 Genetic Engineering in Medicinal Plants Through *Agrobacterium*

The stable introduction of foreign genes into plants represents one of the most significant developments in plant biotechnology. Attempts have been made to manipulate pharmaceutically important medicinal plants for their secondary metabolic pathways by using transgenic technique. Since secondary products are often biosynthesized in mulit-step enzymatic reactions in specifically differentiated cells, manipulations of such pathways to alter metabolic production is complex, complicated and unpredictable [110].

Transformation has many advantages over conventional cell culture systems that may include fast growth and stable high level production of secondary metabolites making them favourable for biotechnological exploitation. *Agrobacterium tumefaciens* and *A. rhizogenes* have proved efficient and have provided highly versatile vehicles for introduction of genes into the desired

Table 5.   Application of molecular markers in some medicinal plants

Plant	Marker	Application	Reference
Achillea ospenifolia	Oligonucleotide finger printing, RAPD	Stability of micropropagated plants	[198]
Allium sativum	RAPD	Genetic diversity in plants regenerated by somatic embryogenesis from long-term-callus	[199]
Artemisia annua	OPGMA-RAPD	Artemisinin and chemotypic variants	[200]
Azadirachta indica	AFLP	Genetic diversity	[201]
A. indica	AFLP	Clonal fidelity in tissue culture raised plants	[202]
Cichorium sp.	AFLP	Diagnostic marker for endive and chicory group	[203]
Codonopsis pilosula	RAPD	Geographic variation	[204]
Datura sp.	AFLP	Genetic diversity	[205]
Digitalis obscura	RAPD	Genetic variation	[206]
Dioscorea bulbifera	RFLP	Linkage (physical) map	[207]
	RAPD	Genetic variability and relationship within the species	[208]
D. rotundata and *D. cayenensis*	AFLP	Genetic diversity	[209]
Duboisia	RFLP	Hybrid origin identification	[210]
Moringa oliefera	AFLP	Genetic variation	[211]
Panax ginseng	RAPD	Genetic stability in micropropagated plants	[212]
P. ginseng and *P. quniquefolium*	RFLP	Ginseng drug	[213]
Plantago major	RAPD	Identifying subspecies	[214]
Rehmannia sp.	RAPD	Homogenity	[215]
Tylophora indica	RAPD	Genetic variation	[216]

plant genome. As a consequence of transfer and integration of genes through plasmids into the plant DNA, the transformed tissues and hairy roots have provided encouraging results. These transformed tissues have thus become potential sources for stable production of plant metabolites (Table 6).

Hairy root cultures of *Trigonella-foenum-graecum* L. produced twice the amounts of diosgenin than the non-transformed roots [111]. Several studies have indicated that *Agrobacterium rhizogenes* affects the levels of polyamine in transformed plants [112] that may influence growth and the production of secondary metabolites. *Atropa baetica* hairy roots synthesized and accumulated a conspicuously high amount of tropane alkaloids [113].

In *Hyoscyamus albus*, hyoscyamine content was more in the transformed roots followed by stem and leaves [114]. Doerk-Schmitz et al. [114], however, reported low proportion of scopolamine in hairy roots of *Hyoscyamus albus* but there was high content of hyoscyamine even after several

Table 6. Some examples of *Agrobacterium*-mediated transformation in medicinal plants

Plant	Strain	Result	Reference
Ammi majus	*Agrobacterium rhizogenes* A4 (20233)	Hairy roots produced higher content of visnagin	[217]
Artemisia annua	*A. rhizogenes* LBA 9402	1-month-old transgenic produced more artemisnic acid and arteannuin B	[218]
A. annua	*A. tumefaciens* C58 , N2 73	Artemisinin content was slightly higher in shoots	[219]
Atropa belladonna	*A. rhizogenes* A4 , TR 105	Higher atropine levels	[220]
A . belladonna	*A. tumefaciens* LBA 4404	Higher scopolamine	[221]
A . belladonna	*A. rhizogenes* 15834	Increased scopolamine content	[222]
A . belladonna	*A. rhizogenes* 15834 and *A. tumefaciens rol ABC* genes	Higher alkaloid	[223]
Catharanthus roseus	*A. rhizogenes*	At reduced pH more alkaloid released	[224]
Cinchona ledgeriana	*A. tumefaciens* A6	Five times more alkaloids (cinchonine and cinchonidine)	[225]
C. ledgeriana	*A. rhizogenes* LBA 9402	Quinine, cinchonidine and quinidine reached a maxima after 45 days	[226]
Datura candida hybrid (*D.candida* × *D.candida*)	*A. rhizogenes*	Scopolamine and hyoscyamine showed increase	[227]
D. innoxia *D. stramonium* *D. ferox* *D. wrightii*	*A. rhizogenes* LBA 9402	Maximum hyoscyamine content in *D. stramonium* and scopolamine in *D. innoxia*	[24]
D. innoxia	*A. rhizogenes* A4, 15834 and A4–24 A4 and 15834 strains more effective	Higher hyoscyamine content	[228]
D. innoxia	*A. rhizogenes* LBA9402, A41027, R1601 R1601 gave best response	Hyoscyamine and scopolamine content was higher and among the two alkaloids hyoscyamine content was much higher	[229]
D. innoxia	*A. rhizogenes*	Permeabilization with Tween 20 for 30 hr period increased the alkaloid concentration in the medium	[230]
D. stramonium	*A. rhizogenes* TR-105	Heat shock given to the cultures resulted in higher hyoscyamine release in the medium	[231]

(Contd)

Table 6.   *(Contd)*

Plant	Strain	Result	Reference
D. stramonium	*A. rhizogenes* TR-105, ATCC 15834, A4, 1855, A41027, ATCC 13333 Among these strains TR-105 proved most effective	Hyoscyamine and scopolamine bioproductivity was higher in hairy root culture	[232]
D. stramonium	*A. rhizogenes*	Release of alkaloids increased at low pH (3.5)	[224]
D. stramonium	*A. rhizogenes* LBA 9402	Live fungal pellets caused enhanced hyoscyamine production	[233]
D. stramonium	*A. rhizogenes* A4 *rol ABC* and *tms* gene	Higher hyoscyamine production	[234]
D. stramonium	*A. rhizogenes* TR-105	An inverse relation between alkaloid accumulation and growth, hyoscyamine content showed an increase	[235]
D. stramonium	*A. rhizogenes* ATCC 15834	Highest hyoscyamine yield with culture medium in which SO_4^{2-} and K^+ was dominant	[236]
D. stramonium	*A. rhizogenes* A4	Lower calcium concentrations reduced the hyoscyamine synthesis	[237]
D. quercifolia	*A. rhizogenes* LBA 9402	5% sucrose in Gamborg B5 medium proved best for growth and higher hyoscyamine accumulation	[238]
Hyoscyamus albus, H. desertorum, H. muticus	*A. rhizogenes* LBA 9402	Hyoscyamine and scopolamine content was highest in *H. albus*	[24]
H. albus	*A. rhizogenes* MAFF 03-01724	Higher yield of hyoscyamine	[239]
H. muticus	*A. rhizogenes* LBA 9402, C58CI, pRTGUS 104	High hyoscyamine content at 3% sucrose in two of the clones, high nitrogen content had negative effect on hyoscyamine production and growth. Copper (11 µM) stimulated hyoscyamine production	[240]
Hyoscyamus × gyorffyi (H. niger × H. albus)	*A. rhizogenes* LBA 9402, A41027, R 1601 Among these strains R 1601 gave the best response	In 14 clones of *H. gyorffyi* hyoscyamine percentage being much higher than scopolamine	[241]
Panax ginseng	*A. rhizogenes*	Produced saponin, and ginsenosides more effectively	[242]
P. ginseng	*A. rhizogenes* A4, 15834 A4 proved more effective	Higher content of glycosides	[243]
P. ginseng	*A. rhizogenes* A4	Faster growth of callus and higher yield of ginsenosides	[244]

Pgq (Panax hybrid) (*P. ginseng* × *P. quinquefolium*)	*A. rhizogenes* ATCC 15834	Ginsenoside content was higher	[245]
Rauwolfia serpentina	*A. rhizogenes* 15834	Increased levels of ajmaline and serpentine	[246]
Solanum eleagnifolium	*A. tumefaciens* T 37	Transgenic shoots showed higher solasodine content	[247]
Scopolia lurida and *S. stramonifolia*	*A. rhizogenes* LBA 9402	Produced little alkaloids	[24]
Withania somnifera	*A. rhizogenes* LBA 9402	Productivity of withanolide D was higher	[248]

subcultures and the transgenic plants could be regenerated directly from such roots via organogenesis [115]. Strains of *Agrobacterium* are reported to affect the growth behaviour and production of secondary metabolites. Influence of *A. rhizogenes* strains on biomass and alkaloid prductivity in hairy root lines of *Hyoscyamus muticus* and *H. albus* was also studied by Zehra et al. [116]. A4 induced hairy root lines of *H. albus* and *H. muticus* were faster growing than those induced by strain LBA 9402. The atropine yield of A4 induced lines of *H. albus* was significantly higher (3.5-fold) than the LBA 9402 induced lines [116].

Cu^{2+} enhanced both, the growth and the alkaloid yield in *Hyoscyamus albus* hairy roots. Similar results have been obtained in the production of shikonin derivative by cell suspension cultures of *Lithospermum erythrorhizon* [117]. Copper concentration up to 11 µM stimulated hyoscyamine production but had no influence on growth of hairy root cultures of *Hyoscyamus*.

Two-year-old transformed root cultures of *Catharanthus roseus* accumulated higher ajmalicine and catharanthine than the non-transformed cultures. Addition of MeJa increased the yield of both the alkaloids [118]. A positive correlation between STR activity and alkaloid accumulation has also been found in tissues of *Cinchona ledgeriana* and *C. roseus* seedlings [72, 119]. TDC (trytophane decarboxylase) activity in developing *Cinchona* seedlings increased after a large pool of tryptophan was formed. It fell to undetectable levels once the tryptophan was converted into tryptamine [119].

Serotonin content enhanced if the hairy root cultures of *Peganum harmala* were fed with tryptamine. But the alkaloid content was not affected [120]. In *Panax ginseng* roots were transformed with *A. rhizogenes*. Inomata et al. [121, 122] found that periodic changes of medium maintained the high growth rate and the ginsenoside production varied during different stages of growth. Mallol et al. [122] also observed that the capacity to produce and accumulate ginsenoside is associated with biomass. The results concur with other investigations as well as our results with *Datura*. Transformation stimulated increased biomass and tropane alkaloid production in axenic root cultures of *Calystegia sepium* and *Atropa belladonna* [123]. Compared to transformed plants, non-transformed plants contained low amounts of tropane alkaloids, especially 6 β-hydroxy hyoscyamine and scopolamine in the roots [124].

In transgenic lines of *Nicotiana tabacum* feeding of lysine to root cultures with low LDC (lysine decarboxylase) activity enhanced cadaverine and anabasine levels [125]. Several hairy root cultures of *N. tabacum* having lDC gene increased cadaverine levels and this was used for the formation of anabasine to obtain a 3-fold increase of this alkaloid. Transformation has indeed helped in the enhancement of secondary metabolites in a number of cases though not to commercial levels. In *Artemisia*, use of arnesyl diphosphate synthase gene promoted artemisinin 3-4 times higher in hairy roots [126]. In *Catharanthus roseus* where *str* (strictocidine synthase) is highly desirable for increased terpenoid indole alkaloid (TIA) production, high STR activity positively influenced the flow of metabolites through the indole pathway [127].

Subroto and Pauline [128] reported the production of steroidal alkaloids in *Solanum aviculare* that was growth associated. It has been shown by Schaller et al. [129] that the introduction of extra copies of a chimeric *hmgr* gene (obtained from *Hevea brasiliensis*) increased the accumulation of sterols by 6-fold in tobacco plants. HMGR (3-hydroxy-3-methylglutaryl-coenzyme A reductase) plays a major role in the regulation of sterol biosynthesis in plants. In *Solanum aviculare*, Cavalcante Argôlo [130] obtained transgenic hairy root clones that grew faster and accumulated up to 4.2 times more solasodine when grown under dark. Upregulation of the *hmgr* gene in tobacco has also been shown to give rise to highly significant increase in sterol accumulation [131].

odc and *adc* genes play important role in the biosynthetic pathway of alkaloids. A stable transformation system has been developed by us for *Datura innoxia* using androgenic callus that was transfected with *Agrobacterium tumefaciens* strain LBA4404 carrying *odc* and *adc* genes. Transformed cultures showed higher amounts of hyoscyamine and early regeneration.

Tiburcio and Galston [132] reported that in *Nicotiana tabacum* ODC pathway is important for cell division and growth, the effects of inhibitors on alkaloid biosynthesis show that ADC is more important in synthesis of pyrrolidine alkaloids.

Imanishi et al. [133] in tobacco and Robins et al. [134] in *Datura stramonium* also found that arginine decarboxylase activity is more important for hyoscyamine formation. However, expression of yeast *odc* gene in *Nicotiana rustica* roots induced an increased accumulation of nicotine [135] that shows that plant secondary products can be elevated by means of genetic manipulation.

4. Epilogue

Biotechnology of medicinal plants has not received the attention it deserves. While other group of plants have been the area of concern for yield improvement including disease and pest resistance, no organised technology has been adopted for improving the yield of medicinal plants. The methodology should include credible selection, micropropagation and studies on abiotic stress related yield as stress has been implicated in the biosynthesis of secondary metabolites. Recombinant DNA technology, molecular biology and metabolic engineering need to be integrated to make a combined concerted effort that will help in improving the productivity of drug components.

References

1. V.P. Kamboj, Herbal medicine, Curr. Sci. 78 (2000) 35–39.
2. J. Holley, K. Cherla, 1998. The medicinal plants sector in India, MAPPA: IDRC/South Asia Regional Office, New Delhi, India.
3. S. Natesh, The changing scenario of herbal drugs: role of botanists, Phytomorphology Golden Jubilee Issue (2001) 75–96.
4. M. Purohit, D. Pande, A. Datta, P.S. Srivastava, Enhanced xanthotoxin content in regenerating cultures of *Ammi majus* and micropropagtion, Planta Med. 61 (1995a) 481–482.

5. M. Purohit, D. Pande, A. Datta, P.S. Srivastava, *In vitro* flowering and high xanthotoxin content in *Ammi majus* L., J. Plant Biochem. Biotechnol. 4 (1995b) 73–76.

6. G. Ali, M. Purohit, M.H. Mughal, M. Iqbal, P.S. Srivastava, A rapid protocol for micropropagation of *Bacopa monniera* L Wettst. a medicinal plant, Plant Tiss. Cult. Biotech. 2 (1996) 208–211.

7. A. Gulati, S. Bharel, M.Z. Abdin, S.K. Jain, P.S. Srivastava, In vitro micropropagation and flowering in *Artemisia annua*, J. Plant Biochem. Biotechnol. 5 (1996) 31–35.

8. S.T. Sai, C.L. Keng, N. Pargini, K.H. Teo Chris, *In vitro* propagation of *Typhonium flagelliforme* (Lodd) Blume, In Vitro Cell. Dev. Biol. Plant. 36 (2000) 402–406.

9. M. Jayanthi, P.K. Mandal, Plant regeneration through somatic embryogenesis and RAPD analysis of regenerated plants in *Tylophora indica* (Burm.F. Merrill), In Vitro Cell. Dev. Biol. Plant. 37 (2001) 576–580.

10. Mc R. Cardell, J.J. Frett, Effect of cytokinin on shoot production in vitro from *Petunia* leaf explants, Hort. Sci. 25 (1990) 627.

11. R.B. Malabadi, K. Nataraja, Shoot regeneration in leaf explants of *Clitorea ternatea* L. cultured in vitro, Phytomorphology 51(2001) 169–172.

12. D.W. Uander, Callus induction in. *Phyllanthus* species and inhibition of viral DNA polymerase and reverse transcriptase by callus extracts, Plant Cell Rep. 10 (1991) 461–466.

13. A. Datta, P.S. Srivastava, Variation in vinblastine production by *Catharanthus roseus* during in vivo and in vitro differentiation, Phytochemistry 46 (1997) 135–137.

14. R. Kaur, M. Sood, S. Chander, R. Mahajan, V. Kumar, D.R. Sharma, *In vitro* propagation of *Valeriana jatamansi*, Plant Cell Tiss. Org. Cult. 59 (1999) 227–229.

15. F. Fracaro, S. Echeverrigaray, Micropropagation of *Cunila galioides*, a popular medicinal plant of south Brazil, Plant Cell Tiss. Org. Cult. 64 (2001) 1–4.

16. P.S. Srivastava, M Purohit, D. Pande, A. Datta, Phenotypic variation and alkaloid content in the androgenic plantlets of *Datura innoxia*, Phytomorphology 43 (1993) 209–216.

17. C.G. Sudha, P.N. Krishnan, S. Seeni, P. Pushpangadan, Regeneration of plants from *in vitro* root segments of *Holostemma annulare* (Roxb.) K Schum., A rare medicinal plant, Curr. Sci. 78 (2000) 1532–1535.

18. A.L. Butiuc-Keul, C. Deliu, Clonal micropropagation of *Arnica montana* L., a medicinal plant, In Vitro Cell. Dev. Biol. Plant. 37 (2001) 581–585.

19. V. Selvakumar, P.R. Anbudurai, T. Balakumar, *In vitro* propagation of the medicinal plant *Plumbago zeylanica* L. through nodal explants, *In Vitro* Cell. Dev. Biol. Plant. 37 (2001) 280–284.

20. A. Narula, S. Kumar, P.S. Srivastava, *In vitro* response of explants of *Abrus precatorius* in different media, J. Trop. Med. Plants. 2 (2001) 57–65.

21. P. Perez-Bermudez, H.U. Seitz, I. Gavidia, A protocol for rapid micropropagation of endangered *Isoplexis*, In Vitro Cell. Dev. Biol. Plant. 38 (2002) 178–182.

22. M.W. Fowler, Plant cell cultures: fact and fantasy, Biochem. Soc. Trans. 11 (1982) 23–28.

23. C.M. O' Neill, A.E. Arthur, R.J. Mathias, The effect of proline thioproline and methyl glyoxalbis (guanylhydrazone) on shoot regeneration frequencies from stem explants of *B. napus*, Plant Cell Rep. 15 (1996) 695–698.

24. A.J. Parr, J. Payne, J. Eagles, B.T. Chapman, R.J. Robins, M.J.C. Rhodes, Variation in tropane alkaloid accumulation within the solanaceae and strategies for its exploitation, Phytochemistry 29 (1990) 2545–2550.

25. M.H. Zenk, H. El-shagi, H. Arens, J. Stockigt, E.W. Weiler, B. Deus, Formation of indole alkaloids serpentine and ajmalicine in cell suspension cultures of *Catharanthus roseus*, in: W. Barz, E. Reinhard, M.H. Zenk (Eds), Plant Tissue Culture and its Biotechnological Applications, Springer, Berlin Heidelberg New York, 1975, pp. 27–43.

26. N. Uozumi, K. Kohketsu, O. Kondo, H. Honda, T. Kobayashi, Fed batch cultures of hairy roots using fructose as a carbon source, J. Ferment. Bioeng. 72 (1991) 457–460.

27. G. Roja, P.S. Rao, Biotechnological investigations in medicinal plants for the production of secondary metabolites, in: I.A. Khan, A. Khanum (Eds), Role of Biotechnology in Medicinal and Aromatic Plants, Ukaaz Publication, Andhra Pradesh, 1998, pp. 95–115.

28. M.W. Fowler, Commercial applications and economic aspects of mass plant cell culture, in: S.H. Mantell, H. Smith (Eds), Plant Biotechnology, Cambridge University Press, Combridge, London, New York, Rochelle, Melbourne, Sydney, 1983, pp. 3–37.

29. L. Nettleship, M. Slayter, Effect of phosphate on secondary metabolism in *Peganum*, J. Exptl. Bot. 25 (1994) 1114–1119.

30. Y. Fujita, Y. Hara, C. Suga, T. Morimoto, Production of shikonin derivatives by cell suspension cultures of *Lithospermum erythrorhizon* II. A new medium for the production of shikonin derivatives, Plant Cell Rep. 1 (1981) 61–63.

31. S.F. Chandler, J.H. Dodds, Effect of nitrogen on secondary metabolite production, Plant Cell Rep. 2 (1983) 205–208.

32. Jimenez-Aparicio, G. Guteierrez-Lopez, Production of food related colorants by culture of plant cells. The case of betalains, in: S.P. Fereidon, J. Kolodziejezyk, J.R. Whitaker, A. Lopez-Munguia, G. Fuller (Eds), Chemicals via Higher Plant Bioengineering, Kluwer Acad. Press, New York, 1999, pp. 195–210.

33. R. Verpoorte, R. Van der Heijden, H.J.G. Ten Hoopen, J. Memelink, Metabolic engineering of plant secondary metabolites pathway for the production of fine chemicals, Biotechnol. Lett. 21 (1999) 467–479.

34. J.M. Furze, M.J.C. Rhodes, A.J. Parr, R.J. Robins, I.M. Whitehead, D.R. Threlfall, Abiotic factors elicit sesquiterpenoid phytoalexin production but not alkaloid production in transformed roots of *Datura stramonium*, Plant Cell Rep. 10 (1991) 111–114.

35. V. Sri Andrijany, G. Indrayanto, L.A. Sochono, Simultaneous effect of calcium, magnesium, copper and cobalt ions on sapogenin steroids content in callus cultures of *Agave amaniensis*, Plant Cell Tiss. Org. Cult. 55 (1999) 103–108.

36. G. Trejo-Tapia, A. Jimenez-Aparicio, M. Rodriguez-Monroy, A. De. Jesus-Sanchez, G. Gutierrez-Lopez, Influence of cobalt and other microelements on the production of betalains and the growth of suspension cultures of *Beta vulgaris*, Plant Cell Tiss. Org. Cult. 67 (2001) 19–23.

37. Saba, D. Pande, M. Iqbal, P.S. Srivastava, Effect of $ZnSO_4$ and $CuSO_4$ on regeneration and lepidine content in *Lepidium sativum*, Biol. Plant. 43 (2000) 253–256.

38. B. Chakarvarty, S. Srivastava, Effect of cadmium and zinc on metal uptake and regeneration of tolerant plants in linseed, Agric. Ecosyst. Environ. 61 (1997) 45–50.

39. G. Ali, P.S. Srivastava, M. Iqbal, Morphogenic response and proline content in *Bacopa monniera* cultures grown under copper stress, Plant Sci. 138 (1998) 191–195.

40. T. Murashige, F. Skoog, A revised medium for rapid growth and bioassays with tobacco tissue cultures, Physiol. Plant. 15 (1962) 473–497.

41. R.U. Schenk, A.C. Hildebrandt, Medium and techniques for induction and growth of monocotyledonous and dicotyledonous plant cell culture, Can J. Bot. 50 (1972) 199–204.

42. A. Brane, W. Urbach, K.J. Dietz, Compartmentation and transport of zinc in barley primary leaves as basic mechanism involved in zinc tolerance, Plant Cell Environ. 17 (1994) 153–162.

43. H.B. Jumin, Plant regeneration via somatic embryogenesis in *Citrus* and its relatives. Phytomorphology 45 (1995) 1–8.

44. P.S. Srivastava, G. Ali, M. Iqbal, A. Narula, N. Bharti, Micropropagation of *Bacopa* and effect of heavy metals on growth performance, in: S.K. Nandi, L.M.S. Palni, A. Kumar (Eds), Role of Plant Tissue Culture in Biodiversity Conservation and Economic Development, Gyanodaya Prakashan, Nanital, India, 2002, pp. 325–344.

45. C. De Backer-Royer, A. Vannereau, S. Duret, R. Villavigens, L. Cosson. Action de metaux lourds Cu et Cd surdes cellules de *Catharanthus roseus cultivees in vitro* adaption et production alcaloidique. Col. Inst. Natl. Res. Alim. 51 (1990) 247–251.

46. H. Mizukami, M. Konoshima, M. Tabata, Effect of nutritional factors on shikonin derivative formation in *Lithospermum* callus cultures, Phytochemistry 16 (1977) 1183–1186.

47. R. Endress, Betacyanin akkumulation in kallus von *Portulacca grandiflora* var JR unter dem eintluss Vonphytohormonem und Cu^{2+} ionen auf unterschiedlichen grund medien, Biochem. Physiol. Pflanz. 169 (1976) 87–98.

48. S. Obrenovic, Effect of Cu (11) D-penicillanine on phytochrome mediated betacyanin formation in *Amaranthus caudatus* seedlings, Plant Physiol. Biochem. 28 (1990) 639–646.

49. G. Indrayanto, L. Rahayu, A. Rahman, P.E. Noerani, Effect of calcium, strontium and magnesium ions on the formation of phytosteroids in callus cultures of *Agave amaniensis*, Planta Med. 59 (1993) 97–98.

50. G. Indrayanto, B. Pratnaningsih, M.H. Santosa, The influence of molybdate and calcium ions on the sapogenin content in callus cultures of *Agave amaniensis*. Indonesian J. Pharm. 7 (1996) 21–27.

51. W.G.W. Kurz, Plant Cell Culture: a potential source of pharmaceuticals, Adv. Appl. Microbiol. 25 (1979) 209–212.

52. M.H. Zenk, H. El-Shagi, U. Schulte, Anthraquinone production by cell suspension culture of *Morinda citrifolia*, Planta Med. Suppl. 79 (1975) 101.

53. D.P. Fulzele, R.K. Satdive, B.B. Pol, Growth and production of camptothecin by cell suspension cultures of *Nothapodytes foetida*, Planta Med. 67 (2001) 150–152.

54. S. Ray, S. Jha, Production of Withaferin A in shoot cultures of *Withania somnifera*, Planta Med. 67 (2001) 432–436.

55. Alain Decendit, Di Liu, Lazhar Ouelhazi, Pierre Doireau, Jean-Michel Mérillon, Marc Rideau, Cytokinin enhanced accumulation of indole alkaloids in *Catharanthus roseus* cell cultures—the factors affecting the cytokinin response, Plant Cell Rep. 11 (1992) 400–403.

56. P.N. Bhatt, D.P. Bhatt, I. Sussex, Studies on some factors affecting solasodine contents in tissue cultures of *Solanum nigrum*, Physiol. Plant. 57 (1983) 159–162.

57. J.G. Marshall, E.J. Staba, Hormonal effects on diosgenin biosynthesis and growth in *Dioscorea deltoidea* tissue cultures, Phytochemistry 15 (1976) 53–55.

58. D. Zhao, J. Xing, Maoyin Li, Dongping Lu, Qiao Zhao, Optimization of growth and jaceosidin production in callus and cell suspension cultures of *Saussurea medusa*, Plant Cell Tiss. Org. Cult. 67 (2001) 227–234.

59. T. Matsumoto, K. Nishida, M. Noguchi, E. Tamaki, Some factors affecting the anthocyanin formation by *Populus* cells in suspension culture, Arg. Biol. Chem. 37 (1973) 561–567.

60. C.A. Hay, L.A. Anderson, M.F. Roberts, J.D. Phillipson, Effect of precursors on the yield of secondary products, Plant Cell Rep. 5 (1986) 1–4.

61. K. Lindsey, Precursor caused higher capsaicin production by cell cultures of *Capsicum frutescens*, Phytochemistry 25 (1986) 2793.

62. T. Tanahashi, N. Nagakura, H. Inouye, M. Zenk, Radioimmunoassay for determination of loganin and biotransformation of loganin to secologanin by plant cell cultures, Phytochemistry 23 (1984) 1917–1922.

63. M. Boitel-Conti, J.C. Laberche, A. Lanoue, C. Ducrocq, B.S. Sangwan-Norreel, Influence of feeding precursors on tropane alkaloid production during an abiotic stress in *Datura innoxia* transformed roots, Plant Cell Tiss. Org. Cult. 60 (2000) 131–137.

64. N.T. Keen, J.E. Patridge, A.I. Zaki, Pathogen-induced elicitor of a chemical defense mechanism in soybean monogenically resistant to *Phytophthora megasperma* var. Sojae, Phytopathology 62 (1972) 768.

65. I.M. Whitehead, D.R. Threlfall, Production of phytoalexins by plant tissue cultures, J. Biotech. 26 (1992) 63–81.

66. J.S. Rokem, J. Schwarzberg, J. Goldberg, Autoclaved fungal mycelia increase diosgenin production in cell suspension cultures of *Dioscorea deltoidea*, Plant Cell Rep. 3 (1984) 159–160

67. C. Funk, K. Gugler, P. Brodelius, Increased secondary product formation in plant cell suspension cultures after treatment with yeast carbohydrate preparation (Elicitor), Phytochemistry 26 (1987) 401–405.

68. D.J. Kim, H.N. Chang, Increased shikonin production in *Lithospermum erythrorhizon* suspension cultures with *in situ* extraction and fungal cell treatment (Elicitor), Biotech. Lett. 12 (1990) 443–446.

69. R. Wijsnama, A. Go JTK, I.N. Van Weerden, P.A. Harkes, B. Veerpoorte, S.A. Barheim, Anthraquinones as phytoalexins in cell and tissue cultures of *Cinchona* sp., Plant Cell Rep. 4 (1985) 241–244.

70. M.M. Yeoman, C.L. Yeoman, Manipulating secondary metabolism in cultured plant cells, New Phytol. 134 (1996) 553–569.

71. T. Hashimoto, Y. Yamada, Alkaloid biogenesis: Molecular aspects, Ann. Rev. Plant Physiol. Plant. Mol. Biol. 329 (1994) 257–285.

72. R.J. Aerts, D. Gisi, E. De Carolis, V. De Luca, T.W. Baumann, Methyl jasmonate vapor increases the developmentally controlled synthesis of alkaloids in *Catharanthus* and *Cinchona* seedlings, Plant J. 50 (1994) 635–642.

73. I. Zabetakis, R. Edwards, D. O'Hagan, Elicitation of tropane alkaloid biosynthesis in transformed root cultures of *Datura stramonium*, Phytochemistry 50 (1999) 53–56.

74. R.E.B. Ketchum, M. Tandon, T.P. Begley, D.M. Gibson, R. Croteau, M.L. Shuler, The production of paclitaxel and other taxanes in *Taxus canadensis* suspension cell cultures elicited with methyl jasmonate, Plant Physiol. 114 (1997) 232.

75. E. Szabo, A. Thelen, M. Petersen, Fungal elicitor preparations and methyl jasmonate enhance rosmarinic acid accumulation in *in vitro* cultures of *Coleus blumei*, Plant Cell Rep. 18 (1999) 485–489.

76. J. Schmitt, M. Petersen, Influence of methyl jasmonate and coniferyl alcohol on pinoresinol and matairesinol accumulation in a *Forsythia* x intermedia suspension culture, Plant Cell Rep. 20 (2002) 885–889.

77. N.E. Delfel, L.J. Smith, The importance of culture conditions and medium component interactions in the growth of *Cephalotaxus harringtonia* tissue cultures, Planta Med. 40 (1980) 237–244.

78. A.B. Ohlsson, L. Bjork, S. Gatenbeck, Effect of light on cardenoline production by *Digitalis lantana* tissue cultures, Phytochemistry 17 (1983) 1907–1910.

79. K. Hirata, M. Horiuchi, T. Ando, M. Asada, K. Miyaomoto, Y. Miura, Effect of near ultra violet light on alkaloid production in multiple shoot cultures of *Catharanthus roseus*, Planta Med. 57 (1991) 499–500.

80. J. Sengupta, G.C. Mitra, A.K. Sharma, Steroid formation during morphogenesis in callus cultures of *Dioscorea floribunda*, J. Plant Physiol. 135 (1989) 27–30.

81. R.M. Buitelaar, J. Tramper, Strategies to improve the production of secondary metabolites with plant cell cultures: a literature review, J. Biotechnol. 23 (1992) 111–141.

82. C. Veeresham, Immobilization of plant cells, in: I.A. Khan, A. Khanum (Eds), Role of Biotechnology in Medicinal and Aromatic plants, Ukaaz Publications, Andhra Pradesh, 1998, pp. 304–319.

83. K. Redenbaugh, J.W. Nichol, M.E. Kossler, B.D. Paash, Encapsulation of somatic embryos for artificial seed production, *In Vitro* Cell. Dev. Biol. Plant. 20 (1984) 256.

84. J. Fujii, D. Slade, K. Redenbaugh, Maturation and green house planting of alfalfa synthetic seeds, *In Vitro* Cell. Dev. Biol. Plant. 25 (1989) 1179–1182.

85. M. Sanada, Y. Sakamoto, M. Hayashi, T. Mashiko, A. Okamoto, N. Ohnishi, Celery and lettuce, in: K. Redenbaugh (Ed.) Synthetic Seeds: Application of Synthetic Seeds to Crop Improvement, CRC Press, Boca Raton, USA, 1993, pp. 305–327.

86. P.S. Rao, T.R. Ganapathi, P. Suprasana, V.A. Bapat, Synthetic seed technology as a method of plant propagation and delivery of tissue cultured plants, in: L.K. Pareek (Ed.), Trends in Plant Tissue Culture and Biotechnology, Agrobotanical Publishers, Jaipur, India, 1996, pp. 47–52.

87. P.S. Ahuja, J. Mathur, N. Lal, A. Mathur, A.K. Mathur, A.K. Kukreja, Towards developing artificial seeds by shoot bud encapsulation, in: A.K. Kukreja, A.K. Mathur, P.S. Ahuja, P.S. Thakur (Eds) Tissue Culture and Biotechnology of Medicinal and Aromatic Plants, Paramount Publishing House, New Delhi, India, 1989, pp. 22–28.

88. B.B. Mandal, K.P.S. Chandel, S. Dwivedi, Cryopreservation of yam (*Dioscorea* spp.) shoot apices by encapsulation-dehydration, Cryo Lett. 17 (1996) 165–174.

89. R.B. Malabadi, K. Nataraja, *In vitro* storage of synthetic seeds in *Clitoria ternatea* Linn., Phytomorphology 52 (2002) 231–243.

90. E. Maruyama, I. Kinoshita, K. Ishii, K. Ohba, A. Saito, Germplasm conservation of the tropical trees, *Cedrela odorata* L., *Guazuma crinita* Mort. and *Jacaranda mimosaefolia* D. Don by shoot tip encapsulation in calcium alginate and storage at 12–25°C, Plant Cell Rep. 16 (1997) 393–396.

91. S. Ahuja, S. Dixit, B.B. Mandal, P.S. Srivastava, Cryopreservation of medicinal plants: prospects and potentials, J. Trop. Med. Plants 3 (2002) 113–120.

92. B. Diettrich, A.S. Popov, B. Pfeiffer, D. Neumann, R.G. Butenko, M. Luckner, Cryopreservation of *Digitalis lanata* cell cultures, Planta Med. 46 (1982) 82–87.

93. B. Vanadenbussche, M.A.C. Demeulemeester, M.P. De Proft, Cryopreservation of alginate coated *in vitro* grown shoot tips of chicory (*Cichorium intybus* L.) using rapid freezing, Cryo Lett. 14 (1993) 259–266.

94. S.W. Decruse, S. Seeni, P. Pushpagadan, Cryopreservation of alginate coated shoot tips of in vitro grown *Holostemma annulare* (Roxb.) K. Schum, and endangered medicinal plant: Influence of pre-culture and DMSO treatment on survival and regeneration, Cryo Lett. 20 (1999) 243–250.

95. L.E. Towill, Cryopreservation of isolated mint shoot tips by vitrification, Plant Cell Rep. 9 (1990) 178–180.

96. Y.P.S. Bajaj, Storage and cryopreservation of *in vitro* cultures, in: Y.P.S. Bajaj (Ed.) Biotechnology in Agriculture and Forestry, vol. 17. Springer, Berlin Heidelberg New York, 1991, pp. 361–381.

97. Y.P.S. Bajaj, Regeneration of plants from pollen embryos of *Arachis, Brassica* and *Triticum* spp. cryopreserved for one year, Curr. Sci. 52 (1983) 484–485.

98. Y.P.S. Bajaj, Regeneration of plants from frozen (–196°C) protoplasts of *Atropa belladonna, Datura innoxia and Nicotiana*, J. Exptl. Bot. 16 (1988) 947–953.

99. G. Weber, E.J. Roth, H.F. Schmeiger, Storage of cell suspension and protoplasts of *Glycine max* (L.) Merri., *Brassica napus* (L.), *Datura innoxia* (Mill.) and *Daucus carota* (L.) by freezing, Z. Pflanzenphysiol. 109 (1983) 29–39.

100. P.C. Debergh, P.E. Read, Micropropagation, in: P.C. Deberg, R.H. Zimmerman (Eds), Micropropagation, Technology and Application, Kluwer Academic, Dordrecht, Boston, London, 1990, pp. 1–13.

101. V. Rani, S.N. Raina, Genetic fidelity of organized meristem-derived micropropagated plants: a critical reappraisal, *In Vitro* Cell. Dev. Biol. Plant. 36 (2000) 319–330.

102. V.B. Shenoy, I.K. Vasil, Biochemical and molecular analysis of plants derived from embryogenic cultures of napier grass (*Pennisetum purpureum* K. Schum.), Theor. Appl. Genet. 83 (1992) 947–954.

103. B.J. Mazur, S.V. Tingey, Genetic mapping and introgression of genes of agronomic importance, Current Opinions in Biotechnology, 6 (1995) 175–182.

104. P. Vos, R. Hogers, M. Bleeker, M. Reijans, T. van de Lee, M. Hornes, A. Frijters, J. Plot, J. Peleman, M. Kuiper, M. Zabeau, AFLP: a new technique for DNA fingerprinting, Nucl. Acids Res., 23 (1995) 4407–4414.

105. P. Breyne, W. Boerjan, T. Gerats, M. Van-Montagu, A. Van-Gysel, Applications of AFLP in plant breeding, molecular biology and genetics, Belg. J. Bot. 129 (1997) 107–117.

106. W.A. Vendrame, G. Kochert, H.Y. Wetzstein, AFLP analysis of variation in pecan   s o m a t i c embryos, Plant Cell Rep. 18 (1999) 853–857.

107. A. Singh, M.S. Negi, V.K. Moses, B. Venkateswarlu, P.S. Srivastava, M. Lakshmikumaran, Molecular analysis of micropropagated neem plants using AFLP markers for ascertaining clonal fidelity, In Vitro Cell. Dev. Biol. Plant. 38 (2002) 519–524.

108. A.M. Kiers, T.H.M. Mes, R. van der Meijden, K. Bachmann, A search for diagnostic AFLP markers in *Cichorium* species with emphasis on endive and chicory cultivar groups, Genome 43 (2000) 470–476.

109. M.S. Negi, A. Singh, M. Lakshmikumaran, Genetic variation and relationship among and within *Withania* species as revealed by AFLP markers, Genome 43 (2000) 975–980.

110. K. Saito, Genetic engineering in tissue culture of medicinal plants. Plant Tiss. Cult. Lett. 10 (1993) 1–8.

111. A. Merkli, P. Christen, I. Kapetanidis, Production of diosgenin by hairy root cultures of *Trigonella foenum-graecum* L., Plant Cell Rep. 16 (1997) 632–636.

112. J. Martin-Tanguy, L.Y. Sun, D. Burtin, R. Vernoy, N. Rossin, D. Tepfer, Attenuation of the phenotype caused by the root inducing, left-hand, transferred DNA and its *rol* gene-correlations with changes in polyamine metabolism and DNA methylation, Plant Physiol. 111 (1996) 259–267.

113. R. Zárate, Tropane alkaloid production by *Agrobacterium rhizogenes* transformed hairy root cultures of *Atropa baetica* Willk (Solanaceae), Plant Cell Rep. 18 (1999) 418–423.

114. K. Doerk Schmitz, L. Witte, A.W. Alfermann, Tropane alkaloid patterns in plants and hairy roots of *Hyoscyamus albus*, Phytochemistry 35 (1994) 107–110.

115. N. Sevon, C.K.M. Oksman and R. Hiltunen, Efficient plant regeneration from hairy root derived protoplasts of *Hyoscyamus muticus*, Plant Cell Rep. 14 (1995) 738–742.

116. M. Zehra, S. Banerjee, S. Sharma, S. Kumar, Influence of *A. rhizogenes* strains on biomass and alkaloid productivity in hairy root lines of *Hyoscyamus muticus* and *H. albus*, Planta Med. 65 (1999) 60–63.

117. P. Christen, T. Aoki, K. Shimomura, Characteristics of growth and tropane alkaloid production in *Hyoscyamus albus* hairy roots transformed with *Agrobacterium rhizogenes* A4. Plant Cell Rep. 11 (1992) 597–600

118. F. Vazquez-Flota, O. Moreno-Valenzuela, M.L. Miranda-Ham, J. Coello-Coello, V.M. Loyola-Vargas, Catharanthine and ajmalicine synthesis in *Catharanthus roseus* hairy root cultues, Plant Cell Tiss. Org. Cult. 38 (1994) 273–279.

119. R.J. Aerts, L.T. Van der, H.R. Van der, R. Verpoorte, Developmental regulation of alkaloid production in cinchona seedlings, J. Plant Physiol. 13 (1990) 86–91.

120. J. Berlin, C. Rugenhagen, N. Greidziak, I.N. Kuzovkina, L. Witte, V. Wray, Biosynthesis of serotonin and β-carboline alkaloids in hairy root cultues of *Peganum harmala*, Phytochemistry 33 (1993) 593–597.

121. S. Inomata, M. Yokoyama, Y. Gozu, T. Shimizu, M. Yanagi, Growth pattern and ginsenoside production of *Agrobacterium* transformed *Panax ginseng* roots, Plant Cell Rep. 12 (1993) 681–686.

122. A. Mallol, R.M. Cusido, J. Palazon, M. Bonfill, C. Morales, M.T. Pinol, Ginsenoside production in different phenotypes of *Panax ginseng* transformed roots, Phytochemistry 57 (2001) 365–371.

123. G. Jung, D. Tepfer, Use of genetic transformation by the Ri T-DNA of *Agrobacterium rhizogenes* to stimulate biomass and tropane alkaloid production in *Atropa belladonna* and *Calystegia sepium* roots grown *in vitro*, Plant Sci. 50 (1987) 145–151.

124. T. Aoki, H. Matsumoto, Y. Asako, Y. Matsunaga, K. Shimomura, Variation of alkaloid productivity among several clones of hairy roots and regenerated plants of *Atropa belladonna* transformed with *Agrobacterium rhizogenes* 15834, Plant Cell Rep. 16 (1997) 282–286.

125. F. Fecker Lothar, C. Rugenhagen, J. Berlin, Increased production of cadaverine and anabasine in hairy root cultures of *Nicotiana tabacum* expressing a bacterial *lysine decarboxylase* gene, Plant Mol. Biol. 23 (1993) 11–21.

126. Da-Hua Chen, Chang-Jun Liu, He-Chun Ye, Guo-Feng Li, Ben-ye Liu, Yu-Ling Meng, Xiao-Ya Chen, Ri-mediated transformation of *Artemisia annua* with a recombinant farnesyl diphosphate synthase gene for artemisinin production, Plant Cell Tiss.Org. Cult. 57 (1999) 157–162.

127. C. Canel, M.I. Lopes Cardoso, S. Whitmer, L. Fits, G. Pasquali, R. Van der Heijden, J.H.C. Hoge, R. Verpoorte, Effect of over expression of strictosidine synthase and tryptophan decarboxylase on alkaloid production by cell cultures of *Catharanthus roseus*, Planta, 205 (1998) 414–419.

128. M.A. Subroto, M.D. Pauline, Production of steroidal alkaloids by hairy roots of *Solanum aviculare* and the effect of gibberellic acid, Plant Cell Tiss. Org. Cult. 38 (1994) 93–102.

129. H. Schaller, B. Grausem, P. Benveniste, M.L. Chye, C.T. Tan, Y.H. Song, Expression of the *Hevea brasiliensis* Mull. Arg. 3-hydroxy-3-methyl glutaryl coenzyme A reductase 1 in tobacco results in sterol overproduction, Plant Physiol. 109 (1995) 761–770.

130. A.C. Cavalcante Argôlo, B.V. Chaarlwood, M. Pletsch, Regulation of solasodine production by *Agrobacterium rhizogenes*-transformed roots of *Solanum aviculare*, Planta Med. 66 (2000) 448–451.

131. Chappell, F. Wolf, J. Proulx, R. Cuellar, C. Saunders, Is the reaction catalyzed by 3-hydroxy-3methyl glutaryl coenzyme-A reductase a rate limiting step for isoprenoid biosynthesis in plants, Plant Physiol. 109 (1995) 1337–1343.

132. A.F. Tiburcio, A.W. Galston, Arginine decarboxylase as the source of putrescine for tobacco alkaloids, Phytochemistry 25 (1986) 107–110.

133. S. Imanishi, K. Hashizume, M. Nakakita, H. Kojima, Y. Matsubayashi, T. Hashimoto, Y. Sakagami, Y. Yamada, K. Nakamura, Differential induction by methyl jasmonate of genes encoding ornithine decarboxylase and other enzymes involved in nicotine biosynthesis in tobacco cell cultures, Plant Mol. Biol. 38 (1998) 1101–1111.

134. R.J. Robins, A.J. Parr, J.N. Walton, Studies on the biosynthesis of tropane alkaloids in *Datura stramonium* L. transformed root culture. On the relative contributions of L-arginine and L-ornithine to the formation of the tropane ring. Planta 183 (1991) 196–201.

135. J.D. Hamill, R.J. Robins, A.J. Parr, D.M. Evans, J.M. Furze, M.J.C. Rhodes, Over expressing a yeast *ornithine decarboxylase* gene in transgenic roots of *Nicotiana rustica* can lead to enhanced nicotine accumulation, Plant Mol. Biol. 15 (1990) 27–38.

136. D. Pande, M. Purohit, P.S. Srivastava, Variation in xanthotoxin content in *Ammi majus* L. cultures during *in vitro* flowering and fruiting, Plant Sci. 162 (2002) 583–587.

137. C. Wawrosch, P.R. Malla, B. Kopp, Micropropagation of *Allium wallichii* Kunth, a threatened medicinal plant of Nepal, *In Vitro* Cell. Dev. Biol. Plant. 37 (2001) 555–557.

138. A.L. Butiuc-Keul, C. Deliu, Clonal propagation of *Arnica montana* L., a medicinal plant, *In Vitro* Cell. Dev. Biol. Plant. 37 (2001) 581–585.

139. A. Ahuja, M. Sambyal, S. Koul, *In vitro* propagation and conservation of *Atropa acuminata* Royle ex Lindl—An indigenous threatened medicinal plant, J. Plant Biochem. Biotech. 11 (2002) 121-124.

140. S. Baburaj, P. Ravichandran, M. Selvapandian, *In vitro* adventitious shoot formation from leaf cultures of *Clerodendrum inerme* (L) Gaertn., Indian J. Exptl. Biol. 38 (2000) 1274-1276.

141. R. Bhattacharya, S. Bhattacharya, *In vitro* multiplication of *Coleus forskohlii* Briq.: An approach towards shortening the protocol, In Vitro Cell. Dev. Biol. Plant. 37 (2001) 572–575.

142. A. Narula, P.S. Srivastava, N.S. Rangaswamy, *In vitro* culture studies on *Dioscorea* species, J. Trop. Med. Plants 1 (2000) 60–74.

143. S. Raha, S.C. Roy, *In vitro* plant regeneration in *Holorrhena antidysenterica* Wall., through high frequency axillary shoot proliferation, *In Vitro* Cell. Dev. Biol. Plant. 37 (2001) 232–236.

144. S.R. Thengane, D.K. Kulkarni, V.A. Shrikhande, K.V. Krishnamurthy, Effect of thidiazuron on adventitious shoot regeneration from seedling explants of *Nothapodytes foetida*, In Vitro Cell. Dev. Biol. Plant. 37 (2001) 206–210.

145. R. Saini, P.K. Jaiwal, *In vitro* multiplication of *Peganum harmala*—an improtant medicinal plant, Indian J. Exptl. Biol. 38 (2000) 499–503.

146. E. Catapan, M.F. Otuki, A.M. Viana, *In vitro* culture of *Phyllanthus caroliniensis* (Euphorbiaceae), Plant Cell Tiss. Org. Cult. 62 (2000) 195–202.

147. M. Das, S.S. Raychaudhuri, Enhanced development of somatic embryos of *Plantago ovata* Forsk. by additives, *In Vitro* Cell. Dev. Biol. Plant. 37 (2001) 568–571.

148. G. Das, M.N. Rashmi, G.R. Rout, *In vitro* clonal propagation of *Plumbago rosea* L.: a potential medicinal plant, Plant Cell Biotech. Mol. Biol. 2 (2001) 105–112.

149. V. Selvakumar, P.R. Anbudurai, T. Balakumar, *In vitro* propagation of the medicinal plant *Plumbago zeylanica* L. through nodal explants, *In Vitro* Cell. Dev. Biol. Plant. 37 (2001) 280–284.

150. M. Jeyakumar, N. Jayabalan, An efficient method for regeneration of plantlets from nodal explants of *Psoralea corylifolia* Linn., Plant Cell Biotech. Mol. Biol. 1 (2000) 37–40.

151. A.R. Sehrawat, Sanjogta Uppal, J.B. Chowdhury, Establishment of plantlets and evaluation of differentiated roots for alkaloids in *Rauwolfia serpentina*, J. Plant Biochem. Biotech 11 (2002) 105–108.

152. S. Anitha, T. Pullaiah, *In vitro* propagation of *Sterculia foetida* Linn (Sterculiaceae), Plant Cell Biotech. Mol. Biol. 2 (2001) 139–144.

153. L. Sivaram, U. Mukundan, Feasibility of commercial micropropagation of *Stevia rebaudiana* in India, J. Trop. Med. Plants 3 (2002) 97–103.

154. Saba, M. Iqbal, P.S. Srivastava, *In vitro* micropropagation of *Silybum marianum* L. from various explants and silybin content, J. Plant Biochem. Biotech. 9 (2000) 81–87.

155. S.R. Bhalsing, V.L. Maheshwari, *In vitro* culture and regeneration of *Solanum khasianum* and extraction of solasodine, J. Plant Biochem. Biotech. 6 (1997) 39–40.

156. M. Jayanthi, P.K. Mandal, Plant regeneration through somatic embryogenesis and RAPD analysis of regenerated plants in *Tylophora indica* (Burm.F. Merrill), *In Vitro* Cell. Dev. Biol. Plant. 37 (2001) 576–580.

157. M. Thiruvengadam, N. Jayabalan, *In vitro* flowering of *Vitex negundo* L. a medicinal plant, Plant Cell Biotech. Mol. Biol. 2 (2001) 67–70.

158. S. Ray, S. Jha, Regeneration of *Withania somnifera* plants, J. Trop. Med. Plants 3 (2002) 89–95.

159. Y. Miura, K. Hirata, N. Kurano, K. Miyatnoto, K. Uchida, Formation of vinblastine in multiple shoot cultures of *Catharanthus* roseus, Planta Med. 54 (1988) 18–20.

160. A. Decendit, Di Liu, L. Ouelhazi, P. Doireau, Jean-Michel Mérillon, M. Rideau, Cytokinin enhanced accumulation of indole alkaloids in *Catharanthus roseus* cell cultures- the factors affecting the cytokinin response, Plant Cell Rep. 11 (1992) 400-403.

161. L.A. Anderson, A.T. Keene, J.D. Phillipson, Alkaloid production by leaf organ, root organ and cell suspension cultures of *Cinchona ledgeriana*, Planta Med. 46 (1982) 25–27.

162. E. Miraldi, A. Masti, S. Ferri, I.B. Comparini, Distribution of hyoscyamine and scopolamine in *Datura stramonium*, Fitoterapia 72 (2001) 644–648.

163. G. Suvarnalatha, L. Rajendran, G.A. Ravishankar, L.V. Venkataraman, Elicitation of anthocyanin production in cell cultures of carrot (*Daucus carota* L.) by using elicitors and abiotic stress, Biotechnol. Lett. 16 (1994) 1275–1280.

164. A.B. Ohlsson, T. Berglund, Effect of high $MnSO_4$ levels on cardenolide accumulation by *Digitalis lanata* tissue cultures in light and darkness, J. Plant Physiol. 135 (1989) 505–507.

165. N.A. O' Dowd, P.G. Mc Cauley, H.S. Richardson David, G. Wilson, Callus production, suspension culture and *in vitro* alkaloid yields of *Ephedra*, Plant Cell Tiss. Org. Cult. 34 (1993) 149–155.

166. R. Verpoorte, V.R. Heijden, J. Schripsema, Plant cell biotechnology for the production of alkaloids. Present status and prospects, J. Nat. Prod. 56 (1993) 186–207.

167. S.B. Rech, C.V.F. Batista, J. Schripsema, R. Verpoorte, A.T. Henriques, Cell cultures of *Rauwolfia sellowii*: growth and alkaloid production, Plant Cell Tiss. Org. Cult. 54 (1998) 61–63.

168. G. Roja, M.R. Heble, Indole alkaloids in clonal propagules of *Rauwolfia serpentina* benthe ex kurz, Plant Cell Tiss. Org. Cult. 44 (1996) 111–115.

169. S. Endreb, H. Takayama, S. Suda, M. Kitajima, N. Aimi, S.I. Sakai, J. Stockigt, Alkaloids from *Rauwolfia serpentina* cell cultures treated with ajmaline, Phytochemistry, 32 (1993) 725–730.

170. P.S. Srivastava, Saba, A. Narula, S. Malik, T. Srivastava, Biotechnological studies to improve silybin content in *Silybum* a hepato-protective herb, in: I.A. Khan, A. Khanum (Eds), Role of Biotechnology in Medicinal and Aromatic Plants, Ukaaz Publications, Andhra Pradesh, 1998, pp. 50–66.

171. P. Khanna, G.L. Sharma, A.K. Rathore, S.K. Manot, Effect of cholesterol on *in vitro* suspension tissue cultures of *Costus speciosus* (Koen) Sm., *Dioscorea floribunda* Mart. & Gal., *Solanum aviculare* Forst. and *Solanum xanthocarpum* Schard & Wendl. Ind. J. Exptl. Biol. 15 (1977) 1025–1027.

172. G. Indrayanto, E. Tristiana, H.S. Mulija, Effect of L-arginine, casein hydrolysate, banana powder and sucrose on growth and solasodine production in shoot cultures of *Solanum laciniatum*, Plant Cell Tiss. Org. Cult. 43 (1995) 237–240.

173. H.A. El-Ashaal, S.A. Ghanem, F.R. Melek, M.A. Kohail, S.H. Hilal, Alkaloid production from regenerated *Solanum* plants, Fitoterapia 70 (1999) 407–411.

174. S. Huang, S. Chen, K. Wu, W. Tang, Strategy for inducing pertinent cell line and optimization of the medium for *Stizolobium hassjoo* producing L-DOPA, J. Ferment. Bioeng. 79 (1995) 342–347.

175. S. Ray, S. Jha, Production of withaferin A in shoot cultures of *Withania somnifera*, Planta Med. 67 (2001) 432–436.

176. U. Eilert, F. Constabel, W.G.W. Kurz, Elicitor stimulation of monoterpene indole alkaloid formation in suspension cultures of *Catharanthus roseus*, J. Plant Physiol. 126 (1986) 11–12.

177. U. Eilert, F. Constabel, W.G.W. Kurz, Elicitor mediated induction of tryptophan decarboxylase and strictosidine synthase in cell suspension cultures of *Catharanthus roseus*, Arch. Biochem. Biophys. 254 (1987) 491–497.

178. F.L.H. Menke, S. Parchmann, M.J. Mueller, J.W. Kune, J. Memelink, Involvement of the actadecanoid pathway and protein phosphorylation in fungal elicitor-induced expression of terpenoid indole alkaloid biosynthetic genes in *Catharanthus roseus*, Plant Physiol. 119 (1999) 1289–1296.

179. S.Y. Byun, H. Pedersen, C.K. Chin, Two-phase culture for the enhanced production of benzophenanthridine alkaloids in cell suspensions of *Eschscholtzia californica*, Phytochemistry 29 (1990) 3135–3139.

180. M.G. Miguel, J.G. Barroso, Accumulation of stress metabolites in cell suspension cultures of *Hyoscyamus albus*, Phytochemistry 42 (1994) 35371–35375.

181. M. Tani, K. Takeda, K. Yazaki, M. Tabata, Effects of oligogalacturonides on biosynthesis of shikonin in *Lithospermum* cell cultures, Phytochemistry 34 (1993) 1285–1290.

182. H. Doernenburg, D. Knorr, Elicitation of chitinase and anthraquinones in *Morinda citrifolia* cell cultures, Food Biotech. 8 (1994) 57–65.

183. S.D. Cline, C.L. Coscia, Stimulation of sanguinarine production by combined fungal elicitation and hormonal deprivation in cell suspension cultures of *Papaver bracteatum*, Plant Physiol. 86 (1988) 161–165.

184. S.D. Cline, R.J. Mc Hale, C.J. Coscia, Differential enhancement of benzophenanthridine alkaloid content in cell suspension cultures of *Sanguinaria canadensis* under conditions of combined hormonal deprivation and fungal elicitation, J. Nat. Prod. 56 (1993) 1219–1228.

185. U. Eilert, W.G.W. Kurz, F. Constabel, Stimulation of sanguinarine accumulation in *Papaver somniferum* cell cultures by fungal elicitors, J. Plant Physiol. 119 (1985) 65–76.

186. P.J. Facchini, A.G. Johnson, J. Poupart, V.D.E. Luca, Uncoupled defense gene expression and antimicrobial alkaloid accumulation in elicited opium poppy cell cultures, Plant Physiol. 111 (1996) 687–697.

187. J.P. Kutney, M.D. Samija, G.M. Hewitt, E.C. Bugante, H. Gu, Anti-inflammatory oleanane triterpenes from *Triterygium wilfordii* cell suspension cultures by fungal elicitation, Plant Cell Rep. 12 (1993) 356–359.

188. Zheng Guang-Zhi, He Jing-bo, Wang Shiling, Cryopreservation of calli and their suspension culture cells of *Anisodus acutangulus*, Acta Bot. Sin. 25 (1983) 512–517.

189. T.H.H. Chen, K.K. Kartha, N.L. Leyng, W.G.W. Kurz, K.B. Chatson, F. Constabel, Cryopreservation of alkaloid-producing cell cultures of periwinkle (*Catharanthus roseus*), Plant Physiol. 75 (1984) 726–731.

190. B. Malaurie, M.F. Trouseat, F. Englemann, N. Chalrillange, Effect of pretreatment conditions on the cryopreservation of *in vitro* cultured yam (*D. alata* and *D. bulbifera*) shoot apices by encapsulation-dehydration, Cryo lett. 19 (1998) 15-26.

191. S. Dixit, S. Ahuja, B.B. Mandal, P.S. Srivastava, Medicinal yams: Potential application of biotechnology for conservation, characterization and use of germplasm, in: I.A. Khan, A. Khanum (Eds) Role of Biotechnology in Medicinal and Aromatic Plants, vol IV, Ukaaz Publications, Hyderabad, India, 2001, pp 65–80.

192. D. Blackesley, R.J. Kierman, Cryopreservation of axillary buds of a *Eucalyptus grandis* × *Eucalyptus camaldulensis* hybrid, Cryo lett. 22 (2001) 13–18.

193. J.C. Pennycooke, L.E. Towil, Cryopreservation of shoot tips from *in vitro* plants of sweet potato (*Ipomea batatas* (L.) Lam.) by vitrification, Plant Cell Rep. 19 (2000) 733–737.

194. K. Redenbaugh, B. Paasch, J. Nichol, M. Kossler, P. Viss, K. Walker, Somatic seeds: encapsulation of asexual plant embryos, Bio/Technology, 4 (1986) 797–801.

195. D. Hirai, A. Sakai, Cryopreservation of *in vitro* grown shoot tip meristems of mint (*Mentha spicata* L.) by encapsulation vitrification, Plant Cell Rep. 19 (1999) 150–155.

196. R.G. Butenko, A.S. Popov, L.A. Volkova, N.D. Chernyak, A.M. Nosov, Recovery of cell cultures and their biosynthetic capacity after storage of *Dioscorea deltoidea* and *Panax ginseng* cells in liquid nitrogen, Plant Sci. Lett. 33 (1984) 285–292.

197. T. Yamada, A. Sakai, T. Matsumara, S. Higuchi, Cryopreservation of apical meristems of white clover (*Trifolium repens* L.) by vitrification, Plant Sci. 63 (1991) 460–467.

198. E. Wallner, K. Weising, R. Kompf, G. Kahl, B. Kopp, Oligonucleotide finger printing and RAPD analysis of *Achillea* species: Characterization and long term monitoring of micropropagated clones, Plant Cell Rep. 15 (1996) 647–652.

199. M.A. Al-Zahim, B.V. Ford-Lloyd, H.J. Newbury, Detection of somaclonal variation in garlic (*Allium sativum* L.) using RAPD and cytological analysis, Plant Cell Rep. 18 (1999) 473–477.

200. R.S. Sangwan, N.S. Sangwan, D.C. Jain, S. Kumar, S.A. Ranade, RAPD profile based genetic characterization of chemotypic variants of *Artemisia annua* L., Biochemistry and Molecular Biology International, 47 (1999) 935–944.

201. A. Singh, M.S. Negi, J. Rajagopal, S. Bhatia, U.K. Tomar, P.S. Srivastava, M. Lakshmikumaran, Assessment of genetic diversity in *Azadirachta indica* using AFLP markers, Theor. Appl. Genet. 99 (1999) 272–279.

202. A. Singh, M.S. Negi, V.K. Moses, B. Venkateswarlu, P.S. Srivastava, M. Lakshmikumaran, Molecular analysis of micropropagated neem plants using AFLP markers for ascertaining clonal fidelity, *In Vitro* Cell. Dev. Biol. Plant. 38 (2002) 519–524.

203. A.M. Kiers, Mes T.M.H., M.R. Van der, K. Bachmann, A search for diagnostic AFLP markers in *Cichorium* species with emphasis on endive and chicory cultivar groups, Genome 43 (2000) 470–476.

204. Zhang YanBo, F. Ngan, T.Z. Wang, TziBun Ng, P. H. But, PanoChui Shaw, J. Wang, Random primed polymerase chain reaction differentiates *Codonopsis pilosula* from different localities, Planta Med. 65 (1999) 157–160.

205. E.S. Mace, R.N. Lester, C.G. Gebhradt, AFLP analysis of genetic relationship in the tribe *Datureae* (Solanaceae), Theor. Appl. Genet. 99 (1999) 642–648.

206. S.G. Nebauer, L. del Castillo-Agudo, J. Segura, RAPD variation within and among natural populations of outcrossing willow-leaved foxglove (*Digitalis obscura* L.), Theor. Appl. Genet. 98 (1999) 985–994.

207. R. Terauchi, T. Terachi, K. Tsunewaki, Physical map of chloroplast DNA of aerial yam, *Dioscorea bulbifera* L., Theor. Appl. Genet. 78 (1989) 1–10.

208. J. Ramser, C. Lopez-Peralta, R. Wetzel, K. Weising, G. Kahl, Genomic variation and relationships in aerial yam (*Dioscorea bulbifera* L.) detected by random amplified polymorphic DNA, Genome 39 (1996) 17–25.

209. H.D. Mignouna, R. Asiedu, N.Q. Ng, M. Knox, N.T.H. Ellis, Analysis of genetic diversity in Guinea yams (*Dioscorea* spp.) using AFLP fingerprinting, in: 11th Symp. Soc. Trop. Root Crops. Trinidad and Tobago, (1997) pp. 70.

210. H. Mizukami, K. Ohbayashi, Y. Kitamura, T. Ikenga, Restriction fragment length polymorphism (RFLP's) of medicinal plants and crude drugs-I. RFLP probes allow clear identification of *Duboisia* interspecific hybrid genotypes in both fresh and dried tissues, Biological and Pharmaceutical Bulletin 16 (1993) 388–390.

211. G.M. Muluvi, J.I. Sprent, N. Soranzo, J. Provan, D. Odee, G. Folkard, J.W. McNicol, W. Powell, Amplified Fragment Length Polymorphism (AFLP) analysis of genetic variation in *Moringa oleifera* Lam., Mol. Ecol. 8 (1999) 463–470.

212. Y. Shoyama, X.X. Zhu, R. Nakai, S. Shiraishi, H. Kohda, Micropropagation of *Panax notoginseng* by somatic embryogenesis and RAPD analysis of regenerated plantlets, Plant Cell Rep. 16 (1997) 450–453.

213. H. Fushmi, K. Komatsu, M. Isobe, T. Namba, Application of PCR-RFLP and MASA analysis on 18S ribosomal RNA gene sequence for the identification of three ginseng drugs, Biological and Pharmaceutical Bulletin 20 (1997) 765–769.

214. K. Wolff, M. Morgan-Richards, PCR markers distinguish *Plantago major* subspecies, Theor. Appl. Genet. 96 (1998) 282–296.

215. M. Hatano, Genetic diagnosis of *Rehmannia* species micropropagated by tip tissue culture and an F_1 hybrid by RAPD analysis, Plant Breeding 116 (1997) 589–591.

216. M. Jayanthi, P.K. Mandal, Plant regeneration through somatic embryogenesis and RAPD analysis of regenerated plants in *Tylophora indica* (Burm. F. Merrill.), In Vitro Cell. Dev. Biol. Plant. 37 (2001) 576–580.

217. J. Troilina, E. Szoke, L. Kursinszke, A. Neszmelyi, Biologically active compounds from genetically transformed hairy root cultures of *Ammi visnaga*, Gyogyszereszet 10 (1996) 212.

218. S. Banerjee, M. Zehra, M.M. Gupta, S. Kumar, *Agrobacterium rhizogenes*-mediated transformation of *Artemisia annua*. Production of transgenic plants. Planta Med. 63 (1997) 467–469.

219. B. Ghosh, S. Mukherjee, S. Jha, Genetic transformation of *Artemisia annua* by A. *tumefaciens* and artemisinin synthesis in transformed cultures, Plant Sci. 122 (1997) 193–199.

220. J.M. Sharp, P.M. Doran, Characteristics of growth and tropane alkaloid synthesis in *Atropa belladonna* roots transformed by A. *rhizogenes*, J. Biotechnol. 16 (1990) 171–186.

221. D.J. Yun, T. Hashimoto, T. Yamada, Metabolic engineering of medicinal plants: transgenic *Atropa belladonna* with an improved alkaloid composition, Proc. Natl. Acad. Sci. USA, 89 (1992) 11799–11803.

222. T. Hashimoto, D.J. Yun, Y. Yamada, Production of tropane alkaloids in genetically engineered root cultures, Phytochemistry 32 (1993) 713–718.

223. V. Bonhomme, D.L. Mattar, J. Lacoux, M.A. Fliniaux, A. Jacquin-Dubreuil, Tropane alkaloid production by hairy roots of *Atropa belladonna* obtained after transformation with *A. rhizogenes* 15834 and *A. tumefaciens* containing *rol A,B,C* genes only, J. Biotechnol. 81 (2000) 151-158.

224. L.A. Saenz Carbonell, I.E. Maldonado Mendoza, O. Moreno Valenzula, R. Ciau Uitz, M. Lopez Meyer, V.M. Loyola Vargas, Effect of the medium pH on the release of secondary metabolites from roots of *Datura stramonium, Catharanthus roseus* and *Tagetes patula* cultured *in vitro*, Appl. Biochem. Biotechnol. 34 (1993) 257–267.

225. J. Payne, M.J.C. Rhodes, R.J. Robins, Quinoline alkaloid production by transformed cultures of *Cinchona ledgeriana*, Planta Med. 53 (1987) 367–372.

226. J.H. Hamill, R.J. Robins, M.J.C. Rhodes, Alkaloid production by transformed root cultures of *Cinchona ledgeriana*, Planta Med. 55 (1989) 354–357.

227. P. Christen, M.F. Robert, J.D. Phillipson, W.C. Evans, Alkaloid of hairy root cultures of a *Datura candida* hybrid, Plant Cell Rep. 9 (1990) 101–104.

228. V.P. Bulgakov, Y.N. Zhuravlev, J.V. Toroptseva, The culture of *Datura innoxia* Mill. transformed roots as a producer of tropane alkaloids, Prikl. Biokhim. Mikrobiol. 27 (1991) 286–291.

229. I. Ionkova, L. Witte, A.W. Alfermann, Spectrum of tropane alkaloids in transformed roots of *Datura innoxia* and *Hyoscyamus × györffyi* cultivated in vitro, Planta Med. 60 (1994) 382–384.

230. M. Boitel-Conti, E. Gontier, J.C. Laberche, C. Ducrocq, B.S. Sangwan-Norreel Permeabilization of *Datura innoxia* hairy roots for release of stored tropane alkaloids. Planta Med. 61 (1995) 287–290.

231. A.M. Escalante-Mane, J.E. Maldonado-Mendoza, V.M. Loyola-Vargas, Effect of heat shock in hairy root cultures of *Datura stramonium*, Plant Physiol. 99 (1992) 49–51.

232. I.E. Maldonado-Mendoza, T. Ayora-Talavera, V.M. Loyola-Vargas, Establishment of hairy root cultures of *Datura stramonium* characterization and stability of tropane alkaloid production during long periods of subculturing, Plant Cell Tiss. Org. Cult. 33 (1993) 321–329.

233. P. Holmes, S.L. Li, N.H. Thomas, B.V. Ford Lloyd, Fungal co-culture for enhanced tropane alkaloid from hairy roots of *Datura stramonium* and *Hyoscyamus albus*, Biotechnol. 100 (1994) 2.

234. M.T. Pinol, J. Palazon, P.R. Cusido, M. Scrrano, Effect of Ri T-DNA from *Agrobacterium rhizogenes* on production in *Datura stramonium* root cultures, Acta Botanica, 109 (1996) 133–138.

235. A.M. Baíza, A. Quiroz, A. Ruíz José, I. Maldonado-Mendoza, V.M. Loyola-Vargas, Growth patterns and alkaloid accumulation in hairy root and untransformed root cultures of *Datura stramonium*, Plant Cell Tiss. Org. Cult. 54 (1998) 123–130.

236. N.N. Sikuli, K. Demeyer, Influence of ion composition of the medium on alkaloid production by "hairy roots" of *Datura stramonium*, Plant Cell Tiss. Org. Cult. 47 (1997) 261–267.

237. M. Teresa-Pinol, J. Palazon, R.M. Cusido, M. Ribo, Influence of calcium ion-concentration in the medium on tropane alkaloid accumulation in *Datura stramonium* hairy roots, Plant Sci. 141 (1999) 41–49.

238. J.M. Dupraz, P. Christen, I. Kapetanidis, Tropane alkaloid production in *Datura quercifolia* hairy roots, Planta Med. (*Supplement Issue*) 59 (1993) A 659.

239. K. Shimomura, M. Sauerwein, K. Ishimaru, Tropane alkaloids in the adventitious and hairy root cultures of solanaceous plants, Phytochemistry 30 (1991) 2275–2278.

240. N. Sevon, T. Varjonen, R. Hiltunen, M.K. Oksman-Coldentey, Effect of sucrose, nitrogen and copper on the growth and alkaloid production of transformed root cultures of *Hyoscyamus muticus*, Planta Med. 58 (1992) A609–A610.

241. I. Ionkova, L. Witte, A.W. Alfermann, Spectrum of tropane alkaloids in transformed roots of *Datura innoxia* and *Hyoscyamus × gyorffyi* cultivated *in vitro*, Planta Med. 60 (1994) 382–384.

242. T.Yoshikawa and T. Furuya, Saponin production by cultures of *Panax ginseng* transformed with *Agrobacterium rhizogenes*, Plant Cell Rep. 6 (1987) 449–453.

243. Yu N. Zhuravlev, V.P. Bulgakov, L.A. Moroz, N.I. Uvarova, Kov V.V. Makhan, G.V. Malinovskaya, A.A. Artyukuv, G.B. Elyakov, Accumulation of paraxosides in a culture of cells of *Panax ginseng*

C.A. Moy transformed with the aid of *A. rhizogenes*, Doklady, Botanical Sciences, 310–312 (1990) 22–23.

244. V.P. Bulgakov, Yu N. Zhuravlev, M.M. Kozyrenko, E.N. Babkina, N.I. Uvarova, Kov V.V. Makhan, The content of dammaran glycosides in different callus lines of *Panax ginseng* C.A. Mey., Rastitel nye-Resursy, 27 (1991) 94–100.

245. D. Washida, K. Shimomura, Y. Nakjima, M. Takido, S. Kitanaka, Ginsenosides·in hairy roots of a *Panax* hybrid, Phytochemistry 49 (1998) 2331–2335.

246. B.D. Benjamin, G. Roja, M.R. Heble, *Agrobacterium rhizogenes* mediated transformation of *Rauwolfia serpentina*: regeneration and alkaloid synthesis, Plant Cell Tiss. Org. Cult. 35 (1993) 253–257.

247. M.A. Alvarez, J.R. Talou, N.B. Paneiego, A.M. Giulietti, Solasodine production in transformed organ cultures (roots and shoots) of *Solanum eleagnifolium* Car. Biotech. Lett. 16 (1994) 393–396.

248. S. Ray, B. Ghosh, S. Sen, S. Jha, Withanolide production by root cultures of *Withania somnifera* transformed with *A. rhizogenes*, Planta Med. 62 (1996) 571–573.

Plant Biotechnology and Molecular Markers
P.S. Srivastava, Alka Narula and Sheela Srivastava (Editors)
Copyright © 2004 Anamaya Publishers, New Delhi, India

7. Production of Phytochemicals in Plant Cell Bioreactors

Saurabh Chattopadhyay, A.K. Srivastava and V.S. Bisaria

Department of Biochemical Engineering & Biotechnology, Indian Institute of Technology, Delhi,
New Delhi 110016, India

Abstract: Plant cell culture provides a viable alternative over whole plant cultivation for the production of useful phytochemicals. In order to successfully cultivate the plant cells at large scale, some engineering parameters such as cell aggregation, mixing, aeration and shear sensitivity are taken into account for selection of a suitable bioreactor. Increased productivity in a bioreactor can be achieved by selection of a proper cultivation strategy (batch, fed-batch, two-stage, etc.), feeding of metabolic precursors and extraction of intracellular metabolites. Proper understanding and rigorous analysis of these parameters would pave the way towards the successful commercialization of plant cell bioprocesses.

1. Introduction

Higher plants are inexhaustible sources of a wide range of biochemicals such as flavors, fragrances, natural pigments, pesticides and pharmaceuticals. Currently many of these compounds are isolated by solvent extraction from the naturally grown whole plants. This continued destruction of plants has posed a major threat to the plant species getting extinct over the years. Clearly, the development of alternative methods to whole plant extraction for the production of these compounds, especially of medicinal value, is an issue of considerable socio-economic importance. These factors have generated considerable interest in the use of plant cell culture technologies for the production of phytochemicals [1]. In plant cell culture, the isolated cells from the whole plant (or parts derived thereof) are cultivated under appropriate physiological conditions and the desired product is extracted from the cultured cells. The recent developments in plant tissue culture techniques and their processing have shown promising results to improve the productivity by many folds.

2. Cell Suspension Cultures

The first step in plant tissue culture is to develop a callus culture from the whole plant. A callus can be obtained from any portion of the whole plant containing dividing cells. To maximize the formation of a particular compound, it is desirable to initiate the callus from the plant part that is known to be a high producer. However, from an engineering perspective, cell suspension cultures have more immediate potential for industrial application than plant tissue and organ cultures, due to extensive expertise which has been amassed for submerged microbial cultures. While tissue and root cultures offer genetic stability as well as, in some instances, superior metabolic performances over suspension cultures of the cell lines, the development of appropriate bioreactors and operating techniques for these systems involve high investment and laborious experimentation [2]. Accordingly, most of the research efforts have been directed towards commercialization of plant cell suspension cultures. A suspension culture is developed by transferring the relatively friable portion of a callus into liquid medium and is maintained under suitable conditions of aeration, agitation, light, temperature and other physical parameters. However, various strategies may have to be adopted to obtain a fairly homogeneous suspension culture.

Mitsui Petrochemical Industry, Japan was the first to produce shikonin (a dyestuff) on commercial scale. While the large-scale cultivation of plant cell suspension cultures is desirable for industrial production of plant-derived biochemicals, the production technology comparable to that used for microbial systems needs to be further developed. Although the basic equipment- and process-related requirements for suspension cultures of plant cells are similar to those of submerged microbial cultures, some of the features used for microbial cultures are not suitable for plant cell cultures because of striking differences in the nature and growth pattern of the two types of cells. The implications of these differences on culturing of the plant cells are summarized in Table 1 [3]. Certain engineering considerations are normally addressed before embarking on the mass scale propagation of plant cells.

3.   Engineering Considerations

Plant cell suspensions can now be successfully cultivated in bioreactors of various configurations. However, many of the unique properties of plant cells in culture such as sensitivity to shear, slow growth rates, and low oxygen requirements are manifested in complex ways at large scale cultivation. As the scale of operation increases, mixing inside the bioreactor vessel becomes difficult, resulting in non-uniform concentration of the nutrients and limited oxygen transfer to respiring cells. Changes in the rheological nature of the fluid, wall growth, and clumping of cells resulting in sedimentation, lead to suboptimal utilization of the bioreactor. These problems

Table 1.   Differences between the characteristic features of plant and microbial cells and their implications for bioreactor design

Characteristic features of a typical plant cell	Implications for reactor design
Lower respiration rate	Lower oxygen transfer rates required
More shear sensitive	May require operation under low-shear conditions by, for example, employing low-shear impellers and bubble-free aeration
Growth as aggregates	May have mass transfer limitations that limit the availability of nutrients to cells within the aggregates
Aggregation important for secondary metabolism	An optimal aggregate size may be required for product synthesis by manipulation of media constituents and environmental conditions
Volatile compounds (e.g. CO_2 or ethylene) may be important for cell metabolism	May need to sparge gas mixtures containing them
Product synthesis may be non-growth-associated	May require a two-step cultivation system for maximal product synthesis

Adapted from [3].

necessitate a more rigorous analysis of bioreactors to be used for the large scale cultivation of plant cells for metabolite production [4–7].

3.1   Aggregation

Plant cells are significantly larger and slower growing cells than most microbial organisms. Aggregation is common, largely due to failure of the cells to separate after division, although the secretion of extracellular polysaccharides, particularly in the later stages of growth, may contribute to increased adhesion. This tendency of the plant cells to grow in clumps results in sedimentation, insufficient mixing and diffusion-limited biochemical reaction. This so-called cell-cell contact is desirable for the biosynthesis of many secondary metabolites by the plant cells. Hence controlled aggregation of plant cells is of interest from process engineering point of view.

3.2   Mixing

Mixing promotes better growth by enhancing the transfer of nutrients from liquid and gaseous phases to cells and the dispersion of air bubbles for effective oxygenation. Although plant cells have higher tensile strength in comparison to microbial cells, their shear sensitivity to hydrodynamic stresses restricts the use of high agitation for efficient mixing. Plant cells are, therefore, often grown in stirred tank bioreactors at very low agitation speeds. Mixing of plant cells grown on a large scale is also hampered by the rheological characteristics of the culture broth [6]. Plant cell suspensions are viscous at high concentrations and behave like non-Newtonian fluids. Non-Newtonian behavior of the culture broth also restricts effective mass and heat transfer inside the bioreactor, leading to non-uniform nutrient concentration and temperature, and the development of dead zones inside the culture vessel. Excretion of polysaccharides at the later stages of cell growth, the extent and nature of which depend on the nature of the plant cells and the carbohydrate source used for growth, also results in a rapid increase in viscosity. Inadequate mixing may lead to clumping of cells, thereby complicating the nature of the reacting system; also the inner cells of the clumps become nutrient deficient, which may have either an adverse or a positive effect on the cell growth and product formation [4]. Adequate mixing can be achieved by proper design of the impeller; helical-ribbon impeller has been reported to enhance mixing at the high density of plant cell suspension cultures [8].

3.3   Oxygen and Aeration Effects

Oxygen requirements of plant cells are comparatively lower than that of microbial cells due to their low growth rates. In some cases, high oxygen concentration is even toxic to the cells' metabolic activities and may strip nutrients such as carbon dioxide from the culture broth. Hence, effective oxygen transfer in plant cell cultures must be carefully analyzed when a bioreactor system is being selected. The intensity of culture broth mixing, the degree of air bubble dispersion, the culture medium's capacity for oxygen, and the hydrodynamic stress inside the culture vessel affect proper aeration of the culture. Effects of aeration on plant cell suspension cultures have focused largely on the influence of $k_L a$, the mass transfer coefficient, in which the aeration and agitation are linked. The $k_L a$ value gives a direct measure of effective oxygenation of culture fluid and helps one choose a suitable bioreactor to cultivate plant cells. Increased viscosity of the culture broth decreases $k_L a$ and signals the need for intensive agitation of the culture for better mixing and oxygen transfer. A balanced analysis of mixing and oxygen transfer as reflected in

$k_L a$ value is, therefore, required to achieve reasonable cell yield and product formation: The effect of initial $k_L a$ on growth and alkaloid production by suspension cultures of *Catharanthus roseus* was studied in 12.5 liter stirred tank bioreactor using either a cross sparger or a sinter sparger, and a 6-bladed Rushton impellor for agitation [9]. It has been observed that, at higher $k_L a$ values, serpentine was produced when the cells were in the log phase, whereas production of serpentine and ajmalicine was maximum at $k_L a$ values of 16 h^{-1} and 4.5 h^{-1}, respectively [9].

High aeration may lead to severe foaming, which has considerable influence on the cell growth and secondary metabolite production [10]. A number of antifoams such as polypropylene glycol 1025 and 2025, Pluronic PE 6100, and Antifoam-C have often been employed to control foaming; however, in some cases this resulted in reduction in cell growth and product formation [11].

3.4 Shear Sensitivity

The sensitivity of plant cells to hydrodynamic stress associated with aeration and agitation can be attributed to the physical characteristics of the suspended cells, viz. their size, the presence of thick cellulose based cell wall, and existence of large vacuoles. Mechanically agitated vessels lead to damaging and breaking the cells through the hydrodynamic stress generated by aeration, agitation, and other operations. The air-lift bioreactor has also been used to achieve better oxygen transfer and good growth. Bubble-free aeration of the culture fluid through a moving membrane provided another suitable alternative for transferring gas without inducing cell damage through shear stress.

The immediate consequence of the shear effect on plant cells is cell damage, which has been quantitatively measured by using a number of system responses such as reduction in cell viability [12], release of intracellular compounds [13], changes in morphology and/or aggregate patterns [14], and changes in metabolism [15]. The effects of hydrodynamic and interfacial stress on plant cell suspension cultures with various modes of quantitative analysis of system response at shake flask as well as bioreactor levels have recently been reviewed by Kieran and co-workers [16].

4. Plant Cell Bioreactors

A suitable bioreactor can be designed for a specific plant cell system from the following considerations [4–7]:

- optimum aeration-agitation with respect to capacity of oxygen supply and intensity of hydrodynamic stress effects on the plant cells.
- intensity of culture broth mixing and air-bubble dispersion.
- control of temperature, pH and nutrient concentration inside the bioreactor.
- control of aggregate size (which may be important to enhance secondary metabolite production).
- maintenance of aseptic conditions for relatively longer cultivation period.

A number of different types of bioreactors (Fig. 1) have been used for mass cultivation of plant cells taking the above considerations into account. Stirred tank bioreactors have been most extensively applied in order to achieve the optimum process parameters. In spite of the fact that stirred tank reactors exert more hydrodynamic stress on plant cells, they have great potential when used with low agitation speed and modified impeller. The first commercial application of large scale cultivation of plant cells was carried out in stirred tank reactors of 200 and 750 liter capacities to produce shikonin by cell cultures of *Lithospermum erythrorhizon* [3]. Cells of *Catharanthus roseus* [17], *Digitalis lanata* [18], *Panax notoginseng* [19], *Taxus baccata* [20] and *Podophyllum hexandrum* [21] have been cultured in stirred tank bioreactors with suitable modifications for production of phytochemicals. Another type of reactor known as bubble column reactor has also been used for large scale cultivation of plant cells. The major advantages of this reactor are the absence of moving parts and ease of maintaining sterile environment, as no sealing parts are required. *Cudrania tricuspidata*, being highly shear sensitive, was cultivated in bubble column reactor [22]. A modification of bubble column reactor 'balloon type bubble bioreactor' has been recently adopted for the production of taxol by *Taxus cuspidata* [23]. The major disadvantage of this reactor is insufficient mixing. A reactor having more uniform flow pattern with slight modification of stirred tank reactor (a draught tube is inserted instead of the impeller) is air-lift bioreactor. The cells of *Catharanthus roseus* [24], *Digitalis lanata* [25], *Cudrania tricuspidata* [22, 26], *Lithospermum erythrorhizon* [26] and *Taxus chinensis* [27] have been successfully cultivated in air-lift bioreactors for production of secondary metabolites. The major disadvantages of this reactor are the development of dead zones inside the bioreactor, insufficient mixing at high cell densities and rupture of cells due to collision between air bubbles

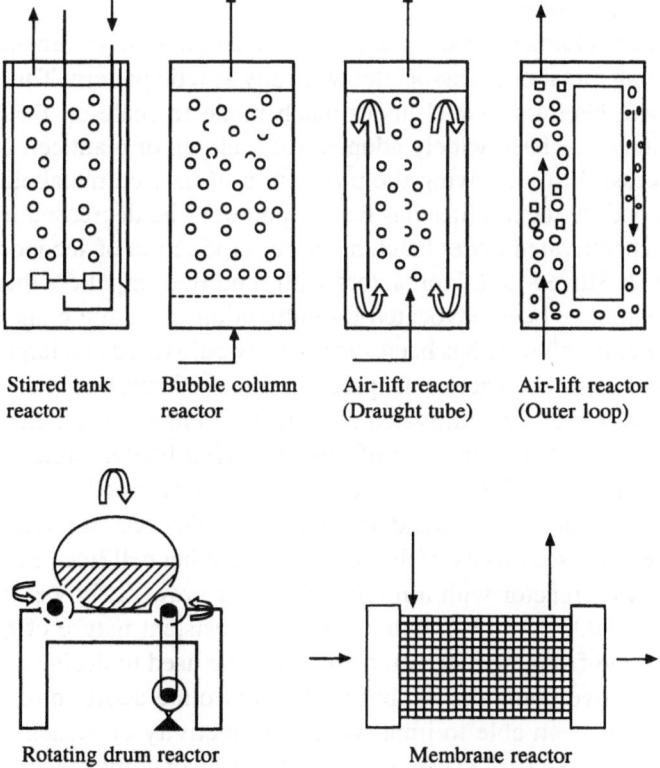

Fig. 1. Configurations of different bioreactors used for plant cell cultivation.

and the cells. Another type of reactor used in plant cell cultivation is rotating drum reactor, which has higher oxygen transfer ability and relatively lower hydrodynamic stress. This consists of a horizontally rotating drum on rollers connected to a motor. Rotary drum reactor has been shown to be superior over other reactors for the cultivation of *Vinca rosea* [28] and *Lithospermum erythrorhizon* [26]. The performance of the different bioreactors has been summarized by Panda and co-workers [4].

5.   Process Strategies

5.1   Selection of Cultivation Techniques

Various modes for culturing plant cells have been employed in suspension culture in order to maximize product formation. The fed-batch mode is used in cases where the addition of a high concentration of substrate affects the growth. The technique of repeated batch cultivation (semi-continuous mode) provides an appropriate approach towards the continuous cultivation of plant cells when the rate of product synthesis (or biotransformation of added precursors) parallels the rate of growth. For non-growth-associated products, the use of a two-stage culture, where cells are propagated in a growth medium and then transferred to a production medium, would be the ideal choice for maximizing product synthesis. Obviously, it would be important to recognize the best physiological state of the cell for maximal product accumulation. Once the type of bioreactor is selected for a specific plant cell process, the mode of operation will depend on the dynamics of the specific culture [29].

Batch cultivations are characterized by constantly changing environmental conditions, and are capable of producing metabolites associated with any kinetic pattern. Therefore, a number of plant cell systems have been cultivated under batch mode to scale-up the process. Although batch cultivation strategy has been widely adopted for scale-up of plant cell bioprocesses, it has not always been successful in improving the production of desired metabolites; in many cases the production of secondary metabolites has been reported to be decreased in scale-up process. A variety of plant cells cultivated under batch mode for production of secondary metabolites are summarized in Table 2. Stirred-tank bioreactors with modified impellers that impart improved mixing under low shear have been advocated for cultivation of fragile plant cells in large scale suspension cultures. *Panax ginseng* has been successfully cultivated at a large scale in 2000 liter and 20000 liter stirred tank bioreactors to produce 500–700 mg/l of ginseng saponins [30]. *Panax ginseng* cell lines have been cultivated in both stirred tank and air-lift bioreactors for the production of gingenoside. Different types of impellers (flat-blade, angled-blade disc turbine, anchor impeller) at various impeller speeds have been used for the cell growth of *Panax ginseng* and it has been observed that angle-blade disc impeller at 100–150 rpm resulted in highest cell growth, indicating the shear sensitivity of the cells [31]. Another cell line of *P. ginseng* cultivated in a 2 liter stirred tank bioreactor with a marine propeller grew fairly well up to an unusually high agitation speed of 1000 rpm [31], indicating shear-resistant nature of the cell line.

The conditions derived from the batch cultivation can be used to design suitable fed-batch or continuous operation to overcome the inhibitions by controlled addition of a limiting nutrient. Fed-batch cultivation has been able to improve the productivity of ginseng by *Panax ginseng* [32], and taxane by *Taxus chinensis* [33]. Fed-batch cultivation of *Coptis japonica* had a significant effect on production of berberine at high cell density, as batch cultivation in stirred tank bioreactor

Table 2. Production of secondary metabolites by plant cell suspension cultures under different modes of cultivation

Plant cell	Product	Bioreactor type, capacity and mode of cultivation	Product (mg/l)	Reference
Anchusa officinalis	Rosmarinic acid	Stirred tank bioreactor, 2.5 liter, batch	3500	[36]
Aralia cordata	Anthocyanin	Jar culture vessel, 500 liter, continuous	1090	[37]
Catharanthus roseus	Ajmalicine	Air-lift bioreactor,	6.4	[38]
	Catharanthine	20 liter, batch	3	[38]
	Serpentine		1.6	[38]
	Tryptamine		16.1	[38]
Coptis japoinca	Berberine	Stirred tank bioreactor, 2.5 liter		
		batch	800	[34]
		Fed-batch	2320	[34]
		Continuous	3500	[34]
Holarrhena antidysenterica	Conessine	Stirred tank bioreactor, 6 liter, batch	106	[39]
Lithospermum erythrorhizon	Shikonin	Stirred tank bioreactor, 200 and 750 liter, two-stage culture	4000	[40]
Nicotiana tabacum	*Cinnamoyl putrescines*	Stirred tank bioreactor		
		batch	160	[35]
		Fed-batch	400	[35]
Panax notoginseng	*Ginseng saponin*	Centrifugal impeller bioreactor, 2.5 liter, batch	800	[41]
		Turbine bioreactor, 2.5 liter, batch	490	[41]
		Air-lift bioreactor, 1 liter, batch	3120	[42]
		Erlenmeyer flask, 0.25 liter, batch	1570	[43]
Perilla frutescens	Anthocyanin	Erlenmeyer flask, 0.5 liter	5800	[44]
Podophyllum hexandrum	Podophyllotoxin	Stirred tank bioreactor, 3 liter		
		batch	13.8	Authors' work
		Fed-batch (intermittent feeding)	43.2	Authors' work
		Continuous with cell retention	48.8	Authors' work
Taxus chinensis	Taxane	Erlenmeyer flask, 0.25 liter	274.4	[33]
Taxus cuspidata	Taxol	Wilson type bioreactor	22	[45]

damaged the cells due to high osmotic pressure of the culture medium. Further, the biomass concentration was reduced due to accumulation of inhibitory products during cell growth. This problem was resolved by suitable fed-batch cultivation, which enhanced both cell growth and berberine production [34]. An increased production of cinnamoyl putrescines has also been observed by fed-batch cultivation of *Nicotiana tabacum* in stirred tank bioreactor [35]. Production of podophyllotoxin has been enhanced to 43.2 mg/l by fed-batch cultivation of *Podophyllum hexandrum* in stirred tank bioreactor as compared to 13.8 mg/l in batch cultivation (Table 2).

Steady state continuous flow or chemostat operation, with a constant withdrawal of culture medium and cells is commonly used for the production of growth-associated products, typically primary metabolites and biomass. It also provides a system to eliminate product inhibition, if any. The continuous culture technique has also been adopted for the cultivation of several plant cells such as, *Coptis japonica* [34], *Catharanthus roseus* [46], and *Nicotiana tabacum* [47]. A high cell density of *Coptis japonica* produced 3500 mg/l berberine when cultivated in continuous mode in 2.5 l stirred tank bioreactor [34]. However, the cellular content of berberine in continuous culture decreased to less than 50% of that observed in batch cultivation because the production of berberine in *C. japonica* was a part of the non-growth-associated kinetics (Table 2).

Cell retention systems have been occasionally employed for the enhancement of growth and product yield in various microbial systems for their ability to achieve high cell density in continuous cultivation. *In situ* cell retention systems have been particularly successful in improving the productivity of product inhibited cultivations, mainly because the bioreactor could be operated at high dilution rates to flush out inhibitory products and at the same time the cells could be retained by the filtration device [48]. Spin filter device has been applied for the somatic embryogenesis of plant cell suspension cultures and for industrial plant propagation [49]. *Podophyllum hexandrum* has been cultivated in stirred tank bioreactor in continuous mode using cell retention device and this further enhanced the production of podophyllotoxin to 48.8 mg/l (authors' work) (Table 2).

5.2 Precursor Feeding

Precursors of biosynthetic pathways have been used in various plant cell suspension cultures to improve the production of secondary metabolites. Factors such as the concentration and the time of addition of the precursor are to be considered when applying the precursor to the cell culture medium. The addition of loganin, tryptophan and tryptamine enhanced the production of secologanin [50], and indole alkaloids [51] by *Catharanthus roseus* suspension cultures. Paclitaxel yields in the cell culture of *Taxus cuspidata* were improved up to six times by feeding phenylalanine and other potential paclitaxel side-chain precursors (e.g. benzoic acid, N-benzoylglycine and serine) [52]. Cholesterol, a precursor of alkaloid biosynthesis, was found to have a strong effect on the production of conessine by *Holarrhena antidysenterica* cell suspension culture [39]. The time of addition of cholesterol as well as its concentration had a significant effect on alkaloid synthesis. A step feeding strategy, in which 50 mg/l cholesterol was added in 4 installments during different phases of growth, enhanced the production of the alkaloid from 63 mg/l to 106 mg/l in 6 liter stirred tank bioreactor; this study highlighted the importance of the physiological state of the culture for effective transformation of the precursor to alkaloid [39].

5.3 Permeabilization of Plant Cells

Plant secondary metabolites are normally produced intracellularly which adds up to the cost of downstream processing of a specific product. It is, therefore, desirable to extract the products into the culture medium. Removal of secondary metabolites from the vacuoles of the cells would also reduce the product inhibition and increase the productivity. Many attempts have been made to permeabilize the plant cell membranes in a reversible manner with organic solvents. Dimethylsulfoxide (DMSO) has been used in many cases, because it is known to extract sterols from the membranes of the eukaryotic cells. Of various cells tested, only *Catharanthus roseus* survived the treatment of DMSO [53]. Taxol has recently been extracted by various organic solvents such as hexadecane, decanol and dibutylphthalate, in the range of 5–20% (v/v), in the culture medium of *Taxus chinensis* [54]. Selection of a specific solvent system with due consideration to its effect on cell growth may lead to substantial release and increase in the production of secondary metabolites.

6. Future Prospects

Plant cell cultivation is a suitable alternative to whole plant cultivation for the production of desired compounds. However, due attention must be given to the relevant engineering parameters influencing cell growth and secondary metabolite production. The inherent difficulties associated with *in vitro* plant cell cultivation, e.g., genetic variation of plant cell lines, sensitivity to shear stress, complex regulatory mechanism etc. are to be properly addressed for a specific cell line. Design of a suitable bioreactor with low-shear impeller, and selection of an appropriate mode of cultivation is required for increased metabolite production. Selection of suitable metabolic precursors, extraction of intracellular metabolites by organic solvents can also lead to significant enhancement in productivity of secondary metabolites.

The third author Prof. V.S. Bisaria along with Prof. Saroj Mishra and Dr. A.K. Panda of National Institute of Immunology, New Delhi, initiated a collaborative project on production of alkaloids by cell cultures of *Holarrhena antidysenterica*, a plant growing in the lawns of Department of Botany, University of Delhi, Delhi, with the cooperation of Prof. S.S. Bhojwani about 15 years ago. Since then the activity at IIT Delhi has increased many folds with substantial inputs of modeling and simulation expertise of the second author Dr. A.K. Srivastava, into the cultivation strategies of plant cells in bioreactors. During the period of association, the authors were greatly benefited by the vast expertise of Prof. Bhojwani. The authors, therefore, feel honored to dedicate this article to Prof. Bhojwani at the time of his superannuation from a very fulfilling and academically rewarding experience.

References

1. F. DiCosmo, P.J. Facchini, M.M. Kraml, Cultured plant cells—the chemical factory within, Chemistry in Britain 25 (1989) 1001–1004.
2. H.E. Flores, W.R. Curtis, Approaches to understanding and manipulating the biosynthetic potential of plant roots, Ann. New York Acad. Sci. 665 (1992) 188–209.
3. G.F. Payne, M.L. Shuler, P. Brodelius, Plant cell culture, in: B. K. Lydensen (Ed), Large Scale Cell Culture Technology, Hanser Publishers, New York, 1987, pp. 193–229.

4.   A.K. Panda, S. Mishra, V.S. Bisaria, S.S. Bhojwani, Plant cell reactors—a perspective, Enz. Microbial Technol. 11 (1989) 386–397.

5.   V.S. Bisaria, A.K. Panda, Large scale plant cell culture methods, applications and products, Curr. Opin. Biotechnol. 2 (1991) 370–374.

6.   A.H. Scragg, The problems associated with high biomass levels in plant cell suspensions, Plant Cell Tissue Organ Cult. 43 (1995) 163–170.

7.   P.M. Kieran, P.F. MacLoughlin, D.M. Malone, Plant cell suspension cultures: some engineering considerations, J. Biotechnol. 59 (1997) 39–52.

8.   M. Jolicoeur, C. Chavarre, P.J. Carreau, J. Archambault, Development of helical-ribbon impeller bioreactor for high density plant cell suspension culture, Biotechnol. Bioeng. 39 (1992) 511–521.

9.   F. Leckie, A.H. Scragg, K.C. Cliffe, An investigation into the role of initial $k_L a$ on the growth and alkaloid accumulation by cultures of *Catharanthus roseus*, Biotechnol. Bioeng. 37 (1991) 364–370.

10. J.J. Zhong, T. Seki, S. Kinoshita, T. Yoshida, Effects of surfactants on cell growth and pigment production in suspension cultures of *Perilla frutescens*, World J. Microbiol. Biotechnol. 8 (1992) 106–109.

11. M. Wongasmuth, P.M. Doran, Foaming and cell floatation in suspended plant cell cultures and the effect of chemical antifoams, Biotechnol. Bioeng. 44 (1994) 481–488.

12. A.H Scragg, E.J. Allan, F. Leckie, Effect of shear on the viability of plant cell suspensions, Enz. Microbial Technol. 10 (1988) 361–367.

13. J.J. Meijer, H.J.G. ten Hoopen, K.C.A.M. Luyben, K.R. Libbenga, Effects of hydrodynamic stress on cultured plant cells: a literature survey, Enz. Microbial Technol. 15 (1993) 234–238.

14. P.M. Kieran, H.J. O'Donnell, D.M. Malone, P.F. MacLoughlin, Fluid shear effects on suspension cultures of *Morinda citrifolia*, Biotechnol. Bioeng. 45 (1995) 415–425.

15. J.J. Zhong, K. Fujiyama, T. Seki, T. Yoshida, A quantitative analysis of shear effects on cell suspension and cell cultures of *Perilla frutescens* in bioreactors, Biotechnol. Bioeng. 44 (1994) 649–654.

16. P.M. Kieran, D.M. Malone, P.F. MacLoughlin, Effect of hydrodynamic and interfacial forces on plant cell systems, in: T. Scheper, K. Schugerl, G. Kretzmer (Eds), Adv. Biochem. Eng./Biotechnol., Vol 67, Springer-Verlag, Berlin, 2000, pp. 139–177.

17. H.J.G. ten Hoopen, W.M. van Gulik, J.E. Schlatman, P.R.H. Moreno, J. Vinke, J.J. Heijnen, R. Verpoorte, Ajmalicine production by cell cultures of *Catharanthus roseus*: from shake flask to bioreactor, Plant Cell Tissue Organ Cult. 38 (1994) 85–91.

18. D. Fulzele, W. Kreis, E. Reinhard, Cardenolide biotransformation by cultured *Digitalis lanata* cells: semi-continuous cell growth and production of deacetyllanatoside-C in a 40 liter stirred tank bioreactor, Planta Med. 58 (1992) A601–A602.

19. J.J. Zhong, F. Chen, W.W. Hu, High density cultivation of *Panax notoginseng* cells in stirred bioreactors for the production of ginseng biomass and ginseng saponin, Process Biochem. 35 (2000) 491–496.

20. V. Srinivasan, L. Pestchanker, S. Moser, T.J. Hirasuma, R.A. Taticek, M.L. Shuler, Taxol production in bioreactors: kinetics of biomass accumulation, nutrient uptake, and taxol production by cell suspensions of *Taxus baccata*, Biotechnol. Bioeng. 47 (1995) 666–676.

21. S. Chattopadhyay, A.K. Srivastava, S.S. Bhojwani, V.S. Bisaria, Production of podophyllotoxin by plant cell cultures of *Podophyllum hexandrum* in bioreactor, J. Biosci. Bioeng. 93 (2002) 215–220.

22. H. Tanaka, Technological problems in cultivation of plant cells at high density, Biotechnol. Bioeng. 67 (2000) 1203–1218.

23. S.H. Son, S.M. Choi, Y.H. Lee, K.B. Choi, S.R. Yun, J.K. Kim, H.J. Park, O.W. Kwon, E.W. Noh, J.H. Seon, Y.J. Park, Large-scale growth and taxane production in cell cultures of *Taxus cuspidata* (Japanese Yew) using a novel bioreactor, Plant Cell Rep. 19 (2000) 628–638.

24. J. Zhao, W.H. Zhu, Q. Hu, Enhanced catharanthine production in *Catharanthus roseus* cell cultures by combined elicitor treatment in shake flasks and bioreactors, Enz. Microbial Technol. 28 (2001) 673–681.

25. E. Reinhard, W. Kreis, U. Barthlen, U. Helmbold, Semicontinuous cultivation of *Digitalis lanata* cells: production of β-methyldigoxin in a 300 liter airlift bioreactor, Biotechnol. Bioeng. 34 (1989) 502–508.

26. H. Tanaka, Large-scale cultivation of plant cells at high density: a review, Process Biochem. 22 (1987) 106–113.

27. Z.W. Pan, H.Q. Wang, J.J. Zhong, Scale-up study of suspension culture of *Taxus chinensis* cells for production of taxane diterpene, Enz. Microbial Technol. 27 (2000) 714–723.

28. W.E. Goldstein, L.L. Lasure, M.B. Ingle, Product cost analysis, in: E.J. Staba (Ed), Plant Tissue Culture as a Source of Biochemicals, CRC Press, Boca Raton, 1980, pp. 191–234.

29. L. Sajc, D. Grubisic, G.V. Novakovic, Bioreactors for plant engineering: an outlook for further research, Biochem. Eng. J. 4 (2000) 89–99.

30. T. Furuya, Saponins (ginseng saponins), in: I.K. Vasil (Ed.), Cell Culture and Somatic Cell Genetics of Plants, Vol 5, Academic Press, CA, 1988, pp. 213–134.

31. J. Wu, J.J. Zhong, Production of ginseng and its bioactive components in plant cell culture: current technological and applied aspects, J. Biotechnol. 68 (1999) 89–99.

32. J. Wu, K.P. Ho, Assessment of various carbon sources and nutrient feeding strategies for *Panax ginseng* cell culture, Appl. Biochem. Biotechnol. 82 (1999) 17–26.

33. H.Q. Wang, J.T. Yu, J.J. Zhong, Significant improvement of taxane production in suspension cultures of *Taxus chinensis* by sucrose feeding strategy, Process Biochem. 35 (1999) 479–483.

34. K. Matsubara, S. Kitani, T. Yoshioka, T. Morimoto, Y. Fujita, Y. Yamada, High density culture of *Coptis japonica* cell increases berberine production, J. Chem. Technol. Biotechnol. 46 (1989) 61–69.

35. O. Schiel, J.K. Redecker, G.W. Piehl, J. Lehmann, J. Berlin, Increased formation of cinnamoyl putrescines by fed-batch fermentation of cell suspension cultures of *Nicotiana tabacum*, Plant Cell Rep. 3 (1984) 18–20.

36. W.W. Su, F. Lei, N.P. Kao, High density cultivation of *Anchusa officinalis* in a stirred tank bioreactor with *in situ* filtration, Appl. Microbiol. Biotechnol. 44 (1995) 293–299.

37. Y. Kobayashi, M. Akita, K. Sakamoto, H. Liu, T. Shigeoka, T. Koyano, Large-scale production of anthocyanin by *Aralia cordata* cell suspension cultures, Appl. Microbiol. Biotechnol. 40 (1993) 215–18.

38. J. Zhao, W.H. Zhu, Q. Hu, Enhanced catharanthine production in *Catharanthus roseus* cell cultures by combined elicitor treatment in shake flasks and bioreactors, Enz. Microbial Technol. 28 (2001) 673–681.

39. A.K. Panda, V.S. Bisaria, S. Mishra, Alkaloid production by plant cell cultures of *Holarrhena antidysenterica*: II Effect of precursor feeding and cultivation in stirred tank bioreactor, Biotechnol. Bioeng. 39 (1992) 1052–1057.

40. M. Tabata, Y. Fujita, Production of shikonin by plant cell cultures, in: M. Zaitlin, P. Day, A. Hollaender (Eds), Biotechnology in Plant Science, Relevance to Agriculture in the Eighties, Acedemic Press, San Diego, 1985, pp. 207–218.

41. J.J. Zhong, F. Chen, W.W. Hu, High density cultivation of *Panax notoginseng* cells in stirred tank bioreactors for the production of ginseng biomass and ginseng saponin, Process Biochem. 35 (1999) 491–496.

42. K. Woragidbumrung, P.S. Tang, H. Yao, J. Han, S. Chauvatcharin, J.J. Zhong, Impact of conditioned medium on cell cultures of *Panax notoginseng* in an airlift bioreactor, Process Biochem. 37 (2001) 209–213.

43. Y.H. Zhang, J.J. Zhong, Hyperproduction of ginseng saponin and polysaccharide by high density cultivation of *Panax notoginseng* cells, Enz. Microbial Technol. 21 (1997) 59–63.

44. J.J. Zhong, T. Yoshida, High-density cultivation of *Perilla frutescens* cell suspensions for anthocyanin production: effects of sucrose concentration and inoculum size, Enz. Microbial Technol. 17 (1995) 1073–1079.

45. L.J. Pestchanker, S.C. Roberts, M.L. Shuler, Kinetics of taxol production and nutrient use in suspension cultures of *Taxus cuspidata* in shake flasks and a Wilson-type bioreactor, Enz. Microbial Technol. 19 (1996) 256–260.

46. H.H. Park, S.K. Choi, J.K. Kang, H.Y. Lee, Enhancement of producing catharanthine by suspension growth of *Catharanthus roseus*, Biotechnol. Lett. 12 (1990) 603–608.

47. T. Hashimoto, S. Azechi, S. Sugita, K. Suzuki, Large scale production of tobacco cells by continuous cultivation, in: A. Fujiwara (Ed), Plant Tissue Culture. Tokyo, 1982, pp. 403–404.

48.   V. Goswami, A.K. Srivastava, Propionic acid production in an *in situ* cell retention bioreactor, Appl. Microbiol. Biotechnol. 56 (2001) 676–680.

49.   D. Wheat, R.P. Bondaryk, J. Nystrom, Spin filter bioreactor technology as applied to industrial plant propagation, Hort. Sci. 21 (1986) 819.

50.   A. Contin, R. van der Heijden, R. Verpoorte, Effects of alkaloid precursor feeding and elicitation on the accumulation of secologanin in a *Catharanthus roseus* cell suspension culture, Plant Cell Tissue Organ Cult. 56 (1999) 111–119.

51.   P.R.H. Moreno, R. van der Heijden, R. Verpoorte, Effect of terpenoid precursor feeding and elicitation on formation of indole alkaloids in cell suspension cultures of *Catharanthus roseus*, Plant Cell Rep. 12 (1993) 702–705.

52.   A.G. Fett-Neto, F. DiCosmo, Production of paclitaxel and related taxoids in cell cultures of *Taxus cuspidata:* perspectives for industrial applications, in: F. DiCosmo, M. Misawa (Eds.), Plant Cell Culture Secondary Metabolism Toward Industrial Application, CRC Press, New York, 1996, pp. 139–166.

53.   P.E. Brodelius, Permeabilization of plant cells for release of intracellularly stored products: viability studies, Appl. Microbiol. Biotechnol. 27 (1988) 561–566.

54.   C. Wang, J. Wu, X. Mei, Enhanced taxol production and release in *Taxus chinensis* cell suspension cultures with selected organic solvents and sucrose feeding, Biotechnol. Prog. 17 (2001) 89–94.

Plant Biotechnology and Molecular Markers
P.S. Srivastava, Alka Narula and Sheela Srivastava (Editors)
Copyright © 2004 Anamaya Publishers, New Delhi, India

8. Development of Biotechnology for *Commiphora wightii*: A Potent Source of Natural Hypolipidemic and Hypocholesterolemic Drug

Sandeep Kumar, S.S. Suri, K.C. Sonie and K.G. Ramawat
Laboratory of Bio-Molecular Technology, Department of Botany, M.L. Sukhadia University,
Udaipur 313001, India

Abstract: *Commiphora wightii* has become an endangered species due to its overexploitation for its gum-resin. Guggulsterones present in gum-resin are potent lipid and cholesterol lowering natural agents. Drugs based on these are currently used clinically in India and Europe. The plant is endemic to Indian subcontinent, therefore major contributions on its biology, chemistry, pharmacology and biotechnology have been made by Indian scientists. Biotechnological approaches made for guggulsterone production by cell cultures and for its micropropagation are reviewed.

1. Introduction

Commiphora wightii (Arnott.) Bhandari, commonly known as 'Indian bdellium', or 'guggul', is an important medicinal plant of the herbal heritage of India. For centuries, guggul has been used extensively by Ayurvedic physicians to treat a variety of afflictions, including arthritis, inflammation, bone-fractures, obesity and disorders of lipid metabolism. It provides 'guggul', an oleogum-resin whose medicinal and curative properties are mentioned in the classic Ayurvedic medical text, the Sushruta Samhita 3000 years ago. The plant has become endangered because of over exploitation for its gum-resin, associated with slow growth of the plant, poor seed set and excessive tapping for gum-resin, which causes mortality of the plant. Gum-resin yields guggulsterones effective against high blood cholesterol and lipids.

2. Distribution

Commiphora is widely distributed in tropical regions of Africa, Madagascar and Asia. It is generally distributed in arid regions and is particularly widespread on the Indian side of Thar Desert. In the Indian subcontinent *Commiphora* species occur in Pakistan, Baluchistan and India. Of the total 185 species, only three (*C. wightii, C. stocksii* and *C. berryi*) have been found in India. *C. wightii* occurs in Rajasthan, Gujarat and Maharashtra [1].

3. Biology

Commiphora wightii (Arnott.) Bhandari (syn. *C. mukul, C. roxburghii, Balsamodendron mukul*) belongs to the family Burseraceae. A characteristic feature of the family is the presence of resin-ducts in the parenchymatous bark.

The plant is a shrub reaching 3 m in height with crooked, knotty branches ending in sharp spines. The papery bark peels in flakes from the older parts of the stem, whereas younger parts is pubescent and grandular leaves are trifoliate. The flowers are sessile and single or in groups of 2-3. The fruit, 6-8 mm in diameter is a drupe, which becomes red on ripening. Fruit yield and seed set is low (about 16% in Aravalli ranges). In drier parts it is even lower. The chromosome number of *C. wightii* is $2n = 26$ [1]. Recently, Gupta et al. [2, 3] reported apomictic seed development associated with polyembryony in guggul. Female plants set seeds irrespective of the presence or absence of pollen. Hand pollination experiments and embryological studies have confirmed the occurrence of non-pseudogamous apomixis, nucellar polyembryony and autonomous endosperm formation. It was inferred that apomixis may have a significant role in the speciation of tropical trees. Apomixis may be favoured by natural selection if the population densities are low and distance between individual trees is greater than the permissible cross-pollination range. Multiple sapling formation by the germination of polyembryonic seeds of *C. wightii* has also been observed [4] thereby confirming the multiple embryo formation in seeds. In another study, Gupta et al. [3] described the cause of low seed set in *C. wightii* on the basis of pollen-stigma interaction in the non-pseudogamous apomictic plants. They observed that although pollen grain germinated on stigma, pistil did not support pollen tube growth perhaps due to changed orientation of the cells of transmitting tissue and absence of proteins in the intercellular matrix. This results in poor seed set.

C. *wightii* is an excellent fuel wood and burns even wet due to the presence of resin in the stem. The plant is cut mercilessly by villagers for cooking the food and used with other wet woods to facilitate burning [5]. Due to abovementioned inherent biological and social problems the plant has become an endangered species. An interesting biological property of ecological significance of resin has also been reported. Essential oil constituents of resin have been shown to enhance sexual maturation of immature adults of the desert locust [6].

4.   Chemistry of Gum-Resin

The presence of guggulsterones differentiates *C. wightii* from 184 other *Commiphora* species. Gum-resin obtained from *Boswellia serrata*, another tree from the family Burseraceae and common in the same regions, is also known locally as salai guggul or white guggul, but *B. serrata* does not contain guggulsterones. It is used as an anti-inflammatory drug.

Phytochemical investigation of guggul gum-resin has been carried out by the group of Dev [7, 8]. Guggul (oleogum-resin) of *C. wightii* is a mixture of 38.5% resins, 32.3 % gum, 1.45% volatile oil, 19.5% minerals, 3.2% organic foreign matter and 3.6% other impurities. During the separation of various products from the complex mixture, the neutral fraction was reported to contain ketonic compounds (5.13%). It is this ketonic fraction that contains biologically important active principles of C_{21} or C_{27} steroid, viz. Z-guggulsterol (0.01%), guggulsterol–VI (0.02%), Z-guggulsterone (1.6%), E-guggulsterone (0.4%), guggulsterol-III (0.03%), guggulsterol-I (0.8%), guggulsterol-IV, guggulsterol-V (Fig. 1) and some defence related secretory ketones [9–12]. Oleogum-resin is a complex mixture and needs stepwise separation [7]. Purification involves separating soluble (45%) and insoluble (55%) components with the aid of ethylacetate, alcohol or petroleum-ether. Insoluble fraction is associated with toxic effects while the soluble fraction contains the guggulsterones and other constituents that are thought to impart the hypolipidemic and anti-inflammatory effects. A method of high performance liquid chromatographic (HPLC) separation was proposed. Recently a HPLC method for quantitative determination of E- and Z-guggulsterones in *C. mukul* resin [13], serum [14, 15] and diet supplement [16] has been developed.

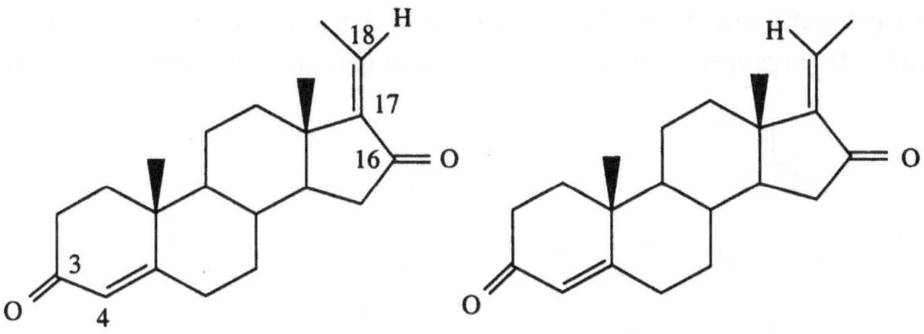

Structure of E- and Z-guggulsterones

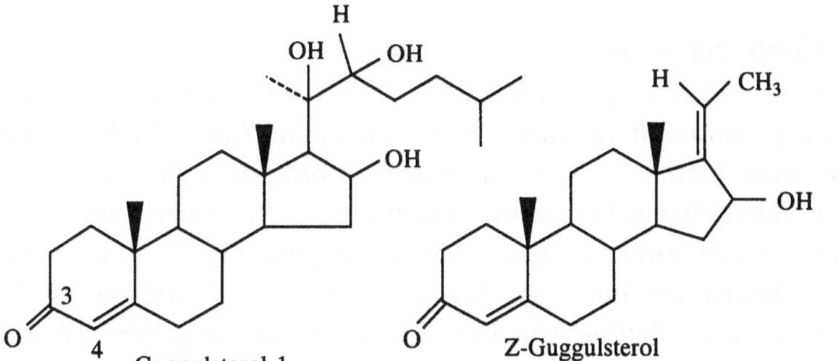

Guggulsterol-1

Z-Guggulsterol

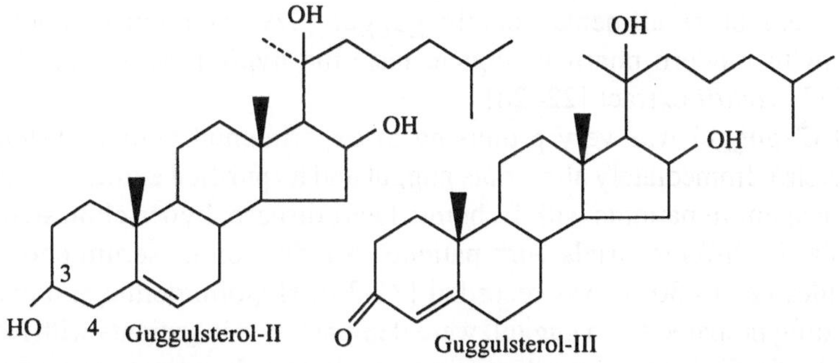

Guggulsterol-II

Guggulsterol-III

Fig. 1.   Structure of guggulsterones and guggulsterols isolated from *C. wightii*.

In addition to these steroids, the gum-resin of *C. wightii* contains diterpenoids (combrene-A and mukulol), steroids derived from pregnane and cholestane, and various carbohydrate derivatives [17]. Upon steam distillation, the gum-resin furnishes an aromatic essential oil. The oil contains the monoterpenes—myrcene, camphorene, polymyrcene and caryophyllene [9]. The aerial parts of *C. wightii* contain β-sitosterol, myricyl alcohol and amino acids [18]. The flowers are rich in flavonoids, most notably quercetin [19].

From the resin of *C. tenuis* growing in Ethiopia, 37 mono- and sesqui-terpenes were detected and identified by GLC and GLC-MS [20]. The main components of the monoterpenoid fraction were α-pinene (60.8%), β-pinene (8.8%), sabinene (6.3%), α-thujene (8.9%), limonene (5.5%), 3-carene (3.7%), β-myrcene (1.8%) and β-elemene (1.1%) constituting 97% of the oil. Identified sesquiterpenoid components constituted approximately 1.6% of the oil (Fig. 2). Oleanolic acid acetate and three other triterpenes were also identified in this oil obtained by wounding the plant.

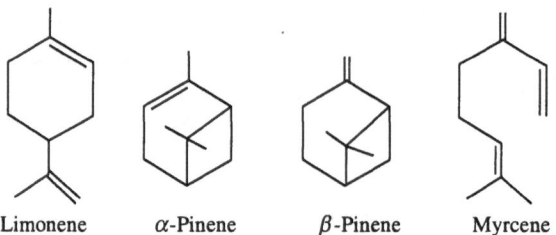

Fig. 2.   **Structure of various monoterpenes isolated from *C. wightii*.**

5.   Medicinal Properties and Pharmacology

Many Indian medicinal plants have come under scientific scrutiny since the middle of the nineteenth century [8]. *Commiphora wightii* is one such plant from which a modern medicine for hyperlipidemia has been prepared based on ancient information. In ancient times, guggul was used primarily as treatment for inflammatory conditions, including arthritis. The development of gum guggul as a potent hypolipidemic agent was first reported by Satyavati working at Banaras Hindu University, Varanasi, leading to the discovery of new anti-cholesterol drug from a plant source [21, 22]. This work was initiated on the basis of information gathered from ancient concept of the pathogenesis of atherosclerosis and obesity described in Sushruta Samhita.

In Ayurveda, guggul is highly valued for the treatment of several ailments including rheumatoid arthritis, lipid disorder and obesity. Currently several formulations of Ayurveda for arthritis, joint pain, sciatica and other ailments contain guggul. Several reports conclusively established scientifically with modern pharmacological tests the hypolipidemic and hypocholesterolemic properties of *C. wightii* extract [22–28].

The effect of guggul was very promising in experimental animal systems [24, 29, 30 and references therein]. Immediately after trials guggul and its purified extract was established effective hypolipidemic agent in patients with ischemic heart disease, hyper-cholesterolemia, obesity and hyperlipidemia. In different trials with patients, a reduction of serum cholesterol (24 to 59%) and triglycerides (22 to 30%) was recorded [31-33]. Hypolipidemic and antioxidant effects of guggulipid, a drug prepared from guggul, were demonstrated in patients with hypercholesterolemia [21, 28, 29, 34, 35]. Guggulipid used as adjunct to dietary therapy decreased the total cholesterol levels by 11.7%, low density lipoprotein (LDL) cholesterol by 12.5%, triglycerides by 12% and the total cholesterol/high density lipoprotein (HDL) cholesterol ratio by 11% [28]. 'Guggulipid', a purified ketonic fraction, is presently used in India and Europe for hyperlipidemia and hypercholesterolemia.

In addition to its lipid lowering activity, guggul may also promote cardiovascular health through its ability to act as an antioxidant and to inhibit platelet aggregation. The guggulsterones inhibited the oxidative modifications of lipid and protein components of LDL induced by copper (Cu) *in vitro* in a concentration dependent manner. Furthermore, guggulsterones also inhibited the formation of hydroxyl (OH$^-$) free radicals created in a non-enzymatic system in a concentration dependent manner [36]. Myocardial necrosis is associated with increased levels of lipid peroxides, xanthine oxidase activity and a lowering of superoxide dismutase, which may lead to increased formation of free radicals with subsequent cardiac cell damage. Guggulsterones, in a manner similar to two other cardioprotective drugs (propranolol and nifedipine), reversed this elevation of lipid peroxides and xanthine oxidase and the decrease in superoxide dismutase activity [37]. The extract of *C. wightii* along with that of *Terminalia arjuna, Inula racemosa* showed protection against isoproterenol induced myocardial necrosis in rats [38] or that with *Allium sativum* and *A. cepa* showed protection against increased cholesterol and blood serum triglycerides, thereby, confering protection against atherosclerosis and myocardial infraction [27]. Mester et al. [39] reported total inhibition of platelet aggregation *in vitro* induced by adenosine diphosphate, serotonin and adrenaline by isolated E- and Z-guggulsterones.

Guggulsterone-Z and guggulsterone-E are responsible for lipid lowering properties in human blood and at least four mechanisms have been proposed to explain their activity. First, guggulsterones might interfere with the formation of lipoproteins by inhibiting the biosynthesis of cholesterol in the liver [40]. Second, guggulsterones have been shown to enhance the uptake of LDL by the liver through stimulation of the LDL receptor binding activity in the membranes of hepatic cells [41]. Third, guggulsterones increase the fecal excretion of bile acids and cholesterol resulting in a low rate of absorption of fat and cholesterol in the intestine [40]. Finally, guggulsterones directly stimulate the thyroid gland [42, 43]. Guggul induced triiodothyronine production with possible involvement of lipid peroxidation, demonstrated thyroid stimulatory effect of guggul administration in the experimental mice [44]. Because serum lipids, including cholesterol, are reduced in response to increased levels of circulating thyroid hormones, the effect of guggulsterones on the thyroid gland might explain the hypolipidemic activity and weight loss property of guggul.

The anti-inflammatory effect of guggul from *C. wightii* on osteoarthritis [21, 45–50] has also been established. Anti-inflammatory effect is common in other plants of the family Burseraceae, viz., *Boswellia dalzielli, B. carteri, B. serrata* and *C. incisa* [48, 51] and references therein).

The extract of *C. molmol*, another species from middle east, possesses anti-thrombosis activity [52], anti-ulcer and cyto-protective property [53], anti-inflammatory effect [54] and cytotoxic and anti carcinogenic effect [55, 56]. On the basis of non-mutagenic, antioxidative and cytotoxic potential of *C. molmol* extract, its use in cancer therapy was recommended [55]. Similarly, sesquiterpenes responsible for hypoglycemic activity were isolated from *C. myrrha* [57].

Dev [7] reported that the steroid profile of *C. wightii* parallels the catabolism of cholesterol to C_{21} steroids. Thus, there is sufficient evidence that both in mammalian tissues and in plants, the catabolism of cholesterol to pregnane derivatives proceeds by either of the two major pathways as shown in Fig. 3. It is because of this property and increased demand for the natural product for hyperlipidemia, the plant has attained great importance in recent years [8].

6. Gum-Resin Production

In *C. wightii* the balsam (oleogum-resin) is present in 'balsam canals' in the phloem of larger veins of the leaf and in the soft base of the stem. The development and widening of gum-resin canal in young stem occurs schizogenously. The lumen of canal is surrounded by an epithelial layer of parenchyma containing dense cytoplasm and shows the presence of gum and resin droplets [58]. The walls of epithelial cells facing the lumen are thin and of fibrillar mesh. The resinous material is synthesized within the epithelial cells and is presumably transported into the canal lumen through the relatively porous wall [59]. Among various plant growth regulators applied on stem with lanolin paste, only kinetin increased the lumen size, while auxin and morphactin had adverse effect causing increase in number of epithelial cells [60].

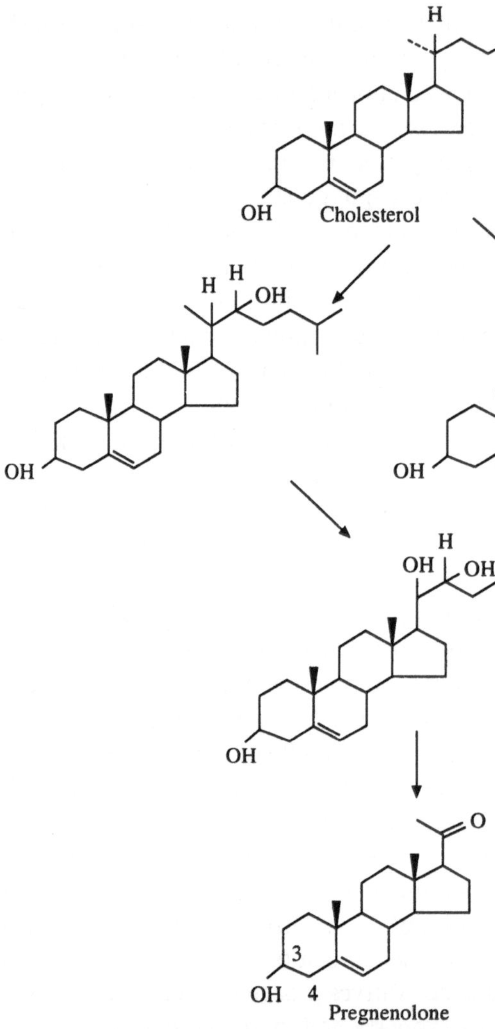

Fig. 3.   Metabolism of cholesterol to pregnenolone.

Gum is tapped in the winter season. Plants over 5 years old with a basal diameter more than 7.5 cm are suitable. Circular incisions of 1.5 cm deep are made on the main branches and stem at a uniform distance of 30 cm apart and at an angle of 60° with the stem. The yellow, fragrant latex oozes out through the incisions and slowly solidifies into vermicular or stalactitic pieces which are collected manually. Subsequent collections of gum-resin are made at an intervals of 10–15 days. About 200–500 g dry guggul is usually obtained from a plant in one season. Application of ethephon on the cuts enhances guggul production 22 times over that obtained in control. This technique developed by Bhatt et al. [61] is inexpensive, safe and requires no skills, and hence can be used by the tribals very easily. They established that guggul production is maximum with the onset of summer (a stress induced secondary product formation) as supported by observations with bright field and fluorescence microscopy. But in the long term, excessive production through ethephon application exhausts the plant and resultantly, kills the plant.

7. Vegetative Propagation

Fruit set and yield of fruits per plant are very low in natural conditions. Poor seed set, poor seed viability and harsh arid conditions are responsible for complete failure of plant establishment in nature from seed. Plants bear fruits in April to May and August to October. About 27% fruits contained single embryo and 7% fruits contained 2 embryos while 66% fruits were without embryo [62]. Therefore, attempts were made to propagate the material by conventional methods of stem cuttings and attempts are being made to propagate the plants through non-conventional biotechnological methods.

7.1 Conventional Methods

Rooting of stem cuttings has its own drawbacks in the arid environment like termites attack, desiccation and heat adversely affecting rooting. Rooting response of stem cuttings was shown to be improved by application of plant growth regulators [63, 64], by selecting cuttings of suitable length and diameter [65] and treating them with potassium salts [66]. However, such methods are not suitable for large-scale multiplication as stock material with sufficient biomass is not available as well as % response of the cutting is variable and affected by seasons [1].

7.2 Biotechnological Methods

Biotechnological research on *C. wightii* has been supported by central funding agencies since 1979 but nothing concrete has come out of these programmes which shows the difficult nature of the material. In nature, the plant is a very slow growing woody shrub. Explants obtained from the mature plants (stem, leaf or petiole) produce fast growing, white and amorphous callus on MS medium containing kinetin and 2,4-dichlorophenoxy acetic acid (2,4-D). Resin exudation from the explants makes the process of sterilization difficult. It is equally difficult to find tender stem explants for *in vitro* growth. Due to these reasons detailed investigations using explant as source material are hampered [67].

7.2.1 Clonal Propagation

Clonal propagation as a biotechnological approach is commonly applied for vegetative propagation of selected materials. Barve and Mehta [68] described a method for clonal propagation of *C. wightii* using stem explants grown on Murashige and Skoog medium [69] containing benzyladenine (BA, 4.0 mgl^{-1}), kinetin (4.0 mgl^{-1}), glutamine 100 mgl^{-1}, thiamine HCl 10 mgl^{-1} and activated charcoal 0.3%. Shoots obtained from explants were incubated to elongate on medium containing lower concentration of BA (0.40 mgl^{-1}) and kinetin (0.4 mgl^{-1}). These elongated shoots were rooted by treating them with IAA and IBA for 24 h in dark and then transferred onto low salt basal medium with activated charcoal. Six-week-old plants (5–6 cm in height) from half strength White's modified medium were used in transplantation. At hardening stage 60% of the transferred plants survived. Micropropagated *C. wightii* plants once established in soil showed vigorous and uniform growth with no morphological abnormalities. The lack of selected high yielding plants and limited number of plants produced by this method are limiting factors for use of this technique for large scale multiplication.

7.2.2 Somatic Embryogenesis

Somatic embryogenesis in callus cultures of *C. wightii* has been achieved. Somatic embryo formation was first observed in immature zygotic explants or intact ovules transferred on B5 medium [62]. Though the frequency of explants producing embryonic culture was low, immature zygotic embryos were the only suitable explants to produce embryonic callus after reciprocal transfers between B5 medium [70] containing 0.1 mgl^{-1} 2,4,5-trichlorophenoxy acetic acid and 0.1 mgl^{-1} kinetin and that devoid of it. All other media failed to produce embryonic callus. Further, somatic embryogenesis in callus obtained from immature zygotic embryos was possible because of selection of embryonic cells. Embryonic cells were small, densely filled with cytoplasm and isodiametric (Fig. 4) as compared to non-embryonic cells, which were large, elongated and vacuolated [67]. Maximum growth of embryonic callus was recorded on MS-2 medium supplemented with 0.25 mgl^{-1} BA and 0.1 mgl^{-1} IBA. MS-2 salts supported higher growth of callus as compared to tissues grown on B5 medium containing same concentrations of plant growth regulators.

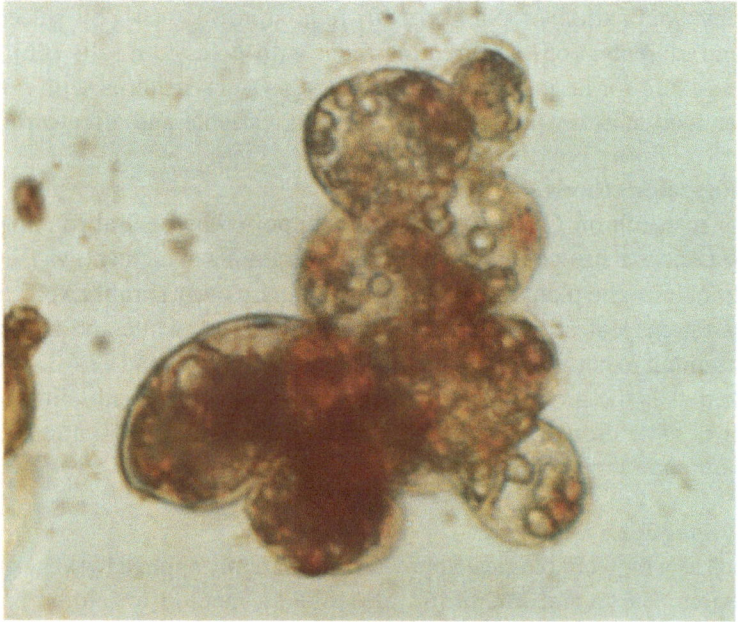

Fig. 4. Embryonic cells of *C. wightii*.

Embryonic callus transferred on MS-2 medium containing various combinations of IAA and BA produced globular, torpedo and a few early cytoledonary stage embryos (Fig. 5). Maximum number of somatic embryos was observed on the medium containing 0.1 mgl^{-1} IAA and 0.25 mgl^{-1} BA. Torpedo and cotyledonary stage embryos obtained from experiments for development of somatic embryos of previous experiments were used. Activated charcoal, ABA, and agar-agar were incorporated in MS-2 medium to generate stress and enhance maturation in somatic embryos. Maximum number of cotyledonary stage embryos were produced on the medium containing 0.5 gl^{-1} activated charcoal and 10 gl^{-1} sucrose. In this experiment embryos were placed on filter

Fig. 5.   Somatic embryos formed from embryonic callus of *C. wightii*.

paper-bridge using liquid medium. Cotyledonary stage somatic embryos kept on various maturation media were transferred onto MS-2HF medium. Somatic embryos grown on media containing plant growth regulators, irrespective of their concentration and combination, produced callus. A high percentage of embryos remained ungerminated, while about 10–25% produced secondary somatic embryos. Therefore, proper selection of mature embryos was required for high percentage of germination. Somatic embryos showed precocious germination and callusing except those grown on MS-2HF medium, which could be maintained for several months. By using static medium or liquid medium with filter paper bridge, about 25% torpedo staged embryos matured into cotyledonary stage embryos and out of these about 25% were converted into plantlets. Thickening of hypocotyl was observed in such plantlets without elongation of internodes (Fig. 6). MS-2 medium containing 20 μg1^{-1} gibberellic acid was most effective for shoot elongation in such plantlets. About 200 plantlets were successfully established in garden soil (Fig. 7) to verify the survivability of regenerants. Survivability was 95% for the plantlets [71–75].

The embryo formation from zygotic embryo and ovule explants may be a case of induced polyembryony as already reported in this plant [2, 4]. This achievement opened new avenues of research on *C. wightii* like cell culture and embryogenesis in bioreactor, formation of resin canals *in vitro* and production of guggulsterones in these organized cultures.

Fig. 6.   Young planlets formed after germination of somatic embryos.

7.3   Guggulsterone Production

Unavailability of sufficient guggul from natural sources and destruction of plants from most of the localities, initiated the search for alternative methods of guggulsterone production. Cell culture is an excellent alternative to produce secondary metabolites. Cell suspension cultures of *C. wightii* were derived from leaf callus in MS medium containing 0.15 mg l^{-1} each of 2,4-D and kinetin. Cells were immobilized in calcium alginate beads. The immobilized cells in stationary phase suspension cultures were less viable but they were active in the synthesis of guggulsterols. Guggulsterol production was 1.97% in stem explants, 0.22% in callus culture (2-month-old) and 0.32% in cell suspension culture (25-day-old) [76]. Guggulsterone is produced in resin canals and hence unorganized cultures proved unproductive for this purpose. However, use of organized cultures, *in vitro* produced embryos and hypocotyls may prove better sources and are being evaluated.

8.   Prospects and Research Need

Commiphora wightii has become an endangered plant species because of high demand for its gum-resin. This plant has become a torchbearer of efficacy of plants used in the Indian system of medicine for treatment of complex human syndrome. Therefore, many more plants are being reinvestigated for their properties mentioned in the ancient system of medicine. The complexity of molecules in mixture warrants their production by natural sources using biotechnological methods.

Development of somatic cell cultures in static and liquid medium opened new avenues of research on this material because large quantities of aseptic material in organized form can be

Fig. 7.   Potted plantlets of *C. wightii*, about 4 months old.

obtained. It is known that the organized material produces several-folds higher amount of active principle as compared to unorganized cultures. This can effectively be used for immobilization of embryos, germinated seedlings, seedling parts, and so on. All these are available in aseptic condition in a material which is otherwise difficult to sterilize in large quantities due to presence of resin and hence sticky nature of the explants. Large amount of aseptic material is required for bioreactor culture and failure of the system due to contamination adds a lot to the cost factor of running the system. Besides, the advantages of mass propagation and development of artificial seeds are evident from the results and need not to be emphasized again (Fig. 8). The other new approach is the development of hairy root culture system using *Agrobacterium rhizogenes* for the production of active principle again on the lines as described above for the organized cultures.

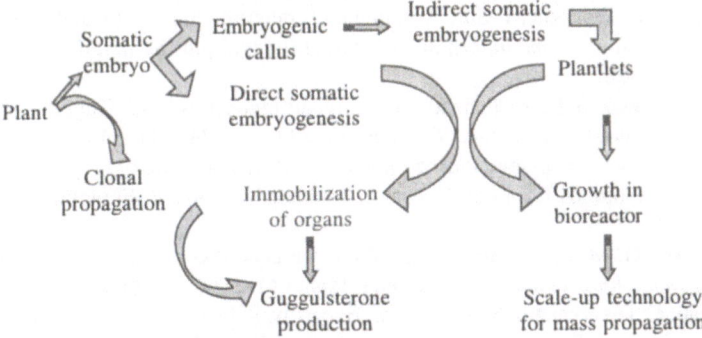

Fig. 8.   Schematic presentation of approaches for guggulsterones production and mass propagation of plants

Acknowledgement

This work was supported by grants from Department of Biotechnology, Government of India, New Delhi (grant No.BT/R&D/08/23/95) to K.G. Ramawat.

References

1. S. Kumar, V. Shankar, Medicinal plants of the Indian desert: *Commiphora wightii* (Arnott.) Bhand. J. Arid Environ. 5 (1982) 1–11.
2. P. Gupta, K.R. Shivanna, H.Y. MohanRam, Apomixis and polyembryony in the guggul plant, *Commiphora wightii*. Ann. Bot. 78 (1996) 67–72.
3. P. Gupta, K.R. Shivanna, H.Y. MohanRam, Pollen-pistil interaction in a non-pseudogamous apomict, *Commiphora wightii*. Ann. Bot. 81 (1998) 589–594.
4. J. Prakash, P. Kasera, D.D. Chawan, A report on polyembryony in *Commiphora wightii* from Thar desert India. Curr. Sci. 78 (2000) 1185–87.
5. K.G. Ramawat, L. Bhardwaj, M.N. Tewari, Exploitation of India desert medicinal plants through tissue culture. Indian Rev. Life Sci., 11 (1991) 3–27.
6. Y.O. Assad, B. Torto, A. Hassanali, P.G.N. Njagi, N.H.H. Bashir, H. Mahamat, Seasonal variation in the essential oil composition of *Commiphora quadricincta* and its effect on the maturation of immature adults of the desert locust, *Schistocera gregaria*. Phytochem. 44 (1997) 833–841.
7. S. Dev, A modern look at an age-old Ayurvedic drug–guggulu. Sci. Age, July 1978 (1987) 13–18.
8. S. Dev, Ancient-modern concordance in Ayurvedic plants: sum examples. Environ. Health Perspect. 107 (1999) 783–789.
9. V.D. Patil, U.R. Nayak, S. Dev, Chemistry of Ayurvedic crude drugs—I. Guggulu-1, Steroidal constituents. Tetrahedron., 28 (1972) 2341–2352.
10. V.D. Patil, U.R. Nayak, S. Dev, Chemistry of Ayurvedic crude drugs—III. Guggulu-3, long chain aliphatic sterols, a new class of naturally occurring lipids. Tetrahedron. 29 (1973) 1595–1598.
11. K.K. Purushothaman, S. Chandrasekharan, Guggulsterols from *Commiphora wightii* (Burseraceae). Ind. J. Chem. Section B: Organic chemistry including medical chemistry, 14 (1976) 802–804.
12. R.S. Prasad, S. Dev, Chemistry of Ayurvedic crude drugs: Guggulu (resin from *Commiphora wightii*)-4, Absolute Steriochemistry of Mukulol. Tetrahedron. 32 (1976) 1437–1441.
13. B. Mesrob, C. Nesbitt, R. Misra, R.C. Pandey, High-performance liquid chromatographic method of fingerprinting and quantitative determination of E- and Z-guggulsterones in *Commiphora mukul* resin and its products. J. Chromatogr. Biomed. Sci. Appl. 720 (1998) 189–196.
14. S.K. Singh, N. Verma, R.C. Gupta, Sensitive high performance liquid chromatographic assay method for the determination of guggulsterone in serum. J. Chromatogr. Biomed. Appl. 670 (1995) 173–176.
15. N. Verma, S.K. Singh, R.C. Gupta, Simultaneous determination of the stereoisomers of guggulsterone in serum by high-performance liquid chromatography. J. Chromato. B708 (1998) 243–248.
16. M. Nagarajan, T.W. Waszkuc, J. Sun, Simultaneous determination of E- and Z-guggulsterones in diet supplements containing *Commiphora mukul* extract (guggulipid) by liquid chromatography. J. AOAC Int., 84 (2001) 24–28.
17. V.D. Patil, U.R. Nayak, S. Dev, Chemistry of Ayurvedic crude drugs—II: Guggul resin from *Commiphora mukul*—II: Diterpenoid constituents. Tetrahedron. 29 (2) (1974) 341–348.
18. H.K. Kakrani, Guggul: A review. Indian Drugs, 18 (1981a) 417–421.
19. H.K. Kakrani, Flavonoids from the flowers of *Commiphora mukul*. Fitoterapie, LII (5) (1981b) 221–222.
20. K. Asres, A. Tei, G. Moges, F. Spoker, M. Wink, Terpenoid composition of the wound-induced bark exudate of *Commiphora tenuis* from Ethiopia. Planta Med. 64 (1998) 473–475.
21. G.V. Satyavati, C. Dwarkanath, S.N. Tripathi, Experimental studies on the hypocholesterolemic effects of *Commiphora mukul* Engl. (Guggul). Indian J. Med. Res. 57 (1969) 1950–1962.

22. G.V. Satyavati, Gum guggul (*Commiphora mukul*)—the success story of an ancient insight leading to a modern discovery. Indian J. Med. Res. 87 (1988) 327–335.

23. R.B. Arora, D. Das, S.C. Kappor, R.C. Sharma, Effect of some fractions of *Commiphora mukul* on various serum lipids in hyperchloresterolemic chicks and their effectiveness in myocardial infraction in rats. Indian J. Expt. Biol. 11 (1973) 166–168.

24. V.S. Baldwa, V. Bhasin, P.C. Ranka, K.M. Mathur, Effects of *Commiphora mukul* (guggul) in experimental induced hyperlipidemia and atherosclerosis. J. Assoc. Physicians India, 29 (1981) 13–17.

25. S.K. Verma, A. Bordia, Effect of *Commiphora mukul* (gum guggulu) in patients of hyperlipidemia with special reference to HDL-cholesterol. Indian J. Med. Res. 87 (1988) 356–360.

26. K. Kuppurajan, S.S. Rajgopalan, T.K. Rao, R. Sitaraman, Effect of guggulu (*Commiphora mukul* Engl.) on serum lipids in obese, hypercholesterolemic and hyperlipemic cases. J. Assoc. Physicians India, 26 (1978) 367–373.

27. S. Lata, K.K. Saxena, V. Bhasin, R.S. Saxena, A. Kumar, V.K. Srivastava, Beneficial effects of *Allium sativum, A. cepa* and *Commiphora mukul* on experimantal hyperlipidemia and atherosclerosis—A comparative evaluation. J. Postgrad. Med. 37 (1991) 132–135.

28. R.B. Singh, M.A. Naiz, S. Ghosh, Hypolipidemic and antioxidant effects of *Commiphora mukul* as an adjunct to dietary in patients with hypercholesterolemia. Cardiovasc Drugs Ther. 8 (1994) 659–664.

29. V.L. Mehta, C.L. Malhotra, N.S. Kalrah, The effect of various fractions of gum guggul on experimentally produced hypercholesteraemia in chicks. Indian J. Physiol Pharmacol.,12 (1968) 91.

30. G.V. Satyavati, Standardized extract: a promising hypolipidemic agent from gum guggul (*Commiphora wightii*), in: Economic and Medicinal Plant Research, Vol. 5. H. Wagner and N.R. Farnsworth (eds). Academic Press (1991) 47–79.

31. S.N. Tripathi, B.N. Upadhyay, A clinical trial of *Commiphora mukul* in the patients of ischaemic heart disease. J. Mole. Cell. Cardio. 10 (Supp. 1) (1978) 124.

32. K. Gopal, Clinical trial of ethyl acetate extract of gum gugulu (Standardized extract) in primary hyperlipidemia. J. Assoc. Physicians India. 34(4) (1986) 249–251.

33. S. Nityanand, J.S. Srivastava, O.P. Asthana, Clinical trials with gugulipid: a new hypolipademic agent. J. Assoc. Physicians India. 37(5) (1989) 323–328.

34. S.N. Tripathi, V.V.S. Sastri, G.V. Satyavati, Experimental and clinical studies of the effect of guggulu (*C. mukul*) in hyperlipidemia and thrombosis. J. Res. Indian Med. 2 (1968) 2.

35. S.C. Malhotra, M.M. Ahuja, K.R. Sundaram, Long-term clinical studies on the hypolipidemic effect of *Commiphora mukul* (guggulu) and clofibrate. Indian J. Med. Res. 65 (1977) 390–395.

36. K. Singh, R. Chander, N.K. Kapoor, Guggulsterone, a potent hypolipidemic, prevents oxidation of low-density lipoprotein. Phytother Res. 11 (1997a) 291–294.

37. S. Kaul, N.K. Kapoor, Reversal of changes of lipid peroxide, xanthine oxidase and superoxide dismutase by cardio-protective drugs in isoproterenol induced myocardial necrosis in rats. Indian J. Exp. Biol. 27 (1989) 625–627.

38. S.D. Seth, M. Maulik, C.K. Katyar, S.K. Maulik, Role of lipistat in protection against isoproterenol induced myocardial necrosis in rats: a biochemical and histopathological study. Indian J. Physiol. Pharmacol. 42 (1998) 101–106.

39. L. Mester, M. Mester, S. Nityanand, Inhibition of platelet aggregation by "guggulu" steroids. Planta Med., 37 (1979) 367–369.

40. A. Gupta, N.K. Kapoor, S. Nityanand, Mechanism of hypolipidemic action of standardized extract. Indian J. Pharmacol. 14(1) (1982) 65.

41. V. Singh et al., Stimulation of low-density lipoprotein receptor activity in liver membrane of guggulsterone treated rats. Pharmacol. Res. 22(1) (1990) 37–43.

42. Y.B. Tripathi, O.P. Malhotra, S.N. Tripathi, Thyroid stimulating action of Z-guggulsterone obtained from *Commiphora mukul*. Planta Med. (1984) 78–80.

43. Y.B. Tripathi, O.P. Malhotra, S.N. Tripathi, Thyroid stimulatory action of Z-guggulsterone: mechanism of action. Planta. Med. 54(4) (1988) 271–277.

44. S. Panda, A. Kar, Gugulu (*Commiphora mukul*) induced triiodothionine production: possible involvement of lipid peroxidation. Life Sci. 65 (1999) 137–141.

45.   R.B. Arora, V. Kapoor, S.K. Gupta, R.C. Sharma, Isolation of a crystalline steroidal compound from *Commiphora mukul* and its anti-inflammatory activity. Indian J. Exp. Biol. 9 (1971) 403.

46.   R.B. Arora, L. Gupta, R.C. Sharma, S.K. Gupta, Standardisation of Indian indigenous drugs and preparations—II. Chemical and biological standardisation of *Commiphora mukul* (Guggul). J. Res. Indian Med. 7 (1972) 20–24.

47.   J.N. Sharma, J.N. Sharma. Comparison of the anti-inflammatory activity of *Commiphora mukul* (indigenous drug) with those of phenylbutazone and ibuprofen in experimental arthritis induced by mycobactrial adjuvant. *Arzneimittelforschung* 27: 1455–1457.

48.   M. Duwiejua, I.J. Zeitlin, P.G. Waterman, J. Chapman, G.J. Mhango, G.J. Provan, Anti-inflammatory activity of resins from some species of the plant family Burseraceae. *Planta Med.* 59 (1993) 12–16.

49.   B.B. Singh, L. Mishra, N. Aquilina, F. Kohlbeck, Usefulness of guggul (*Commiphora mukul*) for *oesteoarthritis*: An experimental case study. Altern. Ther. Healthg Med. 120 (2001) 112–114.

50.   T.G. Fourie, F.O. Snyckers, A pentacyclic triterpene with anti-inflammatory and analgesic activity from the roots of *Commiphora merkeri*. J. Nat. Prod. 52 (1989) 1129–1131.

51.   S.C. Taneja, K.L. Dhar, Studies towards development of a new anti-inflammatory drug from *Boswellia serrata* gum-resin. In: Supplement to culivation and utilization of medicinal plants. S.S. Handa, M.K. Kaul (Eds) RRL, Jammu Tawi, India, (1996) pp. 525–536.

52.   O.A. Olajide, Investigation of the effects of selected medicinal plants on experimental thrombosis. Phytother. Res. 13(3) (1999) 231–232.

53.   M.M. Al-Harbi, S. Qureshi, M. Raza, M.M. Ahmad, A.H. Shah, Gastric antiulcer and cytoprotective effect of *Commiphora molmol* in rats. J. Ethnopharmacol. 55 (1997) 141–150.

54.   A.H. Atta, A. Alkofahi, Anti-nociceptive and anti-inflammatory effects of some Jordanian medicinal plants extracts. J. Ethanopharmacol. 60 (1998) 117–124.

55.   S. Qureshi, M.M. Al-Harbi, M.M. Ahmed, M. Raja, A.B. Giangreco, A.H. Shah, Evaluation of the genotoxic, cytotoxic and antitumour properties of *Commiphora molmol* using normal and Ehrlich ascites carcinoma cell-bearing Swiss albino mice. Cancer Chemother. Pharmacol. 33 (1993) 130–138.

56.   M.M. Al-Harbi, S. Qureshi, M.M. Ahmad, S. Rafatullah, A.H. Shah, Effect of *Commiphora molmol* (oligo-gum-resin) on the cytological and biochemical changes induced by cyclophosphamide in mice. Am. J. Clin. Med. 22 (1994) 77–82.

57.   R.P. Ubillas, C.D. Mendez, S.D. Jolad, J. Luo, S.R. King, D.M. Fort, Antihyperglycemic furanoses-quiterpenes from *Commiphora myrrha*. Planta Med. 65 (1999) 778–779.

58.   R.C. Setia, M.V. Parthsarathy, J.J. Shah, Deveolpment, histochemistry and ultrastructure of gum-resin ducts in *Commiphora mukul*. Englr. Ann. Bot. 41 (1977) 999–1004.

59.   G.M. Nair, K.R. Patel, S.V. Subramanium, J.J. Shah, Secretion of resin across the wall of the epithelial cells in the gum-resin canal of *Commiphora mukul* Engl. Ann. Bot. 47 (1981) 419–421.

60.   R.C. Setia, J.J. Shah, Histological, histochemical and ultrastructure aspect of gum and gum-resin producing structures in plants. Ann. Rev. Plant Sci., CP Malik (Ed) Kalyani Pub. New Delhi (1979) 315–332.

61.   J.R. Bhatt, N.M.B. Nair, H.Y. MohanRam, Enhancement of oleo-gum resin production in *Commiphora wightii* by improved tapping technique. Curr. Sci. 58 (1989) 349–46.

62.   A.K. Singh, Gum-resin production associated cellular and organ differentiation in *Commiphora wightii*. Ph.D. Thesis, M.L. Sukhadia University, Udaipur, (1995) 1–119.

63.   P. Singh, M.L. Sharma, S. Mukherjee, Effect of indole butyric acid on sprouting in plant cuttings of *Commiphora wightii* (Arnott.) Bhand. Indian Drugs 26 (1989) 515–516.

64.   R.R. Shah, D.B. Patel, D.H. Patel, K.C. Dalal, Harmonal effect on germination of guggul cuttings. Indian Drugs 20 (1983) 435–437.

65.   D.N. Puri, R.N. Kaul, Effect of size of stem cuttings on rooting in *Commiphora mukul*. Indian For. 98 (1972) 252–257.

66.   S. Kshetrapal, R. Sharma, Studies on the effect of various plant extracts in sprouting behaviour of *Commiphora wightii* (Arnott.) Bhand. and *C. agallacha*. J. Indian Bot. Soc. 72 (1992) 73–75.

67.   S. Kumar, K.G. Ramawat, Somatic embryogenesis in callus and cell cultures of *Commiphora wightii*:

problems, perseverance and prospects. National Symp. Prospects and Potential of Plant Biotech in India. J. N. Vyas Univ. Jodhpur, pp. (2000) 12.

68.   D.M. Barve, A.R. Mehta, Clonal propagation of mature elite trees of *Commiphora wightii*. Plant Cell Tiss. Org. Cult. 35 (1993) 237–244.

69.   T. Murashige, F. Skoog, A revised medium for rapid growth and bioassays with tobacco tissue cultures. Physol. Plantarum., 15 (1962) 473–497.

70.   O.L.Gamborg, R.A. Miller, K.Ojima, Nutrient requirements of suspension cultures of soybean root cell. Exp. Cell Res. 50 (1968) 51–158.

71.   S. Kumar, S.S. Suri, K.C. Sonie, K.G. Ramawat, Establishmennt of embryogenic cultures and somatic embryogenesis in callus cultures of guggul-*Commiphora wightii* (Arnott.) Bhand. Indian J. Exp. Biol. 41 (2003) 69–77.

72.   S. Kumar, Cell, Callus and tissue culture of *Commiphora wightii* for developing technology for its micropropagation. Ph.D. Thesis, M.L. Sukhadia University, Udaipur, (2002) 1–114.

73.   R. Sharma, S.S. Suri, K.G. Ramawat, K.C. Sonie, Biotechnological approaches to the medicinal plants of Aravalli Hills with special reference to *Commiphora wightii*, in: Role of Biotechnology in Medicinal and Aromatic plants. I.A. Khan and A. Khanum (Eds). Ukaz Pub. Hyderabad, (1999) 140–164.

74.   A.K. Singh, S.S. Suri, K.G. Ramawat, Somatic embryogenesis from immature zygotic embryos of *Commiphora wightii*, a woody medicinal plant. Gartenbauwissen. 62 (1997b) 44–48.

75.   S.S. Suri, K.G. Ramawat, Factors affecting somatic embryogenesis in callus and cell cultures of *Commiphora wightii*. National Symp. Commercial Aspect Plant Tissue Cult., Mol. Biol. & Medicinal Plant Biotechnology. Jamia Hamdard Univ., New Delhi (1998) pp.12.

76.   P. Phale, J. Subramani, P.N. Bhatt, A.R. Mehta, Viability and guggulsterol production in immobilized tissue cultured cells of *Commiphora wightii*. Indian J. Exp. Biol. 27 (1989) 338–340.

Plant Biotechnology and Molecular Markers
P.S. Srivastava, Alka Narula and Sheela Srivastava (Editors)

9. Biotechnology in Quality Improvement of Oilseed Brassicas

Abha Agnihotri[1], Deepak Prem[2] and Kadambari Gupta[2]

[1]Bioresources and Biotechnology Division, TERI, Habitat Place, Lodhi Road, New Delhi 110 003, India

[2]Centre for Bioresources and Biotechnology, TERI-School of Advanced Studies, Habitat Place, Lodhi Road, New Delhi 110 003, India

Abstract: The review presents a comprehensive description of the advances made towards desired quality improvements in rapeseed mustard. The two nutritionally undesired elements, namely erucic acid in the seed oil and glucosinolates in the deoiled meal, are discussed in reference to their nutritional implications and inheritance. The genetic manipulation of fatty acid biosynthetic pathway for diversified uses is also presented. It elaborates the use of biotechnological methods in terms of their conjugation with various other conventional approaches for genetic enhancement and value addition by developing the desired superior genotypes of brassicas for nutritional and industrial purposes.

1. Introduction

Production of crop varieties with increased sustainable production is the most challenging task facing the plant breeders in the current century. Considering the limited resources such as diminishing and deteriorating cultivable land, water supply, fertilizers etc., improvement in terms of yield and quality within a limited time frame, is the demand of the present and future generations. Since the increased yield alone may not sustain the needs of human nutrition, improvement of nutritional quality and value addition for diversified uses are of prime importance.

The oilseeds form the second largest agricultural commodity in India. Among the nine annual oilseed crops grown in the country, oilseed brassica rank second in importance contributing about 30% to the total oilseed produced. It is one of the best edible oils available, having lowest amount of saturated fats as compared to other vegetable oils, provides both essential fatty acids and also the animal feed through oil free meal rich in protein having well balanced aminogram.

The oleiferous Brassicas being the provider of edible oil to a major proportion of our population are prime targets for quality improvement. The presence of high amounts of two nutritionally undesired elements in Indian varieties (40–50% erucic acid in the seed oil and upto 300 µm/g glucosinolates in the deoiled meal) pose a huge challenge to plant breeders working on improving oil and deoiled meal quality in Brassicas. The canola quality exotic rapeseed cultivars, commonly known as double low or 'oo', having less than 2% erucic acid in the seed oil and less than 30 µm glucosinolate/g oil free meal [1] were not found suitable for cultivation under Indian agroclimatic conditions. Since, among the oilseed Brassicas, *B. juncea* acquires the maximum share of cultivated area in our country, the improvement of nutritional quality in *B. juncea* is most desired to suit our needs. The brassica fatty acid profile is also amenable to alterations for developing designer crops for specific food or non-food industrial applications thus having prospects of diversified uses.

2. Seed Oil Quality

The nutritional quality of vegetable oils is considered significant in modern living. Oil quality is described in terms of saturated, monounsaturated and polyunsaturated fatty acids. Mustard oil contains the lowest amounts of saturated fatty acids as compared to other vegetable oils and also has a very good proportion of $n3$ and $n6$ polyunsaturated fatty acids, thus considered beneficial for food consumption. In most vegetable oils, the unsaturated fatty acids consists mainly of oleic and linoleic acid. However, mustard oil is an exception since in addition to oleic acid (8–15%), linoleic acid (13–20%) and linolenic acid (6–14%), it also contains erucic acid (41–50%), and palmitic and stearic acids in trace amounts [2, 3]. Erucic acid contributes approximately 50% of the total fatty acids in mustard oil. However, it is nutritionally undesirable and the high erucic acid *B. napus* oil is reported to be less metabolisable [4]. High erucic acid content is also known to cause impaired myocardial conductance, increased blood cholesterol and cardiac lipidosis with accumulation of erucic acid in mammalian system [5, 6].

High concentration of oleic acid is preferred for cooking purposes since it is thermostable. Both linoleic and linolenic acids are essential fatty acids that need to be supplied in diet from external sources. However, high linolenic acid in the oil being prone to peroxydation causes flavor revision and oil deterioration [7] and therefore 3 to 5% is preferred to meet the dietary requirement. Linoleic and linolenic acid are both produced by a common biosynthetic desaturation pathway [8]. Therefore, selection for high linoleic acid has tended to increase the level of linolenic acid, while selection for low linolenic acid tends to decrease the level of linoleic acid also. For this reason selection within the same germplasm may not be able to meet the breeding objectives. Therefore, the fatty acid scenario in mustard oil implies that efforts should be made towards development of cultivars having low levels of erucic and linolenic acids, high levels of oleic and moderate linoleic acids.

The success of developing *B. juncea* with low erucic acid suitable for Indian agroclimatic condition has been limited due to search for appropriate gene pool. Stefansson et al. [9] and Downey [10] identified genotypes with a genetic block in the biosynthesis of eicosenoic and erucic acid in summer rape (*B. napus*) and summer turnip rape (*B. campestris*), respectively. Studies for inheritance of erucic acid content have shown that it is controlled by multiple genes and the seed erucic acid level is controlled by the embryo genotype in *B. napus* [11, 12]. Kirk and Hurlstone [13] have reported two genes showing dominance and acting in an additive manner for erucic acid biosynthesis in *B. juncea*. Kirk and Oram [14] identified zero erucic genotypes of *B. juncea* and following this low erucic acid genetic stock among Indian accessions of *B. juncea* was also identified [15]. Recently the development of early maturing, low erucic acid strains of *B. juncea* and *B. napus*, suitable to grow under Indian agroclimatic conditions, have been reported and are under the advance stages of testing [16, 17]. In *B. juncea*, the predominantly grown oilseed brassica in India, the main emphasis has been on successful reduction of erucic acid and the work in the direction of developing cultivars with variable fatty acid profile for edible or industrial purposes is just beginning.

3. Deoiled Meal Quality

The defatted Brassica meal contains about 40% protein with a well balanced aminogram and is used as animal feed [18]. Brassica oil meal is particularly rich in lysine and methionine, which are essential amino acids not found in cereal grains. For this reason, Brassica oil meal has been used for animal feed. However, the feeding value of rapeseed-mustard meal has been limited because of the presence of sulfur containing compounds called glucosinolates present in the vegetative tissues and seeds of cruciferous plants [19]. The various kinds of glucosinolates present in *B. juncea*, in decreasing order of their abundance are gluconapin, sinigrin, progoitrin, napoleiferin and glucobrassicanapin. At cellular level, glucosinolates are stored in the vacuole [20, 21] and myrosinase, a glycoprotein enzyme responsible for hydrolysis of glucosinolates is stored in a tonoplast-like membrane bound organelle called the idioblast [22]. On mechanical injury, myrosinase catalyzed hydrolysis of glucosinolates occur to form thiocyanates, isothiocyanates and nitriles. Although glucosinolates as such do not cause much harm, their breakdown products are undesirable in animal feeds. These compounds impart a characteristic flavor and odor to Brassica vegetables and condiments but may reduce palatability and adversely affect iodine uptake by the thyroid glands in non-ruminant animals such as swine and poultry. Thus they reduce the feed efficiency in terms of development and weight gain [19, 23].

To avoid glucosinolate hydrolysis products to accumulate in Brassica oil meal, the myrosinase enzyme is heat inactivated as one of the first steps in oil extraction process. However, this in turn also causes the breakdown of other proteins, which may adversely affect the nutritional value of the oil meal quality. In India heat treatment of seed before oil extraction is usually not done, therefore the extracted edible oil has relatively large proportion of glucosinolates breakdown products imparting the characteristic pungency in the oil. In view of these facts, a prime breeding objective for *B. juncea* quality breeders is to develop low glucosinolate varieties.

Work in the direction of glucosinolate inheritance in Brassicas was started in 1970's and was revolutionised by the discovery of a low glucosinolate *B. napus* cultivar 'Bronowski' from Poland [24]. Since then, this genotype has provided the source of low glucosinolate gene for practically all cultivated Brassicas. Kondra and Stefansson [25] had proposed that in *B. napus* the maternal genotype rather than the embryo genotype controlled seed glucosinolates. They have proposed low glucosinolate level to be controlled by as many as 11 recessive alleles that do not show independent segregation. Rather than linkage to be operative, a simultaneous action of genes for a common biosynthetic pathway for synthesis of all glucosinolates has been suggested. The formation of individual glucosinolates is thought to occur through a chain break at the end of the biosynthetic pathway [26]. Further, Lein [27] has also determined an additional influence of cytoplasm on glucosinolate synthesis.

Till date no germplasm source for low glucosinolate genes has been reported in *B. juncea*. However, Love et al. [28] developed the low glucosinolate *B. juncea* line BJ-1058 using interspecific hybridization between Indian mustard and a low glucosinolate strain of *B. campestris* having the Bronowski gene block. Glucosinolates also show tissue level variation within the same plant [29-31] and the leaf glucosinolate quantity and profile can be correlated to the seed glucosinolate level only in small seedlings suggesting that the glucosinolate content in the leaves and seeds may be under different genetic control [32, 33]. The glucosinolate profile at the seedling level may serve as a tentative tool to predict glucosinolate profile of seeds, but its authenticity is doubtful since differentiation is not clearly understood and may produce unpredictable and drastic changes [34]. There have been some reports indicating that the genes for glucosinolate contents in vegetative tissue are pleiotropic or linked with the grain filling stage [35, 36], and that glucosinolates may contribute towards resistance to insect pests and pathogens [32, 33].

4. Quality Status in *B. juncea*

Most of the work related to quality improvement has been globally concentrated on *B. napus* and the work on genetic enhancement of *B. juncea* quality, the predominant species of Asian subcontinent, is somewhat limited. The facts discussed above suggest that an ideal genotype of mustard from the point of view of nutritional quality would be one having low erucic acid in the seed oil, low levels of glucosinolate in the seed or reproductive tissue and high glucosinolate content in the vegetative tissue. This may seem to be a mammoth task but it has already been accomplished in *B. napus* [33] and is being extensively researched on for *B. juncea*.

Several double low strains/cultivars of *B. napus* are available globally, however, the progress of work to develop double low *B. juncea* has not been very successful as yet, primarily due to the lack of suitable donor germplasm. Following the successful introgression of low glucosinolate genes, the double low *B. juncea* strains have been developed in Canada through cross breeding of BJ-1058 and LDZ (a zero erucic acid, high oil content *B. juncea* strain). The progeny of this cross was backcrossed to the *B. juncea* var. Cutlass, in order to incorporate white rust resistance genes. This material has shown promising results in field trials, and is being improved for its fatty acid profile [37]. In India also, attempts have been made to introgress the double low characteristics in various Indian mustard cultivars [38, 39] but the desired success is yet to be achieved and these low erucic/low glucosinolate/double low strains are being improved for agronomic characteristics.

5. Conventional Approaches for Development of Double Low Cultivars

The conventional breeding techniques for quality improvement vary greatly and have evolved from simple mass selection to hybrid cultivar development. The breeding strategies depend on the objective and practical scientific considerations such as inheritance pattern of genes responsible for a particular trait. The backcrossing approach has been successfully used to transfer simply inherited traits such as low erucic acid. The erucic acid content of the seed is controlled by the genotype of the embryo, that is, the individual F_2 seeds borne on F_1 plants have different erucic acid level. This fact led to the development of the half seed technique [40]. This approach has been used by Kirk and Hurlstone [13] to develop low erucic *B. juncea* lines. The use of backcross technique for development of low glucosinolate *B. juncea* is limited due to the non-availability of any natural low glucosinolate source and polygenic inheritance, but it has been successfully used for development of low glucosinolate *B. napus* [41].

Both *B. napus* and *B. juncea* are predominantly self-pollinated species [42] and thus the commonly used breeding tool of pedigree selection can be employed for cultivar development. A clearly defined breeding objective and identified suitable parent is a pre-requisite to start a pedigree selection program [40]. Backcrossing in conjugation with pedigree selection has been used successfully for the development of early maturing canola quality *B. napus* cultivars in India [17, 43]. Further, modifications in the methods can be done as per available germplasm resources or breeding objectives. However, due to the involvement of multiple recessive genes, development of double low *B. juncea* by conventional methods alone is proving to be a lengthy process, thus necessitating the need for incorporation of suitable biotechnological tools such as doubled haploids production, mutagenesis and molecular approaches to facilitate the quality improvement in a targeted manner.

6.   Quality Improvement Using Doubled Haploids

The efficient production of doubled haploid plants from anther or microspore cultures has become an important new tool for *Brassica* breeders [44]. After initiation of work for this school of thought in late 1970's, efficient protocols to induce embryogenesis in isolated microspore cultures of *B. napus* have been reported by several workers [45-48]. A promising double low variety of *B. napus* (cyclone) developed through doubled haploid technique using isolated microspore culture, is being commercially cultivated in Canada.

Doubled haploids provide several advantages over conventional breeding approaches. The selection of desired genotypes at F_1 haploid level followed by diploidization fixes the desired genes i.e., leads to the production of pure lines that do not segregate further. Hence homozygosity can be achieved in one step equivalent to repeated in-breeding for several generations (8–10 years). Doubled haploids being homozygous, also considerably reduce the time required for parental identification for hybridization programmes [49].

This technique also offers the advantage of significantly smaller population size required to find the least likely recombinants particularly when several genes are involved [50]. Since both glucosinolates and erucic acid are multiple recessive gene governed traits, doubled haploids offer a powerful tool to reduce the perfect population size to approximately 70 to 80 fold than what would be needed to be handled via conventional methods. An integrated approach involving application of doubled haploid technique with early selection for high erucic acid in the cotyledons of microspore derived embryos obtained from the F_1 hybrids of winter oilseed rape established a positive correlation in the erucic acid content of embryos and the seeds derived from them, thus elucidating the efficiency of selection at microspore embryo stage [51].

In addition to the complicated inheritance of glucosinolates, additional effects of maternal inheritance [25] and cytoplasmic influence [27] on seed glucosinolates have been reported. Since glucosinolates in seed are governed by maternal genotype rather than the seed embryo's genotype [52], the F_1 seeds would show glucosinolate content as per maternal parent genotype. This means that effective selection can only be done in F_3 seeds produced by the F_2 population and in order to follow this procedure under field conditions at least 3 years would be required to reach initial screening. However, the pollen/microspore is not the target site for glucosinolate storage [53] and has minimal cytoplasm, therefore seeds produced by the doubled haploids would reflect the genotype of the doubled haploid plant rather than the parent. Moreover, since haploids would express recessive genes, transgressive segregants for recessive traits can efficiently be recovered through diploidization of recessive haploids.

7.   Quality Improvement Through Mutagenesis

Mutagens are also being effectively employed to generate considerable variability in fatty acid composition [54-57]. Chemical mutagenesis has been used to produce *B. napus* lines with reduced linolenic and increased linoleic acid contents [58]. The low linolenic acid lines have been used as genetic base for the development of *B. napus* with less than 3% linolenic acid and more than 22% linoleic acid [1]. Doubled haploid lines of *B. carinata* with modified erucic acid content have also been identified through chemical mutagenesis by EMS treatment of isolated microspores [55, 59]. Mutagens have not only been used to produce variable fatty acid composition but also to substantially increase the oil content. Kumar et al. [56] have reported an increase in oil content upto 4.55% using gamma irradiation in *B. juncea*. An increase in oleic acid content coupled with decrease in erucic acid was also observed in the treated varieties. Wong and Swanson [60] and Auld et al. [61] recovered high oleic acid producing doubled haploids through chemically induced mutagenesis in microspore cultures of *B. napus*.

The use of induced mutations for altering fatty acid profile allows for the selection of variants with either complete or incomplete sets of functionally altered genes responsible for fatty acid synthesis [62]. However, a major limitation of mutagenesis is that apart from the genes controlling the target trait, it may cause several changes in the genetic background thus affecting the non-target traits. For example, mutants for the high oleic acid content have been shown to be associated with undesirable agronomic characteristics [63]. It is considered likely that several genes code for $\Delta12$ desaturase enzymes (that are responsible for conversion of oleic to linoleic) in *B. napus* seed and that some of these genes also regulate production of $\Delta12$ desaturase in the vegetative tissue. Mutation exposure would lead to non-tissue specific changes in both the seed and vegetative tissue $\Delta12$ desaturase genes, which could have detrimental effects on the vegetative tissue where the correct fatty acid composition is required for normal membrane structure and function [63].

As discussed above, mutagenesis aims at altering the existing fatty acid profile and does not have the ability to add genes for new biosynthetic pathways for production of novel fatty acids. Although mutagenesis has been fairly successful for producing genotypes with altered fatty acid compositions, its non-tissue specific action restricts its use for cultivar development. Nevertheless, the lines derived from mutation breeding programs serve as important donor material in forthcoming crop improvement programmes, and the abovementioned bottlenecks can be quite satisfactorily overcome by using the transgenic approach.

8. Quality Improvement Through Genetic Engineering

The potential use of genetic engineering to modify plant seed oil composition has been recognized for a number of years. The oilseed crops have the potential to produce high quality edible oils as well as speciality oils having commercial applications. For instance jojoba is a rich source of wax esters, coconut is rich in capric, lauric and myristic fatty acid, palm has a high proportion of palmitic, oleic and stearic acid, whereas linseed is a rich source of linoleic acid [64]. These fatty acids are used in a wide range of products ranging from the production of soaps, detergents, cosmetics, surfactants, lubricants, plastics, varnishes and pharmaceuticals. Due to the non-domestication of most of the potential sources and their restricted availability, at present the fatty acids for industrial applications are mostly derived from petrochemicals. However, in the near future, with the biased use of global reserves of fossil derived hydrocarbons alternative sources of industrial fatty acids from the environment friendly oil crops are sought after [65]. This can be achieved either by altering the existing fatty acids profile or by adding new genes for synthesis of novel fatty acids.

In most of the oil bearing crops, the biosynthetic pathway of fatty acid synthesis is similar [66] and their differential accumulation in the seed is genetically controlled depending upon the species. During the seed development process, photosynthetically fixed carbon is imported into the seed in the form of sucrose, and is converted into the storage products with the help of enzymes present in the seed. The seed contains all the enzymes that are required for the conversion of sucrose into any of the storage products. However, it is the rate of sucrose uptake by the various biosynthetic pathways that lead to the differential accumulation of a particular storage product in the seed [67]. Thus the genetic manipulation of any of the biosynthetic pathway can lead to a specific ratio of seed storage product, according to the end use of the seed. This can be done either by modifying the length of the existing hydrocarbons in fatty acid chain (modifying the chain elongation enzymes) or by changing the position of double bonds (modifying desaturase enzymes).

The seed specific or tissue specific genetic modifications may be used for creating changes in endogenous fatty acid biosynthesis pathway or addition of new biosynthetic pathways. The use of seed specific antisense technology has allowed for the selective modulation of key enzyme activities in the developing seed, while keeping the rest of the genetic background of the plant absolutely constant [68]. Co-suppression based on post-transcriptional gene silencing of endogenous desaturase gene has shown promising results in developing high oleic acid genotypes of rapeseed-mustard [69]. The recently derived RNAi approach has also shown a great potential for endogenous desaturase silencing. Using this concept total silencing of the $\Delta12$ desaturase gene in *B. napus* has been acheived, resulting in the production of genotypes accumulating 89% oleic acid in the seed oil [70].

The rapeseed oil normally contains low levels of lauric acid (C12) and stearic acid (C18) at a concentration of 1–2% and 0.1–0.2%, respectively. High lauric rapeseed can be used as a substitute in detergent markets, leading to displacement of conventional lauric oils derived from coconut or palm kernel, whereas high stearic rapeseed is a useful substitute in margarine markets and replaces conventional hydrogenated rapeseed oil. The two most notable achievements in oil modification to-date are the 40% stearic and 40% lauric rapeseed varieties (laurical) first produced and entered in field trials by Calgene in 1993–94 [71]. Thus laurical was the first genetically manipulated rapeseed variety given permission for commercial cultivation in 1995 in US. The $\Delta9$ stearoyl ACP desaturase gene which normally converts stearic to oleic acid was partially inactivated in rapeseed using antisense technology, resulting in the accumulation of a seed oil containing up to 40% stearic acid [68]. This high stearic variety contains an antisense copy of a *Brassica* stearate desaturase gene which inhibits the function of the normal rapeseed stearic desaturase gene, resulting in an accumulation of stearic acid, rather than their saturation to oleate. The resulting high stearic oil has many advantages over the normal rapeseed oil for the production of certain solid fats, such as margarines.

With the advent of transgenic technology, the genes coding for enzymes that synthesize industrially important fatty acids can be transferred from non-traditional crops into more important oil crops. The canola oil having low erucic acid has food applications in margarine, salad and salad dressings while the high erucic rapeseed has industrial application. The canola quality rapeseed has also been genetically modified for containing high levels of β-carotene. This high carotenoid canola oil may prove very beneficial to combat the vitamin A deficiency in developing world [73]. Various species of Brassicaceae have been transformed with mutated Sn-2 acyltransferase gene from yeast and have been reported to show increase in seed oil content, seed weight and erucic acid content [72].

Lauric oils are mainly used in soaps and detergents although their use in confectionary fats and milk formulas is also being investigated. Lauric acid which is present at insignificant levels in rapeseed is found at high levels in the seed oil of the California Bay plant, *Umbelluria californica*, due to the presence in the latter species of a lauryo-ACP thioesterase. This gene has been cloned from the Bay plant and inserted into rapeseed causing premature chain-termination, resulting in a novel variety with a seed oil containing almost 25% lauric acid [74]. Following this an Sn-2 acyl transferase gene (LPAAT) from coconut has been introduced in lauric rapeseed to increase the accumulation of lauric acid in the seed triacyl glycerol molecules [75].

Similar to the development of lauric acid producing rapeseed, several novel genes coding for altered fatty acid synthesis have been used for altering seed fatty acid profile. Some worthy examples are Caprilic and Capric acid (from *Cuphea* spp.), myristic acid (from *Myristica fragrans*), Crepenylic acid (from *Crepis alpira*), Richinolic acid (from Castor), Vernolic acid (from *Crepis palaestina*) and petroselenic acid (from *Coriandrum sativum*) [64]. Thus, in future, plant derived oils may be an important source of industrial oil derived chemical or oleo chemicals.

8. Conclusion

Genetic enhancement for improvement in the quality of rapeseed-mustard is a prime breeding target for *Brassica* breeders all over the world. In addition to being the second most important edible oilseed crop in India, the rapeseed-mustard oil also finds its use in industrial applications [76, 77]. The advent of biotechnology has provided the plant breeders with new and more accurate tools that have the ability to compress the time taken in directed evolution of crop species. Efforts towards developing improved quality rapeseed varieties have been consolidated and expedited at global level through the use of biotechnological techniques such as doubled haploid and mutagenesis in conjugation with the conventional methods. The advances in molecular approaches, the genetic engineering and transformation, has made it possible to develop many designer rapeseed varieties with specific fatty acids profile for edible and industrial purposes. The work is in progress towards such quality improvement in Indian mustard, *B. juncea* also. Such targeted value addition in the quality of oilseed Brassicas will not only provide for an improved source of human nutrition but also for non-food, fuel/non-fuel industrial products that could reduce the load on ever depleting natural oil and gas resources.

References

1. Downey, R.K. 1990. *Brassica* oilseed breeding—achievements and opportunities. Plant Breeding Abstracts. 60(10): 1165–1170.
2. Banga, S.K. 1996. Breeding for oil and meal quality, in: Chopra V.L., Prakash S. (Eds). Oilseed and Vegetable Brassicas: Indian Perspective, Chapter 11, pp. 234–249.
3. Prakash, S., Kumar, P.R., Sethi, M., Singh, C., Tandon, R.K. 2000. Mustard Oil: The Ultimate Edible Oil. The Botanica 50: 94–101.
4. Sauer, F.D., Kramer, J.K.G. 1983. The problems associated with the feeding of high erucic acid rapeseed oils and some fish oils to experimental animals, in: High and Low Erucic Acid Rapeseed Oils (Kramer J.K.G., Sauer F.D., Figden W.J. (Eds)), Toronto, Canada. Academic Press. pp 254–292.
5. Renarid, S., McGregor, L. 1976. Antithrombogenic effects of erucic acid poor rapeseed oils in the rats. Rev. Fr. Crops. 23: 393–396. Cited in: Chopra P.K., Kirti P.B. 1996. Biotechnology in: Chopra V.L., Prakash S. (Eds). Oilseed and Vegetable Brassicas: Indian Perspective, Chapter 12, pp. 250–269.
6. Gopalan, C.D., Krishnamurthi, D., Shenolikar, I.S. and Krishnamachari, K.A.V.R. 1974. Myocardial changes in monkey fed on mustard oil. Nutr. Metab. 16: 352–365.
7. Axtell, J.D. 1981. Breeding for improved feeding quality, in: Frey K.J. (Ed) Plant Breeding II, Iowa State Univ. Press. pp 365–414.
8. Diepenbrock, W. 1983. Genotypic differences between the contents of linolenic acid in galactolipids

and triacylglycerol from seeds of rape plants, in: Proc. 6th Internat. Rapeseed Conf. Paris, France. pp 321–362.

9. Stefansson, B.R., Hougen, F.W., Downey, R.K. 1961. Note on the isolation of rape plants with seed oil free from erucic acid . Can. J. Plant Sci. 41: 218–219.

10. Downey, R.K. 1964. A selection of *Brassica campestris* L. containing no erucic acid in its seed oil. Can. J. Plant Sci. 44: 295.

11. Jonsson, R. 1977. Erucic acid heredity in rapeseed (*Brassica napus* and *Brassica campestris*) Heridetas. 86: 159–170.

12. Anand, I.J., Downey, R.K. 1981. A study of erucic acid alleles in digenomic rapeseed (*Brassica napus* L). Can. J. Plant Sci. 61: 199–203.

13. Kirk, J.T.O., Hurlstone, C.J. 1983. Variation and inheritance of erucic acid content in *Brassica juncea*. Z Pflanzenzuchtg 90: 331–338.

14. Kirk, J.T.O., Oram, R.N. 1981. Isolation of erucic acid free lines of *Brassica juncea*: Indian mustard now a potential oilseed crop in Australia. J. Austral. Inst. Agr. Sci. 47: 51–52.

15. Gupta, M.L., Banga, S.K., Banga, S.S., Sandha, G.S., Ahuja, K.L., Raheja, R.K. 1994. A new genetic stock for low erucic acid in Indian mustard. Cruciferae Newsletter. 16: 104–105.

16. Agnihotri, A., Kaushik, N. 1998. Transgressive segregation and selection of zero erucic acid strains from intergeneric crosses of Brassica. Ind. J. Plant Genet. Res. 11(2): 251–255.

17. Anonymous 2001, in: Annual Progress Report of the AICRP on Rapeseed-Mustard, Sewar, Bharatpur, ICAR, Govt. of India.

18. Miller, R.W., Van Etten, C.H., Mc Grew, C.E., Wolf, I.A., Jones, Q. 1962. Amino acid composition of seed meals from 41 species of cruciferae. J. Ag. Fd. Chem. 10: 426–430. Cited in: Banga S.K. 1996. Breeding for oil and meal quality, in: Chopra V.L. and Prakash S. (Eds). Oilseed and Vegetable Brassicas: Indian Perspective, Chapter 11, pp. 234–249.

19. Fenwick, G.R., Heaney, R.K., Mullin, W.J. 1983. Glucosinolates and their breakdown products in food and food plants. Critical Rev. Food Nutrition. 18: 123–201.

20. Matile, P. 1980. Die Senfolbombe: zur komparumentierung des myrosinase systems. Biochem. Physiol. Pflanzen. 175: 722–731. Cited in: Mitten R. 1992. Leaf glucosinolate profiles and their relationship to pests and disease resistance in oilseed rape. Euphytica 63: 71–83.

21. Luthy, B., Matile, P. 1984. The mustard oil bomb: Rectified analysis of the subcellular organisation of the myrosinase system. Biochem. Physiol. Pflanzen. 179: 5–12. Cited in: Mitten R. 1992. Leaf glucosinolate profiles and their relationship to pests and disease resistance in oilseed rape. Euphytica 63: 71–83.

22. Thangstad, O.P., Evjen, K., Bones, A. 1991. Immunogold-EM localization of myrosinase in Brassicacae. Protoplasma. 161: 85–93. Cited in: Mitten R. 1992. Leaf glucosinolate profiles and their relationship to pests and disease resistance in oilseed rape. Euphytica 63: 71–83.

23. Bille, N., Eggum, B.O., Jacobsen, I., Olseno, O., Sorensen, N. 1983. Antinutritional and toxic effects in rats of individual glucosinolates (+) myrosinases added to a standard diet. I. Effects on protein utilization and organ weight. Tierphysiol Tierer nahar Futter-mittelkd. 49: 195–210.

24. Finlayson, A.J., Kryzmanski, J., Downey, R.K. 1973. Comparison of chemical and agronomic characteristics of two *Brassica napus* L. cultivars, Bronowski and Target J. Am. oil Chem. Soc. 50: 407–410.

25. Kondra, Z.P., Stefansson, B.R. 1970. Inheritance of the major glucosinolates of rapeseed (*Brassica napus*) meal. Can. J. Plant Sci. 50: 643–647.

26. Lichter, R. DeGroot, E., Fiebig, D., Schweiger, R., Gland, A. 1988. Glucosinolates determined by HPLC in the seeds of microspore-derived homozygous lines of rapeseed (*Brassica napus* L.). Plant Breeding 100: 209–221.

27. Lein, K.A. 1970. Quantitative Bestimmungsmethoden for samenglucosinolate in *Brassica* Arten and Ihre Anwendung in der Zuchtung Von Glucosinolataramen Raps. Z. Pflanzenzucht 63: 137–154.

28. Love, H.K., Rakow, G., Raney, J.P., Downey, R.K. 1990. Development of low glucosinolate mustard. Can. J. Plant Sci. 70: 419–424.

29. Milford, G.F.J., Fieldsend, J.K., Porter, A.J.R., Rawlinson, C.J., Evans, E.J., Bilsborrow, P.E. 1989. Changes in glucosinolate concentration during the vegetative growth of single and double low cultivars

of winter oil seed rape. Aspects of Applied Biology 23. Production and Protection of Oilseed Rape and Other Brassica Crops. pp 83–90.

30. Inglis, I.R., Wadsworth, J.T., Meyer, A.N., Feare, C.J. 1992. Vertibrate damage to 'OO' and 'O' varieties of oilseed rape in relation to SMCO and glucosinolate concentration in the leaves. Crop Protection 11: 64–68.

31. Schilling, W., Friedt, W. 1992. Breeding 'OO' rapeseed (*Brassica napus*) with differential glucosinolate content in the leaves. Proc.8th Internat. Rapeseed Cong. Saskatoon, Canada. pp 250–255.

32. Glen, D.M., Jones, H., Fieldsend, J.K. 1990. Damage to oilseed rape seedlings by the field slug *Deroceras reticulatum* in relation to glucosinolate concentration. Ann. Appl. Biol. 117: 197–207.

33. Mitten, R. 1992. Leaf glucosinolate profiles and their relationship to pests and disease resistance in oilseed rape. Euphytica 63: 71–83.

34. Razin, A., Cedar, H. 1991. DNA methylation and gene expression. Microbiol. Rev. 55: 451–458.

35. Uppstrom, B. 1983. Glucosinolate pattern in different growth stages of high and low glucosinolate varieties of *Brassica napus*. Sveriges utsadesforening. Tidskrift. 93:331-336. Cited in: Milford G.F.J., Fieldsend J.K., Porter A.J.R., Rawlinson C.J., Evans E.J., Bilsborrow P.E.. 1989. Changes in glucosinolate concentration during the vegetative growth of single and double low cultivars of winter oil seed rape. Aspects of Applied Biology 23. Production and Protection of Oilseed Rape and Other Brassica Crops. pp 83–90.

36. Olivieri, A.M., Parrini, P. 1986. Relationship between glucosinolate content and yield components in rapeseed. Eucarpia 11: 126–127.

37. Rakow, G., Raney, J.P. 1995. Field performance of canola quality *Brassica juncea,* in: Rapeseed Today and Tomorrow. Proc. 9th GCIRC Internat. Rapeseed Cong. Cambridge, UK. 2: 428.

38. Malode, S.N., Swamy, R.V., Khalatkar, A.S. 1995. Introgression of 'OO' quality characters in *Brassica juncea* cv. Pusa bold. In: Proc. 9th GCIRC, Internat. Rapeseed Cong. Cambridge, UK, 4–7 July, pp 431–438.

39. Abha Agnihotri, Nutan Kaushik 2000. Incorporation of superior nutritional quality traits in Indian *B. juncea*. Indian Journal of Plant Genetic Resources 12(3): 352–358

40. Downey, R.K., Harvey, B.L. 1963. Method of breeding for oil quality in rape. Can. J. Plant Sci. 43: 271–275

41. Morice, J. 1974. Selection d'une variete de colza sans acide erucique et sans glucosinolates. In. Proc. 4th Internat. Rapeseed Conf. Giessen, West Germany. pp 31–47. Cited in: Banga S.K. 1996. Breeding for oil and meal quality, in: Chopra V.L., Prakash S. (Eds). Oilseed and Vegetable Brassicas: Indian Perspective, Chapter 11, pp. 234–249.

42. Hinata, K., Nishio, T. 1980. Self incompatibility in cruciferae, in. *Brassica* Crops and Wild allies: biology and breeding (Eds) Tsunoda S., Hinata K., Gomez-Campo C. Japan Sci. Press Tokyo. pp 223–234.

43. Abha Agnihotri, Nutan Kaushik 1999. Transfer of double low characteristics in Indian *B. napus*. Journal of Oilseeds Research 16(2): 227–229.

44. Downey, R.K., Rimmer, S.R. 1993. Agronomic improvements in oilseed *Brassica*s. Advances in Agronomy. 50: 1–66.

45. Lichter, R. 1982. Induction of haploid plants from isolated pollen of *Brassica napus*. Z. Pflanzenphysiol. 105: 427–434.

46. Chuong, P.V., Beversdorf, W.D. 1985. High frequency embryogenesis through isolated microspore cultures in *Brassica napus* L. and *B. carinata* Braun. Plant Sci. (Limerich, Irel.) 39: 219–226.

47. Keller, W.A., Arnison, P.G., Cardy, B.J. 1987. Haploids from gametophytic cells – Recent developments and future prospects. Plant Tissue Cell Cult. Proc. Int. Congr., 6th, 1986, 223–241. Cited in: Downey R.K., Rimmer S.R. 1993. Agronomic improvements in oilseed *Brassica*s. Advances in Agronomy. 50: 1–66.

48. Huang, B., Bird, S., Kemble, K., Simmonds, D., Keller, W.A., Miki, B. 1990. Effects of culture density, conditioned medium and feeder cultures on microspore embryogenesis in *Brassica napus* L. cv. Topas. Plant Cell Rep. 8: 594–597.

49. Ulrich, A., Furtan, W.H., Downey, R.K. 1984. Biotechnology and rapeseed breeding: some economic considerations. Sci. Counc. Can. Rep., 1–67. Cited in: Downey R.K., Rimmer S.R. 1993. Agronomic improvements in oilseed *Brassica*s. Advances in Agronomy. 50: 1–66.

50. Rajhathy, T. 1976. Haploid flax revisited. Z Pflanzenzuecht 76: 1–10. Cited in: Downey, R.K., Rimmer, S.R. 1993. Agronomic improvements in oilseed *Brassica*s. Advances in Agronomy. 50: 1–66.

51. Cegielska-Taras, T., Szala, L., Naleczynska, A., Kolodziej, K., Ogrodowczyk, M. 1999. Selection for high erucic acid content in winter oilseed rape (*Brassica napus* L.) on microspore-derived embryos. Journal of Applied Genetics 40(4) 305–315.

52. Downey, R.K., G. Rakow, 1987. Rapeseed and mustard, in: Principles of Cultivar Development (Ed) Fehr W.R. Macmillan, New York. Vol. 2 pp 437–486.

53. Drozdowska, L., Rogozinska, J. 1982. The occurrence of glucosinolates during the flowering and maturation of oilseed rape (*Brassica napus* L.). Acta Agrobotanica 35(1): 25–29.

54. Ali, I., Shah, S.A., Shah, S.J.A., Ahmad, M. Kafyat-ur-Rehman 1999. Direction and magnitude of variability in fatty acid composition of fast neutrons and gamma radiation induced *Brassica juncea*. Cruciferae Newsletter 21: 93–94.

55. Barro, F., Fernandez-Escobar, J., Vega, M., de la, A. Martin, 2001. Doubled haploid lines of *Brassica carinata* with modified erucic acid content through mutagenesis by EMS treatment of isolated microspore. Plant Breeding 120 (3): 262–264.

56. Kumar Arun, Haider, Z.A., Varade, P.B. 2000. Influence of gamma irradiation on oil quality and quantity in *Brassica juncea* L. Journal of Research, Birsa Agricultural University 12(1): 7–10.

57. Maluszynski, M., Nichterlen, K., Vanzanten, L., Ahloowalia, B.S. 2000. Officially released mutant varieties—The FAO/IAEA Database. Mutation Breeding Review. 12: 1–88.

58. Rakow, G. 1973. Selection for content of linoleic and linolenic acid in rapeseed after mutagenic treatment. Z Pflanzenzuecht 69: 62–68. Cited in: Banga S.K. 1996. Breeding for oil and meal quality, in: Chopra V.L., Prakash S. (Eds). Oilseed and Vegetable Brassicas: Indian Perspective, Chapter 11, pp. 234–249.

59. Velasco, L. 1998. Increasing erucic acid content in Ethiopian mustard through mutation breeding. Plant Breeding 17(1): 85–87.

60. Wong, S.C., Swanson, E. 1991. Genetic modification of canola oil: High oleic acid canola, in: Fat and Cholesterol Reduced Food. (C. Haberstroh, C.E. Morris, Eds) pp 154–164. Gulf, Houstan, Texas. Cited in: Downey R.K., Rimmer S.R. 1993. Agronomic improvements in oilseed *Brassica*s. Advances in Agronomy. 50: 1–66.

61. Auld, D.L., Heikkinen, M.K., Ericson, D.A., Sernyk, J.K., Romero, J.E. 1992. Rapeseed mutants with reduced levels of polysaturated fatty acid and increased levels oleic acid. Crop Science 32: 657–662.

62. Murphy. D.J. 1995. The use of conventional and molecular genetics to produce new diversity in seed oil composition for the use of plant breeders—Progress Problems and Future Prospects. Euphytica 85: 433–440.

63. Kinney, A.J. 1997. Development of genetically engineered oilseed: from molecular biology to agronomics pp. 298–301, in: Physiology, Biochemistry and Molecular Biology of Plant Lipids. J.P. Williams, M.U. Khan, N. Wonlem (Eds) Kluwer Acad Press, Netherlands.

64. Singh, Green 1999. Oleochemicals from genetically engineered oil crops. Malasian Oil Science and Technology 8(2): 1–4.

65. Kerr, R.A. 1998. The next oil crisis looms large; and perhaps close. Science. 281: 1128–1131.

66. Brown, P., Shanklin, J., Whittle, E., Somerville, C. 1998. Catalytic plasticity of fatty acid modification enzymes underlying chemical diversity of plant lipids. Science 282: 1315–17.

67. Murphy, D.J., Rawathorne, S., Hills, D.J. 1993. Storage lipid formation in seeds. Seeds Sci. 3: 79–95.

68. Knutzon, D.S., Thomson, G.A., Radke, S.E., Johnson, W.B., Knauf, V.C., Kridl, J.C. 1992. Modification of Brassica seed oil by antisense expression of a stearoyl-acid carrier protein desaturase gene. Proc. Natl. Sci. USA 89: 2624–2628.

69. Stoutjesdijk, P.A., Hurlstore, C., Singh, S.P., Green, A.G. 1999. Genetic manipulation for altered oil quality in Brassica. Proc. Xth Int. Rapeseed Conf, Canberra.

70. Smith, N.A., Singh, S.P., Wang, M.B., Stoutjesdijk, P.A., Green, A.G., Waterhouse, P.M. 2000. Total silencing by intron-spliced hairpin RNAs. Nature 407: 319-320.

71. Murphy, D.J. 1995. Designer oilseed rape—Biotechnological possibilities and commercial realities, in: Rapeseed Today and Tomorrow. Proc. 9th Internat. Rapeseed Cong. Cambridge UK 4: 1322–1334.

72. Marillia, E.F., Zou, J.T., Katavic, V., Qi, Q., Jako, C., Barton, D.L., Friesen, W., Giblin, E.M., Gossen, K.K., Kumar, A., Covello, P.S., Keller, W.A., MacKenzie, S.L., Taylor, D.C. 1999–2000. Metabolic engineering of brassica seeds oils; improvement of oil quality and quantity and alteration of carbon flux, in: Arencibia. A.D. (Ed) Plant Genetic Engineering: Towards the Third Millennium; Proceedings of the International Symposium on Plant Genetic Engineering, Havana, Cuba, 6–10 December, 182–188.

73. Shewmaker, C., Sheehy, J., Daley, M., Colburys, Ke. D.Y. 1999. Seed specific over expression of phytoene synthase: increase in carotenoids and other metabolic effects. The Plant Journal 20(4): 401–412.

74. Voelkar, T.A., Worell, A.C., Anderson, L., Bleibaum, J., Ian, C., Hawkins, D.J. 1992. Fatty acid biosynthesis redirected to medium chains in transgenic oilseed plants. Science 257: 72–73.

75. Voelkar, T.A. et al 1996. Genetic engineering of a quantitative trait: metabolic and genetic parameters influencing the accumalation of laurate in rapeseed. 9(2): 229–241.

76. Strayer, R.C., Blake, J.A., Craig, W.K. 1983. Canola and high erucic rapeseed oil as substitutes for diesel fuel—Preliminary tests. J. Am. Oil Chem. Soc. 60: 1587–1592.

77. Harold, S., Lal, K., Lawate, S. 1995. Varieties of rapeseed oil and derived products used in fuels and lubricants, in: Rapeseed Today and Tomorrow, Proc. 9th GCIRC Internat. Rapeseed Cong. Cambridge, UK. 4: 1341–1344.

Plant Biotechnology and Molecular Markers
P.S. Srivastava, Alka Narula and Sheela Srivastava (Editors)
Copyright © 2004 Anamaya Publishers, New Delhi, India

10. Role of Biotechnology for Incorporating White Rust Resistance in Brassica Species

Kadambari Gupta[2], Deepak Prem[2] and Abha Agnihotri[1]

[1]Bioresources and Biotechnology Division, TERI, Habitat Place, Lodhi Road, New Delhi 110 003, India.

[2]Centre for Bioresources and Biotechnology, TERI–School of Advanced Studies, Habitat Place, Lodhi Road, New Delhi 110003, India

Abstract: The productivity of the oilseed brassica species, the second most important edible oilseed crop in India, is adversely affected by several fungal diseases, white rust being one of the them. White rust caused by *Albugo candida* may cause 17–34% yield losses which may enhance upto 60% under environmental conditions favourable to disease infestation. While most of the cultivated species of brassicas are susceptible to this fungal disease, some sources of white rust resistance have been reported in widely related species. Apart from conventional methods of selection and hybridization, several *in vitro* techniques have been used to utilize these genes for incorporating resistance/tolerance in the cultivated varieties. The article presents a comprehensive status of Indian cultivated brassicas vis-a-vis white rust resistance, and the use of biotechnological tools such as embryo rescue, somatic hybridization, somaclonal variation and molecular techniques for incorporation of disease resistance.

1. Introduction

Oleiferous Brassicas are the important cash crops of India and stands second only to groundnut among the nine annual oilseeds being cultivated. It contributes 27 and 25.3% to the total oilseed production and hectarage, respectively, and is cultivated in about 6.81 million ha with 6.96 m tons production of oilseed [1]. The productivity level of 1022 kg ha^{-1} is far below that of the developed countries (2500–3000 kg/ha) and the world average of 1500 kg/ha (Economic Survey, 2000–2001, GOI). This is mainly due to certain abiotic and biotic factors that adversely influence the average yield of cultivated varieities.

Amongst the abiotic factors, drought, frost and salinity are a major cause of concern since they may cause yield losses to the magnitude of 20–70% [2]. The important biotic factors are weeds and insects-pests causing 17–41% and upto 60% loss, respectively, whereas fungal diseases alone can cause major damage to the crop contributing to a yield loss of upto 70% under favourable conditions for the disease infestation.

The fungal diseases that attack Brassica species in India throughout the cultivated areas include *Alternaria* blight caused by *Alternaria brassicae* and *Alternaria brassicicola*, white rust caused by *Albugo candida* and Downy mildew caused by *Peronospora parasitica*. The other diseases that attack Brassicas, but are distributed more commonly in the temperate regions include sclerotinia stem rot caused by *Sclerotinia sclerotiorum*, blackleg by *Leptosphaeria maculans,* club root by *Plasmodiophora brassicae* and powdery mildew caused by *Erysiphe cruciferarum.*

Among the abovementioned fungal diseases white rust has emerged as a major limiting factor in production of Brassicas causing a loss of 17–34% [3, 4] which may reach upto 60% depending upon the severity of infection and environmental conditions [3, 5-9]. In addition, Downy mildew that alone does not cause much of damage, when combined with white rust causes synergistic damage resulting upto 35% yield loss [10].

B. juncea, the most predominantly grown Brassica species in India, is highly susceptible to white rust disease. Although *B. nigra, B. oleracea, B. napus, B. carinata* and some species of *B. campestris* have been reported to be comparatively tolerant to this disease, adequate amount of resistance is not available in cultivated Brassica species.

This article discusses the white rust disease in terms of its symptoms, effect on plant system, its physiology, inheritance, available sources of resistance along with the biotechniques utilized to achieve the adequate amount of resistance in Brassicas.

2.   White Rust

2.1   Symptoms
White rust caused by fungal pathogen *Albugo candida* (Pers.) Kunzee belonging to family Albuginacae appears in almost all rapeseed mustard growing states of India. *A. candida* can infect all above ground parts of the plant, producing characteristic white blisters known as sori [10, http: // www. extento. hawaii. edu/kbase/crop/Type/a_candi.htm]. The fungal pathogen attacks the plant at both vegetative and reproductive phase. In the vegetative phase the fungal pathogen infects leaves and cotyledons causing local infection resulting in the appearance of white to creamy yellow pustules on the abaxial (lower) surface corresponding to tan yellow pustules on the adaxial (upper) surface of the leaves such that disease can be easily recognized from the upper surface of the affected leaves. The pustules rupture after maturity and release white coloured dust of spores known as sporangia. With the increase in duration of disease, tissues around the pustules become necrotic and lead to senescence of leaves.

At the flowering stage the fungus causes systemic infection, leading to extensive distortion, hypertrophy, hyperplasia and sterility resulting in severe inflorescence malformation known as staghead [10, 11]. This leads to early foliar infection and abnormalities in reproductive organs leading to complete sterility. This systemic staghead infection of the inflorescence is often in association with *Peronospora parasitica* [4, 12, 13]. However in a recent report it has been elucidated that the inflorescence malformation in *B. juncea* is due to *A. candida* and not because of *P. parasitica* [14].

2.2   Effect on Plant
White rust has a significant impact on the yield and quality of seeds. It also has a profound effect on important end products such as total oil content, fatty acid composition and seed protein content. The fungal infection tends to decrease dry matter, and increase erucic acid [15]. High proportion of erucic acid is reported to cause impaired myocardial conductance and increased blood cholesterol and is thus nutritionally undesirable [16]. A positive correlation exists between the amount of chlorophyll, sugars, flavonoids, waxy deposition on leaves, total phenols and the extent of infection by *Albugo candida*. The moderately resistant cultivars contain higher amount of the abovementioned biochemicals than the susceptible cultivars at all stages of growth

[17-21]. Phenols, in particular, have been reported to impart resistance whereas more proteins led to higher disease severity [22, 23]. Thus, white rust infection not only damages the plant morphology but also disrupts its physiological metabolism.

2.3   Biology of the Fungus

The biggest challenge in breeding white rust resistant brassicas is in the fact that as many as 13 pathotypes of *Albugo* parasitize different cruciferous plant species [24-28]. Besides infecting cruciferous plant species, *Albugo* also finds its host in several wild species of different families such as *Portulacaceae, Chenopodiaceae, Amaranthaceae, Convolvulaceae, Boraginaceae* to name a few. The white rust races are classified based upon their ability to infest different host species. The different races that infect Brassica species are given in Table 1. However, these races at times, may not retain their species specificity and can also attack the related species, i.e. host specificity in *Albugo candida* is not an absolute adaptation to a particular species especially when the races are from hosts sharing a common genome.

Table 1.   Various pathotypes of white rust and their specific host species

White Rust Pathotypes	Host Species
Race 1	*Raphanus sativus*
Race 2	*Brassica juncea*
Race 3	*Armoracia rusticana*
Race 4	*Capsella bursa pastoris*
Race 5	*Sisymbium officinale*
Race 6	*Rorrippa islandica*
Race 7	*Brassica campestris*
Race 8	*Brassica nigra*
Race 9	*Brassica oleracea*
Race 10	*Sinapis alba*
Race 11	*Brassica carinata*
Race 12	*Brassica juncea* (Indian isolates)
Race 13	*Brassica campestris* var. *toria* (Indian isolates)

Source: Singh et al. [28].

2.4   Inheritance of Resistance

The information on the genetics of host parasite interaction for white rust has centered on the level of specificity between the races of pathogen and genotypes of related host species [29]. Genetic analysis of available white rust resistance through biometrical techniques has elucidated a digenic mode of inheritance with duplicate gene action in *B. napus* [9, 30] and monogenic dominant resistance in *B. juncea* [31-38]. Cheung et al [39] and Prabhu et al. [40] have confirmed this through gene mapping. The white rust resistance in the three Brassica species, *B. campestris, B. nigra* and *B. carinata*, is reported to be under the control of a single dominant gene [27, 41-44]. It is suggested that a few major genes in *Brassica* are responsible to initiate the disease resistance whereas other minor genes may be involved in the control of the intensity of sporulation of the fungus in the plant [45]. However, additive genetic variations were also found to be predominant for the intensity of white rust resistance [35, 46-50] and thus differential expressions have been obtained.

2.5 Sources of Resistance

The traits for resistance to white rust are found to be present in some species of Brassica as well as in related weedy and wild speices. Among the various *Brassica* species grown in India, *B. napus* and most cultivars of *B. oleracea* [41, 51, 52], some species of *B. campestris* [53-57] and *B. carinata* [58] have been reported to exhibit moderate resistance and thus utilized as a source of resistance to white rust. Recently moderate resistance has also been found in *B. tournefortii* [59, 60] and in certain species of related genus *Diplotaxis* [61] and *Sinapis alba* [58]. *Eruca sativa* has been identified as a potential source of white rust resistance and all the accessions of this genera are reported to be resistant to race 2 that attacks *B. juncea* [62].

3. Disease Control Strategies

The different strategies adopted to control plant diseases include non-chemical and chemical control. The non-chemical control includes hot water treatment and biological control [63, 64]. The chemical control, though found to be effective, results in development of resistance in the pathogens and residual toxicity [10], thus having detrimental effect on non-target species. Besides this, the fungicide sprays affect crop physiology independent of disease occurrence. These may decrease the triacylglycerol fraction of the oil and increase the diacylglycerol fraction [15], thus affecting the oil quality. Owing to the abovementioned problems associated with chemical control, focus has been to develop new biotechnological techniques for crop protection and production.

An upcoming alternative approach to crop protection is the use of externally applied biotic and abiotic stress inducers that activate plant's natural defence mechanism. They create a hypersensitive response in the plant thus leading to systemic acquired resistance. For instance, actigard, an isonicotinic acid derivative identified by Syngenta has been recently commercially utilized to prevent downy mildew on spinach [65]. Under such biological stresses, plant synthesize a variety of compounds that include phytoalexins and pathogenesis-related proteins [66-68]. However, work in this direction is still in infancy and commercial products are yet to be realized on a large scale.

Among the biological approaches, exploitation of genetic resistance, present in the existing plant species for its incorporation in the cultivated varieties, is seen as the eco-friendly and environmentally safe approach [10, 69]. This includes sexual hybridization, wide hybridization, somatic hybridization, somaclonal variations and genetic engineering. The possibility of using these biotechniques to enhance the scope and efficiency of transfer of desired traits with special emphasis on white rust for improvement of crop Brassicas are briefly discussed.

3.1 Selection and Hybridization

Several germplasm lines and popular cultivars have been screened for white rust resistance and varying degree of response to *A. candida* has been reported in *B. juncea* [58, 70–75]. However, it was found that the selected lines are not stable for the trait, and breakdown of resistance occurs in successive years [74]. The breakdown of resistance may be due to mutations in the existing pathotypes leading to new pathotypes. This indicates that there is a continuous need to expand the genetic base for white rust resistance. Intervarietal transfer of disease resistance has been attempted and resistant F_2 generation plants have been selected by Chauhan et al. [76]. However, response of the plants to pathogen is yet to be studied under field conditions.

B. juncea cultivars have also been hybridized with other species of *Brassica* such as *B. napus* [52, 77, 78] and advanced progenies of the cross were identified to possess similar degree of response to white rust as the resistant donor. *B. carinata* has also been utilized for transfer of disease resistance in *B. juncea* and moderate disease tolerance was observed in the hybrids [51, 79]. However, *B. juncea* accessions having resistance to white rust are under different stages of development.

3.2   Wide Hybridization

Genes conferring resistance to biotic and abiotic stress are frequently scattered in weedy and widely related species that can be used for the incorporation of resistance in cultivated varieties [80]. However, majority of these belong to secondary and tertiary gene pools. Their exploitation is problematic because of the difficulties in obtaining hybrids and subsequent gene transfer in desirable genetic background due to pre- and post-fertilization barriers.

Pre-fertilization considerations include spatial separation, asynchrony of flowering, pollination system, floral characteristics and competitiveness of pollen whereas post-fertilization considerations include genetic/sexual compatibility, hybrid viability, fertility of progeny and successful introgression. For successful gene introgression all pre- and post-fertilization requirements must be met, failure of any one requirement will lead to non-introgression of the gene, and thus would not produce the desirable results [81, 82]. A large number of interspecific and intergeneric sexual hybrids have been produced using *in vitro* techniques to study the compatability barriers that exists between the species [83]. These techniques have been scantily utilized in Brassicas for transfer of disease resistance traits, however, there are a few reports for transfer of *Alternaria* blight resistance. Chevre et al [84] generated hybrids between *B. napus* and *S. alba* through ovary culture to transfer resistant traits for *Alternaria* blight. Similar study was made utilizing *B. campestris* and *B. spinenscens* through sequential embryo rescue technique by Agnihotri et al. [85]. However, *Alternaria blight* resistant genotypes are yet to be realized and only a preliminary report by Gupta and Agnihotri [86] is available for transfer of resistance to white rust in *Brassica* species utilizing this technique.

3.3   Somatic Hybridization

Somatic hybridization involves enzymatic removal of the cell wall and the resulting spherical protoplasts are fused together. Fusion of protoplasts at the level of plasma membrane is non-specific, and there is no barrier to interspecific, intergeneric or even intertribal fusion of cells. The resulting hybrid cells are cultured and subsequently regenerated to give rise to somatic hybrids. The use of this technique can bypass both pre- and post-fertilization barriers [87, 88]. Among the cultivated *Brassica* species, the main focus so far has been on *B. napus* and *B. oleracea* and this technique has not been utilized effectively in other *Brassica* species. Attempts have been made to transfer resistance traits to some fungal diseases in Brassicas through somatic hybridization such as *Leptosphaeria maculans* (Black leg) in *B. napus* [89-96] and *B. olearacea* [97]; *Plasmodiophora brassicae* in *B. napus* [98] and *B. oleracea* [99]; *Alternaria* blight in *B. napus* [98, 100] and *B. oleracea* [101-105]. Therefore, as apparent several studies have been undertaken to transfer black leg, downy mildew and *Alternaria* blight disease resistance, mainly in *B. napus* and *B. oleracea*, and only a few reports are available for transfer of white rust resistance in *B. oleracea* [106, 107] and *B. juncea* [108]. However, in both the cases the somatic hybrids obtained were sterile and could not be utilized further. Hence, production of hybrids that are either sterile or do not survive, mainly due to meiotic irregularities, is the major drawback of this technique.

3.4 Somaclonal Variation

Somaclonal variation, regarded as the spontaneous epigenetic variations that occur *in vitro,* have been a source of genetic variation suitable for crop improvement. Somaclonal variations have been utilized for many abiotic stress resistance traits, however, it has been scantily used for disease resistance. Somaclones have been selected for salt tolerance [109], high yield [110] and for transfer of *Alternaria* blight disease resistance in *B. juncea* [111]. So far only one study has been published for resistance to white rust in *B. juncea* through generation of somaclones [111], reporting a stable and heritable resistance till R_2 generation in the field.

3.5 Molecular Techniques

Recent developments in DNA marker technology has led to a better understanding of the complex genome of various crop plants. Molecular markers that are tightly linked to the trait of interest, besides helping in identifying the desired species at any growth stage of the plant [112], also helps to select for the trait under strict quarantine laws [113-115]. However, even then the crop plants have to be tested for virulence against the pathogen to confirm the effectiveness of the marker associated with the resistant gene [116]. Marker assisted selection, or MAS as it is commonly known, has been successfully utilized in identifying oil quality in *B. napus* [117-119] and *B. campestris* [120, 121], seed coat colour in *B. napus* [122] and *B. juncea* [123], and for fungal disease *Leptosphaeria maculans* in *B. napus* [44, 124].

Work is in progress to identify the genes responsible for resistance to white rust for use in molecular assisted selection. In *B. juncea,* resistance to white rust race 2 was observed to be controlled by a single dominant allele. With the help of restriction fragment length polymorphism (RFLP) a locus Acr [39] and Ac21 [40] have been identified in *B. juncea.* Recently, flanking markers have been identified for a white rust resistant locus, AcAl in *B. napus* [34] and in *B. campestris* [45], and Ac2t in a Polish *B. juncea* accession [125]. Similarly, 3 genes namely Ac71, Ac72 and Ac73 have been identified for resistance to white rust race 7 in *B. campestris* [43]. However, the use of these markers in molecular assisted selection has not been successful as yet [39, 40, 43] mainly because of their specificity to their respective host species.

Work has been undertaken to identify the molecular markers that could be used in precise and efficient screening. Two markers, WR2 and WR3 [40] and OPNO11000 and OPBO61000 [125] linked to white rust resistance have been identified which flank the resistant locus. Prabhu et al. [40] have reported that although these markers were effective in identifying the presence or absence of the resistance gene in the population of the cultivars, these are specific to the Russian source of white rust resistance. Work is in progress to study the mechanism of resistance response and mapping of the loci responsible for resistance to white rust [43, 39]. Although, a few markers linked to white rust resistance locus have been identified in some species of Brassica, work has to be consolidated to employ these markers in routine marker-assisted selection for efficient utilization. Furthermore, the focus has now been shifted from identifying trait linked markers to the mapping of the genes to utilize them more efficiently in developing new cultivars [126].

During the past decade, different strategies have been used to produce transgenic plants that are less susceptible to disease caused by phytopathogenic fungi and bacteria [127]. For achieving transgene derived resistance, genes from organisms other than plants and endogenous plant genes have been utilized. The basic concept of utilizing these genes revolves around induction of systematically acquired resistance utilizing transgene mediated production of pathogenesis related (PR) proteins [127]. Almost 14 distinct PR-protein groups have been identified from different plants [68, 128] and have been utilized for transgene mediated production of PR-proteins. This technique has been successfully utilized in cereals [68, 129, 130, 131]. However, transgene mediated resistance for white rust has not been exploited in crop Brassicas because of poor understanding of pathogenesis related proteins and their role in trigerring systematically acquired resistance towards *Albugo candida*.

4. Conclusion

Oilseed Brassicas contribute to about 30% of the edible oilseeds being produced in India. However, in spite of the horizontal increase in the area and production, vertical increase in productivity per unit area has remained far below the yield potential of presently cultivated varieties. This is mainly due to various abiotic and biotic stresses, fungal diseases being one of them. The major fungal diseases affecting oilseed Brassicas are *Alternaria* blight, white rust and downy mildew, which together cause severe yield losses under environmental conditions favouring disease infestation. White rust caused by *Albugo candida* is reported to be under the control of digenic inheritance with duplicate gene action in *B. napus* and monogenic dominant resistance in *B. juncea*. Upto 13 races of *Albugo candida* affecting different *Brassica* species have been reported that are host specific. Strategies for controlling white rust have mainly focussed on development of disease resistant cultivars, exploiting the resistance available within crop species and also utilizing the modern biotechnological tools to tap resistance from secondary or tertiary gene pool. Utilization of resistance genes from wide species has been a promising proposition since some of the widely related species of crop Brassicas such as *Eruca sativa*, species of *Diplotaxis* and *B. tournefortii* have been reported to have resistance/tolerance to white rust. These sources of resistance have been utilized by various scientists for introgression of resistance genes in the cultivated varieties. Some work has also been undertaken to identify and clone the resistance genes and develop molecular markers for precise selection. Systematic characterization of the Indian gene pool of Brassicas and its related species for identification of white rust resistant genes is important and work in this direction is being pursued at various national and international institutions. Utilization of these genes through the use of biotechnological tools will help in expediting the development of varieties having resistance to white rust.

References

1. Kumar, P.R., Chauhan, J.S., Singh, A.K. 2000. Rapeseed mustard genetic resources: status and priorities. Indian J Plant Genet Res 13(3): 207–218.
2. Chopra, V.L., Prakash, S. 1996. Oilseeds and Vegetable Brassicas. An Indian Perspective. Oxford & IBH Publishing Co. Pvt. Ltd. New Delhi.
3. Saharan, G.S., Kaushik, C.D., Gupta, P.P., Tripathi, N.N. 1984. Assessment of losses and control of white rust of mustard. Indian Phytopathol 37: 397.

4. Bisht, I.S., Agrawal, R.C., Singh, R. 1994. White rust (*Albugo candida*) severity in mustard (*Brassica juncea*) varieties and its effects on seed yields. Plant Var Seeds 7(2): 85–89.

5. Bernier, C.C. 1972. Diseases of Rapeseed in Manitoba in 1971. Can. Plant. Dis. Surv. 52: 108.

6. Harper, F.R., Pittman, U.J. 1974. Yield loss in *Brassica campestris* and *Brassica napus* from systemic stem infection by *Albugo cruciferatum*. Phytopathol 64: 408–410.

7. Petrie, G.A., Vanterpool, T.C. 1974. Fungi associated with hypertrophies caused by infection of Cruciferae by *Albugo cruciferarum*. Can Plant Dis Surv 54: 37–42.

8. Bains, S.S., Jhooty, J.S. 1979. Mixed infection by *Albugo candida* and *Peronospora parasitica* on *Brassica juncea* inflorescence and their control. Indian Phytopathol 32: 268–271.

9. Verma, V., Bhowmik, T.P. 1989. Inheritance of resistance to a *Brassica juncea* pathotype of *Albugo candida* in *Brassica napus*. Can. J. Plant Pathol 11: 443–444.

10. S.J. Kolte, 1996. Diseases, pp. 184–207, in: Oilseeds and Vegetable Brassicas. An Indian Perspective (Eds) Chopra V.L., Prakash S., Oxford & IBH Publishing Co Pvt Ltd New Delhi 293.

11. Verma, P.R., Petrie, G.A. 1980. Effect of seed infestation and flower bud inoculation on systemic infection of turnip rape by *Albugo candida*. Can J Plant Sci 60: 267–271.

12. Petrie, G.A. 1988. Races of *Albugo candida* (white rust) and staghead on cultivated cruciferae in Saskatchewan. Can J Plant Pathol 10: 142–150.

13. Awasthi, R.P., Nashaat, N.I., Heran, A., Kolte, S.J., Singh, U.S. 1997. The effect of *Albugo candida* on the resistance to *Peronospora parasitica* and vice versa in rapeseed mustard, in: ISHS Symposium Brassicas: 10th Crucifer Genet Workshop, ENSAR-INRA, Rennes, France, 49.

14. Nashaat, N.I., Heran, A., Mitchell, S.E., Awasthi, R.P. 1997. New genes for resistance to downy mildew (*Peronospora parasitica*) in oilseed rape (*Brassica napus* ssp. *oleifera*). Plant Pathol 46 (6): 964–968.

15. McCartney, H.A., Doughty, K.S., Morton, G., Booth, E.J., Kightey, S.P.J., Landon, G., West, G., Walker, K.C., Thomas, J.E. 1999. A study of the effect of disease on seed quality parameters of oilseed rape, in: Proceedings of 10th International Rapeseed Congress 26–29 Sept Canberra, Australia.

16. Kramer, J.K. 1998. Hematological and lipid changes in new born piglets fed milk replacer diets containing erucic acids. Lipids 33: 1–10.

17. Yadav, O.P., Yadav, T.P., Kumar, P. 1994. Inheritance of white rust resistance in Indian mustard. Indian Phytopathol 47(2): 159–163.

18. Rao, M.V.B., Raut, R.N. 1994. Inheritance of resistance to white rust (*Albugo candida*) in an interspecific cross betwen Indian mustard (*Brassica juncea*) and rapeseed (*B. napus*). Indian J Agri Sci 64 (4): 249–251.

19. Gupta, M.L., Singh, G., Ahuja, I., Raheja, R.K., Ahuja, K.L., Banga, S.K. 1997. Chlorophyll content in relation to white rust (*Albugo candida*) resistance in Indian mustard. Cruciferae Newslett 19: 105–106.

20. Singh, G., Gupta, M.L., Ahuja, I., Raheja, R.K. 1998. Biochemical traits in relation to white rust resistance in Indian mustard (*B. juncea* L. Coss). Crop Improv 25(1): 48–52.

21. H.V. Singh, 2000. Biochemical basis of resistance in Brassica species against downy mildew and white rust of mustard. Plant Dis Res 15(1): 75–77.

22. O.P. Yadav, T.P. Yadav, P. Kumar, S.K. Gupta, 1996. Inheritance of phenols and protein in relation to white rust (*Albugo candida*) resistance in Indian mustard. Indian J Genet Plant Breed 56(3): 256–261.

23. V. Pruthi, H.K.L. Chawla, G.S. Saharan, 2001. *Albugo candida* induced changes in phenolics and glucosinolates in leaves of resistant and susceptible cultivars of *Brassica juncea*. Cruc Newslett 23: 61–62.

24. Pound, G.S., Williams, P.H. 1963. Biological races of *Albugo candida*. Phytopathol 53: 1146–1149.

25. Willams, P.H. 1985. CRGC Resource Book. Deptt. of Plant Pathology, Univ. of Wisconsin, Madison, USA, pp. 1–7.

26. Hill, C.B. Crute, I.R. Sherriff, C., Williams, P.H. 1988. Specificity of *Albugo candida* and *Peronospora parasitica* pathotypes toward rapid-cycling crucifers. Cruc Newslett 13: 112–113.

27. Liu, Q., Rimmer, S.R. 1991. Inheritance of resistance in *Brassica napus* to an Ethiopian isolates of *Albugo candida* from *Brassica carinata*. Can J Plant Pathol 13(3): 197–201.

28. Singh, R., Bartaria, A.M., Singh, U.C., Pandya, R.K. (1998). White rust of rapeseed and mustard problems and progress—A review. *Agric. Rev* 19(4): 211–226.

29. Saharan, G.S. 1992. Disease Resistance, 1992. In: Breeding Oilseed Brassicas (Eds) Labana, K.S., Banga, S.S., Banga, S.K. Narosa Publishing House, New Delhi, pp. 181–200.

30. Fan, Z., Rimmer, S.R., Stefansson, B.R. 1983. Identification of resistance to *Albugo candida* in rape (*Brassica napus* L.). Can J Genet Cytol 25: 420–424.

31. Tewari, A.S., Petric, G.A., Downey, R.K. 1988. Inheritance of resistance to *A. candida* race 2 in mustard (*B. juncea* (L.) Czern). Can J Plant Sci, 68: 297–300.

32. Paladhi, M.M., Prasad, R.C., Dass, B. 1993. Inheritance of field reaction to white rust in Indian mustard. Indian Phytopathol 41: 96–99.

33. Rimmer, S.R., Buchwaldt, L. 1995. Diseases, in: Brassica Oilseeds Production and Utilization. (Eds) Kimber D., Mcgregor D.I. CAB International, Oxford, UK, 111–140.

34. Kole, C., Teutonico, R., Mengistu, A., Williams, P.H., Osborn, T.C. 1996. Molecular mapping of a locus controlling resistance to *Albugo candida* in *Brassica rapa*. Genetics 6(4): 367–369.

35. Sridhar, K., Raut, R.N. 1998. Differential expression of white rust resistance in Indian mustard (*Brassica juncea*). Indian J Genet Plant Breed 58(3): 319–322.

36. Bansal, V.K., Thiagarajah, M.R., Stringam, G.R., Tewari, J.P. 1999. Inheritance of partial resistance to race 2 of *Albugo candida* in canola-quality mustard (*Brassica juncea*) and its role in resistance breeding. Plant Pathol 48(6): 817–822.

37. Sachan, J.N., Kolte, S.J., Singh, B. 2000. Inheritance of white rust (*Albugo candida* race 2) in *Brassica juncea*. Indian Phytophythol 53(2): 206–209.

38. Chauhan, S.K., Sharma, J.B. 2001. Inheritance of white rust resistance in Indian mustard incorporated from *Brassica napus*. Indian J Genet Plant Breed 61(3): 250–252.

39. Cheung, W.Y., Gugel, R.K., Landry, B.S. 1998. Identification of RFLP markers linked to the white rust resistance gene (*Acr*) in mustard (*Brassica juncea* (L.) Czern. And Coss.). Genome 41(4): 626–628.

40. Prabhu, K.V., Somers, D.J., Rakow, G., Gugel, R.K. 1998. Molecular markers linked to white rust resistance in mustard *Brassica juncea*. Theor Appl Genet 97(5/6): 865–870.

41. Delwiche, P.A., Williams, P.H. 1974. Resistance to *Albugo candida* race 2 in *Brassica* sp. Proc Am Phytopathol Soc 1: 66.

42. Delwiche, P.A., Williams, P.H. 1981. Thirteen marker genes in *Brassica nigra*. J Hered 72: 289–190.

43. Liu, J.Q., Parks, P., Rimmer, S.R. 1996. Development of monogenic lines for resistance to *Albugo candida* from a Canadian *Brassica napus* cultivar. Phytopathol 86(9): 1000–1004

44. Tanhuanpaa, P., Vilkki, J. 1999. Tagging of a locus for resistance to *Albugo candida* in *Brassica rapa* ssp. *oleifera*, in: Proceedings of 10th International Rapeseed Congress, 26–29 September, Canberra, Australia.

45. Ferreira, M.E., Williams, P.H., Osborn, T.C. 1995. Mapping of a locus controlling resistance to *Albugo candida* in *Brassica napus* using molecular markers. Phytopathol 85(2): 218–220.

46. Yashpal, Singh, H. 1991. Genetic component of white rust resistance and seed yields in Indian mustard. J Oilseed Res 8: 259–261.

47. Yadav, O.P., Yadav, T.P., Kumar, P., Gupta, S.K. 1994. Inheritance of sugars in relation to white rust resistance in Indian mustard. Indian Phytopathol 47(1): 56–59.

48. Pal, R., Kumar, P. 1996. Genetics of white rust (*Albugo candida*) resistance in Indian mustard. Indian J Agri Res 30(2): 85–90.

49. Mani, N., Gulati, S.C., Raman, R. 1996. Breeding for genetic resistance to white rust in Indian mustard. Crop Improv 23(1): 75–79.

50. Sheikh, I.A., Singh, J.N. 2001. Genetics of resistance to white rust in Indian mustard (*Brassica juncea* L. Czern & Coss). Crop Res 21(3): 341–344.

51. Singh, D., Singh, H. (1987). Genetic analysis of resistance to white rust in Indian mustard, in: 7th International Rapeseed Congress, Pozman, Poland, 11–14 May, pp. 126.

52. Gugel, R.K., Raney, J.P., Duczek, L.J., Rakow, G.F.W., Olson, T.V. 1999. Resistance to white rust in *Brassica juncea* from an interspecific cross with *B. napus*, in: Proceedings of 10th International Rapeseed Congress, 26–29 September, Canberra, Australia.

53. Ebrahimi, A.G., Delwich, P.A., Williams, P.H. 1976. Resistance in *B. juncea* to *Peronospora parasitica* and *Albugo candida* Race 2 Proc Am Phytopathol Soc 3: 273 (Abstr.).

54. Katiyar, R.K., Chopra, V.L. 1990. Somaclonally induced earliness in a *B. juncea* germplasm accessions with field resistance to important disease. Plant Breeding. 104: 262–264.

55. Kolte, S.J. 1987. Importance diseases of rapeseed and mustard in India: Present research progress and future research needs, in: Proc IDRC Third Oilcrops Network Workshop, Addis Ababa, Ethiopia, October 6–10, 1986, pp. 91–106.

56. Kolte, S.J., Tewari, A.N. 1980. Note on the susceptibility of certain oleiferous Brassicae to downy mildew and white blister disease. Indian J Mycol Plant Pathol 10: 191–192.

57. Parui, N.R., Bandyopadhyay, D.C. 1973. A note on the screening of rai (*B. juncea*) against white rust (*Albugo candida*). Curr Sci 42: 798–799.

58. Saharan, G.S., Kaushik, C.D., Kaushik, J.C. 1988. Sources of resistance and epidemiology of white rust of mustard. Indian Phytopathol 41(1): 96–99.

59. Yadav, R.C., Sareen, P.K., Chowdhury, J.B. 1991. Interspecific hybridisation in *Brassica juncea* × *Brassica tournefortii* using ovary culture. Cruc Newslett 14/15: 84.

60. Ljungberg, A., Cheng, B.F., Heneen, W.K. 1993. Investigation of hybrids between *Brassica tournefortti* Gouan and *B. alboglabra* Bailey. Sveriges Utsadesforenings Tidskrift. 4: 191–197.

61. Gupta, S., Sharma, T.R., Chib, H.S. 1995. Evaluation of wild allies of *Brassica* under natural conditions. Cruc Newslett 17: 10–11.

62. Bansal, V.K., Tewari, J.P., Tewari, I., Gomez-Campo, C., Stringam, G.R. 1997. Genus Eruca: a potential source of white rust resistance in cultivated Brassicas. Plant Genet Res Newslett 109: 25–26.

63. Sharma, A.K., Gupta, J.S., Maheshwari, R.K. 1984. The relationship of *Streptomyces arabicus* to *Alternaria brassicae* (Berk) Sacc and *Alternaria brassicicola* (Schew) Wiltshire on the leaf surface of yellow sarson and taramira. Geobis New Reports 3: 83–84.

64. Sharma, A.K., Gupta, J.S., Singh, S.P. 1985. Effect of temperature on the antifungal activity of *Streptomyces arabicus* against *Alternaria brassicae* (Berk) Sacc and *A. brassicicola* (Schew) Wiltshire. Geobis 12: 168–169.

65. Moffat, A.S. (2001). Finding new ways to fight plant diseases. Science 292: 2270–2273.

66. Collinge, D.B., Slusarenke, A.J. 1987. Plant gene expression in response to pathogens. Plant Mol Biol 9: 389–401.

67. Linthorst, H.J.M. 1991. Pathogenesis-related proteins of plants. Crit Rev Plant Sci 10: 123–150.

68. Muthukrishnan, S., Liang, G.H., Trick, H.N., Gill, B.S. 2001. Pathogenesis-related proteins and their genes in cereals. Plant Cell Tiss Org Cult 64: 93–114.

69. Dhillon, S.S., Singh, G., Singh, S., Singh, K. 1992. Variability of *Alternaria* blight resistance in newly developed strain of *Brassica juncea*. Plant Dis Res 7(1): 48–53.

70. Khetmalas, M.B., Lambhate, S.S., Tendulkar, A.V., Rodage, R.G. 1994. Reaction of mustard cultivars against white rust. J Maharashtra Agri Univ 19(2): 288–289.

71. Kolte, S.J., Awasthi, R.P., Vishwanath. 1994. Divya mustard: a unique plant type and its developmental traits in disease management. Cruc Newslett 16: 128–129.

72. Katiyar, R.K., Chamola, R. 1995. Useful end products from *Brassica juncea* × *B. carinata* and *Brassica juncea* × *B. campestris* crosses. Cruc Newslett 17: 20–21.

73. Pandey, I.D., Singh, B., Sachan, J.N. 1999. Brassica hybrid research in India: status and prospects, in: Proceedings of 10th International Rapeseed Congress, 26–29 September, Canberra, Australia.

74. Sheikh, I.A., Singh, J.N. 1999. The inheritance of resistance to white rust in Indian mustard (*Brassica juncea* L. Czern & Coss.). In: Proceedings of 10th International Rapeseed Congress, 26–29 September, Canberra, Australia.

75. Yadav, M.S., Dhillon, S.S., Singh, K., Brar, K.S. 1999. Screening of Indian mustard germplasm for resistance to *Alternaria* blight and white rust. Plant Dis Res 14(1): 70-72.

76. Chauhan, Y.S., Kumar, K., Maurya, K.N., Singh, R.B. 1999. Development of NDM 87-1, a white rust resistance strain of mustard (*Brassica juncea* L. Czern & Coss.). Cruc Newslett 21: 117–118.

77. Franke, C., Pottes, D.A., Males, D.R. 1999. The development and genetics of white rust (*Albugo candida*) race 2V resistant canola quality *Brassica juncea*, in: Proceedings of 10th International Rapeseed Congress, 26–29 September, Canberra, Australia.

78. Pal, S.S., Gupta, T.R., Vineet Kumar, Dhaliwal, H.S. 1999. Transfer of white rust resistance from *Brassica napus* to *B. juncea* cv. RLM 198. Crop Improv 26(2): 249–251.

79. H. Singh, D. Singh, T.P. Yadav, 1988. Comparative performance of the genotypes of Indian and Ethiopian mustard under semi-arid regions of India. Cruc Newslett 13: 36–37.

80. Tsundo, S. 1980. Eco-physiology of wild and cultivated forms in Brassica and allied apecies, in: Brassica crops and wild allies, Biology and Breeding (Eds) Tsunoda S, Hinata K, Gomez-Campo C, 109–120, Japan Scientific Society Press, Tokyo.

81. Raghavan, V. 1986. Variability through wide crosses and embryo rescue. In: Cell culture and somatic cell genetics of plants. (Ed) Vasil I.K. Academic Press, Oriando 2: 613–633.

82. Bing, D.J., Downey, R.K., Rakow, G.F.W. 1995. An evaluation of the potential gene transfer between *Brassica napus* and *Sinapis arvensis*. Plant Breeding 114: 481–484.

83. Shivanna, K.R. 1996. Incompatability and wide hybridization, in: Oilseed and Vegetable Brassicas: Indian perspective. Eds, V.L. Chopra, S Prakash, Oxford & IBH Publishing Co. Pvt. Ltd., New Delhi, 77–102.

84. Chevre, A.M., Eber, F., Margale, E., Kerlan, M.C. 1997. Comparison of somatic and sexual *Brassica napus-Sinapis alba* hybrids and their progent by cytogenetic studies and molecular characterization. Genome 37: 367–37.

85. Agnihotri, A., Laxmikumaran, M., Shivanna, K.R., Jagannathan, V. 1990. Embryo rescue of interspecific hybrids of *Brassica spinencens* × *Brassica campestris* and DNA analysis, in: Current Plant Science and Biotechnology, in Agriculture; Progress in Plant Cellular and Molecular Biology, Eds, Niikamp et al., Kluwer Academic Publisher, 270–274.

86. Gupta, K., Agnihotri, A. 2003. Interspecific hybridization and embryo rescue for combining white rust resistance in *Brassica* species, in Proc. National Symposium on Plant Biotechnology, Role in Sustainable Development, Feb. 17–19, 2003, Jaipur, pp. 16.

87. Glimelius, K., Fahleson, J., Landgren, M., Sjodin, C., Sundberg, E. 1995. Improvements of the Brassica crops by transfer of genes from alien species via somatic hybridization, in: Progress in Plant Cellular and Molecular Biology, (Eds) Nijkamp H.J.J., Vander Plas L.H.W., Van Aartrijk J., Kluwer Academic Publishers, Dordrecht, The Netherlands. 299–304.

88. Waara, S., Glimelius, K. 1995. The potential of somatic hybridization in crop breeding. Euphytica 85: 217–223.

89. Liu, J.H., Dixelius, C., Eriksson, I., Glimelius, K. 1995. *B. napus* (+) *B. tournefortii*, a somatic hybrid containing traits of agronomic importance of rapseed breeding. Plant Sci 109: 75–86.

90. Gerdemann-Knorck, M., Sacristan, M.D., Braatz, C., Schieder, O. 1994. Utilization of asymmetric somatic hybridization for the transfer of disease resistance from *B. nigra* to *B. napus*. Plant Breeding 113: 106–113.

91. Gerdemann-Knork, M., Nielen, S., Tzcheetzch, C., Iglish, J., Schleder, O. 1995. Transfer of disease resistance within the genus *Brassica* through asymmetric somatic hybridization. Euphytica 115: 480–483.

92. Brever, E.P., Saunders, J.A., Angle, J.S., Chaney, R.L., McIntosh, M.S. 1997. Somatic hybridization between the zinc accumulator *Thlaspi caerulescens* and *B. napus*. Theo Appl Genet 99: 761–71.

93. Stieve, G., Robbelen, G. 1994. Establishing cytoplasmic male sterility in *B. napus* by mitochondrial recombination with *B. tournefortii*. Plant Breeding 113: 294–304.

94. Sjodin, C., Glimelius, K. 1989. Transfer of resistance against *Phoma lingam* to *B. napus* by asymmetric somatic hybridization combined with toxin selection. Theor Appl Genet 78: 513–520.

95. Plumper, B., Scristan, M.D. 1995. Asymmetric somatic hybrids *Sinapis alba* + *Brassica napus*, in Proc. 9th Int. Rapeseed Congress, Cabridge, UK, pp. 3.

96. Sundberg, E., Langercrantz, U., Glimelius, K. 1991. Effects of cell type used for fusion on chromosome elimination and chloroplst segregation in *B. oleracea* + *B. napus* hybrids. Plant Sci 78: 89–98.

97. Ryschka, U., Schuman, G., Klocke, E., Scholza, P., Herimann, M. 1996. Somatic hybridization in Brassicaceae. Acta Horticulturae, 407: 201–208.

98. O'Neill, C.M., Murata, T., Morgan, C.L., Mathia, R.J. 1996. Expression of the C3-C4 intermediate character in somatic hybrids between *B. napus* and the C3-C4 species *Moricandida arvensis*. Theor Appl Genet 93: 1234–1241.

99. Hagimori, M., Nagaoka, M., Kato, N., Yoshikawa, H. 1992. Production and characterization of somatic hybrids between the Japanese radish and cauliflower. Theor Appl Genet 84: 819–824.

100. Primard, C., Vedel, F., Mathieu, C., Pelletier, G., Chevre, A.M. 1988. Interspecific somatic hybridization between *Brassica napus* and *Brassica hirta* (*Sinapis alba* L.). Theor Appl Genet 75: 546–552.

101. Hansen, L.N., Earle, L.D. 1997. Somatic hybrids between *B. oleracea* and *Sinapis alba* with resistance to *Alternaria* blight. Theor Appl Genet 94: 1078–1085.

102. Jourdan, P.S., Salazae, E. 1993. *B. carinata* resynthesized by protoplast fusion. Theor Appl Genet 86: 567–572.

103. Hansen, L.N. 1998. Intertribal somatic hybridization between rapid cycling *Brassica oleracea* and *Camelina sativa*. Euphytica 104: 173–179.

104. Sharma, T.R., Singh, B.M. 1992. Transfer of resistance of *Alternaria brassicae* in *Brassica juncea* through interspecific hybridization among Brassicas. J Genet and Breed 46: 373–378.

105. Kirti, P.B., Mohapatra, T., Khanna, H., Prakash, S., Chopra V.L. (1995). *Diplotaxis catholica* + *Brassica juncea* somatic hybrids: molecular and cytogenetic characterization. Plant Cell 11: 5–6.

106. Toriyama, K., Hinata, K., Kameya, T. 1987. Production of somatic hybrid plants "Brassicomoricandia", through protoplast fusion between *Moricandida arvensis* and *Brassica oleracea*. Plant Sci, 48: 123–128.

107. Toriyama, K., Yanagino, T., Razmjoo, K., Ishii, R. and Hinata, K. 1988. Chloroplast DNA and CO_2 compensation point of somatic hybrids plants between *B. oleracea* and *Moricanida arvensis*. Jap J Genet 63: 543–547.

108. Kirti, P.B., Prakash, S., Chopra, V.L. 1991. Interspecific hybridization between *Brassica juncea* and *Brassica spinescens* through protoplast fusion. Plant Cell Reports, 9: 639–642.

109. Jain, S., Narnawatee, H.S., Jain, R.K., Chowshary, J.B. 1991. Protein status of genetically stable salt tolerant *Brassica juncea* L. somaclones and their parent cv. Prakash. Plant Cell Rep 9(12): 684–687.

110. Katiyar, R.K., Chopra, V.L. 1990. Somatically induces earliness in *Brassica juncea* germplasm accessions with field resistance to important diseases. Plant Breeding 104(3): 262–264.

111. Sharma, T.R., Singh, B.M. 1995. Generations and evaluations of somaclones of *Brassica juncea* for resistance to *Albugo candida* and *Alternaria brassicae*, in: Proc. Indian National Science Academy, Part B, Biological Sciences 61(2): 155–161.

112. Winter, P., Kahl, G. 1995. Molecular marker technologies for plant improvement. J Microbial Biotech 11: 438–448.

113. Melchinger, A.E. 1990. Use of moelcular markers in breeding for oligogenic disease resistance. Plant Breeding 104: 1–19.

114. Kelly, J.D. 1995. Use of random amplified polymorphic DNA markers in breeding for major gene resistance to plant pathogens. Horticulturae 30(3): 461–465.

115. Lanridge, P., Chalmers, K. 1998. Techniques for marker development, in: Proc. 9th Internations Wheat Genetics Symposium 1(4): 107–117.

116. Michelmore, R.W. 1995. Molecular approaches to manipulation of disease resistance genes. Annual Rev of Phytopathol 15: 393–427.

117. Ecke, W., Uzunova, M., Weissleden, K. 1995. Mapping the genome of rapeseed (*Brassica napus* L.) II. Localization of genes controlling erucic acid synthesis and oil content. Theor Appl Genet 91: 972–977.

118. Jourden, C., Simmonneaux, D., Renard, M. 1996. Selection of pollen for linolenic acid content in rapeseed *Brassica napus* L. Plant Breeding 115(1): 11–15.

119. Hu, J. Li, G. Struss, D., Quiros, C.F. 1999. SCAR and RAPD markers associated with 18-carbon fatty acids in rapeseed, *Brassica napus*. Plant Breeding 118: 145–150.

120. Teutonico, R.A., Osbrown, T.C. 1994. Mapping of RFLP and quantitative trait loci in *Brassica rapa* and comparison to the linkage maps of *B. napus*, *B. oleraceae* and *Arabidopsis thaliana*. Theor Appl Genet 89: 885–894.

121. Tanhuanpaa, P.K., Vikki, J.P., Vikki, H.J. 1996. Mapping of a QTL for oleic acid concentration in spring turnip rape (*Brassica rapa* ssp. *oleifera*). Theor Appl Genet 92: 952–956.

122. Van Deynze, A.E., Landry, B.S., Pauls, K.P. 1995. The identification of restriction fragments length polymorphisms linked to seed colur genes in *Brassica napus*. Genome 38: 534–542.

123.   Negi, M.S., Devic, M., Delseny, M., Laxmikumaran, 2000. Identification of AFLP fragments linked to seed coat colour in *Brassica juncea* and conversion to SCAR marker for rapid selection. Theor Appl Genet 101: 146–152.

124.   Dion, Y., Gugel, R.K., Rakow, G.F.W., Seguin-Swartz, G., Landry, B.S. 1995. RFLP mapping for resistance to the blackleg disease [Causal agent *Leptosphaeria maculans* (Desm.) Ces. Et al Not.] in canola (*B. napus*). Theor Appl Genet 91: 1190–1194.

125.   Mukherjee, A.K., Mohapatra, T., Varshney, A., Sharma, R., Sharma, R.P. 2001. Molecular mapping of a locus controlling resistance to *Albugo candida* in Indian mustard. Plant Breeding 120: 483–487.

126.   Botha, A.M., Venter, E. 2000. Molecular marker technology linked to pests and pathogens resistance in wheat breeding. South African J Science, 96: 233–240.

127.   Herrera-Estrella, L., Simpson, J. 1995. Genetically engineered resistance to bacterial and fungal pathogens. World J Micro Biotech Vol. 11, pp. 383–392.

128.   Van Loon, L.C., Van Strein, E.A. 1999. The families of pathogenesis-related proteins, their activities and comparative analysis of PR-I type proteins. Physiol Mol Plant Pathol 55: 85–97.

129.   Anuratha, C.S., Zen, K.C., Cole, K.C., Mew, T., Muthukrishnan, 1996. Induction of chitinases and β-glucanases in *Rhizoctonia solani* infected rice plants: Isolation of an infection related chitinase cDNA clone. Physiol Plant 97: 39–46.

130.   Repellin, A., Baga, M., Jauhar, P.P., Chibbar, N. 2001. Genetic enrichment of cereal crops via alien gene transfer: New Challenges. Plant Cell Tiss Org Cult 64: 159–183.

131.   Toyoda, H., Matsuda, Y., Yamaga, T., Ikeda, S., Morita, M., Tamai, T., Ouchi, S. 1991. Suppression of the powdery mildew pathogen by chitinase microinjected into barley coleoptile epidermal cells. Plant Cell Rep 10: 217–220.

11. Current Trends in Forest Tree Biotechnology

E.M. Muralidharan and Jose Kallarackal

Division of Genetics and Division of Plant Physiology, Kerala Forest Research Institute,
Peechi 680653, Kerala, India

Abstract: Modern tools of biotechnology offer a variety of options through which it is possible to match the strides made in crop improvement in agriculture and horticulture. Current trends in forest tree biotechnology indicate that this is indeed happening and that some of the hurdles of conventional forest tree improvement are no longer a serious bottleneck. The progress made in *in vitro* culture of forest trees and the current status of application of the technology is discussed. The trends in use of molecular tools particularly the wide variety of DNA markers available and the identification of novel genes controlling traits of interest are examined. The current status of the technology in genetic transformation of forest trees is also reviewed. The bio-safety issues in forest biotechnology especially those relating to transgenic trees are presented without bias to either side of the ongoing debate.

1. Introduction

With growing realization of the ecological role of forests in sustaining life on earth and the consequences of indiscriminate exploitation being felt in several parts of the world, afforestation and reforestation have in recent decades been high on the agenda of most nations. While agriculture and horticulture advanced by leaps and bounds in bringing about a quick domestication of a large number of plant species that supply the world most of its food, fodder and fibre, another primary need, namely, shelter, was based on wood biomass that continued to be taken out from the natural forests. Productive lands being always chosen for growing food crops, the trees have been relegated to the fringes except where agroforestry has been the tradition. Until the latter part of the last century, much of the wood based produce had been taken out of the natural forests, which was considered inexhaustible. It has become clear that this activity is no longer sustainable or environmentally sensible. Unbridled clearing of tropical forests has been particularly severe in several of the poorer nations that happen to be the richest in biodiversity. The loss of biodiversity is often inestimable in some of the biodiversity hot spots.

In consonance with the traditional emphasis given to agriculture, in plant biotechnology too, the accent has always been on crop plants. The relatively short history of domestication, the long lifecycles and the large size of most trees and inaccessibility in the wild have always been a disadvantage to researchers.

However, intensive forestry has become the order of the day when the availability of land and other factors are making traditional forestry practices increasingly unsustainable. Trees with shorter rotations, and genetically improved for disease and pest resistance, superior form etc., have been deployed in plantations in many of the developed nations where genetic improvement programmes has been initiated. The advent of biotechnology in the past two decades, however, broadened the scope of genetic improvement of trees, mainly by removing the hurdles encountered in conventional breeding programmes. Partly due to the environmental concerns and also due to the increasing realization of the advantages of intensively managed plantations of fast growing tree crops, the interest in application of biotechnology to forest crops has been kindled. Mixing of trees with other crop plants (agroforestry) is also gaining more popularity in several countries. Biotechnological research has a major role to play in developing the right kind of trees for inter-planting among annual crops [1]. While the role of traditional knowledge and conventional technology in conservation and utilization of plant resources cannot be underplayed, it is clear that to keep pace with the growth and development of human civilization, strategic changes in our approach will be necessitated.

This article takes a perspective view of the area of forest tree biotechnology and evaluates the progress made to date and examines the current trends in four areas of research and application, namely, *in vitro* culture techniques and micropropagation, application of molecular biological tools to forestry, genetic transformation of trees and biosafety vis-à-vis forestry.

2. *In vitro* Culture of Forest Trees

More than half a century of developments in *in vitro* culture of trees has not significantly changed the empirical methods used to induce morphogenesis in cultures. The latter half of the last century saw the development of protocols for *in vitro* culture and plantlet regeneration in several of the important tree crops. But in spite of the volume of work done, it is still rare to find protocols that are dependable in terms of efficiency and repeatability. Additions have been made to the list of plant growth regulators found to be useful for culture, with TDZ and ancymidol, being used with success in several plant species including trees.

The factors underlying the maturation of trees, which is a great hurdle in the propagation of important species, are still imperfectly understood. Two approaches to circumvent this problem are (i) maintenance of juvenility over long periods through cryopreservation or long term *in vitro* storage, while field trials are on with clones of juvenile origin, so that promising clones can be mass multiplied easily at the end of the testing period, and (ii) the induction and use of juvenile material from mature trees. The problem of physiological aging which has been noticed in micropropagated plantlets of juvenile origin and which eventually results in losses, e.g. in volume, is an issue of great concern. Rejuvenation of 20-year-old Radiata Pine through somatic embryo-genesis induced from vegetative apical meristems [2] does indicate a possible solution to this problem. Aderkas and Bonga [3] have reviewed the factors influencing rejuvenation in trees using methods of enhancing micropropagation through manipulations that involve application of osmotic, temperature or hormonal stress. There is a need to re-evaluate the morphogenetic competence of tissues from different levels of the tree. No significant differences were found in bud break, shoot multiplication or callus derived from cambium taken from top branches and epicormic shoots of *Robinia pseudoacacia* [4]. Epicormic shoots are considered more juvenile compared to the top branches and generally believed to be better explants for *in vitro* morphogenesis.

In tree species of the humid tropics the presence of endophytes within the tissues pose a serious problem in establishment of sterile cultures. Although this fact is rarely highlighted in literature, it is without doubt a vexing problem that needs to be tackled with a combination of techniques such as proper selection of explant type, prophylactic treatments of plants, use of antimicrobial agents for pre-treatment of explants and for inclusion in the culture media.

Browning of explants, contaminations and the effect of seasons still remain major obstacles for establishment of cultures from mature trees. In mature female trees of *Ceratonia siliqua* [5], shoot culture initiation was greatly influenced by season, with the highest survival percentage observed in spring.

Some definite advantages have been shown for micropropagated forest trees. A comparison was made between the growth of trees produced by micropropagation from nodal stem sections or callus tissue of a 20-year-old silver birch (*Betula pendula*) tree with that of seedlings [6]. Micropropagated trees were more uniform in height and trunk girth than seedling trees, and more than 80% of the trees flowered within three years of field planting, whereas only 39% of seedling trees flowered within the same period. Besides, micropropagated trees had lesser bark fissuring (a desirable character) than seedling trees. In Loblolly pine [7] the early reduced growth and mature morphology observed in plantlets derived from cotyledons does not occur in micropropagation from fascicular and axillary shoots or epicotyls. Increased biomass production was observed in micropropagated plants when compared to seedling progenies of the plus trees of two species of eucalyptus [8].

The recent advancements made in plant regeneration through somatic embryogenesis in several hardwood and conifer trees [9, 10] will greatly facilitate efficient mass propagation, conservation and genetic transformation. Eucalyptus are perhaps the most popular among the plantation tree crops around the world. *In vitro* regeneration systems in the different *Eucalyptus* species have been developed particularly using somatic embryogenesis [11-14]. The development of transgenics is thus facilitated. Shoot induction as well as somatic embryogenesis were induced on zygotic embryos in several genotypes. Among the other important tree species where improvement continues to be made in technology is the American chestnut (*Castanea dentata*) a timber and nut-yielding tree that has been on the decline due to the chestnut blight, where use of tissue culture and genetic engineering for restoration is approaching reality. Plantlet regeneration through germinated somatic embryos and microshoots derived from somatic embryos were obtained from developing ovules [15].

Among broad-leaved trees, micropropagation using shoot cultures appears to be the predominant method for cloning. Several important genera of forest trees can now be successfully micropropagated either through multiple shoot induction or somatic embryogenesis e.g. *Acacia, Albizzia, Casuarina, Dalbergia, Prosopis, Eucalyptus, Populus, Ficus, Bambusa, Dendrocalamus, Phyllostachys* besides important tree species like *Azadirachta, Gmelina, Tectona*, and species of *Salix, Shorea, Cassia* etc. [16-18]. India, with its rich biodiversity of tree species and with a long list of successful reports of *in vitro* culture [16], still has very few trees micropropagated on a large scale. However, the Micropropagation Technology Parks set up by the Department of Biotechnology (DBT) have been successful in scaling-up of protocols for forestry species such as eucalyptus, bamboo, poplars and teak used in energy plantations and reforestation [19]. Among the forest trees of the world, teak (*Tectona grandis*) is an important species in which work in standardizing micropropagation, cryopreservation and transformation continues to be carried out in several countries [20, 21]. The absence of an *in vitro* regeneration and transformation method in teak has been a major bottleneck for development of transgenics in this important timber species. Improved cryopreservation of *in vitro* shoot tips were obtained in teak [22]. Somatic embryogenesis has been reported from hypocotyls, endosperm, stem segments and protoplasts of sandalwood. Improvements in the techniques are being made [23-25]. The use of extracts of a cyanobacterium, *Plectonema boryanum* for somatic embryogenesis, in the absence of hormones, and the successful encapsulation in a composite gel of 50% silica gel and 4% sodium alginate and their germination were also reported in sandal [26]. Somatic embryogenesis and proliferation through repetitive embryogenesis were achieved in *Dalbergia sissoo*, an important timber tree of the sub-tropics, from callus derived from immature embryos [27]. Direct organogenesis and plantlet regeneration were also obtained from semi-mature and mature cotyledons of *D. sissoo* [28].

Bamboo and rattan (canes) are forest plants of great importance to the tropical countries especially of South East Asia where a significant proportion of the population utilise them as a source for a wide range of products of both traditional household uses and for food and industrial uses (paper and pulp). Interest in scientific management of bamboo and rattan resources has however been relatively recent. Application of biotechnology to the genetic improvement and conservation of these plants assumes importance since the potential for improvement in productivity is tremendous and resource management through conservation and establishment of plantations is gaining importance in many parts of the world.

In bamboo, tissue culture and micropropagation have been very successful and the technology has been commercialized for several years now. Regeneration has been obtained from a variety of explants [16, 29, 30], but rapid and high rate of multiplication are achieved through sprouting of nodal explants taken from *in vitro* raised seedlings. The formation of shoot cultures from secondary branches of culms of mature culms is much more difficult and reports are still few [31, 32]. Cultures derived from seeds and nodes of *in vitro* raised shoot cultures or mature culms have also produced embryogenic callus and regenerated plantlets [29]. Artificial seeds of *Dendrocalamus strictus* were produced [33] by encapsulation of somatic embryos. Minimal growth and storage up to 8 months was achieved in *Bambusa arundinacea* and *Thyrsostachys siamensis* cultures stored at 5°C or 10°C or on media containing different concentrations of 2-chloroethyltrimethyl ammonium chloride (CCC) or butanedioic acid mono (2, 2-diemthyl-hydrazide) diaminozide. Such *in vitro* methods can therefore be expected to increase the availability of planting material for much longer periods than is possible through seeds.

The results of small-scale field trials using tissue culture plants derived from mature culms and seedlings of bamboo were reported by Mascarenhas et al. [34]. They observed early culms formation and improvement in several other growth parameters in tissue culture raised plants as compared to seed-raised plants.

Although the induction of suspension cultures and the isolation of protoplasts from different species of bamboo have been reported [35, 36], further progress has apparently not been obtained in utilizing the cultures. Virus-free plantlets and salt resistant plantlets have been regenerated through *in vitro* culture [35]. The flowering cycle in bamboo is unique and gregarious flowering resulting in the death of the entire population takes place in cycles of 12–120 years depending on the species. This has been a hurdle in the propagation and breeding of bamboo. *In vitro* flowering is a first step in bringing about a control on the phenomenon so that studies could be carried out. This phenomenon has been reported in several bamboo species [37, 38]. It however appears that an understanding of the factors responsible for flowering both *in vitro* as well as in nature, is still eluding us and the benefits of the procedure cannot yet be realized. The *in vitro* strategies available for genetic improvement of bamboo have been discussed [39].

Rattans are climbing palms, which are the source of rattan or cane, which is an important raw material for wickerwork in several countries of the South East Asia. Overexploitation of the wild resources and lack of sufficient plantations have resulted in several species being endangered. Tissue culture procedures have been standardized for different species in India [40, 41], Thailand [42], Philippines [43] and Malaysia [44]. Maziah [45] found a growth dependency of *in vitro* micropropagated *Calamus manan* on vesicular arbuscular mycorrhiza (VAM) prior to transplanting to field. Multiple shoots and *in vitro* flowering were reported in *C. thwaitesii* [46] in embryo cultures. Regeneration from mature plant tissues is however not very successful in rattans. Evidence of somatic embryogenesis from root tip explants of mature plants of *C. manan* has been reported [47]. Mass multiplication of superior genotypes will depend on perfection of this technique. Until then tissue culture will perhaps be useful for rare species which do not produce enough seeds for meeting the demand for planting stock. Genetic transformation of rattan will also be facilitated if an efficient regeneration system based on somatic embryogenesis is available.

A large volume of literature is available on *in vitro* culture of conifers [48-51]. Somatic embryogenesis and high quality plantlet regeneration have been achieved and several patents attest to the commercial interest in this technology [52-54]. Somatic embryogenesis has been achieved in about 30 species [51] and methods involving immature embryos have been used wherein cleavage polyembryony is induced as in the natural case or the formation of an embryonal suspensor mass (ESM) from the different parts of the embryo is obtained. The different pathways to embryogenic cultures and the importance of osmolarity regulation for normal development and conversion of somatic embryos are now understood [51]. Organogenesis from embryonic cotyledons of Radiata pine is a well-established procedure and field trials have been conducted in New Zealand [55]. Rejuvenation, as evidenced by complete restoration of rooting competence of *Sequoia sempervirens* was achieved [56] through *in vitro* grafting of adult shoot tips onto juvenile rootstocks *in vitro* repeatedly for five times. They also found that rejuvenation was correlated with a disappearance of adult-associated esterase and peroxidase isozymes and an appearance of isoesterases and isoperoxidases that were characteristic of juvenile-phase shoots and hence these isozymes could serve as markers to assist phase-change investigations. The development of efficient *in vitro* regeneration systems in conifers through somatic embryogenesis has facilitated the genetic transformation of a number of conifers. The potential for automated systems for plantlets regeneration and delivery to soil have also been discussed by Gupta et al. [57].

3.   Application of Molecular Biological Tools to Forestry

A wide variety of DNA-based markers have been developed and procedures are getting simpler and inexpensive. Nuclear and chloroplast based Single Sequence Repeat (SSR), interSSR, Random Amplification of Polymorphic DNA (RAPD), SSCP, Amplified Fragment Length Polymorphism (AFLP), microsatellite DNA, Expressed Sequence Tags (EST) and Sequence Tagged Sites (STS) are some of the commonly used markers used for genome and QTL mapping for conservation and understanding of the evolutionary genetics and sequences controlling traits of economic interest of forest trees.

Genetic linkage maps and mapping of QTLs have been prepared for a wide range of trees using RAPD and AFLP markers [58, 59]. Molecular markers linked to specific traits can predict inheritance and is one of the most important applications of biotechnology in tree improvement. Wilcox et al. [60] used genome mapping in loblolly pine and identified a locus behaving as a single dominant gene imparting resistance to fusiform rust disease. Years of conventional genetic analysis had failed to detect any such resistance factor. The characterization of the phytoplasma causing the spike disease of sandalwood and its detection has been aided by the use of PCR [61]. The power and versatility of neutral DNA-based marker technologies allow for flexible high-resolution investigation of genetic variation at different levels of the population [62]. Both nuclear and organelle DNA can be studied with advantage since the rate of evolution of the two are different. Grattapaglia et al. [62], studied several tropical tree species with different DNA technologies to devise strategies for *in situ* and *ex situ* conservation. Markers can be powerful tools for tree improvement at both the early and advanced stages of the breeding programme. Marker assisted selection (MAS) will be particularly useful in studying the inheritance of quality traits in later stages of tree breeding. RAPDs with their dominant inheritance only detect one allele per locus and hence the information content is low, but high quality genetic maps of trees can be generated using simple reagents, if some care is taken to ensure repeatability [63-65]. AFLP markers too are versatile and use simple reagents, but a single assay generates much higher information [66, 67]. Faster and cheaper DNA marker technologies are becoming available. DNA chips that permit simultaneous analysis of thousands of loci are just around the corner and increased accessibility will make their routine use possible for forestry conservation programmes.

Genes associated with control of flowering are of particular interest in trees because of their potential to promote or inhibit flowering or in determination of gender. Ecological effects of transgenic trees could in a large measure be under control through manipulation of flowering. Our understanding of the genetic pathways controlling phase change, flower initiation and flower development has improved in recent years. Homology between the *Arabidopsis* flowering genes, LEAFY and flowering organ identity gene *AGAMOUS* and genes involved in flowering in conifers have been identified [67, 68]. Expressed Sequence Tag (EST) studies have been used to identify genes important in the regulation of flowering. The range of options that are available for control of flowering in transgenic trees [69] is encouraging.

Much research has been done to develop new clones of trees that are resistant or tolerant to different kinds of environmental stress such as salinity, drought, flooding etc. Abiotic stress affecting productivity of tree crops such as drought and heat-shock stress also has attracted the attention of forest biotechnologists. Mayne et al. [70] identified a S-Adenosyl Methionine Synthetase (SAM-S) cDNA by differential screening of a cDNA library constructed from root mRNA from jack pine (*Pinus banksiana*) seedlings exposed to two cycles of drought conditioning. The increase in the rate of SAM-S enzyme activity after drought conditioning was also correlated with increase in rates of ethylene and betaine synthesis. Investigators [71] have cloned and characterised three cDNAs (PgEMB22, 27 and 29) predicted to encode low-molecular-weight (LMW) heat-shock proteins (HSPs) from white spruce (*Picea glauca*) somatic embryos by differentially screening a zygotic embryo cDNA library. They were developmentally regulated during somatic embryo development and germination and also showed strong response to heat-shock stress. Abscisic acid and polyethylene glycol, stimulators for spruce embryo maturation, could also induce the HSP genes. A cDNA clone (pLP6) of a gene, which is repressed under water deficit and by wounding, was isolated from a loblolly pine (*Pinus taeda*) cDNA library and characterized [72]. The predicted polypeptide for pLP6 bears strong resemblance to a number of Class I chitinases although some of the diagnostic domains are absent.

Tolerance to oxidative stress can help the plants to survive in many adverse conditions. To achieve this, stem explants of a poplar hybrid (*Populus tremula* × *P. alba [P. canescens]*) clone were co-cultivated with *Agrobacterium tumefaciens* strain C58pMP90 having binary vectors with constructs with bacterial genes for either glutathione reductase (GR) (gor) or glutathione synthetase (GS) (gshII) [73]. When *gor* was targeted to the chloroplasts, leaf GR activities were up to 1000 times greater than in all other lines. These results suggest that overexpression of GR in the chloroplasts increased the antioxidant capacity of the leaves and that this improved the capacity to withstand oxidative stress. The high chloroplastic GR expressors showed increased resistance to photoinhibition. The herbicide methyl viologen inhibited CO_2 assimilation in all lines, but the increased leaf levels of glutathione and ascorbate in the high chloroplastic GR expressor persisted despite this treatment.

Use of markers has been of indirect application in improving the productivity of forest trees. For the reclamation or reforestation of poor soils the use of nitrogen fixing trees are of great potential. Improved survival and productivity of such trees are obtained if specific strains of nitrogen-fixing bacteria are used for inoculation of the seedlings. The identification of such strains through conventional microbiological means is extremely slow and unreliable and therefore, studies on suitability and nodulation behaviour and persistence in soil will be facilitated by the use of genetic markers. In the identification of *Frankia* strains that are symbionts of *Casuarina* the use of amplified nifH and rDNA segments has been found useful [74, 75].

Lignin, the complex polymer constituent of the secondary cell walls of xylem tracheids and fibres of trees is of great interest to breeders, wood and paper/pulp industry and biotechnologists alike. While playing an essential role in the plant structure and function, lignin causes severe problems in the efficient utilization of biomass for pulp production as well as for food. Modification of the lignin composition and content through suppression of the key enzymes involved in the biosynthesis is the target of several genetic transformation studies around the world. Some of the genes of interest in the lignin biosynthesis pathway are caffeic acid/5-hydroxy ferulic acid *o*-methyltransferase, Phenylalanine ammonialyase (PAL), *p*-Coumaric acid: CoA ligase genes [76-79]. Transgenics with alteration of wood colour phenotypes [79], reduced lignin, repression of lignin biosynthesis, high-cellulose, accelerated growth [80, 81] have been obtained. Sequencing of cDNAs isolated from specialized tissues of wood has been used as a tool to identify genes involved in wood formation [82]. Such studies will help to increase our knowledge of the environmental influence on wood properties.

4. Genetic Transformation of Trees

A significant number of crops such as corn, soybean, tomato and cotton in many developed nations and also in China consist of genetically engineered plants. However, transgenic trees are yet to be released commercially. Progress is being made rapidly and the constraint appears to be the time taken to complete the field trials. Among the important tree crops where transgenics have been reported are eucalyptus [83], quaking aspen [84], sweet gum—*Liquidambar styraciflua* [85], larches, spruces and pines [86]. The particle bombardment and *Agrobacterium* mediated gene transfer systems have both been used successfully.

Advancements in molecular tools have made possible the development of flexible and adaptable expression vector systems for plant transformation utilizing *Agrobacterium* or biolistics. They aim at a cassette system that allows quick and easy replacement of promoter, terminator, markers or the gene of interest. Besides the genes of interest, the promoters are a factor of importance.

Wound induced promoters [87] hold great potential for use with insect resistance genes to avoid constitutive expression, which is potentially a threat to non-target organisms including plant friendly ones.

Genetic engineering for resistance to pests, diseases and environmental stress is a major objective in forestry. As of today the *Bacillus thuringiensis* (Bt) and the Cry family of genes against insect pests are the most widely studied. Although the strategy appears not without drawbacks—effective concentrations of the toxin are difficult to achieve and resistance may develop over time. Further research to modify the genes involved and to engineer multiple genes into trees may find acceptance. Proteinase inhibitors are another choice. Here too, high levels of protein are required for killing the target organisms and there is a need to target expression to specific organs of the tree. Controlling the cytokinin producing genes [88] forms an alternative strategy that is closer to the natural defense mechanisms of plants and therefore safer and more acceptable than Bt genes. As discussed above options are becoming available for preventing any gene flow from transgenics to wild plants like induction of sterility [69].

Agrobacterium-mediated gene transfer is the method of choice for many plant biotechnology laboratories because of the high percentage of single-copy and single-locus insertion events compared to biolistics. In Norway spruce [89], increased transformation efficiencies of 1000-fold from initial experiments were obtained, where little or no transient expression was detected by varying the strain of *Agrobacterium*, source material and co-cultivation conditions. In loblolly pine, transient expression increased 10-fold utilizing modified *Agrobacterium* strains. Both *A. tumefaciens* and *A. rhizogenes* have been used for stable transformation in conifers. The reporter or selectable marker genes most commonly transferred to conifers are the *uid*A and *npt*II genes and among the useful genes, the Bt and genes controlling lignin synthesis [81, 85, 86], herbicide resistance and control of flowering [68].

5.   Biosafety vis-à-vis Forestry

The advocates of a ban on plant genetic engineering cry hoarse over the lack of sufficient testing of the modified organisms. They, in particular, point to the risks involved in escape of genes to wild relatives of the crop plants, resulting in the modified plants turning into weeds or unproductive strains, and the danger of toxic gene products on animal and human systems on ingesting them. The pro-genetic engineering lobby, with scientists and the corporate sector with a stake in the spoils, dismiss the arguments. Admittedly, all the possible risks have not been assessed and the debate is far from over. It would, therefore, be prudent for all concerned in these early years of GMO's to bring in an extra measure of caution lest a Thalidomide or Minimata be repeated.

Regulations have been primarily designed for annual plant species with which agriculturists are familiar. Genetically modified trees have to be treated as a separate category because unlike annuals, a much larger time frame is involved in growing and testing the perennial species, and several aspects of the biology of most tree species are relatively less studied and understood. The possibility of trees modified for fast growth and resistance to stress, turning to weeds or smothering other vegetation in a low intensity management regime is not to be ignored. The interaction of the introduced gene in the genome over a long period of time needs careful monitoring. Danger of silent genes getting activated (atavism) during the different growth phases of the tree is another possibility, given our inadequate understanding of developmental biology. Faster growing trees, transgenic or otherwise, carry a price tag—greater water and nutrient demand and reduced opportunity for nutrient cycling leading to site deterioration over a few rotations. This leads to the use of·fertilizers, or in tropical areas, the increased probability of abandoning the site. In biotechnological applications for pests and disease resistance, Bt genes are known not to discriminate between pests and friendly insects and can also affect the biodiversity of the plantation, which is important although not comparable with that of a natural ecosystem. However, it is possible to restrict the range of insects affected through modification of the genes involved.

Transparency in the regulatory and supervisory processes in testing of GMO's is yet to be evident. The issues relating to transgenic forest plantations has been debated in IUFRO, which represents the professional forest scientists all over the world, and a position statement released to promote informed public debate. The statement calls for a scientific appraisal of the transgenic technology and points out that advantages of the technology also lies in the significant environmental benefits accrued out of increased productivity leading to decreased dependence on natural forests, diminished use of pesticides and pulping chemicals and that options exist for mitigating the risks posed by gene flow to wild relatives [90]. FAO has also addressed the issue by organising a debate on the risks of gene flow from genetically modified organisms including transgenic trees [91].

Biosafety regulations are either non-existent or inadequate in many developing countries and many may not have the capability or desire to implement them. In the absence of an international system of regulations, unscrupulous entrepreneurs can be tempted to take advantage and introduce GMO's commercially or for testing without adequate safeguards. The complacency that characterises testing of GMO's has more to do with the confidence that the scientists have in their understanding of the way genes behave rather than abundant caution regarding the broader environmental consequences. A comprehensive internationally coordinated programme involving governments, non-governmental organisations and biotechnology companies to assess, study and monitor GMO trials and lend confidence to researchers and public is required to remove the stalemate.

6.   Conclusion

Technologies in molecular biology and tissue culture could play an increasing role in the choice of genotypes for successful establishment of plantations and agroforestry practices. Research areas such as micropropagation, somatic embryogenesis, genetic engineering, marker-aided selection and molecular diagnostics are merging with traditional forestry to help identify and produce better-suited trees for plantations and agroforestry. A combination of classical and molecular biological research could be used to improve pest and stress resistance of selected genotypes, modify structure and function, and monitor pests of trees. This merger of approaches, as well as continued technological development, could accelerate the production and selection of suitable tree genotypes for forestry.

References

1.   N.B. Klopfenstein, J.G. Kerl, The potential of biotechnology in temperate agroforestry practices, Agroforestry Systems 32 (1995) 29–44.
2.   D.R. Smith, P.J. Battle, C.P. Holliday, M.J. Walsh, S.A. Merkle, Progress with somatic embryogenesis, cryopreservation and transformation of slash pine. Proceedings of the 25th Biennial Southern Forest Tree Improvement Conference, July 11–14, 1999, New Orleans, LA (2000) pp. 236–238.

3.   P. von Aderkas, J.M. Bonga, Influencing micropropagation and somatic embryogenesis in mature trees by manipulation of phase change, stress and culture environment, Tree Physiol. 20 (2000) 921–928.

4.   K.H. Han, D.I. Shin, D.E. Keathley, Tissue culture responses of explants taken from branch sources with different degrees of juvenility in mature black locust (*Robinia pseudoacacia*) trees. Tree Physiology 17 (1997) 671–675.

5.   A. Romano, S. Barros, M.A. Martins-Loução, Micropropagation of the Mediterranean tree *Ceratonia siliqua*, Plant Cell Tissue and Organ Cult. 68 (1) (2002) 35–41.

6.   O.P. Jones, M. Welander, B.J. Waller, M.S. Ridout, Micropropagation of adult birch trees: production and field performance, Tree Physiology 16 (1996) 521–525.

7.   L.J. Frampton, H.V. Amerson, G.N. Leach, Tissue culture method affects *ex vitro* growth and development of loblolly pine, New Forests 16 (1998) 125–138.

8.   S.S. Khuspe, P.K. Gupta, D.K. Kulkarni, U.J. Mehta, A.F. Mascarenhas, Increased biomass production by tissue culture of *Eucalyptus*, Can. J. For. 17(1987) 1361–1363.

9.   S.M. Jain, P.K. Gupta, R.J. Newton (eds.) Somatic Embryogenesis in Woody Plants, Vol. 1–3, Kluwer Academic Publishers, Netherlands, 1995.

10.   S.M. Jain, K. Ishii, Recent Advances in somatic embryogenesis in forest trees, in: Recent Advances in Biotechnology for Tree Conservation and Management, Proc. of IFS Workshop, Florianoplis, Brazil, 15–19 September 1997, International Foundation for Science. 1998, pp. 214–231

11.   E.M. Muralidharan, A.F. Mascarenhas, Somatic embryogenesis in Eucalyptus, in: S.M. Jain, P.K., Gupta, R.J. Newton (Eds) Somatic Embryogenesis in Woody Plants, Vol. 2, Kluwer Academic Publishers, Netherlands, (1995), pp 23–40.

12.   K.V. Mullins, D.J. Llewellyn, V.J. Hartney, S. Strauss, E.S. Dennis, Regeneration and transformation of *Eucalyptus camaldulensis*, Plant Cell Rep. 16 (1997) 787–791.

13.   C. J. Tsai, Y.N. Wang, C.H. Chiang, Callus induction and plant regeneration from embryos of *Eucalyptus saligna.*, Q. Jour. Chin. For. 25 (3) (1992) 3–14.

14.   M.P. Watt, F.C. Blakeway, C.F. Cresswell, B. Herman, Somatic embryogenesis in *Eucalyptus grandis*, South African Forestry Journal, 157(1991) 159–165.

15.   Z. Xing, W.A. Powell, C.A. Maynard, Development and germination of American chestnut somatic embryos, Plant Cell Tissue Org Cult. 57 (1999) 47–55.

16.   A.F. Mascarenhas, E.M. Muralidharan, Tissue culture of forest trees in India, Curr. Sci. 58 (1989) 606–613.

17.   A.F. Mascarenhas, E.M. Muralidharan, Clonal Forestry with Tropical Hardwoods, in: M.R. Ahuja, W.J. Libby (Eds), Clonal Forestry II: Conservation and Application, Springer-Verlag. (1993) pp 169–187.

18.   M.H. El-Lakhany, Rapid propagation of fast growing tree species in developing countries: Its potentials, constraints and future developments, in: F.W.G Baker (Ed) Rapid Propagation of Fast Growing Woody Species, CAB International, (1992) pp.102–108.

19.   DBT, Plant Tissue Culture: From Research to Commercialisation, A Decade of Support, Department of Biotechnology, New Delhi, 2000, pp. 1–224.

20.   Z. Fadillah, M.Y. Aziah, Collection techniques for *in vitro* propagation of *Tectona grandis* (Teak), in: Paper presented at the 5th Conference on Forestry and Forest Products Research (CFFPR), Series 4–5th, October, 1999, FRIM, Kepong, Malaysia

21.   M.D. Gradaille, L. Ramos, Y. Lezcano, R. Rodrigues, M. Escalona, Algunos elementos en la micropropagacion de la teca, Biotecnologia vegetal, 1 (2000) 39–44.

22.   S.V. Kendurkar, R.S. Nadgauda, S. Von Arnold, Studies on cryopreservation of teak (*Tectona grandis*): a tropical hard wood tree (Abstract), in: International Tree Biotechnology Meeting, NCL, India, 1999, pp. 53–57.

23.   P.S. Rao, V.A. Bapat, Micropropagation of Sandalwood (*Santalum album* L.) and mulberry (*Morus indica* L.), in: M.R. Ahuja (ed.) Micropropagation of Woody Plants. Kluwer Academic Publ. Netherlands, (1993) pp. 317–346.

24.   P.S. Rao and V.A. Bapat, Somatic embryogenesis in Sandalwood (*Santalum album* L.) and mulberry (*Morus indica* L.), in: S.M. Jain, P.K. Gupta and R.J. Newton (Eds), Somatic Embryogenesis in Woody Plants. Kluwer Acad. Publ. Netherlands, (1995) pp. 153–170.

25. G. Lakshmi Sita, A. Bhattacharya, Cell and molecular approaches for obtaining disease resistance plants in sandalwood, in: International Seminar on Sandalwood and its Products, Bangalore, India, 1997, pp. 5.

26. V.A. Bapat, *In vitro* studies on sandal wood (*Santalum album* L.), in: Paper presented at the International Tree Biotechnology Meeting (Nov.17–19) National Chemical Laboratory, Pune, 1999 pp. 19–23.

27. P. Das, S. Samantaray, A.V. Roberts, G.R. Rout, *In vitro* somatic embryogenesis of *Dalbergia sissoo* Roxb.—a multipurpose timber-yielding tree, Plant Cell Rep. 16(8) (1997) 578–582.

28. A.K. Singh, S. Chand, S. Pattnaik, P. K. Chand, Adventitious shoot organogenesis and plant regeneration from cotyledons of *Dalbergia sissoo* Roxb., a timber yielding tree legume, Plant Cell Tissue Org. Cult. 68 (2) (2002) 203–209.

29. P. Das, G.R. Rout, Analysis of current methods and approaches on the micropropagation of bamboo, Proc. Natl. Acad. Sci. (India), LXIV(B) (1994) 235–246.

30. C. X. Zhang, Y.F. Xie, Y.F. Zhang, D.T. He, T.B. Chen, W.W. Wu, The current advances in bamboo tissue culture and its prospects, Journal of Bamboo Research 18(3)(1999) 46–49.

31. H.C. Chaturvedi, M. Sharma, A.K. Sharma, *In vitro* regeneration of *Dendrocalamus strictus* Nees through nodal segments taken from field-grown culms, Plant Science Limerick. 91(1) (1993) 97–101.

32. S. Saxena, S.S. Bhojwani, Towards regeneration and mass propagation of bamboo through tissue culture: Bamboo in Asia and the Pacific, Proceedings of the Fourth International Bamboo Workshop, 27–30 November 1991, Chiangmai, Thailand (1994) 157–164.

33. S. Mukunthakumar, J. Mathur, Artifical seed production in the male bamboo *Dendrocalamus strictus* L. Plant Science (Limerick) 87(1) (1992) 109–113.

34. A.F. Mascarenhas, S.S. Khuspe, R.S. Nadgauda, P.K. Gupta, E.M. Muralidharan, B.M. Khan, Biotechnological application of plant tissue culture to forestry in India, in: V. Dhawan (Ed), Applications of Biotechnology in Forestry and Horticulture, Plenum Press (1989) pp. 73–86.

35. L.C. Huang, B.L. Huang, Bamboo tissue culture. Institute of Botany, Academia Sinica Monograph Series No. 13 (1993) 203–212.

36. J G.N. Que, Q. Zhuge, Study on cell suspension culture and isolation of protoplasts of *Dendrocalamus membranceus*, Forest Research 7(1) (1994) 44–47.

37. H.R. Heuch, *In vitro* flowering of bamboo species—prospects and aims, in: Zhu Shilin, Li Weidong, Zhang Xinping, Wang Zhongming (Eds) Bamboo and its Use, Proc. of International Symposium on Industrial Use of Bamboo, 7–11 December 1992, Beijing, China 1993.

38. R.S. Nadgauda, V.A. Parasharami, A. F. Mascarenhas, Precaucious flowering and seedling behaviour in tissue cultured bamboos. Nature, 344 (1990): 335–336.

39. Gielis, J., Peeters, H., Gillis, K., Oprins, J., Debergh, P.C. Tissue culture strategies for genetic improvement of bamboo. Acta Hort. (ISHS) 552 (2001) 195–204.

40. K. Valsala, E.M. Muralidharan, Plant regeneration from *in vitro* cultures of rattan (*Calamus*), in: A.D. Damodharan (Ed), Proceedings of 10th Kerala Science Congress, Kozhikode, January 1998, pp. 161-163.

41. K. Valsala, E.M. Muralidharan, *In vitro* regeneration in three species of Rattan (*Calamus* spp.), in: P.B. Kavi Kishor (Ed), Plant Tissue Culture and Biotechnology: Emerging Trends, Universities Press. (1999) pp. 118–122.

42. P. Chuthamas, P. Prutpong, I. Vong Kaluang, S. Tantiwiwat, *In vitro* culture of immature embryos of *Calamus manan* Miq, in: A.N. Rao Isara Vongkaluang (Eds) Recent Research on Rattans, Kasetsart University, Thailand and IDRC, Canada (1989). pp144–147.

43. L.F. Patena, M.M.S. Mercado, R.C. Barba, Rapid propagation of rattan (*C. manillensis* H.A. Wendl.) by tissue culture, Philip. J. Crop Sci. 9(1984) 217–218.

44. M.Y.Aziah, M.N. Nur Supardi, Initial growth of tissue culture raised *Calamus manan* seedlings. RIC Bulletin 12 (1/2) (1994) 6–7.

45. Z. Maziah, Preliminary studies on growth dependency of *in vitro* micropropagated *Calamus manan* on vesicular arbuscular mycorrhiza (VAM) prior to transplanting to field, RIC Bulletin 10(1) (1991).

46. S.M.S.D. Ramanayake, Viability of excised embryos, shoot proliferation and *in vitro* flowering in a

species of rattan *Calamus thwaitesii* Becc. Journal of Horticultural Science and Biotechnology 74(5) (1999) 594–601.

47.   D.K.S Goh, N. Michaux Ferriere, O. Monteuuis, M.C. Bon, Evidence of somatic embryogenesis from root tip explants of the rattan *Calamus manan. In Vitro* Cellular and Developmental Biology (Plant) 35(5) (1999) 424–427.

48.   N.B. Jones, J. Van Staden, Somatic embryogenesis in *Pinus patula*. Scheide et Deppe, in: S.M. Jain, P.K. Gupta, R.J. Newton (Eds), Somatic Embryogenesis in Woody Plants. Vol. 4. Kluwer Academic Publishers, Dordrecht, 1999, pp. 431–447.

49.   N.B. Jones J. Van Staden, Improved somatic embryo production from embryogenic tissue of *Pinus patula*. In Vitro Cellular and Development, Biol. Plant, 37 (2001) 543–549.

50.   M.R. Becwar, G.S. Pullman, Somatic embryogenesis in loblolly pine (*Pinus taeda* L.), in: S.M. Jain, P.K. Gupta, R.J. Newton (Eds), Somatic Embryogenesis in Woody Plants Vol. 3, Gymnosperms, Kluwer, The Netherlands, 1995, pp. 287–301.

51.   P.K. Gupta, J.A. Grob, Somatic embryogenesis in conifers, in: S.M. Jain, P.K. Gupta, R.J. Newton (Eds), Somatic Embryogenesis in Woody Plants. Vol. 1. Kluwer Academic Publishers, Dordrecht, The Netherlands. 1995 pp. 81–98.

52.   P.K. Gupta, G.S. Pullman, Method for Reproducing Conifers by Somatic Embryogenesis Using Stepwise Hormone Adjustment, U.S. Patent No. 5236841, August 17, 1993.

53.   P.K. Gupta, G.S. Pullman, Method for Reproducing Douglas fir by Somatic Embryogenesis. U.S. Patent No. 5482857, January 9, 1996.

54.   G.S. Pullman, P.K. Gupta, Method for Reproducing Conifers by Somatic Embryogenesis Using Mixed Growth Hormones for Embryo Culture, U.S. Patent No. 5294549. March 15, 1994.

55.   J.A. Gleed, Development of plantlings and stecklings of radiata pine, in: M.R. Ahuja, W.J. Libby (Eds), Clonal Forestry II: Conservation and Application, Springer-Verlag. (1993). pp. 149–168.

56.   H.J. Huang, Y. Chen, J.L. Kuo, T.T. Kuo, C.C. Tzeng, B.L. Huang, C.M. Chen, L.C. Huang, Rejuvenation of *Sequoia sempervirens in vitro:* changes in isoesterases and isoperoxidases. Plant and Cell Physiology 37 (1996) 77–80.

57.   P.K. Gupta, G. Pullman, R. Timmis, M. Kreitinger, W.C. Carlson, J.E. Grob., E. Welty, Forestry in the 21st Century: The biotechnology of somatic embryogenesis, Biotechnology 11 (1993) 454–459.

58.   International Foundation for Science, Recent Advances in Biotechnology for Tree Conservation and Management, Proceedings of an IFS Workshop, Florianopolis, 15–19 Sept.1997.

59.   D.L. Remington, R.W. Whetten, B.-H. Liu, D.M. O'Malley, Construction of an AFLP genetic map with nearly complete genome coverage in *Pinus taeda*. Theo. Appl. Gen., 98 (1999) 1279–1292.

60.   P.L. Wilcox, H.V. Amerson, E.G. Kuhlman, B.-H. Liu, D.O. O'Malley, R.R. Sederoff, Detection of genetic resistance to fusiform rust disease in loblolly pine by genomic mapping, Proc. Nat. Acad. Sci. 93 (1995) 3859–3864.

61.   S. Thomas, M. Balasundaran, Detection of sandal spike phytoplasma by polymerase chain reaction, Curr. Sci. 76 (1999.) 1574–1576.

62.   D. Grattapaglia, A.Y. Ciampi, F.A. Gaiotto, M.G. Squilassi, R.G. Collevatti, V.J. Ribeiro, A.M. Reis, F.B. Gandara, B.M. Walter, R.P.V. Brondani, in: Recent Advances in Biotechnology for Tree Conservation and Management, Proceedings of an IFS Workshop, Florianopolis, International Foundation for Science 1998, pp. 50–61.

63.   D. Grattapaglia, R.R. Sederoff, Genetic linkage maps of *Eucalyptus grandis* and *E. urophylla* using a pseudo-testcross strategy and RAPD markers, Genetics 137 (1994) 1121–1137.

64.   D. Grattapaglia, F.L. Bertolucci, R. Penchel, R.R. Sederoff, Genetic mapping of quantitative trait loci controlling growth and wood quality traits in *Eucalyptus grandis* using a maternal half-sib family and RAPD markers, Genetics 144 (1996) 1205–1214.

65.   C. Plomion, N. Barhman, C.E. Durel, D. O'Malley, Genomic mapping in *Pinus pinaster* (maritime pine) using RAPD and protein markers, Heredity 74 (1995) 661–668.

66.   C.M. Marques, J.A. Araújo, J.G. Ferreira, R. Whetten, D.M. O'Malley, B. H. Liu, R.R. Sederoff, AFLP genetic maps of *Eucalyptus globulus* and *E. tereticornis*, Theor. Appl. Genet. 96 (1998) 727–737.

67. R. Rutledge, S. Regan, O. Nicolas, P. Fobert, C. Côté, W. Bosnich, C. Kauffeldt, G. Sunohara, A. Séguin, D. Stewart, Characterization of an AGAMOUS homologue from the conifer black spruce that produces floral homeotic conversions when expressed in *Arabidopsis*. The Plant J. 15 (1998) 625–634.

68. W.H. Rottmann, R. Meilan, L.A. Sheppard, A.M. Brunner, J.S. Skinner, C. Ma, S. Cheng, L. Jouanin, G. Pilate, S.H. Strauss. Diverse effects of overexpression of LEAFY and PTLF, a poplar (*Populus*) homolog of LEAFY/FLORICAULA, in transgenic poplar and *Arabidopsis*, Plant J. 22 (2000) 235–246.

69. J.S. Skinner, R. Meilan, A.M. Brunner, S.H. Strauss, Options for genetic engineering of floral sterility in forest trees, in: S.M. Jain, S.C. Minocha (Eds), Molecular Biology of Woody Plants, Vol. 1. Kluwer Academic Publishers, Dordrecht, The Netherlands, (2000) pp. 135–153.

70. M.B. Mayne, J.R. Coleman, E. Blumwald, Differential expression during drought conditioning of a root-specific S-adenosylmethionine synthetase from jack pine (*Pinus banksiana* Lamb.) seedlings, Plant, Cell and Environment 19 (1996) 958–966.

71. J.Z. Dong, D.I. Dunstan, Characterization of three heat-shock-protein genes and their developmental regulation during somatic embryogenesis in white spruce (*Picea glauca* (Moench) Voss), Planta. 200 (1996) 85–91.

72. S.J. Chang, J. Puryear, E.A. Funkhouser, R.J. Newton, J. Cairney, Cloning of a cDNA for a chitinase homologue which lacks chitin-binding sites and is down-regulated by water stress and wounding, Plant-Molecular-Biology 31 (1996) 693–699.

73. C.H. Foyer, N. Souriau, S. Perret, M. Lelandais, K.J. Kunert, C. Pruvost, L. Jouanin, Overexpression of glutathione reductase but not glutathione synthetase leads to increases in antioxidant capacity and resistance to photoinhibition in poplar trees. Plant-Physiology, 109 (1995) 1047–1057.

74. C. Rouvier, S. Nazaret, M.P. Fernandez, B. Picard, S. Simonet, P. Normand, *Rrn* and *nif* intergeneric spacers and isoenzyme patterns as tools to characterize *Casuarina*-infective *Frankia* strains, Acta Oecologica, 13 (1992) 487–495

75. M. Valdes, H. Olivera, L.Vasquez, N.-O. Perez, Developing genetic markers for ecological studies of the symbiosis *Casuarina-Frankia*, in: Recent Advances in Biotechnology for Tree Conservation and Management, Proceedings of an IFS Workshop, Florianopolis, 15–19 Sept.1997 (1998), pp. 308–316.

76. J.J. MacKay, W.W. Liu, R. Whetten, R. Sederoff, D. O'Malley, Genetic analysis of cinnamyl alcohol dehydrogenase in loblolly pine: single gene inheritance, molecular characterization and evolution. Mol. Genet. 247 (1995) 537–545.

77. K.S. Voo, R.W. Whetten, D.M. O'Malley, R.R. Sederoff, 4-Coumarate: CoA ligase from loblolly pine xylem: Isolation, characterization and cDNA cloning, Plant Physiol. 108 (1995) 85–97.

78. C.J. Tsai, M.R. Mielke, J.L. Popko, W.-J. Hu, G.K. Podila, V.L. Chiang Novel wood coloration and altered lignin composition in transgenic aspen through manipulation of caffeic acid/5-hydroxyferulic acid *o*-methyltransferase gene expression, Plant Physiology 114 (3) (1997) 300.

79. C.J. Tsai, J.L. Popko, M.R. Mielke, W.J. Hu, G.K. Podila, V.L. Chiang Suppression of *o*-methyltransferase gene by homologous sense transgene in quaking aspen causes red-brown wood phenotypes, Plant Physiology, 117 (1998) 101–112.

80. W.J. Hu, S.A. Harding, J. Lung, J.L. Popko, A. Kawaoka, Y.Y. Kao, K. Osakabe, H. Suzuki, D.D. Stokke, C.J. Tsai, V.L. Chiang, Repression of lignin biosynthesis in transgenic trees results in pleiotropic effects including high-cellulose and accelerated growth phenotypes, in: Wood and Wood Fibers: Properties and Genetic Improvement, IEG-40 Workshop, July 19–22, 1998, Atlanta, Georgia.

81. W.J. Hu, J.L. Popko, J. Lung, A. Kawaoka, Y.Y. Kao, S. Hideki, D.D. Stokke, C.J. Tsai, V.L. Chiang, Transgenic aspen trees with reduced lignin quantity and increased cellulose content, in: 215th American Chemical Society National Meeting, March 31–April 2, 1998, Dallas, Texas.

82. I. Allona, M. Quinn, E. Shoop, K. Swope, S. St. Cyr, J. Carlis, J. Riedl, E. Retzel, M.M. Campbell, R. Sederoff, R.W. Whetten, Analysis of xylem formation in pine by cDNA sequencing, PNAS USA 95 (1998) 9693–9698.

83. C.K. Ho, S.H. Chang, J.Y. Tsay, C.J. Tsai, V.L. Chiang, Z.Z. Chen, *Agrobacterium tumefaciens*-mediated transformation of *Eucalyptus camaldulensis* and production of transgenic plants, Plant Cell Reports 17 (1998) 675–680.

84. C.-J. Tsai, G.K. Podila, V.L. Chiang, *Agrobacterium*-mediated transformation of quaking aspen (*Populus tremuloides*) and regeneration of transgenic plants, Plant Cell Reports 14 (1994) 94–97.

85. V.L. Chiang, Z.Z. Chen, G.K. Podila, W.Y. Wang, R.C. Bugos, W.H. Campbell, U.N. Dwivedi, J. Yu, C.J. Tsai, J.Y. Tsay, J.C. Yang, Genetic manipulation of lignification in *Liquidambar styraciflua* (sweetgum) by introduction of a chimeric sense or antisense *o*-methyltransferase gene cloned from *Populus tremuloides* (aspen), in: Intl. Conf. on Emerging Technologies for Pulp and Paper Industry, May 18–20, Taipei, Taiwan, 1993, pp. 26–29.

86. R.J. Newton, J.C. Bloom, D.H. Bivans, S.M. Jain, Stable genetic transformation of conifers, Phytomorphology Golden Jubilee Issue 2001 421–434.

87. J.B. Hollick, M.P. Gordon, A poplar tree proteinase inhibitor-like gene promoter is responsive to wounding in transgenic tobacco, Plant Molecular Biology 22 (1993) 561–572.

88. A.C. Smigocki, J.W. Neal, Enhanced insect resistance in plants genetically engineered with a plant hormone gene involved in cytokinin synthesis, United States Patent No. 5496732 (1996).

89. A.R. Wenck, M. Quinn, R.W. Whetten, G. Pullman, R.R. Sederoff. High-efficiency *Agrobacterium*-mediated transformation of Norway spruce (*Picea abies*) and loblolly pine (*Pinus taeda*), Plant Mol. Biol. 39 (1998) 407–416.

90. Position Statement on Benefits and Risks of Transgenic Plantations, IUFRO Working Party on Molecular Biology of Forest Trees, 2.04.06, September 1999. http://users.ox.ac.uk/~dops0022/conference/forest_biotech99_home.html

91. FAO, Summary Document of Conference on Gene flow from GM to non-GM populations in the crop, forestry, animal and fishery sectors, FAO Electronic Forum on Biotechnology in Food and Agriculture, 31 May to 5 July 2002. http://www.fao.org/biotech/logs/C7/summary.htm

Plant Biotechnology and Molecular Markers
P.S. Srivastava, Alka Narula and Sheela Srivastava (Editors)

12. Cloning Forestry Species

Vibha Dhawan and Sanjay Saxena

Bioresources and Biotechnology Division, Tata Energy Research Institute, Habitat Place,
Lodhi Road, New Delhi 110 003, India

Abstract: Over-exploitation of Indian forests has led to progressive decline in the forest cover and its productivity. Since a large number of people depend on forest resources for their livelihoods and many more for meeting their energy needs it is imperative to enhance the productivity levels of our forests for a sustainable harvesting. Way back in 1999, the Department of Biotechnology, Government of India took the initiative of setting-up Tissue Culture Pilot Plants for micropropagation of various plant species. One of these facilities was established at TERI and so far about 12 million plants have been despatched out of which 3.7 million are of forest species alone. The field trials have clearly established clonal uniformity of tissue cultured plants and substantial increase in productivity levels.

1. Introduction

Forest play a major role in maintaining climatic stability, conserving water and soil, housing biological diversity and serve as a valuable source of various timber and non-timber products. India with its wide geographical distribution is endowed with rich forest resource and has two of the eighteen hotspots in biodiversity in the world. Traditionally, communities have lived in close harmony with the forest and their dependence on this important resource has taught them to be caring for sustenance. In the recent past, however, this situation has changed tremendously and the forests are being heavily over-exploited leading to reduction in forest cover, lowering of plant density, and eroding of floral and faunal diversity. On account of intense population pressures (both human and cattle) and heavy reliance of inhabitants on forests to cater their daily needs, the forests are under severe strain. This problem is quite complicated and has no easy solutions given by the fact that forests belong to the State Governments and thus conservation is viewed as the Government's responsibility while harvesting in many of these areas is the privilege of the local communities. There are many industries, which are dependent on forest resource, and at the time of independence many concessions were given to them, which includes making timber available at subsidized rate. Many such actions have caused heavy damage to our forests. Much of this loss is yet to be made up. This problem is further aggravated by the fact that there is no incentive for industries to get themselves involved into growing of raw material required by them. Further, due to fewer job options in the rural parts, cattle rearing and collection of non-timber forest produce remain the favourite revenue-earning activities. Since cattle are left in open for grazing, natural regeneration of the forests has become very difficult. This has resulted in denudation of several areas and spread of wastelands in the country. To repair the damage already caused to our forests and to restrict their future abuse, the Government of India has formulated several plans and policies directed towards large-scale afforestation and using improved planting material for higher yields. Due to increasing human needs, it is not possible to divert agricultural land for forestry purposes and the only option therefore, is to improve the productivity of existing forests. This involves selection of suitable species (matching species with sites), using superior quality planting material, managing the plantations properly, and developing a programme for sustainable harvesting.

While the importance of conventional breeding in improving the productivity cannot be undermined, breeding of trees is technically difficult and time consuming. Since the generation time in most tree species is very long, in many cases one may not even see the results of the experiments in lifetime. It may take several decades to release an improved variety/clone. Fortunately, most trees species are cross-pollinated and consequently lot of genetic variability exists in nature today. For immediate gains, it would be worthwhile to exploit this genetic variability existing in nature by selecting and cloning superior genotypes/individuals from their natural population.

2.   Cloning of Tree Species by Conventional Methods

Compared to horticultural plants, research in tree species has lagged behind. There has been very little interest shown by the private sector in this field and that too mainly in short rotation crops and softwoods. General interest in cloning of hardwood species started with eucalyptus. Aracruz (Brazil) took the lead in raising commercial plantations of eucalyptus through rooting of cuttings. The concept of raising plantations by pulp and paper companies is not new and many countries including United States, Australia and New Zealand are involved in similar kind of activities but largely with pines. Commendable work has also been done in Thailand and Indonesia on hardwood species such as acacia, casuarina, eucalyptus, teak etc. and increase in productivity has been reported.

Of late, in India too, foresters who are the custodians of our forests have realized the importance of planting clonal material. Massive projects have been launched by several state forest departments for raising clonal nurseries. Some initiatives have also been taken by private companies such as ITC Bhadrachalam, JK industries, West Coast Paper Mills and Andhra Pradesh Forest Cooperation. However, the total plantlet production through conventional techniques is far from adequate. This is largely because till date methods largely of rootings of cuttings have been developed only for a few species and that too are effective only when the mother tree is in a juvenile phase.

3.   Major Constraints in Vegetative Propagation

- Because of large size of the propagules only few functional cuttings can be derived from the desired clone/genotype.
- In many tree species the cuttings lose their ability to root by the time a particular clone is evaluated for its useful traits.
- The cutting-raised-plants tend to form adventitious roots which unlike the tap root of seedlings do not penetrate very deep inside the ground, thereby, making the plant highly prone to felling by strong winds.
- Propagation through cuttings also poses a potential risk for spread of various systemic diseases.
- Propagation through cuttings is extremely slow and season specific
- Depending on the species and the efficiency of asexual methods of propagation, the production cost of cutting raised plants is marginally to significantly higher in comparison

to seed-raised plants. This is quite a crucial factor for the foresters as they are often given large target of afforestation and have limited funds at their disposal.

4.   Micropropagation

Tissue culture perhaps is the most commercially exploited field of plant biotechnology. It overcomes many of the constraints that the conventional methods of propagation are inflicted with. Cloning of plants under aseptic conditions, commonly called micropropagation results in mirror images of selected mother plants on a large-scale within a short period of time. Micropropagation assumes greater significance in those species which cannot be regenerated or are difficult to regenerate by conventional methods such as seeds and vegetative propagation, where conventional methods are inadequate to meet the demand of planting material, and vast variability exists in seed-raised progenies.

5.   Advantages of Micropropagation

Rapid multiplication: By using an efficient protocol one may produce over a million plants starting from a single bud.

Saving of space: Unlike conventional cuttings, that measure 8–12 inches in length and occupy lot of space, large number of cultures can be accumulated within a small area.

Production of disease free plants: The plants produced by tissue culture are free of almost all bacterial and fungal diseases. In those species where virus infestation is known to affect the quality of the plant as well as the productivity, virus elimination can be achieved by tissue culture. Since virus elimination is a time consuming and expensive process, it can only be applied if one produces a large number of plants from a single explant, which is free of known viruses.

Clonal uniformity: Unlike seedlings which represent only the half siblings, the tissue culture raised plants are true-to-the-mother type and there is no segregation of genes or change in genetic character/traits in the progeny during the regeneration process.

Independent of seasonal constraint: Since plants are produced in controlled conditions of light, temperature and humidity, there is no effect of the outside environment on the regeneration process. While the plants can be produced independent of season inside the lab, the transplantation process remains dependent on the season.

6.   Tissue Culture of Trees

Tissue culture of woody species was first reported by Gautheret way back in 1933. However, the progress made with trees has been rather slow as compared to herbaceous species. This is largely because tree species have a distinct juvenile and an adult phase and trees, especially in their adulthood, are more recalcitrant to tissue culture technology. Further, they have a long gestation period and thus, field evaluation and commercial exploitation takes much longer vis-a-vis agricultural and horticultural species.

Although micropropagation process has been in use in the developed world for several decades, in India the commercial importance of this technique was realized only in late 1980s and early 1990s when several tissue culture companies were set-up. However, most of these companies were export-oriented units with focus on ornamentals. By and large these companies had a general reluctance to venture into tissue culture of forest trees on account of following reasons :

- *Policy matter*   : Being export-oriented units with a buy-back arrangement with their collaborators, these companies have largely been catering to the demand of their collaborators.
- *Technical problems*   : There are several technical problems associated with tissue culture of forest species and there are very few groups in the country who have the technical capability to deal with them.
- *Commercial factors*   : As compared to ornamentals (most of which are seasonal and demand frequent replacement), the market of forest species is relatively small on account of their perennial nature and selective users. The paucity of funds with most of the State Forest Departments who are expected to be the main users also deterred these companies to take up tissue culture of forest species.

6.1 Tissue Culture Pilot Plant

To meet this challenge, the Department of Biotechnology (DBT), Government of India in 1989 decided to set-up a Tissue Culture Pilot Plant (TCPP) at TERI for mass propagation of forest species using tissue culture technique to augment the biomass production in the country. Located within TERI's 36-hectare campus at Gual Pahari, Gurgaon, Haryana, here all the infrastructural facilities ranging from modern laboratories and greenhouses to nurseries required for mass production of tissue cultured plants, are available. In 1997, this Tissue Culture Pilot Plant was upgraded into a Micropropagation Technology Park (MTP) to provide an effective platform for transfer of the proven tissue culture related technologies to the entrepreneurs. Thus MTP has an annual capacity of over two million plants.

6.2 Objectives of Micropropagation Technology Park

- Propagate superior clones of various plant species on a large-scale using tissue culture technology
- Multiply those species on a mass-scale which are difficult to regenerate by conventional methods of propagation
- Multiply species *in vitro* where conventional methods of propagation are inadequate to meet the demand of planting material
- Enhance further the productivity of *in vitro* raised plants using symbionts such as *rhizobia* and *mycorrhizae*
- Impart training in the field of plant tissue culture
- Technology transfer to new entrepreneurs or industry for commercialization
- Function as a think-tank for the tissue culture industry

6.3 Achievements of TCPP/MTP

- Established a modern, indigenously designed tissue culture laboratory with an annual production capacity of two million plants at Gurgaon (Haryana).
- Developed micropropagation protocols for a large number of species and refined procedures for several others so as to make them suitable for large-scale propagation.
- Supplied over 12 million plants of various species to different state forest departments, non-governmental organization, agro-based companies, private growers etc.
- Successfully demonstrated the application of tissue culture technology at the farmers' field.
- Established high survival, plant uniformity, and better growth rates of tissue culture plants as compared to conventional propagules.
- Successful technology transfer to industry.
- Contractual research/plant production for clients.
- Created awareness about MTP and the tissue culture technology through lectures, demonstrations, seminars/workshops and exhibitions.
- Conducted several training programmes.

6.3.1 *Criteria for the Selection of Hardwood Species*

For the species that can be conventionally propagated through seeds and exhibit wide variability, tissue culture is of immense value if the plus trees are mass multiplied. To achieve this, the selection of the mother tree must be done very carefully. Also it is necessary to select newer and newer clones to avoid monoculture and degeneration of clone. Some of the criteria followed for selections are:

General Criteria
- A tree can be evaluated earliest at half its rotation age. For eucalyptus and populus this age could vary from 4 to 5 years while for species like teak this may be as long as 30 years.
- Superior growth in height and diameter of the bole is judged in relation to neighboring trees of the same or similar age. Isolated trees cannot be marked as plus trees.
- The marked tree should be free of all diseases.

Criteria for Stem Form
- Straight bole.
- Leading shoot must be showing active growth.
- No spiral grain.

Criteria for Crown and Branching Habit
- It should provide dense mass of healthy foliage.
- Good natural pruning and well-healed knot scars.
- Branches should be small in relation to the stem at the point of origin.

6.4 Multiplication of Tree Species at TERI

Following tree species have been/are being multiplied at TERI's production facility:

- *Anogeissus* spp. (*A. pendula* and *A. latifolia*)

- Bamboos (*Bambusa tulda, Bambusa arundinacea* and *Dendrocalamus strictus*)
- Eucalyptus (*E. tereticornis, E. camaldulensis* and *E. citriodora*)
- *Paulownia fortunei*
- Poplars (*P. deltoides* and *P. euphratica*)
- *Leucaena* hybrids

Till March 31, 2002 nearly 3.7 million plants of various forest species alone had been dispatched to various State Forest Departments, NGOs and private growers for field trials and routine plantations. These plants would cover an area of over 4000 hectares (Tables 1 and 2).

Table 1.   Area covered under tissue cultured plants of various forest species till March 31, 2002

S. No.	Species	No. of plants dispatched	Spacing	Approximate area covered (ha)
1.	*Anogeissus pendula*	447144	3 m × 3 m	402.46
2.	*Anogeissus latifolia*	127532	3 m × 3 m	114.79
3.	*Eucalyptus* spp.	1600481	3 m × 2 m	960.67
4.	*Populus deltoides*	1235410	3 m × 5 m	1854.96
5.	*Dendrocalamus strictus*	95385	5 m × 5 m	238.46
6.	*Paulownia fortunei*	220981	4 m × 5 m	441.96
7.	Miscellaneous	11094	3 m × 5 m	16.65
	Total	3738027		4029.95

Table 2.   Number of tissue cultured plants of various forest species dispatched to different states for field trials and routine plantations till March 31, 2002

State	Total number of plants dispatched
Assam	17163
Bihar	20517
Delhi	85113
Gujarat	20905
Haryana	1704507
Himachal Pradesh	1400
Jammu & Kashmir	29700
Madhya Pradesh	45106
Maharashtra	2735
Orissa	4321
Punjab	30075
Rajasthan	616580
Uttar Pradesh	1034635
Tamil Nadu	8404
West Bengal	10779
Kerala	1300
Karnataka	4835
Miscellaneous	99952
Total	3738027

6.5 Field Trials

Although TERI has been into production of forest species by tissue culture since 1991, the evaluation of the tissue cultured plants started much later because it is recommended that the performance of a forest species should be evaluated only after it had completed half its rotation age. The field trial data of various species available thus suggest the following:

- High survival rate of tissue cultured plants in the field at times even when the soil and other growth conditions are not favourable and life saving irrigation facilities are lacking.
- The plants showed high degree of clonal uniformity.
- Most of clones (CPTs) selected and multiplied at TERI outperformed the local clones or seedling raised plants in biomass production. Depending on the nature of the clone and its suitability at a particular location, the gains varied from marginal to significant (upto 200%).
- In the initial trials of *Populus deltoides* some problem of formation of 'kinks' was observed which was later rectified by modifying the regeneration procedures and management practices.

6.5.1 Anogeissus spp.

Anogeissus pendula

A. pendula is a very slow growing tree that grows 9 to 15 m in height and 1 m in girth. The tree is essentially an inhabitant of dry and hot regions of Haryana, Madhya Pradesh, Rajasthan and southern Uttar Pradesh, where the annual rainfall ranges between 400 and 800 mm. It can also withstand a temperature regime of 3°C to 47°C. While the leaves are used as fodder, the timber is valued for its strength and working qualities. It is used extensively for making various items of domestic and agricultural use. The wood is also consumed for making charcoal of high calorific value. The utility of this species makes it highly vulnerable to felling by rural communities and grazing by their livestock. Regeneration through seeds is extremely difficult and is not much in practice as the viability of seeds is very low (0.2–0.4%). Methods of vegetative propagation by cuttings are not yet available.

In the absence of identified plus trees and recalcitrant nature of adult tissues, cultures were established from seeds. For tissue culture work, the mother trees were carefully selected on the basis of their phenotype and the seeds were collected only from tall and healthy looking trees. The shoots obtained from aseptically raised seedlings served as the explant. The shoots were multiplied by axillary branching method and rooted individually on a suitable rooting media. After 4 weeks of hardening inside the greenhouse and polyhouse, the plants were hardened further in the nursery for at least 3 months before transfer to the field.

Till March 31, 2002, over 4.4 lakh plants of *A. pendula* had been lifted from TERI's facilities by various forest departments and other agencies for field trials and routine plantations. Most of these plants were lifted by Haryana Forest Department and planted in Aravalli Hills. The feedback received from the concerned forest departments suggests a transplantation success of over 85%. Since *A. pendula* is a very slow growing species other growth parameters such as height, girth, etc. do not hold much relevance in early years of plantation. Hence, the emphasis has been accorded only to the survival success. In the absence of conventional seedlings no controls were possible. Generally, the tissue cultured plants of a particular genotype exhibited similar growth pattern. However, as expected, there was some variation in the performance of the plants of different genotypes.

In a 36-month-old trial conducted at National Research Centre for Agro-forestry, Jhansi, AP-28 has proved to be the most promising genotype. The tallest plant of this particular genotype attained a height of 365 cm. Since these plants were grown under routine plantations, performance of individual genotypes could not be monitored. Furthermore, in the absence of conventional propagules, the comparison of tissue cultured plants with seedlings was not possible.

The available data clearly suggest that by and large tissue cultured plants of *Anogeissus pendula* have survived well in the field. This is despite the fact that most of these plants were planted in extremely hostile conditions (poor soil and no irrigation) prevailing in the barren hills of Aravalli. Low survival in few trials was mainly due to unsuitability of the site or biotic interference. The plants are constantly being monitored for their growth performance and in coming years more data would be available on the subject. Production of over 4.4 lakh plants and high survival of tissue cultured plants in the field fully testify the utility of *in vitro* technology in mass propagation of a forest species which was almost impossible regenerate by conventional methods.

Anogeissus latifolia

Commonly referred to as 'dhaura', *A. latifolia* is a large tree that grows upto 33 m in height and 2.4 m in girth. It is commonly found in the forests of the sub-Himalayan tract and Shivalik hills. The tree is found at its best in Madurai, Coimbatore, and Salem districts of Tamil Nadu and some parts of Maharashtra, Karnataka and UP. Like *A. pendula, A. latifolia* is also a good fodder and timber tree. The timber is fairly durable and is deployed for making furniture, cart-axles, shafts, poles, tool handles etc. The regeneration problems in conventional methods are similar to that described for *A. pendula* and, therefore, justify the need of *in vitro* techniques for plant propagation. To initiate cultures, seeds were collected from healthy looking trees growing in Udaipur (Rajasthan). After removing the seed coat, the seeds were put for germination and the shoots derived from 3-week-old seedlings were used for further multiplication.

Till date, nearly 1.2 lakh tissue cultured plants have been lifted from TCPP/MTP. Almost 80% of these have been planted in Haryana. Since a majority of plants were transferred to the field only during the last couple of years, it is premature to comment upon the specifics of various growth parameters. However, initial feedback received on the performance of tissue cultured plants suggest high survival rate and vigorous growth. A large percentage of plants were grown as routine plantation for which the forest departments do not maintain any record. In some cases the trials were laid initially but were abandoned later due to heavy biotic pressure or various administrative reasons. Nevertheless, repeated requests for the plants made by the end-users suggest that in *vitro* plants are doing well in the field.

6.5.2 *Bamboos*

Bamboos are one of the fastest growing perennial grasses belonging to the family Poaceae. On account of their versatility and immense utility, bamboos have been used for a variety of purposes since times immemorial. Being straight, light, hard and strong, bamboos are extensively used for construction of houses, scaffoldings, ladders, bridges, fences, furniture, sticks, tool handles, pipes, basket mats and a large number of items of domestic and agricultural use. Bamboo leaves are used for thatching and are also valued as fodder. However, the most important use of bamboo is in the paper and pulp industry to which it serves as the basic raw material.

Over-exploitation of bamboo resources by paper and pulp industry, bad management practices and interference by biotic factors such as grazing and forest fires are some of the major factors that have resulted in scarcity of bamboos. Although propagation of bamboos take place both by seeds as well as vegetatively, however, both the methods of propagation are beset with many problems that restrict their large-scale use. In view of the constant increase in demand, the scarcity of planting material and the problems associated with conventional methods of propagation, development of effective *in vitro* methods of propagation are highly desirable.

TERI scientists have developed *in vitro* regeneration protocol for four bamboo species viz., *Bambusa tulda, Bambusa arundinacea, Dendrocalamus longispathus* and *D. strictus*. However, keeping in view the demand of various species and area of distribution, the emphasis was laid only on mass propagation of *D. strictus*.

Commonly known as 'lathi bamboo', *D. strictus* is a densely tufted bamboo with strong culms that grow 20 to 50 ft in height and 1 to 3 inches in diameter. It is the most widely grown bamboo species in India. *D. strictus* is found in almost all parts of the country except northern parts of West Bengal, Assam and other very moist areas. It grows well on dry, properly drained soil up to a height of 2,000 m. Unlike most other bamboos, culms of *D. strictus* are either solid or have a very narrow lumen. Because of this property, *D. strictus* is relatively harder and stronger than other bamboo species.

For initiation of cultures, seeds were dehusked and after surface sterilization cultured on 2,4-D containing medium for induction of callus and somatic embryos. The somatic embryos were multiplied for several passages on a suitable multiplication medium. On being transferred to a germination medium the somatic embryos formed plantlets. Till March 31, 2002 over 95,000 plants had been dispatched to various states.

6.5.3 *Eucalyptus spp.*

Commonly known as 'safeda', eucalyptus is a versatile tree that grows in almost all parts of the country, from coastal areas to an elevation of 200 m. It can attain a height of 40–50 m and a girth of 1 to 1.4 m. The tree is valued for its fast growth, high adaptability to grow in different kind of soil and climatic conditions, and multiple uses. The wood is heavy, hard and mostly straight grained. In India, the plantation-grown wood is mainly used for scaffolding, construction of houses, making rayon-grade pulp and paper pulp, agricultural implements, furniture, boxes, carts, etc.

Eucalyptus is one of the fastest growing tree species producing large amount of biomass. Because of its rapid and straight growth that casts very little shadow, eucalyptus is extremely popular as an agro-forestry species. It is easy to cultivate and can even be grown in nutritionally deficient soils. One of the major advantages of growing eucalyptus is that the animals do not browse it and therefore, its protection does not pose any problem. Also, after planting once, one can have three harvests without going for re-planting.

Conventionally, eucalyptus is propagated through seeds. However, due to segregation of genes, the seed-raised population is highly heterogeneous. Clonal propagation of eucalyptus by rooting of cuttings has met with limited success. Not only it is difficult to obtain large number of plants of a particular clone by conventional vegetative methods, but also there is a potential risk of spread of various diseases along with the propagules. In contrast, using tissue culture technology, a large number of healthy and disease-free plants of selected clones can be produced within a short span of time.

At TERI, we have been successful in multiplying three species of eucalyptus i.e., *E. tereticornis, E. camaldulensis* and *E. citriodora.* However, most of the production has been of *E. tereticornis* only. Disease-free trees in possession of various desirable traits, such as faster growth, higher biomass, straight bole etc. were selected from the natural population or field plantations of different state forest departments. Referred to as 'Candidate Plus Trees (CPTs)', such superior clones served as the source material for tissue culture work. The CPTs were coppiced in a particular season to obtain juvenile shoots. Single node segments (explants) derived from such coppiced shoots were then used to initiate cultures. Under the influence of the media, the axillary bud present at the node sprouted and formed shoot(s). *In vitro* shoots were then multiplied and rooted separately on well defined media. The tissue-cultured plants were hardened inside the greenhouse before their transfer to the field.

Till March 31, 2002, 1.6 million tissue-cultured plants of eucalyptus had been supplied to various state forest departments, NGOs and private growers for field trials and routine plantations. Field data confirms high transplantation success (more than 90%), uniform and faster growth, higher yields and better timber qualities. In some of the clones selected and multiplied at TERI, the yield is more than twice as compared to conventional (seed raised) plants. Whereas seed-raised plantations have yielded a maximum mean annual increment (MAI) of 20 m^3/ha/year, TERI clones have shown growth with MAI up to 40 m^3/ha/year. Under natural (non-irrigated) conditions, the average yield of TERI clones after 6 years of planting is estimated to be around 120 tons/ha as against only 80 tons/ha in case of seed-raised plants. Higher yields can be expected if the plantations are raised under irrigated conditions.

6.5.4   *Paulownia fortunei*

Paulownia is receiving increasing attention as a short rotation woody species. A species of Chinese origin, it is characterized by fast growth, attractive growth habit and flowers, and biomass production. Besides timber, *Paulownia* leaves are used for fodder and flowers for honey production. The value of *Paulownia* for afforestation, mine site reclamation and inter-cropping systems has also been demonstrated. Despite all its potential uses, the species could not be evaluated at the commercial level in India because of lack of planting material.

Although conventionally *Paulownia* can be propagated through seeds as well as vegetatively from root or shoot cuttings, yet these methods are not adequate to meet the demand of planting material that is required for carrying out extensive field trials. At TERI, success has been achieved in developing a complete micropropagation protocol of *P. fortunei* using adult tissue. Till March 31, 2002, over 2.2 lakh tissue-cultured plants had been dispatched from TERI's MTP. Since *P. fortunei* is an exotic species and its planting material is not so easily available, the ideal edapho-climatic conditions required for its growth in India are not yet known. With the availability of planting material it will now be possible to carry out extensive field trials in different geographical conditions. Based on the performance suitable *Paulownia* growing areas may be identified and industrial plantations be raised.

6.5.5 *Populus spp.*

Populus deltoides

P. deltoides, which was first introduced in eastern UP, has now become a common tree in Tarai region and states of Punjab and Haryana. It thrives well in tropical and sub-tropical regions of India. It is an excellent source of biomass, and as a raw material its wood accounts for 50–60% for plywood and nearly 90% for match stick industry. The wood being light and of low density is an excellent source of packaging material.

To maintain its clonal nature, the species is always propagated through vegetative means and the seeds are mainly used for breeding purposes. For vegetative propagation, stem cuttings derived from superior trees are used.

It is recommended that only the leader shoot are used to derive cuttings because cuttings obtained from the side branches are not successful and the plant tends to die within 2–3 years of raising. However, the number of cuttings that can be obtained from the leader shoot of a particular tree is rather small. Therefore, in order to meet the ever-increasing demand of industry, it would be useful to carry out micropropagation of *P. deltoides*.

Mass propagation of several superior clones of *P. deltoides* such as G-3, G-48, D-121, L-34 and S7C15 using tissue culture technology has been undertaken at TERI and over 1.2 million plants have been dispatched. In the initial lots of tissue cultured plants that were transferred to the field, many of the plants showed kinks/bends in the stem. Sometimes the percentage of such plants was as high as 40%. However, cuttings derived from such kink-bearing plants in the following year produced almost normal plants. More than 95% of the plants were straight and the remaining plants showed decreased degree of bends. Disappearance of bends confirms the fact that this problem was not due to any change in the genetic make-up of the plants during the course of *in vitro* process. The clonal fidelity of the tissue cultured plants was further confirmed by DNA fingerprinting. In order to overcome the problem of kink formation in tissue cultured plants, the regeneration protocol as well as the management practices adopted in the field were modified. Following the same, the frequency of shoots bearing kinks became negligible.

Populus euphratica

P. euphratica is a unique species that can tolerate drought as well as water logging. Besides *P. deltoides*, TERI has also worked out efficient protocol for *in vitro* regeneration of *P. euphratica*. However, due to restricted geographical distribution and limited demand, the emphasis continues to be on the mass propagation of *P. deltoides*.

Conclusions

TERI is one of those few organizations not only in India but in the world that are involved in large-scale production of superior quality planting material of various tree species using tissue culture technology. The dispatch of over 3.7 million tissue cultured plants of tree species clearly demonstrates the technical feasibility of using tissue culture technology for large-scale production of forest species. The clonal uniformity, and thereby increase in productivity, signifies enormous potential the technology has to offer for increasing land productivity and thus face the challenge of meeting biomass needs of the country.

References

1. Saxena, S., Dhawan, V. (1999). Regeneration and large-scale propagation of bamboo (*Dendrocalamus strictus* Nees) through somatic embryogenesis. Plant Cell Reports. 18: 438–443.
2. Saxena, S., Dhawan, V. (2001). Large-scale production of *Anogeissus pendula* and A. *latifolia* by micropropagation. *In Vitro* Cell. Devp. Biol.-Plant. 37: 586–591.

Plant Biotechnology and Molecular Markers
P.S. Srivastava, Alka Narula and Sheela Srivastava (Editors)

13. Micropropagation of Woody Plants

J.S. Rathore, Vinod Rathore, N.S. Shekhawat, R.P. Singh*, G. Liler,
Mahendra Phulwaria and H.R. Dagla

Biotechnology Unit, Department of Botany, JNV University, Jodhpur 342 001, India
E-mail: biotechunit@satyam.net.in

*Department of Botany, Shri B.R.M. Govt. College, Nagaur, Rajasthan, India

Abstract: Micropropagation protocols for cloning of mature trees of *Balanites aegyptiaca,* the Hingota (Balanitaceae); *Citrus limon,* the Nimbu (Rutaceae) and *Syzygium cuminii,* the Jamun (Myrtaceae) have been developed. In order to harvest responsive nodal explants the mother tree(s) were pruned during the winter. Fresh shoot sprouts derived from the trees were used as explants. The nodal explants produced multiple shoots *in vitro* by activation of axillary meristems on MS medium + 0.45 µM BAP. Shoots were further multiplied in culture by (i) repeated transfer of the mother explants and (ii) the subculturing of the nodal segments of *in vitro* differentiated shoots. Shoots multiplication in *Citrus limon* could be achieved by amendment of the nutrient medium. The *in vitro* cloned shoots of the three species were rooted *in vitro* and *ex vitro*. *Ex vitro* root induction was followed to produce plants. Micropropagated plants were hardened in the green house. The hardened and acclimatized plants were transferred to pots and subsequently to field. The cloned plants are growing normal. The protocols defined are reproducible. These can be used for mass multiplication of selected clones and genetic improvement of these species.

1. Introduction

In vitro technologies are continuously expanding in the field of biology. Plant tissue culture has become a general title for a very broad subject. While in the beginning it was possible to culture plant cells either as established organs, such as roots or as disorganized masses, it is now possible to culture plant cell in a variety of ways, individually (as single cells in microculture systems), collectively (as calluses or suspensions, on petri-dishes, in Erlenmeyer flasks or in large-scale fermentors), or as organized units as shoots, roots, ovules, flowers, fruits etc. [1]. In case of *Arabidopsis*, which has been the subject of the most intensive research effort into technology development [2], it is even possible to culture complete plants for generations from seed germination to seed set without having to revert to an *in vivo* phase [3]. In its most general definition plant cell culture covers all aspects of the cultivation and maintenance of plant material *in vitro*. The cultures produced are being put to an ever-increasing variety of uses. At the early stages, *in vitro* cultured systems were developed as experimental tools for basic research and studies on plant cell division, growth, differentiation, physiology and biochemistry [4]. Such systems were seen as ways to reduce the degree of complexity associated with whole plants, providing additional exogenous control over endogenous processes, to enable more reliable conclusions to be made through simpler experimental designs [5]. However, in the recent past tissue culture technology has been increasingly used in highly applied contexts. Successes in a number of areas have been achieved. There has been major change in both the number of people making use of these techniques and also in an enhancement of the degree of sophistication associated with *in vitro* technology. Techniques of micropropagation and production of disease-free plant stocks have been defined and refined to such an extent that they have become standard practice for a range of (usually vegetatively propagating) crop plants. Thus creating what is now a multi-million plant/multi-million dollar industry [3, 6, 7]. Moreover, the discipline within this technology in which advances have been most rapid and will eventually have the greatest impact on both fundamental and applied plant sciences is that of genetic modification of plant cell.

Micropropagation deals with the propagation of plants, *in vitro*, has many advantages over conventional vegetative propagation. Its application in horticulture, agriculture and forestry is currently expanding world-wide. The goal of micropropagation is to mass-produce genetically identical, physiologically uniform, developmentally normal and pathogen-free plantlets, which can be acclimatized in a reduced time period and at a lower cost. Development of both automated environmental control systems and improved *in vitro* culture systems are essential for a significant reduction in production cost [8]. However, commercial use of micropropagation is still limited, because of its relatively high production cost resulting mainly from high labor costs, low growth rate *in vitro*, and poor survival rate of the plantlets during acclimatization [9]. Altman and Loberant [10] have elegantly reviewed principles and practices of micropropagation.

Micropropagation of woody/tree/forest plant is feasible [11-13]. However, with some exceptions traditional *in vitro* methods are not as yet practical or commercially viable for most forest trees. Therefore, improvement in current procedures and their scaling-up is required. Although it has been argued that from the environmental perspective, the genetic diversity of forest should be maintained/conserved, hence the traditional use of mixed population of seedling for forestation be applied and clonal forestry may not be appropriate. The case of cuttage propagation and micropropagation for all types of woody perennials is strongly affected by ontogenetic age [14]. Cloning *in vitro* and *in vivo* of adult and/or mature plants is adversely affected by characteristics accompanying maturation such as reduced growth rate, reduced or total lack of rooting ability or sometimes the unpleasant phenomenon of plagiotropy [15]. Maturation, a complex phenomenon, is the major problem preventing a wider application of tissue culture technology among woody plant species. Micropropagation of woody plants of stressed environments which experience types of (annually recurring) abiotic-stresses, become more difficult as the seasonal and environmental factor influence the behavior of explant(s) in culture to a great extent.

In simple terms, plant tissue culture can be considered to involve three phases. First, isolation of the plant (tissue) from its usual environment. Second, the use of aseptic techniques to obtain clean material free of bacterial, fungal, viral and even algal contaminations. Third, the culture and maintenance *in vitro* in a strictly controlled physical and chemical environment [9, 16, 17]. The components of this environment are then in the hands of the researcher who gains a considerable degree of external control over the subsequent rate of the plant material concerned. Hall [3] suggested an extra fourth phase where recovery of whole plants for rooting and transferring to soil is the ultimate goal. The success of this technology is to a great extent dependent upon abiding by a number of fundamental rules and following a number of basic protocols.

During last four decades a number of plant tissue culture technologies have been developed for a number of plant species in India [18-20]. Govil and Gupta [21] have reviewed commercialization of plant tissue culture in India. It has been suggested that plant tissue culture would play a very important role in conservation, propagation and genetic improvement of plants of our country and also in restoration ecology and restoration of degraded habitats [22].

Since 1980, we have been working on development of tissue culture protocols for application in propagation and genetic improvements of woody plants of arid and semi-arid regions (namely the Indian Thar Desert and the Aravallis). Some of the woody plant species (as important biomass producer) are keystone species of these regions of the country. We developed tissue culture processes for cloning and mass propagation, using nodal shoot explants of mature and selected woody plants namely *Aegle marmelos* [23], *Capparis decidua* [24], *Celastrus paniculatus* [25], *Maytenus emarginata* [26], *Zizyphus* spp. [27]. We also cloned shoots of *Prosopis cineraria* [28] and *Tecomella undulata* [29]. Micropropagated shoots were rooted by pulse treatment with root-inducing auxins. Several species of *Anogeissus* (Combretaceae) were first micropropagated in our laboratory using cotyledonary nodes [30–35]. Later, Saxena and Dhawan [36] of Tata Energy Research Institute (TERI), New Delhi reported the micropropagation of *Anogeissus latifolia* and *A. pendula*, also using juvenile explants.

Now we describe the development of micropropagation protocols for cloning of *Balanites aegyptiaca* (Hingota), *Citrus limon* (Nimbu) and *Syzygium cuminii* (Jamun). These woody species are economically and ecologically important as they yield valued products. *Balanites aegyptiaca* (Balanitaceae) is a tree of arid regions. This has multiple uses particularly for the aboriginals and rural people. The stem-bark is used as a fish-poison, and the pulp of fruit as detergent/soap for washing cloths/hair. Hard and durable timber is utilized for making agricultural appliances and household articles. The powder of mature fruits is taken orally by the women to prevent unwanted pregnancy. The roots and fruits of *B. aegyptiaca* yield 'Diosgenin'—a sapogenin widely used for production of pharmaceutical steroid and oral contraceptives [37]. *Citrus* (Rutaceae) is considered as number one fruit of the world for its nutritional values, the magnitude of fruit production and an array of commercial products which are derived from it. *Citrus limon* is an important horticultural species. Similarly, the Black-plum *S. cuminii* (Myrtaceae) is a tropical fruit tree which has multiple uses [38]. Also this tree has very high water use efficiency and thus is effective biomass producers. We developed cloning processes using nodal segments of rejuvenated (fresh shoot sprouts) shoots of selected mature trees.

2. Materials and Methods

2.1 Source Plants

Selected mature tree(s) of *Balanites aegyptiaca*, *Citrus limon* and *Syzygium cuminii* were pruned during December-January. Shoot sprouts were harvested during the months of February-March-April. Fresh shoot sprouts collected during the months of March/April were used as explants. The nodal explants were dressed and treated with 0.1–0.2% Bavistin and 0.1% Tetracylin for 10–15 min. These were surface sterilized with 0.1% $HgCl_2$ (5–6 min), then with 90% ethanol (60 sec) and were thoroughly washed with sterile water. These were finally treated with chilled sterile antioxidant solution (0.1% ascorbic acid; 0.05% citric acid and 0.1% PVP) for 15–20 min. The explants were inoculated on MS [39] medium supplemented with different concentrations of BAP or kinetin.

2.2   Culture Establishment and Axillary Bud Breaking

The nodal explants of the three species were inoculated in culture tubes on agar-gelled MS medium supplemented with different concentrations of benzylaminopurine (BAP) or kinetin. The cultures were incubated at 28 ± 2°C in a culture room with 10 h per day photoperiod. The responses of the explants were recorded regularly.

2.3   Amplification of Shoots in Culture

Shoot of *B. aegyptiaca* were multiplied by subculturing of nodal shoot segments of *in vitro* generated shoot on MS medium + 0.2 µM BAP. Shoot amplification in *Citrus limon* was achieved when the mother explants were repeatedly transferred or nodal explants subcultured on amended MS (50% of NH_4NO_3, KNO_3) with 0.25 µM BAP. Multiplication of shoots of *Syzygium cuminii* was achieved by (a) repeated transfer of mother explants and (b) subculturing of *in vitro* produced shoots on above mentioned amended medium with K_2SO_4 (100 mgl^{-1}), KCl (70 mgl^{-1}) and ammonium citrate (50 mgl^{-1}). Subculturing was done after 20–25 days. The cultures were amplified in 250 ml flaks or bottles. These were kept under the controlled conditions of temperature (28 ± 2°C), light (40–50 µmol m^{-2} s^{-1} SFP for 12 h/d photoperiod) and 60% RH.

2.4   Rooting of Cloned Shoots

Experiments were conducted to induce the roots *in vitro* and *ex vitro* from the micropropagated shoots. For *in vitro* rooting 4 to 5 cm long shoots were excised and cultured on agar-gelled full, half, one-third and one-fourth strengths of MS medium containing 0.1% of activated charcoal and different concentrations (1.23 to 16.1 µM) of IBA or NAA or NOA. These shoots were cultured at 30°C, under different regimes of light (8–10 h photoperiod per day). For *ex vitro* rooting, the individual shoots were pulse-treated with sub-lethal concentrations of root-inducing auxins and cultured on autoclaved soilrite in glass bottles (jam bottles). These bottles were kept in the green house at 30 ± 2°C.

2.5   Acclimatization of Micropropagated Plants

In vitro rooted plantlets were washed with sterile water to remove adhered nutrient agar and transferred to sterile soilrite in the culture bottles. In case of *ex vitro* rooted plantlets after roots were visible, the culture bottles were shifted from low temperature/high relative humidity (RH) regime of green house to the region which experienced relatively high temperatures and low RH. Also the rooted plantlets were exposed gradually to external environment by loosening/removing the caps of the culture bottles. Micropropagated and hardened plantlets were transferred to polybags containing mixture of organic manure, garden soil and sandy soil. These plants were watered regularly. The green-house-hardened plants were kept in nursery covered with agronet.

3.   Results

3.1   Selection of Explants

Nodal shoot segments harvested from pruned and non-pruned tree(s) were used as explants for establishment of cultures of three species. Explants prepared from fresh (rejuvenated) shoots regenerated from pruned plants during the months of March-April proved to be the most suitable for culture establishment. The explants harvested from non-pruned tree(s) proved to be difficult to surface sterilize as these carried recalcitrant microbial contaminations; these rarely showed bud breaking, caused excessive browning of the culture medium and exhibited browning/darkening of cut ends/explants. Thus management and pruning of mother tree was found to be essential for harvesting shoots to be used as responsive explants.

3.2 Establishment of Shoot Cultures

The surface sterilized nodal explants could be cultured on MS media containing 0.45 μM BAP or higher concentrations of BAP or kinetin. The axillary meristems were activated and bud breaking was observed after 10-15 days of inoculation in 85-90% of the explants of three species. Maximum number of shoots differentiated on MS medium supplemented with 0.45 μM BAP. Shoot differentiated from each node were 2-3 in *Citrus limon*, 1-2 in *B. aegyptiaca* and 3-4 in *S. cuminii*, respectively. Kinetin proved to be less effective as compared to BAP in the activation of axillary buds. More than 0.45 μM of either of BAP or kinetin caused callusing from the explants and proved to be inhibitory.

3.3 Amplification of Shoots *in vitro*

After the activation of meristems, bud breaking and axillary bud differentiation, the shoots were further multiplied on suitable culture media. Shoots of *B. aegyptiaca* were multiplied by sub-culturing of segments of *in vitro* produced shoots (Fig. 1). Shoot amplification occurred on MS + 0.22 μM BAP. Shoots of *C. limon* were multiplied by repeated transfer of mother explants on amended MS medium + 0.22 μM of BAP. About 12-15 shoots differentiated from each mother explant (Fig. 2). Three-fold rate of shoot multiplication was achieved by repeated transfer of the mother explants.

By repeated transfer of mother explants on amended MS medium, shoots of *S. cuminii* multiplied. Two- to three-fold rate of shoot multiplication was achieved (Fig. 3). The cultures were transferred on to fresh media after 20-25 days. The cultures were maintained at high light intensity (50–60 μmol m^{-2}s^{-1}) at 28–30°C. The shoot cultures of all the three species are being multiplied and maintained for the last 3 years.

3.4 Rooting of Cloned Shoots

3.4.1 In Vitro Rooting

Isolated shoots of all the three species rooted on half-strength MS medium with 0.1% activated charcoal. Ninetyfive to 100% of the shoots of *C. limon* rooted on half-strength MS medium + 27.0 μM of NAA. From each shoot six to eight roots regenerated. Rooting was poor on media supplemented with IBA or NOA. Of the shoots of *B. aegyptiaca*, 80-90% rooted *in vitro* on half-strength MS medium + 0.2 μM IBA + 0.1% activated charcoal (Fig. 4). Ninety percent of the shoots of *S. cuminii* rooted on half-strength MS medium + 0.1% activated charcoal + 9.8 μM of IBA.

3.4.2 Ex vitro Rooting

About 90-95% of the *in vitro* amplified shoots of *C. limon* rooted *ex vitro* (Fig. 5) if pulsed with 0.98-2.46 μM IBA. The rooting percentage was 85-90% if the shoots were treated with equimolar NOA. The shoots treated with 1.07-2.68 μM NAA showed maximum rooting. The *ex vitro* roots were visible after 10-12 days of pulse treatment.

Fig. 1. Multiplication of shoots of *Balanites aegyptiaca in vitro* by subculturing; **Fig. 2.** Amplification of shoots of *Citrus limon*; **Fig. 3.** Amplification of shoots of *Syzygium cuminii* by repeated transfer of mother explant; **Fig. 4.** Rooting of shoots of *Balanites aegyptiaca in vitro* on half-strength MS medium + 0.2 μM IBA + 0.1% activated charcoal; **Fig. 5.** *Ex vitro* rooted plantlets of *Citrus limon* being hardened in the green house; **Fig. 6.** *Ex vitro* rooting of shoots of *Syzygium cuminii*; **Figs. 7, 8 and 9.** Cloned plants of *Balanites aegyptiaca, Citrus limon* and *Syzygium cuminii,* respectively.

The shoots of *S. cuminii* could also be rooted *ex vitro*. A pulse treatment with 2.46 μM of IBA for 10–15 min was found to be sufficient to induce *ex vitro* roots from the shoots. Cent-per-cent of the shoots rooted on soilrite in the green house within 20–25 days. If the shoots were pulsed with NAA, 65% of these rooted (Fig. 6) after 30–35 days.

About 70% of the shoots of *B. aegyptiaca* rooted on soilrite after 12–15 days if treated with 1.0–2.5 mM IBA for 2–5 min.

3.5 Hardening of Micropropagated Treelets

In vitro rooted plants were hardened by transfer to soilrite containing bottles in the green house. These were kept near pad section for 8–10 days and gradually shifted towards fan section. After 10 days the caps of culture bottles were loosened and gradually removed. Plantlets rooted *ex vitro* were acclimatized in the green house. After formation of roots the plants were exposed to low RH and high temperatures. *Ex vitro* rooted plantlets were found to be easy to harden and acclimatize than those rooted *in vitro*. Hardened and acclimatized plants were transferred to black bags containing garden soil; sandy soil and organic manure (Figs. 7, 8 and 9). Several plants have been transferred to the field. These are growing normal. Flowering of these is yet to be recorded.

4. Discussion

The research work presented in this article demonstrates that the mature woody plants can be cloned using appropriate *in vitro* methods. We have described the development protocols for cloning of *Balanites aegyptiaca*, *Citrus limon* and *Syzygium cuminii*. These are valuable woody species that yield products of economic value. Selection of the individual plant with desired (superior) characters is possible only after certain age, when reproductive maturity is reached. Such selected and mature plants give high yield of quality product. Once the selection is done it is necessary to maintain genetic fidelity of the clone. This is done by vegetative propagation *in vivo* or *in vitro* (micropropagation). Cloning of mature woody plants *in vitro* and *in vivo* is adversely affected by characteristics accompanying maturation such as reduced growth rate, reduced or total lack of rooting ability or sometimes the unpleasant phenomenon of plagiotropy [14, 15, 40]. Maturation, a complex phenomenon, is the major problem preventing a wide application of tissue culture technology among woody species. Nevertheless, a number of woody species/trees have been micropropagated. Success with several species have been achieved mainly by the use of special starting (explanting) material, by special pre-treatment to mother/source plant(s) *in vivo* or by *in vitro* culture [11, 26, 28]. All of these tricks, which improve clonal propagation are often described by the general term rejuvenation. It is clear that rejuvenation is a pre-requisite for possible cloning of adult trees and that the success in practice mainly depend on the ability to rejuvenate them. We found that in all the three species under investigation, pre-treatment (pruning during winter) of mother plant was desirable otherwise the explant did not respond in culture. The shoot sprouts (flushes) from plants pruned during winter proved to be the only useful (suitable explants) for culture initiation. Rejuvenation (also known as phase reversal or return to the juvenile form) includes the complete reversal of maturation as a result of sexual reproduction or vegetative propagation via shoot formation (through activation of pre-existing axillary- or apical-meristems) or somatic embryogenesis. Re-invigoration is defined as the reversal of ageing (which leads to reduced vigor and rooting ability). Reinvigoration can be used when rooting-ability and vigor are increased as a result of, for example, pruning, hedging, repeated culturing, BAP-treatment and grafting [41]. The nodal explants of *B. aegyptiaca*, *C. limon* and *S. cuminii* derived from fresh shoot sprouts, responded in culture and produced multiple shoots on BAP (0.45 µM) supplemented medium. The shoots could be further amplified *in vitro* by (i) repeated transfer of explants and (ii) subculturing, but on medium with comparatively lower concentration of BAP. In quite a number of plant species repeated subculture/transfer of adult shoots (mother explants) were reported to induce invigoration and complete rejuvenation, by which shoot multiplication and rooting ability are strongly improved. It is suggested that once the cultures/explants were established these become conditioned and they required low cytokinin for further multiplication. In case of *Citrus limon* the cultures could be multiplied by lowering the concentrations of certain salts (NH_4NO_3 and KNO_3). Chaturvedi [42] et al. critically reviewed tissue culture employing vegetative explants in *Citrus* spp. It is stated that maximum tissue culture research has been done in *Citrus* during the last four decades however the results of practical value are meager. We have successfully established procedure for large-scale shoot multiplication of *Citrus limon*. This is important contribution in *Citrus* tissue culture with practical utility.

The micropropagated shoots were rooted *in vitro* on half-strength MS medium + 0.1% activated charcoal supplemented with IBA (*B. aegyptiaca* and *S. cuminii*) and NAA (*Citrus limon*). Bonga and Von Aderkas [43] suggested that roots from rejuvenated shoots of woody plants, is induced *in vitro* by IBA or NAA. Probably the nature of auxin required and the concentration for *in vitro* root regeneration are species specific.

In the present case the micropropagated shoots of all the three species could be rooted *ex vitro*. The main advantage of *ex vitro* over *in vitro* rooting is that root damage during transfer to soil is less likely. Furthermore, rooting rates are often higher and root quality is better when the rooting takes place *ex vitro* [43]. McClelland et al. [44] studied the effect of *in vitro* and *ex vitro* root initiation on subsequent microcutting root quality in three woody plants. They suggested greater resistance of *ex vitro* rooted plants to stress. Arya et al. [25] found that the *ex vitro* rooted plantlets of a woody climber, *Celastrus paniculatus* were easy to harden. The duration of time and cost of plant production are also reduced by switching to *ex vitro* root generation. IBA proved to be more effective auxin for pulsing of the shoots for *ex vitro* root induction. The auxin, most commonly used for root formation is IBA. It is generally assumed that the greater ability of IBA as compared with other auxins to promote rooting is due to its relatively higher stability [45, 46]. It has been possible to induce *ex vitro* induction in number of woody species [35] of stressed environments.

The rooted plantlets of all the species could be hardened in the green house and pot transferred with ease. The survival rates have been satisfactory. Development of protocol for micropropagation of *B. aegyptiaca* is important contribution as this could be applied for cloning of plants selected for higher yield of diosgenin. Selected and tested plants of *Citrus limon* bearing desired attributes of horticultural importance can also be cloned using our protocol. Yadav et al. [38], and Jain and Babbar [47] reported *in vitro* micropropagation of *Syzygium cuminii*. They used explants from young seedlings. This method of cloning is not preferred for fruit trees. Multiple shoot induction from 1- to 2-year-old seedlings of *S. travancoricum* was recorded by Anand et al. [48]. Mathew and Hariharan [49] reported *in vitro* shoot multiplication in *S. aromaticum*. Shah Valli Khan et al. [50] reported *in vitro* micropropagation of mature *S. alternifolium*. In this article we have described a process for cloning of mature tree of Black-plum (*S. cuminii*). This is the most desired level of cloning. The micropropagated plantlets of all the three woody species could be hardened and pot transferred. The processes defined are highly reproducible and efficient and these can be utilized for cloning of selected trees of these species.

Acknowledgements

N.S. Shekhawat is grateful to the Department of Biotechnology (DBT), Government of India for financial assistance for the establishment of Micropropagation Unit and Green House (Grant No. BT/R&D/08/03/93), University Grants Commission (UGC), New Delhi for providing support under UGC-SAP-DSA Programme (1997–2001) to the Department of Botany and the Department of Science and Technology, Government of India for providing support under DST-FIST Programme. We are grateful to the Head, Department of Botany for providing facilities for this work. We also appreciate the technical assistance provided by Shri M.S. Panwar.

References

1. E.F. George, Plant propagation by tissue culture I: The Technology, Exegenetics Ltd., Edington, UK, 1993.
2. J.M. Martinez-Zapater, J. Salinas, *Arabidopsis* protocols. Humana Press, Totowa, New Jersey, USA, 1998.
3. R.D. Hall, Plant Cell Culture Initiation: Practical Tips. Molecular Biotechnology 16 (2000) 161–173.
4. H.E. Street (Ed), Tissue Culture and Plant Science. Academic Press, New York, 1974.
5. G.G. Henshaw, J.F. O'Hara, K.J. Webb, Morphogenetic studies in plant tissue cultures, in: M.M. Yeoman D.E.S. Truman (Eds) Differentiation *in vitro*: British Society for Cell Biology Symposium-4. Cambridge University Press, UK, 1982, pp. 231–251.
6. R.L.M. Pierik, *In vitro* culture of higher plants. Kluwer, Dordrecht, The Netherlands, 1997.
7. K.L. Giles, K.R.D. Friesen, Micropropagation, in: P.D. Shargool, T.T. Ngo (Eds) Biotechnological Applications of Plant Culture. CRC Press, Boca Raton, Florida, USA, 1994, pp. 111–128.
8. J. Aitken-christie, T. Kozai, M.A.L. Smith, Automation and environmental control in plant tissue culture. Kluwer Academic Publishers, Dordrecht, 1995.
9. T. Kozai, C. Kubota, B.R. Jeong, Environmental control for the large-scale production of plants through *in vitro* techniques. Plant Cell, Tissue and Organ Culture 51 (1997) 49–56.
10. A. Altman, B. Loberant, Micropropagation: Clonal propagation *in vitro*, in: A. Altman (Ed) Agricultural Biotechnology. Marcel Dekker, Inc., New York, 1998, pp. 19–42.
11. J.M. Bonga, Clonal propagation of mature trees: problems and possible solutions, in: J.M. Bonga, D.J. Durzan (Eds) Cell and Tissue Culture in Forestry, Vol. I. Martinus Nijhoff Publishers, Dordrecht, 1987, pp. 249–271.
12. B.E. Haissig, N.D. Nelson, G.H. Kidd, Trends in the use of tissue culture in forest improvement. Biotechnology 5 (1997) 52–87.
13. I.S. Harry, T.A. Thorpe, *In vitro* culture of forest trees, in: I.K. Vasil, T.A. Thorpe (Eds) Plant Cell and Tissue Culture. Kluwer Academic Publishers, Dordrecht, 1994, pp. 539–560.
14. W.P. Hacket, Juvenility, maturation and rejuvenation in woody plants. Horticultural Review 7 (1985) 109–155.
15. R.L.M. Pierik, Rejuvenation and micropropagation, in: H.J.J. Nijkamp, L.H.W., Van Der Plas, K. Van Aartrijk (Eds) Progress in Plant Cellular and Molecular Biology. Kluwer Academic Publishers, 1990, pp. 91–101.
16. R.H. Smith, Plant Tissue Culture: Techniques and Experiments. Second edition. Academic Press, San Diego, USA, 2000.
17. Q.T. Nguyen, T. Kozai, Environmental effects on the growth of plantlets in micropropagation. Environment Control in Biol. 36 (1998) 59–75.
18. G. Lakshmi Sita, C.S. Vaidyanathan, T. Ramakrishnan, Applied aspects of plant tissue culture with special reference to tree improvement. Current Science 51 (1982) 88–92.
19. N.S. Shekhawat, B.M. Johri, P.S. Srivastava, Morphogenesis and plant tissue culture, in: B.M. Johri

(Ed) Botany in India: History and Progress, Vol. II. Oxford & IBH Publishing Co. Pvt. Ltd., New Delhi, 1994, pp. 307–356.

20. N.S. Shekhawat, M.H. Mughal, B.M. Johri, P.S. Srivastava, Indian contribution to plant tissue and organ culture, in: P.S. Srivastava (Ed) Plant Tissue Culture and Molecular Biology: Applications and Prospects. Narosa Publishing House, New Delhi, 1998, pp. 751–811.

21. S. Govil, S.C. Gupta, Commercialization of plant tissue culture in India. Plant Cell Tissue and Organ Culture 51 (1997) 65–73.

22. Manju Sharma, Biodiversity conservation and socio-economic development: role and relevance of biotechnology, in: Nandi, S.K., Palni, L.M.S., A. Kumar (Eds) Role of Plant Tissue Culture in Biodiversity Conservation and Economic Development. G.B. Institute of Himalayan Environments and Development, Himavikas Occassional Publication No. 15 Kosi-Katarmal, Almora, India, 2002, pp. 1–9.

23. R. Bhati, N.S. Shekhawat, H.C. Arya, *In vitro* regeneration of plantlets from root segments of *Aegle marmelos*, Indian J. Experimental Biology 30 (1992) 844–845.

24. N.S. Deora, N.S. Shekhawat, Micropropagation of *Capparis decidua* (Forsk.) Edgew—a tree of arid horticulture, Plant Cell Reports 15 (1995) 278–281.

25. V. Arya, R.P. Singh, N.S. Shekhawat, A micropropagation protocol for mass multiplication and off-site conservation of *Celastrus paniculatus*—A vulnerable medicinal plant of India. J. Sustainable Forestry 14 (2002) 107–120.

26. T.S. Rathore, N.S. Deora, N.S. Shekhawat, Cloning of *Maytenus emarginata* (Wild.) Ding Hou—a tree of the Indian Desert, through tissue culture. Plant Cell Reports 11 (1992) 449–451.

27. T.S. Rathore, R.P. Singh, N.S. Deora, N.S. Shekhawat, Clonal propagation of *Zizyphus* species through tissue culture. Scientia Horticulturae 51 (1992) 165–168.

28. N.S. Shekhawat, T.S. Rathore, R.P. Singh, N.S. Deora, S.R. Rao, Factors affecting *in vitro* clonal propagation of *Prosopis cineraria*. Plant Growth Regulation 12 (1993) 273–280.

29. T.S. Rathore, R.P. Singh, N.S. Shekhawat, Clonal propagation of desert teak (*Tecomella undulata*) through tissue culture. Plant Science 79 (1991) 217–222.

30. R. Joshi, N.S. Shekhawat, T.S. Rathore, Micropropagation of *Anogeissus pendula* Edgew—an arid forest tree. Indian J. Experimental Biology 29 (1991) 615–618.

31. G. Kaur, R.P. Singh, T.S. Rathore, N.S. Shekhawat, *In vitro* propagation of *Anogeissus sericea*. Indian J. Experimental Biology 30 (1992) 788–791.

32. T.S. Rathore, N.S. Deora, N.S. Shekhawat, R.P. Singh, Rapid micropropagation of a tree of a rid forestry, *Anogeissus accuminata*. Biologia Plantarum 35 (1993) 381–386.

33. R.P. Singh, N.S.Shekhawat, Micropropagation of *Anogeissus rotundifolia* Blatt. and Hallb.—An endemic and rare tree of the Thar desert. J. Sustainable Forestry 4 (1997) 159–170.

34. N.S. Shekhawat, J. Yadav, V. Arya, R.P. Singh, Micropropagation of *Anogeissus latifolia* (Roxb. Ex. DC.) Wall. Ex. Guill. & Perr.—a tree of fragile ecosystems. J. Sustainable Forestry 11 (2000) 83–96.

35. J. Yadav, N. Choudhary, M.S. Shekhawat, J.S. Rathore, V. Arya, S.R. Vishnoi, V. Rathore, R.P. Singh, N.S. Shekhawat, Micropropagation of plants of arid zones: achievements and limitation, in: Nandi, S.K., Palni, L.M.S., A. Kumar (Eds) Role of Plant Tissue Culture in Biodiversity Conservation and Economic Development. G.B. Institute of Himalayan Environments and Development. Himavikas Occassional Publication No. 15, Kosi-Kartmal, Almora (India), 2002, pp. 11–28.

36. S. Saxena, V. Dhawan, Large-scale production of *Anogeissus pendula* and *A. latifolia* by micropropagation. In Vitro Cell Dev. Biol.-Plant 37 (2001) 586–591.

37. V. Singh, R.P. Pandey, Ethanobotany of Rajasthan, India. Scientific Publishers (India), Jodhpur, 1998, pp. 54–55.

38. U. Yadav, M. Lal, V.S. Jaiswal, *In vitro* micropropagation of the tropical fruit tree *Syzygium cuminii*. Plant Cell, Tissue and Organ Culture 21 (1990) 87–92.

39. T. Murashige, F. Skoog, A revised medium for rapid growth and bioassays with tobacco tissue cultures. Physiol. Plant. 15 (1962) 473–497.

40. W.M. Cheliak, D.L. Rogers, Integrating biotechnology into tree improvement programs. Canadian J. Forest Research 20 (1990) 452–463.

41. M. Boulay, Conifer Micropropagation: Applied research and commercial aspects, in: J.M. Bonga, D.J. Durzan (Eds) Cell and Tissue Culture in Forestry, Vol. I. Martinus Nijhoff, Dordrecht,1987.

42. H.C. Chaturvedi, S.K. Singh, A.K. Sharma, S. Agnihotri, Citrus tissue culture employing vegetative explants. Indian J. Experimental Biology 39(2001) 1080–1095.

43. J.M. Bonga, P. Von Aderkas, *In vitro* culture of trees. Forestry Sciences, Vol 38, Kluwer Academic Publishers, Dordrecht, 1992.

44. M.A.L. McClellan, Z.B. Smith, Carothers, the effect of *in vitro* and *ex vitro* root initiation on subsequent microcutting root quality in three woody plants. Plant Cell, Tissue and Organ Culture, 23 (1990) 115–123.

45. E. Epstein, J. Ludwig-Muller, Indole-3-butyric acid in plants: occurrence, synthesis, metabolism and transport. Physiol. Plant. 88 (1993) 382–389.

46. W.M. Van der Krieken, H. Breteler, M.H.M. Visser, D. Mavridou, The role of the conversion of IBA into IAA on root regeneration in apple: introduction of a test system. Plant Cell Reports 12 (1993) 203–206:

47. N. Jain, S.B. Babbar, Recurrent production of plants of Black-plum, *Syzygium cuminii* (L.) Skeels, a myrtaceous fruit tree, from *in vitro* cultured seedling explants. Plant Cell Reports 19 (2000) 519–524.

48. A. Anand, C. Srinivas Rao, P. Balakrishna, *In vitro* propagation of *Syzygium travancoricum* Gamble— an endangered tree species. Plant Cell, Tissue and Organ Culture 56 (1999) 59–63.

49. K.M. Mathew, M. Hariharan, *In vitro* multiple shoot regeneration in *Syzygium aromaticum*. Ann. Bot. 65 (1990) 277–279.

50. P.S. Shah Valli Khan, E. Prakash, K.R. Rao, *In vitro* micropropagation of an endemic fruit tree *Syzygium alternifolium* (Wight) Walp. Plant Cell Reports 16 (1997) 325–328.

Plant Biotechnology and Molecular Markers
P.S. Srivastava, Alka Narula and Sheela Srivastava (Editors)

14. Biotechnology in Mulberry (*Morus* spp.) Crop Improvement: Research Directions and Priorities

S.B. Dandin and V. Girish Naik

Central Sericultural Research and Training Institute, Srirampura, Mysore 570 008, India

Abstract: Mulberry (*Morus* spp.) is a crop plant of economic importance in sericulture. Mulberry improvement through conventional breeding has substantially contributed to the success of sericulture industry. However, the application of biotechnology in mulberry crop improvement holds a great promise especially in those areas where conventional research has not achieved the desired success. The biotechnological research in genome characterization with isozyme and DNA markers, micropropagation, regeneration from callus, somatic hybridization, *in vitro* conservation technologies like slow-growth storage and cryopreservation, genetic transformation etc., have contributed to the success in mulberry improvement. Besides discussing the progress achieved so far in mulberry biotechnology, the article also emphasizes the future priorities in this direction both in terms of supportive and strategic research.

1. Introduction

Application of biotechnological methods for crop improvement has significantly contributed to the success of modern day agriculture. Enhancement of yield potential, improvement of quality, resistance to pests and diseases, tolerance to abiotic stresses and resistance to herbicides are the main focus of crop improvement in many agricultural crops through biotechnological approach. Mulberry (*Morus* spp.) is a crop of economic importance in the sericulture industry. Its foliage forms the sole source of food for the domesticated silkworm, *Bombyx mori* L. Mulberry is a dioecious, heterozygous and perennial tree. In spite of the problems associated with tree crop improvement, considerable progress has been achieved in mulberry breeding through conventional approaches. However, biotechnology application holds a great promise in further improvement of mulberry crop especially in those areas where conventional research has not achieved the desired success. Already considerable progress has been made in this direction. The article attempts to consolidate the important outcome of the biotechnological applications in mulberry and also discusses the need for future research priorities in mulberry improvement, utilization and conservation.

2. Genome Characterization

Understanding of genetic structure of the plant is very important for crop improvement, utilization and conservation. Mulberry being a perennial, heterozygous tree, traditional methods of analysis have not provided sufficient insight into the genetic architecture. Compared to the phenotypic characters, molecular markers are highly heritable, consistent, fast and easy to measure and evaluate. Among the molecular markers, isozyme and DNA markers are widely employed for genome characterization and analysis of plants and animals.

2.1 Isozyme Markers

Hunter and Markert [1] were first to introduce isozymes as genetic markers in plants. Hirano [2] used peroxidase isozyme technique to evaluate the affinities in mulberry and its relatives and showed that the results supported the conventional view. The study [3] of inheritance of peroxidase isozyme of mulberry was initiated in Japan and established that particular isozyme banding type was significantly correlated with leaf stalk length. Hirano [4] also used isozyme technique to analyze 284 mulberry varieties. He used seven enzyme systems and a sap protein to characterize these varieties. Based on the electrophoretic pattern he categorized 131 varieties into seven groups and established the gentic relationship among them. The study also demonstrated the correlation between amino acid content and peroxidase enzyme in the leaf. Katagiri and co-workers [5] successfully utilized peroxidase isozyme technique to differentiate hexaploid mulberry strains collected from Mexico. In India, peroxidase isozyme studies were reported on introduced species from Indonesia [6], triploids [7] and aneuploids [8] of mulberry. Even though isozyme analysis is comparatively easy, less costly and the markers are codominant in expression, they are less attractive compared to the DNA markers because of lack of sufficient polymorphism.

2.2 DNA Markers

Studies on mulberry genome was first initiated in Japan. Katagiri and coworkers [9] successfully isolated chloroplast DNA from mulberry. Later Machii [10] reported the isolation of total DNA by ultracentrifugation method. Chengfu and coworkers [11] detected DNA marker variation using RAPD technique in 12 mulberry varieties with 24 primers. Relationships among the operational taxonomical units (12 species and 2 varieties) of *Morus* were examined with 20 random decamer primers, generating 238 polymorphic markers [12]. Phylogenetic analysis of RAPD data indicated that grouping so obtained is in conformity with morphological classification. Polymorphism in genomic DNA of five parents and their four resulting hybrids were analyzed by RAPD technique [13]. Of the F_1 patterns, most of the markers appeared were same as their respective parents, however, few were unique not found in their parents. Sharma and coworkers [14] assessed the genetic diversity in *Morus* germplasm collections using fluorescence-based AFLP markers. The wide range in the genetic similarity (0.58–0.99) indicated that the mulberry germplasm collection represents a genetically diverse population. However, the study also concluded that the genetic base of cultivated mulberry is narrow. A recent study [15] showed that as many as five RAPD and one DAMD primers generated profiles can together differentiate all the nine mulberry varieties in terms of unique bands. Central Sericultural Germplasm Resources Centre, Hosur in collaboration with Seribiotech Research Laboratory, Bangalore has characterized number of mulberry germplasm using DNA fingerprinting techniques [16–19]. RAPD analysis of 15 mulberry species revealed few species diagnostic markers indicating the usefulness of the technique in identification. Phylogenetic analysis of RAPD and ISSR markers showed the separation of wild and cultivated mulberry species into a different cluster. Study of 44 cultivated mulberry varieties and 27 *M. laevigata* collections with RAPD marker data has resulted in generation of useful information on genetic diversity and identity. A research project on "Genome analysis of mulberry" [20] is currently underway at Central Sericultural Research and Training Institute at Mysore. The results indicate that RAPD can be effectively used to DNA fingerprint mulberry cultivars and also can be successfully employed to study the inheritance pattern and for the development of molecular linkage map.

3.   Micropropagation by Tissue and Organ Culture

Most of the initial studies on mulberry tissue culture concentrated on the regeneration of complete plantlets from various explants like shoot tip, axillary bud, winter bud, leaf, cotyledon, hypocotyls etc. Ohayama [21] for the first time successfully obtained complete plant from axillary bud of *M. alba* on MS medium supplemented with growth regulators. Since then, shoot proliferation was observed in many mulberry genotypes using different explants and supplementing the media with cytokinins like BAP [22–43]. However, BAP had a negative response at a higher concentration on shoot proliferation of mulberry genotypes [44–45]. Modification of basal media with macro- and micro-salt were tested on different genotypes for shoot proliferation [41]. Micropropagation of shoots of *M. indica* was tested with MS salts and B5 vitamins [46]. The pH level of various media tested ranged from 4 to 5.6 for shoot multiplication in different mulberry genotypes [47–48]. However, the optimum pH level appears to be in the range of 5.6–5.8 for many genotypes. Best shooting response was obtained at 0.8% of agar concentration [41, 46, 48].

Auxin rich media induced rooting within 10–14 days of culture in *M. laevigata* [44]. In several mulberry species, rooting was enhanced by treatment with NAA and IAA [21, 23, 41]. Combination of IBA, IAA and IPA helped root proliferation in *M. australis* [45], *M. lhou*, *M. cathayana* and *M. serrata* [49]. In *M. alba* shoots produced roots in auxin-free media [48].

Hardening of regenerated plantlets is an essential perquisite for successful establishment in the field. Various kinds of potting mixtures like steam sterilized peat and agroperlite (2:1), autoclaved soil, soilrite mixture and vermi-compost have been used for establishment of regenerated plants [39, 49]. Half-strength Hogland's nutrient solution and water was used to irrigate the plantlets [39, 45, 49].

4.   Callus Formation and Differentiation

Induction of callus of mulberry genotypes from different explants sources like stem segments [50], young leaf [51, 52] and hypocotyls segments [53, 54] were successfully attempted on MS media supplemented with 2-4 D. Addition of Kn, IAA and NAA in the media resulted in the better proliferation of calli. Calli can be prolonged in the culture medium up to eight weeks in good conditions by adding ABA and PABA [55]. Calli of *M. bombycis*, *M. alba* and *M. multicaulis* were regenerated in medium supplemented with auxins and cytokinins [56, 57]. Shoot regeneration from callus of *M. alba* was obtained [26, 58] on MS medium supplemented with BAP. In *M. bombycis* shoot bud induction was reported in the callus on LS medium supplemented with BAP [42]. Addition of GA3 and DTT to the culture medium enabled to break the pseudo-dormancy and obtained regeneration in stored calli [59].

Rhizhogenesis of calli was frequently reported from the cultures on media containing auxins [60–61]. Rooting was also obtained from the cell suspension of the callus from hypocotyl tissue [62]. From the callus of internodal segment and leaf explants of *M. laevigata,* rhizogenesis was observed on MS medium supplemented with NAA [63]. Recently, few workers have reported the complete regeneration of mulberry plants from callus culture using TDZ [64, 65].

5.   Development of Haploids

As already discussed, mulberry is a dioecious, outbreeding and heterozygous tree species. Development of homozygous lines through conventional method has not been successful. Homozygous lines are extremely important in genetical studies and exploitation of hybrid vigor in crop improvement programme. In this background, constant efforts have been made in mulberry to develop homozygous lines through the production of haploids. Lin and coworkers [66] successfully reported the regeneration of haploid plants from uninucleate anthers in a Chinese mulberry variety. Venkateshwaralu and Katagiri [67] reported globular and heart-shaped embryoids on B5 medium in a Japanese variety. Similar results were also obtained by Sethi and coworkers [68] in an Indian mulberry variety on MS medium. Katagiri [69] reported the colony formation and induction of callus from pollen culture studies. Katagiri and Venkateswaralu [70–71] observed embryo like structures on culturing the pollen in B5 medium. Addition of fructose to B5 medium [72] resulted in profuse division and obtained a compact calli. On MS media supplemented with glutamine, coconut water and 2–4 D, Tewary and coworkers [73] obtained globular embryoids from pollens isolated from anthers starved at 10–12°C for 72 h in S-1 variety. Lakshmi Sita and Ravindran [74] for the first time reported the gynogenic haploids plants from the ovary culture of mulberry. Dennis Thomas and coworkers [75] developed a reproducible protocol for the production of gynogenic haploids of a female clone of mulberry (*M. alba* L. Cv. K-2) from unpollinated ovary culture.

6. Protoplast Isolation, Culture and Somatic Hybridization

Genetic barrier in hybridization due to sexual incompatibility and other associated problems can be successfully overcome by somatic hybridization of protoplast cells. Methodology for isolation of protoplast, its culture and fusion of cells play a critical role in successful regeneration of plants. Protoplast isolation in mulberry was attempted from callus [76] and mesophyll cells [77–78]. Ohnishi and Kiyama [79] showed that primary callus culture gave a better protoplast yield than secondary callus cultures of mulberry. Tewary and Lakshmi Sita [78] reported the optimized concentration of cellulase (2%), macerozyme (1%) and macerase (0.5%) for better protoplast yield in mulberry. Katagiri [80] observed the colony formation in cultures of mulberry mesophyll protoplasts. Differences in division of mesophyll protoplasts cultured on different media and under different light intensities were studied in few mulberry species. Ming and coworkers [81] demonstrated the regeneration of complete plant from the callus derived from mesophyll protoplast of mulberry through organogenesis and somatic embryogenesis on MS medium. Protoplast fusion in mulberry was successful using chemical fusogen [76] and electro-fusion [82].

7. *In Vitro* Methods for Conservation of Genetic Resources

Conventional approach to germplasm conservation of reclalcitrant seed species as well as vegetatively propagated crops did not overcome the inherent limitation in storage technique. In contrast, *in vitro* conservation methods offered suitable alternative to seed and field gene bank. *In vitro* conservation refers to maintenance of germplasm in a relatively stable form under more or less defined nutrient conditions in artificial environment. Potential advantages of conserving mulberry genetic resources by *in vitro* methods are:

  (i)　can be utilized for germplasm collection in the field
 (ii)　rapid multiplication of germaplsm genotypes
(iii)　pathogen-free germplasm can be maintained
 (iv)　require very small storage space
  (v)　loss due to diseases, pests and natural calamities avoided
 (vi)　germplasm exchange is easier as quarantine requirement is effectively met

Some of the methods which are attempted/employed for conservation of mulberry genetic resources are discussed below:

7.1   Synthetic Seeds

Synthetic seeds, also called artificial seeds, are prepared by encapsulating the apical/axillary buds with 3–5% sodium alginate and 100 mM calcium chloride solution as a complexing agent. Sodium alginate solution is mixed to liquid medium supplemented with all macro- and micronutrients and growth regulators necessary for the development of mulberry plant [83]. About 50 ml of sodium alginate and 120 ml of calcium chloride solution making a total of 170 ml is sufficient to make 200–220 encapsulated beads (artificial seeds). The artificial seeds can be germinated either *in vitro* or *in vivo*. However, success of germination *in vivo* is comparatively less. For *in vitro* germination of artificial seeds of mulberry, MS medium was found suitable. A cytokinin supplement in MS medium enhances the germination but inhibits root formation. Even though, encapsulation of apical buds are of limited value in germplasm conservation, these artificial seeds retain viability upto 45 days at +4°C and for a long period under cryopreservation. These artificial seeds are also useful in germplasm exchange.

7.2   Slow-Growth Storage

Slow-growth condition in *in vitro* provide a secondary storage method for field gene bank, a storage mode for experimental, or a rescue of germplasm for plant distribution [84]. Slow-growth storage may provide short- or medium-term conservation strategy for germplasm materials depending on the period of storage achieved. This is done by maintaining the cultures under growth limiting condition, which reduces the requirements of sub-culturing and associated risks.

Even though, *in vitro* slow-growth storage appears to be good choice for conservation of vegetatively propagated species, the information on germplasm conservation is limited to few genera. *In vitro* technique has been utilized to conserve wide range of species including temperate woody plants, fruit trees, horticultural species, and numerous tropical species. A recent FAO survey indicates that only 37,600 accessions are conserved *in vitro* worldwide. Slow-growth storage is routinely used in the conservation of only few species like banana, potato and cassava.

In mulberry, single shoots of *M. nigra* L. stored on multiplication medium at 4°C for 16-hour photoperiod survived for only six months. Survival was enhanced to 42% at nine months by storing them at 25°C with activated charcoal as supplement [85]. High viability (80%) for six months was observed in 15 genotypes of *M. alba* stored at 4°C and in dark on shoot proliferation medium [86]. Rooting was observed in all the shoots and shoots retained their multiplication potential.

In vitro techniques are becoming increasingly popular in storing and distributing germplasm throughout the world. Certification programme insist on *in vitro* cultures for providing virus-free plants from stock collections. However, additional research is needed to be carried out in the field of genetic stability of *in vitro* grown plants. Field experimentation and molecular analysis are needed to confirm the genetic stability. Additional research to develop standard method along with regular evaluation of culture materials will provide safe storage for *in vitro* cultures.

7.3   Cryopreservation

Cryopreservation (storage in liquid nitrogen at –196°C) is considered an ideal method for long-term germplasm storage. At cryogenic temperature cell divisions and all metabolic activities are stopped, minimizing the possibility of any genetic change. Cryopreservation can be applied to different plant parts/structures including seed, apical or axillary buds, embryos, pollen and *in vitro* cultures. Sakai [87] was first to report the survival of plant tissue exposed to ultra-low temperature, when he demonstrated that very hardy mulberry twigs could withstand freezing in liquid nitrogen (LN) after dehydration mediated by extra-organ freezing. Generally, the technique of preserving at low temperature improvised with chemical cryoprotectant, slow dehydration, cooling followed by rapid immersion in LN, storage in LN, rapid thawing, washing and recovery.

As mentioned earlier; cryopreservation technique possibility was first demonstrated using mulberry. Since then, considerable work on cryopreservation of mulberry has been undertaken especially in Japan. Shoot tips of pre-frozen winter buds of *M. bombycis* Koidz. cv Kenmochi were able to withstand storage in LN [88], however, grafts and cuttings could not on immersing in LN. With modification of this method Wang and coworkers [89] were able to regenerate plants of *M. multicaulis* Loud cv. Lusang through tip culture of frozen winter buds. Shoot segments were prefrozen at –3°C for 10 days, –5°C for 3 days, –10°C for 1 day and –20°C for 1 day before immersion in LN. Buds were cultured on MS medium after thawing in air at 0 to 20°C. Observed survival rate was 55 to 90%. Excised shoot tips from winter buds of *M. bombycis* cv. Kenmochi prefrozen to –20°C at 5°C/day were able to produce more shoots compared to the buds prefrozen at 10°C/day [90]. Prior to prefreezing at –20°C, partial dehydration to 38.5% improved the recovery rates. The survival rates of the winter buds stored in LN from one month to 3.5 years did not change [91]. Direct dehydration with silica gel at 25°C of excised shoot tips (2 mm long) from winter bud could be done before immersion in LN. With decreasing water content shoot formation increased and at about 19% water content, a maximum of 80% survival rate was observed. Encapsulation by alginate coating of winter hardened shoot tips of many *Morus* species had 81% of shoot formation with 22–25% water content [90, 92]. *In vitro* grown shoot tips of thirteen cultivars of mulberry were tested for cryopreservation. Slow freezing (0.5°C/min to –42°C), vitrification (PVS2, 90 min) and air drying (24% water content) or encapsulation-dehydration (33% water content) was tested for survival, which ranged from 40 to 81.3%. Niino and coworkers [93] also reported long-term storage of mulberry winter buds by cryopreservation. Winter buds from *M. bombycis* with about 10 mm vascular tissue were kept at 0°C for 1 day before freezing. Buds were cooled to –10°C steps at daily intervals from 0 to –30°C. They were kept for one day at –30°C prior to immersion in LN or before transfering to –135°C. After storage, buds were rapidly thawed at 37°C in a water bath and then cultured on MS medium supplemented with 2% fructose and 1 mg/l 6-BAP. Rate of shoot formation did not vary much in buds stored in LN or deep freezer at –135°C after a storage period of 3.5 years.

8. Genetic Transformation

Genetic transformation has been successfully attempted in many agricultural crops. According to an estimate about 50 million ha of transgenic crops were cultivated worldwide in 2001. These estimates do not include those cultivated in China. In spite of the resistance to the genetically modified plants (GMPs) from some quarter, the popularity is gaining among the cultivators. "Golden rice" is a remarkable achievement and a major leap. This genetically modified rice is nutritionally enriched with Vitamin A and iron content, which can effectively prevent malnutrition among the population, especially in Asian countries, where rice is a staple food.

Even though, genetically transformed mulberry is yet to be released for cultivation, preliminary work in this direction has been initiated. Machii [94] used *Agrobacterium tumefaciens* LBA 4404 as vector to incorporate a foreign gene into mulberry. He transferred kanamycin resistance gene and β-glucouronidase (GUS) gene through Ti plasmid PB1121 to mulberry leaf discs and showed their expression in transformed plantlets. Oka and Tewary [95] induced hairy roots in *in vitro* grown mulberry (*M. indica* L.) hypocotyls using Japanese wild *Agrobacterium rhizogenes* strains. Specific amplification of DNA fragment by PCR showed that portions of the *rol* genes in the T-DNA core region of the Ri plasmid were integrated into the hairy roots. A genomic clone, *Mahmg 1*, was isolated from *M. alba* and its expression characterized in mulberry and transgenic tobacco [96].

9.   Future Priorities

Biotechnological tools are of immense value in generating genetical information in crops especially in problematic plants like trees. It holds a great promise in mulberry improvement, utilization and conservation.

India has large resources of mulberry, which needs to be characterized unambiguously with DNA marker technology and the total diversity required to be assessed as a supportive research work for breeding. Developing DNA fingerprints of indigenous mulberry cultivars and important genotypes for their individualization will be of immense value to the breeders as well as for the curators of gene banks. DNA fingerprints of mulberry can be successfully used as 'molecular I.D. cards' in context of IPR/patent protection and also protection of Plant Breeders' Rights. Based on the molecular data, a core collection is required to be developed for efficient utilization of mulberry germplasm for crop improvement. There is an urgent need to identify DNA markers for important agronomic traits, resistances to biotic stresses and tolerance to adverse edaphic and climatic conditions, which can be utilized to hasten the mulberry breeding programme and thereby saving considerable physical and financial resources. The major values of molecular markers lie in the long-term strategic research. An important aspect in this direction is the study of quantitative trait loci (QTLs) of mulberry. Absence of any linkage map based on morphological/agronomic traits, necessitates the immediate development of molecular framework linkage of mulberry. The map can be used to tag genes of agronomic importance and to perform map based cloning of target genes. Genetic transformation techniques needs to be further fine tuned for stable expression of cloned genes. The silkworm is completely dependent on mulberry leaves for their entire nutritional requirement. Hence, the development of transgenic mulberry with qualitatively superior protein and carbohydrate content of the other known system, which may have significant impact on silk production, is also an important area.

In the *in vitro* culture front, further refinement is required to obtain a consistent regeneration of plants from callus culture. This will help in a long way in successful regeneration of transformed cells. Further, *in vitro* protocols already developed needs to be practically utilized for medium and long-term conservation of mulberry genetic resources.

References

1. R.L. Hunter, C.L. Markert, Histochemical demonstration of enzymes separated by zone electrophoresis in starch gels, Science 125 (1957) 1294–1295.
2. H. Hirano, Evaluation of affinities in mulberry and its relatives by peroxidase isozyme technique, JARQ 11 (1977) 228–233.
3. H. Hirano, K. Naganuma, Inheritance of peroxidase isozymes in mulberry (*Morus* spp.), Euphytica 28 (1979) 73–79.
4. H. Hirano, Thremmetological studies of protein variation in mulberry, Bull. Sericult. Expt. Sta. 36 (1980) 67–186.
5. K. Katagiri, Y. Kunitomo, R.R. Davalos, E. Iwata, Hexapolids found in mulberry strains collected in Mexico, J. Seric. Sci. Jpn. 63 (1994) 425–426.
6. M. Venkateswaralu, B.N. Susheelamma, N. Suryanarayana, K. Sengupta, Peroxidase isozyme studies in four mulberry species introduced from Indonesia, Ind. J. Seric. 28 (1989) 271–273.
7. M. Venkateswaralu, B.N. Susheelamma, M.V. Rajan, P.K. Tewari, A. Sarkar, Peroxidase isozyme studies in triploids of mulberry (*Morus* spp.), Ind. J. Seric. 34 (1995) 153–155.
8. M. Venkateswaralu, B.N. Susheelamma, N. Suryanarayana, N.K Dwivedi, K. Sengupta, Peroxidase isozyme banding patterns in aneuploids of mulberry, Sericologia 29 (1989a) 99–104.
9. K. Katagiri, H. Hirano, H. Hirai, H. Ichikawa, Isolation of chloroplast DNA in mulberry, J. Seric. Sci. Jpn. 53 (1984) 83–84.
10. H. Machii, Isolation of total mulberry DNA, J. Seric. Sci. Jpn. 58 (1989) 349–350.
11. L. Chengfu, Z.Youzuo, Z.Yaozhou, Studies on RAPD in mulberry, Journal of Zhejiang Agric. Univ. 22 (1996) 149–151.
12. F. Lichun, Y. Guang Wei, Y. MaoDe, Z. Xiaoyong, X. Zhong Huai, Study of relationship among species in *Morus* L. using random amplified polymorphic DNA (RAPD). Scientia Agricultura Sinica 30 (1997) 52–56.
13. C.F. Lou, Y.Z. Zhang, J.M. Zhou, Polymorphisms of genomic DNA in parents and their resulting hybrids in mulberry *Morus*. Sericologia 38 (1998) 437–445.
14. A. Sharma, R. Sharma, H. Machii, Assessment of genetic diversity in a *Morus* germplasm collections using fluorescence-based AFLP markers, Theor. Appl. Genet. 101 (2000) 1049–1055.
15. E. Bhattacharya, S.A. Ranade, Molecular distinction amongst varieties of mulberry using RAPD and DAMD profiles, BMC Plant Biology 1 (2001) 1–8.
16. Annual Report 1997–98, Silkworm and Mulberry Germplasm Station, Hosur.
17. Annual Report 1999–2000, Central Sericultural Germplasm Resources Centre, Hosur.
18. Annual Report 2000–2001, Central Sericultural Germplasm Resources Centre, Hosur.
19. Annual Report 2001–02, Central Sericulural Germplasm Resources Centre, Hosur.
20. Annual Report 2001–2002, Central Sericultural Research and Training Instiute, Mysore.
21. K. Ohyama, Tissue culture in mulberry tree, JARQ 5 (1970) 30–34.
22. K. Ohyama, S. Oka, Effects of absicic acid and gibberellic acid on bud break of mulberry plants. J. Seric. Sci. Jpn. 44 (1975) 321–326.
23. K. Ohyama, S. Oka. Regeneration of whole plants from isolated shoot tips of mulberry tree. J. Seric. Sci. Jpn 45 (1976) 115–120
24. S. Oka, K. Ohyama, *In vitro* initiation of adventitious buds and its modification by high concentration of benzyladenine in leaf tissues of mulberry (*Morus alba*) Can. J. Bot. 59 (1981) 68–74.
25. H.R. Kim, K.R. Patel, T.A. Thorpe, Regeneration of mulberry plantlets through tissue culture, Bot. Gazett. 146 (1985) 335–340.
26. P. Narayanan, S. Chakraborty, G.S. Rao, Regneration of plantlets from the callus of stem segements of mature plants of *Morus alba* L. Proc. Ind. Nat. Sci. Acad. B55 (1989) 469–472.
27. K.K. Sharma, T.A. Thorpe, *In vitro* propagation of mulberry (*Morus alba* L.) through nodal segments, Scientia Hort. 42 (1990) 307–320.
28. P.K. Tewary, G.S. Rao, Multiple shoot formation through shoot apex culture of mulberry, Ind. J. Forestry (1990) 13: 109–111.

29. S. Chattopadhay, S.K. Datta, A rapid clonal propagation of mulberry tree (*Morus alba*) through tissue culture, Proc. Ind. Sci. Cong. Part 3 (1990) 208–209.

30. S. Chattopadhay, S. Chattopadhyay, S.K. Datta, Quick *in vitro* production of mulberry (*Morus alba*) plantlets for commercial purpose. Ind. J. Exp. Biol. 28 (1990) 522–525.

31. M.K. Raghunath, S.P. Chakraborthy, B. Roy, S.K. Sen, Micropropagation of superior hybrids through axillary bud culture in mulberry (*Morus* L.). Cell and Chromosome Res. 15 (1992) 12.

32. K. Kathiravan, A. Shajahan, A. Ganapathi. Regneration of plantlets from hypocotyl derived callus cultures of mulberry *Morus alba* L. Israel J Plant Sci. 43 (1995) 259–262.

33. K. Hayashi, S. Oka, Formation of multiple bud bodies and plant regeneration in mulberry, *Morus alba*, J. Seric. Sci. Jpn. 64 (1995) 117–123.

34. P.K. Tewary, A. Sarkar, V. Kumar, S. Chakraborthy, Rapid *in vitro* multiplication of high yielding mulberry (*Morus* spp.) genotypes V1 and S-34, Ind. J. Seric. 34 (1995) 133–136.

35. B. Verma, V. Patni, U. Kant, H.C. Arya, Direct plantlet regeneration from axillary buds of *Morus alba* L. (grafted var. – White) J. Ind. Bot. Soc. 75 (1996) 45-47.

36. V.A. Bapat, M. Mhatre, P.S. Rao Propagation of *Morus indica* L. (mulberry) by encapsulated shoot buds, Plant Cell. Rep. 6 (1987) 393–395.

37. S.K. Pattnaik, Y. Sahoo, P.K. Chand, Efficient plant retrieval from alginate-encapsulated vegetative buds of mature mulberry trees, Scientia. Hort. 61 (1995) 227–239.

38. Y. Sahoo, S.K. Pattnaik, P.K. Chand, Plant regeneration from callus cultures of *Morus indica* L. derived from seedlings and matured plants, Scientia Hort. 69 (1997) 85–98.

39. J. Ivanica, *In vitro* micropropagation of mulberry, *Morus nigra* L. Scientia. Hort. 32 (1987) 33–39.

40. U. Yadav, M. Lal, V.S. Jaiswal, Micropropagation of *Morus nigra* L. from shoot tip and nodal explants of matured trees, Scientia Hort. 44 (1990) 61–67.

41. A.K. Jain, S.B. Dandin, K. Sengupta, *In vitro* micropropagation through axillary bud multiplication in different mulberry genotypes, Plant Cell Reps. 8 (1990) 737–740.

42. A.K. Jain, R.K. Datta, Shoot organogenesis and plant regeneration in mulberry (*Morus bombycis* Koidz.): factors influencing morphogentic potential in callus cultures, Plant Cell Tiss. Org. Cult. 29 (1992) 43–50.

43. H. Yakua, S. Oka, Plant regeneration through meristem cultures from vegetative buds of mulberry (*Morus bombycis* Koidz.) stored in liquid nitrogen, Ann. Bot. 62 (1988) 79–82.

44. M. Hossain, S.M. Rahman, A. Zaman, O.I. Joarder, R. Islam, Micropropagation of *M. laevigata* Wall. from matured trees. Plant Cell Rep. 11 (1992) 522–524.

45. S.K. Pattnaik, Y. Sahoo, P.K. Chand, Micropropagation of a fruit tree *M. australis* Poir. Syn. *M. acidosa* Griff. Plant Cell Rep. 15 (1996) 841–845.

46. M. Mhatre, V.A. Bapat, P.S. Rao, Regneration; of plants from the culture of leaves and axillary buds in mulberry (*Morus indica* L.), Plant Cell Rep. 4 (1985) 78–80.

47. S. Oka, K. Ohyama, Studies on *in vitro* cultures of excised buds in mulberry tress-III. Effects of agar concentration, pH and sugar of medium on the development of shoot from winter buds, J. Seric. Sci. Jpn 47 (1978) 15–20.

48. S. Enmoto, Preservation of genetic resource of mulberry by means of tissue culture, JARQ 21 (1987) 205–210.

49. S.K. Pattnaik, P.K. Chand, Rapid clonal propagation of three mulberries, *Morus cathyana*, Hemsl., *M. lhou* Koidz. and *M. serrata* Roxb. through *in vitro* culture of apical shoot buds and nodal explants from mature trees. Plant Cell Rep. 16 (1997) 503–508.

50. S. Oka, K. Ohyama, Induction of callus and effects of constituents of medium on callus formation of mulberry trees. J. Seric. Sci. Jpn. 42 (1973) 317–324.

51. K.S. Ogurstov, K.S. Atabekova, M.M. Khalmirzaev, S.H.R. Madyarov, Culturing the meristem of young mulberry on artificial nutrient medium, Shelk 1 (1986) 3.

52. P.K. Tewary, B.K. Gupta, G.S. Rao, *In vitro* studies on the growth rate of callus of mulberry (*Morus alba* L.), Indian. J. Forestry 12 (1989) 34–35.

53. T. Ohnishi, Y. Kobayashi Separation of free cells from the hypocotyl of mulberry. J. Seric. Sci. Jpn (1991a) 117–124.

54. T. Ohnishi, Y. Kobayashi, Suspension culture of cells from hypocotyl of mulberry, J. Seric. Sci. Jpn. 60 (1991b) 320–323.

55. T. Ohnishi, F. Yasukura, J. Tan, Preservation of mulberry callus by the addition of both abscisic acid (ABA) and p-aminobenzoate (PABA) in culture medium, J. Seric. Sci. Jpn. 55 (1986) 252–255.

56. K. Oshigane, On the redifferentiation of shoot callus to different culture in mulberry, J. Fac. Text. Sci. Tech. Shinshu Univ. Agric. Seric 107 (1989a) 38.

57. K. Oshigane, D. Zhuang, Effect of *in vitro* colchicine treatment in the culture of isolated buds in mulberry, J. Seric Sci. Jpn. 59 (1990) 187–195.

58. S. Rao, M.K. Raghunath, Callus initiation and plant regeneration on mulberry (*Morus alba* L.) cultivars, J. Swamy. Bot. Club. 10 (1993) 17–21.

59. F. Yasukura, T. Ohnishi, Effects of regeneration period of mulberry cultures on their growth after their storage and after the pseudo dormancy induced by storage, J. Seric. Sci. Jpn. 59 (1990) 255–258.

60. H. Seki, M. Takeda, K. Trutrumi, Y. Ushiki, Studies on the callus culture of mulberry trees (1) The effect of concentration of auxin and kinetin on the callus culture of mulberry stem J. Seric. Sci. Jpn 40 (1971) 81–85.

61. D.D. Ghugale, D.D. Kulkarni, R. Narasimhan, Effects of auxin and gibberillic acid on growth and differentiation of *Morus alba* and *Populus nigra* tissue *in vitro*, Ind. J. Exp. Biol. 9 (1971) 381–384.

62. T. Ohnishi, Y. Kobayashi. Root formation from the suspension cells of mulberry hypocotyl. J. Seric. Sci. Jpn. 60 (1991c) 505–507.

63. R. Islam, A. Zaman, M. Hossain, A.C. Barmar, O.I. Joarder, A.B.M.B. Hossain, Effects of growth regulators on *in vitro* callogenesis in *Morus laevigata* wall, Bull. Sericult. Res. 3 (1992) 55–58.

64. K.S. Jagadischandra, N. Sathyanarayana, Regeneration of plants in mulberry (*Morus indica*) var. Mysore Local through leaf callus culture, J. Plant Biol. 28 (2001) 147–152.

65. A. Kapur, S. Bhatnagar, P. Khurana, Efficient regeneration from mature leaf explants of Indian mulberry via organogenesis, Sericologia 41 (2001) 207–214.

66. S. Linn, D. Ji, J. Qin, *In vitro* production of haploid plants from mulberry (*Morus*) anther culture, Scientia Sinica 30 (1987) 853–863.

67. M. Venkateswaralu, K. Katagiri, Induction of greenish status and callus in anther culture of mulberry, Proc. Japnese Sericulture. Association (1991) Mito, Japan, Nov. 28–30. P.11

68. M. Sethi, S.Bose, A. Kapur, N.S. Rangaswamy, Embryo differentiation in anther culture of mulberry, Ind. J. Expt. Biol. 30 (1992) 1146–1148.

69. K. Katagiri, Callus induction in cultures of mulberry pollen. J. Seric. Sci. Jpn. 58 (1989b) 267–268.

70. K. Katagiri, M. Venkateswarulu, Induction of callus, embryo like and organ like division in pollen cultures of Indian mulberry variety Behrampur, Proc. Japnese Sericulture Association (1991a) Mito, Japan Nov. 28–30 P.30

71. K. Katagiri, M. Venkateswaralu, Induction and calli and organ-like structures in isolated pollen culture of mulberry, *Morus australis* Poiret. J. Seric. Sci. Jpn. 62 (1993) 1–6.

72. K. Katagiri, M. Venkateswaralu, Effects of sugars and sugar alcohols in the division of mulberry pollen in tissue culture, J. Seric. Sci. Jpn. 60 (1991b) 514–516.

73. P.K. Tewary, S.P. Chakraborthy, S.S. Sinha, R.K. Datta, *In vitro* study on pollen culture in mulberry, Acta Botanica Indica 22 (1994) 186–190.

74. G. Lakshmisita, S. Ravindran, Gynogenic plants from ovary cultures of mulberry (*Morus indica*), in: J. Prakash, RLM Pierik (Eds) Horticulture-New Technologies and applications, pp. 225–229. Kluwer Acad. Publ., Dordrecht.

75. T. Dennis Thomas, A.K. Bhatnagar, M.K. Razdan, S.S. Bhojwani, A reproducible protocol for the production of gynogenic haploids of mulberry, *Morus alba* L., Euphytica 110 (1999) 169–173.

76. T. Ohnish, S. Kiyama, Increase in yield of mulberry protoplasts by treatments with chemical substance, J. Seric. Sci. Jpn. 56 (1987a) 407–410.

77. K. Katagiri, Difference in cell division in mulberry mesophyll protoplast culture due to species and culture conditions, J. Seric.Sci. Jpn 57 (1988) 445–446.

78. P.K. Tewary, G. Lakshmi Sita, Protoplast isolation, purification and cultures in mulberry (*Morus* spp.) Sericologia 32 (1992) 651–657.

79. T. Ohnishi, S. Kiyama, Effects of change in temperature, pH, Ca ion concentration in the solution used for protoplast fusion on the improvement of the fusion ability of mulberry protoplasts, J. Seric. Sci. Jpn 56 (1987b) 418–421.

80. K. Katagiri, Colony formation in culture of mulberry mesophyll protoplasts, J. Seric. Sci. Jpn. 58 (1989a) 267–268.

81. W.Z. Ming, Z.H. Xu, N. Xu, H.M. Ren, Plant regeneration from leaf mesophyll protoplasts of *Morus alba*, Plant Physiol. Commu. 28 (1992) 248–249.

82. T. Ohnishi, K. Tanabe, On the protoplast fusion of mulberry and paper mulberry by electrofusion method, J. Seric. Sci. Jpn. 58 (1989) 353–354.

83. P.K. Chand, Y. Sahoo, S.K. Pattnaik, Artificial seeds: a novel approach to mulberry propogation, Indian Silk Sept. (1994) 33–38.

84. B.M. Reed, Y. Chang, Medium and long-term storage of *in vitro* culture of temperate and nut crops, In: Razdan M.K. and Cocking E.C, Conservation of plant genetic resources *in vitro*: General aspects, Science Publishers Inc. USA, 1 (1997) Pp. 67–106.

85. C.P. Wilkins, H.J. Newburry, J.H. Dodds, Tissue culture conservation of fruit tree FAO/IBPGR, Plant Gen. Res. Newsletter 73/74 (1988) 9–20.

86. K.K. Sharma, T.A. Thorpe, *In vitro* propagation of mulberry, Sci. Hort. 42 (1990) 307–320.

87. A. Sakai, Survival of twigs of woody plants at ‑196°C, Nature 185 (1960) 392–394.

88. T. Yokoyama, S. Oka, Survival of mulberry winter buds at super low temperatures, J. Seric. Sci. Jpn. 52 (1983) 263–264.

89. L. Wang, X. Zhang, Z. Yu, Cryopreservation of winter mulberry buds, Acta. Agric. Boreali-Sinica, 3 (1988) 103–106.

90. T. Niino, A. Sakai, H. Yakawa, Cryopreservation of dried shoot tips of mulberry winter buds and subsequent plant regeneration, Cryo-lett. 13 (1992b) 51–58.

91. T. Niino, Cryopreservation of germplasm of mulberry (*Morus* spp.), in: YPS Bajaj (Ed). Biotechnology in Agriculture and Forestry 32 (1995), Springer-Verlag, Berlin pp. 102–113.

92. T. Niino, A. Sakai, Cryopreservation of alginate coated *in vitro* grown shoot tips of apple, pear and mulberry, Plant Sci. 87 (1992) 199–206.

93. T. Niino, A. Koyama, K. Shirata, S. Ohuchi, M. Suzuki, A Sakai, Long-term storage of mulberry winter buds by cryopreservation, J. Seric. Sci. Jpn. 62 (1993) 431–434.

94. H. Machii, Leaf disc transformation of mulberry plants (*Morus alba* L.) by *Agrobacterium* Ti plasmid, J. Seric. Sci. Jpn. 59 (1990) 105–110.

95. S. Oka, P.K. Tewary, Induction of hairy roots from hypocotyls of mulberry (*Morus indica* L.) by Japanese wild strains of *Agrobacterium rhizogenes*, J. Seri. Sci. Jpn. 69 (2000) 13–19.

96. A.K. Jain, R.M. Vincent, C.L. Nessler, Molecular characterization of hydroxymethylglutaryl-CoA reductase gene from mulberry (*Morus alba* L.), Plant Molecular Biology 42 (2000) 559–569.

Plant Biotechnology and Molecular Markers
P.S. Srivastava, Alka Narula and Sheela Srivastava (Editors)
Copyright © 2004 Anamaya Publishers, New Delhi, India

15. Development of High Efficiency Micropropagation Protocol of an Adult Tree—*Wrightia tomentosa*

S.D. Purohit*, P. Joshi, K. Tak and R. Nagori

Post Box No. 100, Plant Biotechnology Laboratory, Department of Botany,
M.L. Sukhadia University, Udaipur 313 001, Rajasthan, India

*e-mail: sdp_56@hotmail.com

Abstract: Highly efficient and reproducible micropropagation protocol for *Wrightia tomentosa* using sexually adult material has been developed. Multiple shoots were induced from nodal shoot segments through forced axillary branching *in vitro*. Nature and management of the donor tree, season of collecting explant and their orientation on the medium strongly influenced the initial establishment of cultures. Explants collected in April-June period and placed vertically on the MS medium containing 2 mgl^{-1} BAP produced shoots from axillary nodes *in vitro*. Management of donor tree by serial harvesting of explants every fortnight was necessary to obtain vigorous growth of shoots *in vitro*. Explants of the fifth flush (F_5) were found most suitable to obtain more than 7 shoots per node on the above medium. The rate of multiplication in subsequent subcultures was a little more than two and half-fold. Incorporation of phloroglucinol (100 mgl^{-1}) into the multiplication medium containing BAP (2 mgl^{-1}) accelerated the rate of multiplication to 3-fold per subculture. Similar response could be obtained by using 10 mM thidiazuron (TDZ) alone in the multiplication medium. Nodal segments from *in vitro* raised shoots were also used to initiate a new culture cycle. The shoots could be multiplied for at least 24 months without loss of vigor. More than sixty per cent shoots obtained after sixth subculture developed roots when treated with pre-autoclaved indole-3-butyric acid solution (100 mgl^{-1}) for 10 min and implanted on modified MS medium (major salts reduced to $^1/_4$ strength and 400 mgl^{-1} activated charcoal). Successfully rooted plants were hardened *in vitro* in glass bottles containing Soilrite^{TM} irrigated with $^1/_4$ strength MS salt solution (pH 5.0). More than 5,000 plantlets were successfully hardened *in vitro* and transferred to greenhouse for acclimatization. The survival rate of the plants during hardening was more than 95 per cent.

1. Introduction

Wrightia tomentosa (Roxb.) Roem et Schult (Apocynaceae), once a common tree species of Aravallis in Rajasthan (India), has traditionally been exploited for its ivory-white wood in making toys and as fuel. High rate of seedling mortality, lack of suitable method for natural regeneration and overexploitation has reduced its population drastically and the plant has been listed as an endangered species [1]. There is, therefore, a strong need for an alternative method to produce large number of plants of superior types for conservation and regeneration. Micropropagation methods have been widely applied for clonal propagation of tree species for afforestation, woody biomass production and conservation of elite and rare germplasm [2, 3]. These methods have been successfully integrated with modern forest tree management programs for rapid restoration of the degraded lands [4]. A lab-scale protocol for micropropagation of *W. tomentosa* using adult material was reported by Purohit et al. [5]. Only a limited number of plants could be produced by that method with major constraint being in hardening of *in vitro* developed plants. The present article describes a highly efficient and reproducible micropropagation protocol that is being taken up for scaling-up production by this laboratory for large-scale plantation by the foresters.

2.   Materials and Methods

Trees of *W. tomentosa* selected and marked for quality of wood (age of tree more than 30 years) were used as a source of explants. Shoots were harvested from these plants round the year, divided into four distinct periods viz. April-June, July-September, October-December and January-March. Management of donor tree was done by lopping one major fork and juvenile shoots produced near cut ends were collected for explantation. Such newly flushed shoots were serially harvested fortnightly and successive flushes were termed as first (F_1), second (F_2), third (F_3), fourth (F_4) and fifth (F_5). Nodal shoot segments (1.5, 3.0 and 4.5 cm long and 0.2–0.5 cm in diameter) were prepared as explants. Two different orientations of explant on medium was tested. Horizontal placement was done in two ways: (i) explant lying horizontally on medium (H_1) and (ii) one side of node, having axillary bud, inserted in medium while other side exposed to air (H_2). Vertical placement of explant was done in three different ways: (a) node completely immersed in medium (V_1), (b) node on the surface of medium (V_2) and (c) node one centimeter above the medium (V_3).

Explants were washed thoroughly with sterilized distilled water containing few drops of Tween-20 and then surface-sterilized with 0.1% (w/v) mercuric chloride for 5 min followed by thorough washing with sterile distilled water. Surface sterilized explants were inoculated on standard multiplication medium containing MS salts [6] with 2 mg l^{-1} BAP. Explants were also inoculated on MS medium containing different concentrations of Kn (0.5–5.0 mg l^{-1}), TDZ (50–10,000 nM) and GA_3 (1.0–2.0 mg l^{-1}). Proliferated shoots from nodes of F_5 flush were further subcultured on MS medium with various concentrations of TDZ (0.1, 1.0 and 10.0 μM) or BAP (2 mgl^{-1}). Phloroglucinol (50, 100 and 250 mg l^{-1}) was also added to standard shoot multiplication medium. Cultures in conical flasks (100, 150 ml) covered with non-absorbent cotton plugs were kept under controlled conditions of temperature (28 ± 2°C), light (45 μmol m^{-2} s^{-1} for 16 h/day provided by fluorescent tubes) and 60–70% relative humidity. Once culture conditions for optimum shoot induction from explants were established, the shoots produced *in vitro* were subcultured on fresh medium every 3 weeks.

Shoots having passed through three, six and nine passages in multiplication medium were used for rooting. Shoots (2.0–3.0 cm) were excised and their cut ends were dipped in different concentrations of IBA solution (50, 100, 200 and 500 mg l^{-1}) for different duration (5–15 min) followed by their implantation on standard rooting medium containing quarter strength MS salts, sucrose (1%) and agar (0.6%). Activated charcoal (50, 100, 200 and 400 mg l^{-1}) was also tested in standard rooting medium. Initially the culture vessels were kept wrapped with black paper or in darkness for 5–7 days at 30 ± 2°C temperature at 60–70% relative humidity.

Rooted shoots from 3-week-old cultures were hardened prior to *ex vitro* exposure. Hardening was attempted by three different methods. In first method, individual plantlets were planted carefully in Soilrite™ (Karnataka Explosives, Bangalore, India) filled netted pots (3 cm high) and placed in horizontally kept pickle bottles (30 cm long) which accommodated 25 such pots (W_1). In second method, individual plantlets, transferred to netted pots were placed in glass troughs (30 cm diameter) covered with polythene sheets (W_2). Thirdly, autoclaved 400 ml screw cap glass bottles one-fourth filled with soilrite™ irrigated with 40 ml inorganic salt solution (major salts of MS medium reduced to $1/4$ strength, pH 5.0) were used (W_3). Each bottle containing 4 plantlets were kept in culture room for 30 days.

After 30 days, plantlets hardened by methods described as W_1 and W_2 were shifted individually to plastic pots (10 cm high) and covered with polythene bags. Gradually, humidity was lowered by perforating polythene cover, then opening it for 1 h/day and finally completely removing it.

Plantlets hardened *in vitro* (W_3) were kept in closed bottles till they touched the caps of bottles (nearly after 30 days). The caps were loosened and finally opened in misthouse (with 70–85% RH). After one month, plants were transferred to pots and kept under greenhouse conditions where a gradient of humidity (80–40%) was maintained by a Fan-Pad evaporative cooling system.

2.1 Statistical Analyses

Standard analysis procedures [7, 8] were followed for CRD analysis. Abnormality, non-additivity and heterogeneity of variance in raw data of different experiments were minimized using square root ($\sqrt{x}$ and $\sqrt{x} + 0.5$) transformation [9]. In ANOVA, test for significance (F test), standard error of mean and critical difference at 5 and 1 per cent probability was calculated on transformed data which were tabulated along with retransformed values in each experiment. In case of discrimination amongst two treatments 't' test was used [8].

3. Results

3.1 Initiation of Shoot Cultures

Bud break frequency was strongly influenced by the nature and management of donor tree, season of explant collection and their orientation on the medium. April-June was found to be the best period for collection of explant to obtain maximum (98%) bud break response with minimum (5.0%) loss due to contamination (Table 1). The explants collected during the months of July-September developed fungal growth associated with shoot bud proliferation. Least bud break response was observed in the winter months of October-December. Explants prepared from one-year-old mature branches responded poorly in cultures as compared to explants taken from freshly flushed branches (Table 2).

Table 1. Effect of season of harvest on shoot initiation from mature node explants of *W. tomentosa* on standard multiplication medium

Period	Per cent contamination*	Per cent response*	Callus intensity
April–June	5 ± 1.08	98 ± 13.85	++
July–September	90 ± 8.78	78 ± 12.68	++++
October–December	81 ± 7.02	20 ± 4.12	+++++
January–March	72 ± 8.60	45 ± 8.02	+++

*Mean ± SE.

Explants of different size showed varied response in terms of bud break and amount of associated callus during shoot initiation. Maximum bud break was found in explant measuring

Table 2.   Effect of source of explant on shoot initiation in *W. tomentosa* on standard multiplication (SM) medium

Source of explant	Per cent explant sprouted	Mean number of shoots per node	Mean shoot length (cm)
Mature branches	52	2.17	0.85
Juvenile branches	98	3.83*	2.92*

*Significant at 1% level using t test.

3.0 and 4.5 cm in length (Table 3). However, the size of explant did not make significant difference in shoot bud proliferation both in terms of their number and length. In very small explants (1.5 cm) the basal callus developed upto node, posing difficulty in further subculturing.

Table 3.   Effect of length of explant on shoot initiation in *W. tomentosa* on standard multiplication medium

Size of explant	Per cent explant sprouted	Mean number of shoots per node	Mean shoot length (cm)
1.5 cm	82	4.32 (2.08)	1.92
3.0 cm	98	4.64 (2.15)	2.33
4.5 cm	97	4.47 (2.11)	2.00
SEm±	0.07	0.18	
CD.05		NS	NS'

*Figures in parentheses are $\sqrt{\chi}$ transformed values.

Orientation of explant on medium was a significant factor in shoot proliferation from nodal segments. Between horizontal and vertical orientation of explants on medium, the latter was found significantly superior (Table 4). Maximum number of shoots were produced in explants oriented in vertical position V_3 followed by V_2 and horizontal position H_2, both statistically at par in terms of number and length of shoots. When positioned vertically, callusing was associated with lower internodal region of explant only while it extended to whole surface in horizontally placed explants. Nodal explants placed 1.0 cm above the medium provided callus-free shoot proliferation.

Management of donor tree from which the explants were collected was a very important step in accelerating the number of shoots per node during initial phases of cultures establishment. Explants obtained from serially lopped branches producing different flushes of juvenile shoots exhibited graded increase in shoot bud proliferation. An increase in per cent bud break response and number of shoots per explant from first flush (F_1) to fifth flush (F_5) was noted with a concomitant decrease in per cent contamination (Table 5). Explants from F_5 flush exhibited initiation of ca 7.33 axillary shoots per node as compared to 3.98 shoots induced in explants from F_1 flush. However, effect of flushes on length of proliferated shoots was non-significant.

Explants from F_5 flushes responded differently as compared to that of F_1 flush (Table 6). About 90 per cent bud break response was observed when explants from F_5 flush were inoculated on the MS medium containing any of the four growth regulators tested. In explants from F_1

Table 4. **Effect of node orientation on shoot initiation in *W. tomentosa* on standard multiplication medium**

Orientation of nodal explant	Per cent explant sprouted	Mean number of shoots per node*	Mean shoot length (cm)
Horizontal placement			
Both sides of node on the medium (H$_1$)	98	2.45 (1.57) c	1.42 c
One side inserted into the medium (H$_2$)	94	3.29 (1.81) b	2.50 b
Mean		2.85 (1.69)	1.96
Vertical Placement			
Node immersed in the medium (V$_1$)	10	1.30 (1.14) d	1.00 d
Node on the surface of the medium (V$_2$)	95	3.98 (1.99) b	3.83 b
Node nearly 1 cm above the medium (V$_3$)	98	5.31 (2.30) a	3.33 a
Mean		3.28 (1.81)*	2.38*
SEm±		0.09	0.18
CD$_{.05}$		0.25	0.52
CD$_{.01}$		0.35	0.71

*Figures in parentheses are $\sqrt{x}$ transformed values.

  Means followed by different letters in the same column differ significantly.

*Placement of explant (horizontal or vertical) differ significantly.

Table 5. **Effect of serial harvesting on *in vitro* response by MN of *W. tomentosa* on standard multiplication medium**

Flush number	Per cent contamination	Per cent explant sprouted	Mean number of shoots per node*	Mean shoot length (cm)
I flush (F$_1$)	10	88	3.98 (1.99) d	2.54
II flush (F$_2$)	10	90	5.14 (2.27) c	2.18
III flush (F$_3$)	8	91	5.49 (2.34) bc	2.43
IV flush (F$_4$)	5	95	6.51 (2.48) b	2.17
V flush (F$_5$)	1	98	7.33 (2.71) a	2.31
SEm±			0.06	0.17
CD$_{.05}$			0.17	NS
CD$_{.01}$			0.23	NS

*Figure in parentheses are $\sqrt{x}$ transformed values.

  Means followed by different letters in the same column differ significantly.

flush, mean number of axillary shoots initiated were statistically insignificant on tested concentrations of any of the four growth regulators except GA$_3$ with least shoot formation. However, influence of PGRs on shoot initiation was marked in explants from F$_5$ flush, maximum being on BAP (2 mg l^{-1}). None of the other three growth regulators (Kn, TDZ and GA$_3$) at any of the concentrations showed better response in terms of number of axillary shoots induced per node. Numerically, 2 mg l^{-1} BAP produced maximum number of shoots, followed by 250 nM TDZ.

Table 6.   Effect of different PGRs on shoot initiation in _W. tomentosa_

MS + PGR		F$_1$ flush		F$_5$ flush	
		Per cent explant sprouted	Mean number of shoots/node*	Per cent explant sprouted	Mean number of shoots/node
BAP	2 mg l^{-1}	84	3.74 (1.93)a	97	6.52 (2.55)a
Kn	0.5 mg l^{-1}	48	2.91 (1.71)ab	89	3.73 (1.93)b
	2.0 mg l^{-1}	58	2.69 (1.64)ab	90	3.19 (1.79)bc
	2.5 mg l^{-1}	52	3.19 (1.79)ab	88	3.11 (1.79)bcd
	5.0 mg l^{-1}	65	2.44 (1.56)abc	95	2.91 (1.71)bcd
TDZ	50 nM	61	2.44 (1.56)abc	98	2.91 (1.71)bc
	250 nM	60	3.44 (1.85)a	92	3.48 (1.87)bc
	500 nM	52	2.69 (1.64)ab	98	2.91 (1.71)bcd
	1,000 nM	61	2.91 (1.76)ab	97	3.48 (1.87)bc
	10,000 nM	53	2.69 (1.64)ab	92	3.19 (1.79)bc
GA$_3$	mg l^{-1}	35	1.72 (1.31)bc	95	2.23 (1.49)cd
	2 mg l^{-1}	24	1.45 (1.21)c	92	1.93 (1.39)d
SEm±		0.14			0.13
CD.05		0.40			0.38
CD$_{.01}$		NS			0.52

*Figure in parentheses are $\sqrt{\chi} + 0.5$ transformed values.

Means followed by different letters in the same column differ significantly.

3.2   Shoot Multiplication

Shoots after their initial proliferation from F$_5$ explants on medium containing 2.0 mgl^{-1} BAP along with the mother explant were further subcultured onto standard multiplication medium after every 3 weeks. Substitution of BAP with TDZ in subcultures increased shoot multiplication rate, highest being on medium containing 10 µM TDZ (2.92 fold) which was significantly superior to the rate obtained on standard multiplication medium containing 2.0 mg l^{-1} BAP (Table 7). Incorporation of phloroglucinol (PG) in standard multiplication medium increased the rate of shoot multiplication above 3-fold. PG also induced healthy cultures with dark green and lustrous leaves (Table 8).

Table 7.   Effect of different concentrations of TDZ on shoot multiplication in _W. tomentosa_ cultures

MS + TDZ (µM)	Multiplication fold*	Callus intensity
Control (SM medium)	2.42b	+
0.1	1.92c	+
1.0	2.67ab	+++
10.0	2.92a	+++
SEm±	0.14	
CD$_{.05}$	0.42	
CD$_{.01}$	0.58	

*Means followed by different alphabets in the same column differ significantly.

Table 8. **Effect of different concentrations of phloroglucinol on shoot multiplication in *W. tomentosa***

MS + BAP (2 mgl^{-1}) +PG (mgl^{-1})	Multiplication fold*	Callus intensity
Control	2.34 b	+++
50	1.08 c	+++
100	3.33 a	+++
250	1.92 b	++++
SEm±	0.20	
CD$_{.05}$	0.61	
CD$_{.01}$	0.83	

*Means followed by different letters differ significantly.

3.3 Rooting in Shoots

Those shoots having passed through six multiplication cycles, responded to rooting treatments (Table 9). With the increasing number of shoot multiplication cycles, the rooting response was more favorable, showing early root initiation and reduced callusing.

Table 9. **Effect of number of subcultures on rooting in IBA pulse treated shoots of *W. tomentosa* on standard rooting medium**

Shoots harvested after subculture	Per cent rooting response	Mean number of days to rooting	Callus intensity
III subculture	00.0	00.0	+++
VI subculture	15.4	35.6	+++
IX subculture	40.3	18.1	+

* IBA pulse treatment (100 mgl^{-1} for 10 min).

Concentration of IBA and duration of treatment affected the root induction process considerably. Among various IBA concentrations tested for pulse treatment, rooting response was maximum in shoots treated with 100 mg l^{-1} IBA solution for 10 min (Table 10). Such shoots exhibited ca 2.92 roots with 2 cm mean root length. Higher or lower concentrations of IBA did not improve rooting response. Rooting percentage was found to be positively related with concen-tration of activated charcoal (AC) added to rooting medium (Table 11). Maximum rooting response (69.7%) was observed when the IBA-treated shoots were placed in medium containing 400 mgl^{-1} AC. Addition of AC helped in early root initiation, increased root number and reduced callusing at the root-shoot junction.

3.4 Hardening and Acclimatization

Plantlets raised *in vitro* initially posed problems in hardening and acclimatization. Rooted plants when transferred directly to pots without prior hardening started wilting, no sooner they were transferred, and desiccated completely within 24 h. Seedlings were also used in experimentation to understand their requirements for hardening. Even the seedlings were prone to transplantation shock similar to *in vitro* developed plantlets.

Table 10. Effect of IBA pulse treatment on rooting in shoots of *W. tomentosa* on standard rooting medium

Strength of IBA solution (mgl^{-1})	Duration (min)	Per cent rooting response	Mean number of roots*	Mean root length (cm)	Callus intensity
50	5	11	1.30 (1.14) d	0.78 d	–
	10	15	1.49 (1.22) cd	1.25 bc	+
	15	18	1.84 (1.36) bcd	1.50 b	+
100	5	40	2.00 (1.41) bc	1.50 b	++
	10	59	2.92 (1.71) a	2.00 a	++
	15	48	1.90 (1.38) bc	1.85 a	+++
200	5	18	2.37 (1.54) ab	1.85 a	+++
	10	21	1.96 (1.40) bc	1.00 cd	+++
	15	19	1.79 (1.34) bcd	1.00 cd	++++
500	5	0	–	–	
	10	0	–	–	
	15	0	–	–	
SEm$_\pm$			0.08	0.11	
CD$_{.05}$	0.23	0.31			
CD$_{.01}$	0.31	0.42			

*Means followed by different alphabet in same column differ significantly.

Table 11. Effect of activated charcoal (AC) on rooting in pulse treated shoots of *W. tomentosa* on standard rooting medium

Medium + AC (mgl^{-1})		Per cent rooting response	Mean number of roots*	Mean root length (cm)	Mean shoot length (cm)	Callusing intensity
Control		59	2.89 (1.70) b	1.87 b	3.74	+
AC	50	45	2.93 (1.71) b	1.92 b	3.67	–
	100	50	2.76 (1.66) b	2.11 ab	3.56	–
	200	65	3.06 (1.75) b	2.41 a	3.89	–
	400	69.7	3.78 (1.94) a	2.06 b	3.60	–
SEm$_\pm$			0.07	0.11	0.18	
CD$_{.05}$			0.21	0.34	NS	
CD$_{.01}$			0.30	0.46	NS	

*Values in parentheses are $\sqrt{\chi}$ transformed values.

Means followed by different letters in the same column differ significantly. IBA pulse treatment (100 mgl^{-1} for 10 min).

Owing to fast desiccation of rooted plants on direct pot transfer, other methods of hardening were employed. Rooted plants with nearly 4 cm long shoot, 1 to 2 cm long root and 4 to 6 leaves in number were hardened by three different methods as described in materials and methods. Apical growth was visible in more than 95 per cent of plantlets reared through any of the three methods. After 30 days, plantlets attained an average height of 5.4 cm with 2.1 cm long root (root shoot ratio being 0.37) and 6-8 broad leaves. Such plantlets were transferred to individual plastic pots and covered with polythene bags in case of W_1 and W_2 while the caps of glass bottles were loosened in W_3 plantlets. Plantlets exhibited wide variation in growth during this period. Most of the plantlets in all the three methods exhibited good shoot growth while root growth was better only in case of plantlets hardened through W_3 method. Generally, the plants were ca 9.08 cm long with 3.67 cm long roots and 8 to 10 leaves. These plants on an average accumulated 20.22 mg dry matter in shoots (without leaves), 7.95 mg in roots and 13.54 mg in each leaf. During gradual exposure to *ex vitro* conditions in greenhouse, all plants remained green and healthy for initial 15 days. Plants hardened through W_1 and W_2 modes exhibited yellowing of leaves and leaf fall in next 15 days. The rate of survival was 26.8% after 30 days of transfer to pots which declined to 2.4% after 60 days. Plantlets hardened through W_3 mode grew vigorously having rigid and thick stem, highly branched root system and green, lustrous, healthy and broad leaves. Survival rate of such plants was more than 95 per cent. More than 5,000 plantlets have been successfully hardened and acclimatized and are ready for transplantation into field (Fig. 1).

Fig. 1. Hardened plants of *W. tomentosa* kept in a nursery.

4. Discussion

An adult superior tree can be micropropagated for desired attributes by enhanced axillary proliferation. The buds residing in the axil of twigs are induced to proliferate and generate multiple shoot buds *in vitro*. Proliferation of these axillary buds may be difficult due to micro-organism contamination [10], phenolic oxidation [11] and tissue maturity [12]. Maturity of tissue is accompanied with reduced growth rate, reduced/lack of rooting ability and sometimes plagiotrophy [13]. By reverting a part of tree to complete/partial juvenility by *in vivo* and *in vitro* methods problems associated with maturity can be minimized. In *W. tomentosa* the explants collected from previously lopped trees showed better proliferation when shoots were harvested serially. Severe pruning has been found to be an efficient method for rejuvenation [14]. In *Quercus robur* forced flushing method, related to severe pruning, was adopted [15].

The season of explant collection greatly influenced establishment of *W. tomentosa* cultures *in vitro*. Effect of season on bud sprouting was also noted in many tree species viz. *Tectona grandis*

[16], guava [17], *Tecomella undulata* [18], *Prosopis cineraria* [19] and *W. tinctoria* [20]. Vertical orientation of *W. tomentosa* explants was found better than horizontal orientation in terms of number of proliferated shoots. On the contrary, horizontal orientation of explants was found better in *Fraxinus angustifolia* [21] and *Quercus robur* [15].

Cytokinins promote cell division in plant tissues under specific conditions and are found obligatory for shoot differentiation [22]. Hu and Wang [23] reported superiority of BAP among cytokinins in differentiation of shoots from explants of trees. Buising et al. [24] suggested that transient exposure of soybean embryonic axes to BAP interrupted chromosomal DNA replication and reprogrammed the developmental fate of a large number of cells in shoot apex. In our case, maximum number of shoots were obtained in medium containing BAP in comparison to other PGRs. Recently, TDZ has been found to be one of the most active cytokinin-like substances used for woody plant tissue culture [25]. In present study, TDZ did not supercede the response obtained with BAP in shoot induction, but it did enhance multiplication of shoots in subcultures. Incorporation of phloroglucinol in the multiplication medium improved shoot multiplication rate. Similar results have been reported in apple root stock M.7 by Jones [26]. The effect could be related to hastening of rejuvenation process *in vitro* by phloroglucinol.

Rooting by dip treatment of auxin has been recommended by Harry and Thorpe [27]. It is supposed to eliminate the inhibitory effect on root growth when IBA is incorporated in the media [28]. Purohit et al. [5] have recommended IBA pulse treatment for rooting in *W. tomentosa* shoots. This method of root induction has been successfully employed in the present studies also.

Hardening is most critical factor for achieving success in pot transfer of regenerated plantlets. We have observed that a gradual shifting of plants from medium to culture bottles containing low salt concentration without sucrose allowed stress, compelling plants to become partially autotrophic. This step proved useful in achieving more success in hardening.

The results have demonstrated the feasibility of application of this protocol for raising large number of *W. tomentosa* which would greatly help in afforestation programmes in Aravallis in Rajasthan (India). Large numbers of plants are ready for field transfer that can be used for field evaluation studies.

Acknowledgements
Authors thank Dr. N.S. Shekhawat, Incharge, Plant Biotechnology Laboratory, J.N. Vyas University, Jodhpur, India for providing hardening facilities. Thanks are also due to the Department of Biotechnology, Govt. of India, New Delhi, for financial support.

References
1. S. Sharma, A census of rare and endemic flora of south-east Rajasthan, in: Jain, S.K., Rao, R.R. (Eds), Threatened Plants of India, Botanical Survey of India, Howrah, (1983) pp. 63–70.
2. J.M. Bonga, Clonal propagation of mature trees: problems and possible solutions, in: Bonga, J.M. Durzan, D.J. (Eds), Tissue Culture in Forestry. Martinus Niijhoff, The Hague 1982a, pp. 249–271.
3. Y.P.S. Bajaj, Biotechnology of tree improvement for rapid propagation and biomass energy production, in: Bajaj, Y.P.S. (Ed), Biotechnology in Agriculture and Forestry, Vol.1-Tree. Springer-Verlag, 1986, pp. 1–23.
4. J. Aitken-Christie, M. Connett, Micropropagation of forest trees, in: Kurata, K., Kozai, T. (Eds), Transplant Production Systems. Kluwer Academic Publishers, The Netherlands,1992, pp. 163–194.

5. S.D. Purohit, G. Kukda, P. Sharma, K. Tak, *In vitro* propagation of an adult tree *Wrightia tomentosa* through enhanced axillary branching. Plant Sci., 103 (1994) 67–72.

6. T. Murashige, F. Skoog, A revised medium for rapid growth and bioassays with tobacco tissue cultures. Physiol. Plant., 15 (1962) 473–497.

7. O. Kempthorne, Design and analysis of experiments. John Wiley & Sons, New York. 1952.

8. G.W. Snedecor, Statistical methods. Iowa State College Press. Iowa, USA, 1956.

9. M.E. Compton, Statistical methods suitable for analysis of plant tissue culture data 1994, Plant Cell, Tiss. and Org. Cult., 37, 217–242.

10. P.C. Debergh, L.J. Maene, A scheme for commercial propagation of ornamental plants by tissue culture. Sci. Hortic., 14 (1981) 335–45.

11. T.R. Marks, S.E. Simpson, Reduced phenolic oxidation at culture initiation *in vitro* following the exposure of field-grown stock plants to darkness or low level of irradiance. J. Hortic. Sci., 65 (1990) 103–11.

12. J.M. Bonga, Vegetative propagation in relation to juvenility, maturity and rejuvenation, in: Bonga, J.M., Durzan, D.J. (Eds), Tissue Culture in Forestry. Martinus Niijhoff, The Hague, 1982b, pp. 387–412

13. R.L.M. Pierik, Micropropagation: Technology and opportunities. In: Prakash, J., Pierik, R.L.M. (Eds.), Plant biotechnology: Commercial prospects and problems. Oxford & IBH Publishing Co. Pvt. Ltd., New Delhi, 1993, pp 9–22

14. A. Franclet, Manipulation des pieds–meres et amelioration de la qualite des boutures. In: AFOCEL (Ed) AFOCEL Etudes et Recherches. Nangis, France,1977 pp 1–21.

15. A.M. Vieitez, M.C. Sanchez, J.B. Amo-Marco, and A. Ballester, Forced flushing of branch segments as a method for obtaining reactive explants of mature *Quercus robur* trees for micropropagation. Plant Cell, Tiss. Org. Cult., 37(1994) 287–95.

16. P.K. Gupta, A.L. Nadgir, A.F. Mascarenhas, V. Jagannathan, Tissue culture of forest trees: Clonal multiplication of *Tectona grandis* L. (Teak) by tissue culture, Plant Sci. Lett., 17 (1980) 259–68.

17. V.S. Jaiswal, M.N. Amir, *In vitro* propagation of guava from shoot cultures of mature trees. J. Plant Physiol, 130 (1987) 7–12.

18. T.S. Rathore, R.P. Singh, N.S. Shekawat, Clonal propagation of desert teak (*Tecomella undulata*) through tissue culture. Plant Sci., 79 (1991) 217–222.

19. N.S. Shekhawat, T.S. Rathore, R.P. Singh, N.S. Deora, S.R. Rao, Factors affecting *in vitro* clonal propagation of *Prosopis cineraria*. Plant Growth Reg. 12 (1993) 273–280.

20. G. Kukda, 1994. Clonal propagation of toy wood tree *Wrightia tinctoria* through tissue culture biotechnology. Ph.D. Thesis, Mohanlal Sukhadia University, Udaipur, India.

21. M.A. Parez-Parron, M.E. Gozalez-Benito, C. Perez,. Micropropagation of *Frazinus angustifolia* from mature and juvenile plant material. Plant Cell, Tiss. Org Cult. 37 (1994) 297–302.

22. V. Dhawan, Tissue culture in hardwood species, in: "Plant Biotechnology: Commercial Prospects and Problems" J. Prakash, R.L.M. Pierik (Eds). Oxford & IBH Publishing Co. Pvt. Ltd., New Delhi, 1993 pp 43–72.

23. C.Y. Hu, P.J. Wang, Meristem shoot tip and bud cultures, in: "Handbook of Plant Cell Culture". Vol I "Techniques for Propagation and Breeding". D.A. Evans, W.R. Sharp, P.V. Ammirato, Y. Yamada (Eds). MacMillan Publ. Co., New York, 1983, pp 177–227.

24. C.M. Buising, R.C. Shoemaker, R.M. Benbow, Early events of multiple bud formation and shoot development in soybean embryonic axes treated with the cytokinin, 6-benzylaminopurine, Amer. J. Bot., 81(11) (1994) 1435–1448.

25. C.A. Huetteman, J.A. Preece, Thidiazuron—a potent cytokinin for woody plant tissue culture. Plant Cell, Tiss. Org. Cult., 33 (1993) 105–119.

26. O.P. Jones, Effect of phloridzin and phloroglucinol on apple shoots. Nature, 262 (1976) 392–3 Erratum 262–724.

27. I.S. Harry, T.A. Thorpe, Englemann Spruce (*Picea engelmannii* Parry ex. Englem), in: Y.P.S. Bajaj, (Ed), Biotechnology in agriculture and forestry, Vol. 16, Springer-Verlag, 1991, pp. 408–422.

28. H.T. Hartmann, D.E. Kestu, F.T. Davies, Plant propagation: Principles and practices, V edition Prentice Hall, Engle Wood Cliffs, New Jersey, Jr. (Eds) 1990.

Plant Biotechnology and Molecular Markers
P.S. Srivastava, Alka Narula and Sheela Srivastava (Editors)

16. *In Vitro* Regeneration and Improvement in Tropical Fruit Trees: An Assessment

Madhulika Singh, Uma Jaiswal and V.S. Jaiswal

Laboratory of Morphogenesis, Department of Botany, Banaras Hindu University, Varanasi 221 005, India

Abstract: *In vitro* regeneration protocol has been developed for many tropical fruit trees by using juvenile as well as mature explants. Regeneration via somatic embryogenesis have been obtained in a number of cases e.g., while in citrus, sugar apple and papaya, etc. induction of androgenic haploids are successful, in guava and feijoa only callus results in anther cultures. Somaclones have helped in the selection of seedless *Musa*. Synthetic seed technology has aided in raising plantlets from encapsulated embryos of guava, mango, papaya, etc. Gene transfer techniques can further prove to be useful in the improvement of varieties.

1. Introduction

Commercial cultivation of fruits is still in infancy and even at present the yield of fruits in most cases remains low and are not within the means of working classes of the developing world. Many tropical fruits remain under utilized due to the complex circumstances that encompass their production and marketing. Moreover, the green revolution could make a relatively little impact on fruit cultivar development. Increase in fruit production can be achieved by advances in horticultural practices, post harvest handling and disease and pest control. Conventional propagation methods, i.e. grafting, air layering and removal of suckers, for improving the tropical fruit crop trees already exist for many important tropical fruits but the long juvenile period has made these techniques time consuming and cumbersome. Clonal propagation and selection of fruit crops using tissue and organ culture techniques have considerable potential for the improvement of economically important fruit trees that have been under cultivation for many generations. Improvement of plant quality and yield by cell manipulation through the sophisticated methods of genetic engineering has to rely on tissue culture for the final product. Generation of new variability through somaclonal variant selection, production of androgenic and gynogenic haploids to achieve homozygosity, rapid fixation of specific traits in hybrids, freeing plants from disease causing organisms by shoot tip culture and production of industrial compounds by cell culture are some well-known applications of plant tissue culture. Tissue culture technique and other biotechnological intervention have proved fairly successful and could be commercialised for some temperate fruit crops [1]. However, due to difficulty in controlling normal somatic embryo development and achieving high rates of their germination the progress in the application of tissue culture for clonal multiplication of tropical fruit trees and biotechnological tools has been rather slow. The purpose of this review is to present the current status of *in vitro* regeneration and improvement of tropical fruit trees.

2.   *In vitro* Regeneration of Tropical Fruit Trees

One of the earliest attempts to regenerate tropical fruit trees through *in vitro* culture technique was made by Maheshwari and Rangaswamy [2]. Subsequently, several species of tropical fruits have been regenerated through the process of organogenesis as well as somatic embryogenesis. Organogenesis involves adventitious and axillary shoot production. The adventitious shoot production comprises *de novo* shoot meristem formation from callus tissue or directly from organized tissues such as epidermal or subepidermal cells. The axillary shoot production involves shoot formation from axillary buds, shoot tips and meristems. The regenerated shoots are excised and used to produce additional shoots. The axillary shoot production is a direct method involving multiplication of preformed buds, usually without any callus formation and produces in general, genetically stable cultures. It produces the smallest number of plants, since the number of shoots produced is limited by the number of axillary buds placed in culture. Although the initial multiplication rate is low, it increases during the first few subcultures and eventually reaches a steady state, which may be maintained through numerous subcultures. Somatic embryogenesis is the process in which structures are formed containing a shoot and root connected by a closed vascular system (directly analogous to zygotic embryos).

2.1   Regeneration via Organogenesis

The main factors that influence the mode and rate of *in vitro* regeneration are the nature of explant, composition of the medium and the physical conditions in which the cultures grow. Organogenesis has been induced *in vitro* both from seedlings and mature tree explants (Table 1). Adventitious shoots have arisen directly from internode segments without a callus phase in *Citrus* [3]. Direct shoot organogenesis and plant regeneration have also been reported from seedling leaf explants of *Annona squamosa* [4] and *Garcinia mangostana* {5, 6] and from hypocotyl and seedling petioles of *A. cherimola* [7]. Adventitious shoots have differentiated following callus initiation and proliferation in *Citrus* [8, 9]. New vegetative growth that occurs from the base of the main stem during the period of vigorous vegetative growth in guava [10] serves as a reliable source of shoot tip and nodal explants. Papayas have been decapitated in order to stimulate lateral branching and to increase the number of explants from stock plants [11]. Shoot tip culture is the basic technique for *Musa* propagation [12–15]. It has been successfully applied to the rapid propagation of AA and AAA bananas, cooking ABB bananas and to a limited extent to AAB plantains and 'Silk' and 'Pome' AAB dessert bananas.

Organogenesis of tropical fruit species have generally been based on MS medium [16]. In a few cases (mangosteen, *Musa*) other media have been used for optimum morphogenesis (Table 1). In most studies callus initiation and shoot induction have been reported on the same medium which contains cytokinin, BA (Table 1) or a cytokinin together with an auxin. A high cytokinin to auxin ratio favours caulogenesis. Occasionally, shoot formation can occur following subculture of callus initiated on a medium with either a high auxin to cytokinin ratio, or high cytokinin to auxin ratio, or with cytokinin alone [17]. Usually, the auxin, NAA has been preferred for its synergistic effect on shoot induction. Some tropical fruit trees which have been regenerated via organogenesis have been listed in Table 1.

2.2   Regeneration via Somatic Embryogenesis

Somatic embryogenesis has several distinct advantages over organogenesis [17–19]. In woody

Table 1. *In vitro* regeneration of tropical fruit trees: Organogenesis

Species	Explant	Mature/ Juvenile	Medium	Growth Regulator	Reference
Annona cherimola	H, P	J	MS	NAA, BA	[7]
Annona squamosa	L	J	MS	BA	[4]
Annona squamosa	H	J	WPM	BA, NAA, IBA	[61]
Artocarpus heterophyllus	ST	M	MS	BA, KIN, NAA	[62]
Carica papaya	S	J	MS	KIN, IAA	[63]
Carica papaya	C	J	MS	BA, NAA	[64]
Carica papaya	N	J	MS	BA, KIN, IBA, NAA	[65]
Carica papaya	L, P, S, R	M	MS	BA, IBA	[66]
Citrus acida	Ep	J	MS	BA, NAA, IAA, 2, 4–D, GA$_3$	[67]
Citrus aurantifolia	S, R	J	MS	BA	[68]
C. aurantium	S	J, M	MS	NAA, KIN	[69]
C. grandis	S, L	J	MS	NAA, BA	[8]
C. grandis	S, T	J	MS	BA, TDZ, NAA, GA$_3$	[70]
C. halimii	H	J	MS	BA, NAA	[71]
C. jambhiri	S, R	J	MS	BA, KIN, NAA	[72]
C. limetoides	S	J	MS	NAA, KIN	[9]
Citrus limon	S	J, M	MS	NAA, BA	[69]
Citrus limon	S, R	J	MS	BA, KIN, NAA,	[72]
C. madurensis	S	–	MS	BA, NAA	[73]
C. paradisi	S, L	MS	MS	BA, NAA	[74]
C. reticulata	S	J,M	MS	NAA, KIN	[69]
C. reticulata	ST	M	MS	BA, KIN NAA, IBA	[75]
C. sinensis	S, L	M	MS	BA, NAA	[8]
C. sinensis	S	J, M	MS	BA, IBA	[69]
Citrus sinensis × *Poncirus trifoliata*	S	J, M	MS	BA	[3]
C. sinensis × *P. trifoliata*	R	J, M	MS	BA, 2, 4-D	[76]
Garcinia mangostana	L, C	J, M	MS	BA	[5]
Garcinia mangostana	L	M	WPM	BA, IBA	[77]
Garcinia mangostana	L	J, M	*	BA, auxin	[78]
Garcinia mangostana	Seed	–	MS	BA, NAA	[6]
Garcinia mangostana	L	J	MS, WPM	BA, TDZ, NAA	[79]
Litchi chinensis	Seed C	J	MS	BA, IBA	[80]
Mangifera indica	L	M	MS	KIN, IAA, IBA	[81]
Poncirus trifoliata	ST	M	MS	BA, IBA	[82]
Psidium guajava	ST	M	MS	BA, NAA, IBA	[83]
Psidium guajava	N	M	MS	BA, NAA, IBA	[10]
Psidium guajava	N	M	MS	BA, NAA, IBA	[84]
Psidium guajava	N	J	MS	BA	[85]
Psidium guajava	ST	J	MS	BA, NAA, IBA	[86]
Psidium guajava	Seedling	J	MS		[87]
Psidium guajava	S	M	MS	BA, NAA, IBA	[88]
Musa	ST	–	MS, Knop's	BA, NA	[89]
Syzygium cumini	ST, N	J	MS	BA, BAA, IBA	[90]

C = cotyledon, Ep = epicotyl, H = hypocotyl, L = leaf, N = node, P = petiole, R = root, S = stem, ST = shoot tip, M = mature, J = juvenile, MS = Murashige and Skoog, WPM = Woody plant medium [91], Knop's = Knop's medium [92], * = specific formulation, BA = 6-benzylamino purine, KIN = kinetin, TDZ = thidiazuron, NAA = α-naphthalene acetic acid, IAA = Indole-3-acetic acid, IBA = indole-3-butyric acid, Zea = zeatin, 2ip = 2-isopentenyl adenine, GA$_3$ = gibberellic acid.

species somatic embryogenesis is achieved less frequently than other methods of regeneration. However, most of the tropical fruit trees have been regenerated via somatic embryogenesis.

Among the tropical fruit trees *in vitro* somatic embryogenesis was first reported in *Citrus*. The initial attempt on induction of somatic embryogenesis in *Citrus* was made by Stevenson [20]. Later on, Maheshwari and Rangaswamy [2] reported the induction of somatic embryogenesis in *Citrus* by showing the formation of subcuticular globular proembryos from nucellus explants. Since then, the list of species has been extended and numerous publications have appeared on the initiation of somatic embryogenesis (both direct and indirect) using diverse explants (Table 2).

Among the different explants used to induce somatic embryogenesis in tropical fruit trees, nucellus has been the most appropriate. Somatic embryogenesis has been induced directly in cultured nucelli of *Citrus* [21] and indirectly in mango [22–24] and papaya [25]. Immature zygotic embryo has also proved to be regenerable tissue for many species (Table 2). The culture of zygotic embryo is a relatively easy *in vitro* procedure. Embryo culture has been used to multiply *Litchi* which is one of the most recalcitrant tropical fruit species.

Somatic embryogenesis is reported to follow two different patterns [26]. In the first, embryogenesis proceeds from the cells that are embryogenic in origin [27, 28] and in the second, embryogenesis is induced in highly differentiated tissues such as leaf, stem, nucellus and inflorescence [29, 30]. Embryogenesis from proembryogenic determined cells (PEDC) requires only an *in vitro* environment to follow the requisite pattern of cell division [19]. Since mature tissues are highly differentiated than those of proembryos, embryogenesis from the former tissues proceeds via the other route described by Sharp et al. [26], i.e. through induced embryogenic determined cells (IEDC) [31]. These highly differentiated tissues must undergo major epigenetic changes to initiate somatic embryogenesis. Therefore, IEDC requires an *in vitro* environment initially to dedifferentiate and then to redifferentiate quiescent cells to an embryogenic state. Direct and indirect embryogenesis, are two additional terms used to describe PEDC and IEDC embryogenesis respectively [17].

A number of media have been used for the induction of embryogenic cultures. However, most of the successful reports are based on Murashige and Skoog's (MS) medium (Table 2). The effect of medium composition and strength on induction of somatic embryogenesis has been demonstrated in some species, e.g. *Citrus* [30], papaya [32] and mango [22, 23, 33]. Generally, embryogenic callus has been obtained following explanting onto the medium containing 2,4-D or other synthetic auxins, like dicamba, NAA, etc. (Table 2). The requirement of exogenous auxin for the induction of somatic embryogenesis depends on the nature of the explant used. Although cytokinins have sometimes been incorporated into the induction medium, they are probably not critical for induction. But, in a few cases, e.g. in longan [34], induction of embryogenic callus has been shown to be cytokinin-dependent. Nitrogen in the form of glutamine has been shown to be essential for somatic embryogenesis in mango [22–24, 33]. Addition of polyamines to the culture media promoted somatic embryogenesis in coconut and papaya [35]. The complex organic nutrients such as coconut water, casein hydrolysate, malt extract, etc., have also been used in the induction medium for some species [36, 37]. Sucrose is the commonly used carbon source and a relatively high concentration of sucrose (5–6%) is optimum for somatic embryogenesis in guava [38], *Citrus* [39], mango [22–24, 33] and longan [34]. In addition to culture medium and explants, different genotypes of a species influence the ability of somatic embryogenesis [23, 37, 40].

Table 2.   *In vitro* regeneration of tropical fruit trees: *Somatic embryogenesis*

Species	Explant	Mature/ Juvenile	Medium	Growth Regulator	Reference
Carica papaya	S	J	MS	KIN, IAA	[63]
Carica papaya	ZE	J	MS	2, 4–D, KIN	[32]
Carica papaya	Protoplasts isolated somatic embryos	J			[93]
Carica papaya	P	J	MS	2, 4-D, BA	[94]
Carica papaya	H	J	MS	2, 4-D, IBA	[95]
Carica papaya	H	J	MS	2, 4-D, ABA	[25]
C. papaya × *C. cauliflora*	ZE	J	MS	BA, ABA	[96]
Citrus aurantifolia	Nu	M	MS	IAA, KIN, GA$_3$	[97]
Citrus aurantium	Nu	M	MT		[98]
Citrus clementina	Nu	M	MS		[99]
Citrus grandis	Nu	M	MS		[21]
Citrus jambhiri	Nu	M	MT		[98]
Citrus limon	Nu	M	MS		[21]
Citrus limon	In (style)	M	MS	BA, NAA	[29]
Citrus limon	ST	M	MS	BA, KIN, NAA, IBA	[75]
Citrus limon	Nu	M	MS	–	[100]
Citrus microcarpa	Nu	M	W	–	[2]
Citrus nobilis	Nu	M	MT	–	[101]
Citrus paradisi	Nu	M	MT	–	[102]
Citrus reticulata	Nu	M	W	–	[103]
Citrus reticulata	ST	M	MS	BA, KIN, NAA, IBA	[75]
Citrus reticulata	L, E, C, R	M	MS	KIN, NAA	[30]
Citrus sinensis	Nu	M	MS	–	[21]
Citrus sinensis	Nu	M	MS	–	[100]
C. unshiu	Juice vesicle	M	MS	KIN, GA, NAA	[104]
Cocos nucifera	Inf	M	MS	2, 4-D, BA, zip	[105]
Cocos nucifera	Inf	M			[106]
Eugenia spp.	ZE	M	MS	2,4-D	[114]
Euphoria longan	L	M	B5	2, 4-D, KIN	[34]
Eriobotrya japonica	Nu	M	MS	2, 4-D, BA	[107]
Feijoa sellowiana	ZE	J	MS	2, 4-D, KIN	[108]
Mangifera indica	Nu, ZE	M, J	MS	2, 4-D	[109]
Mangifera indica	Nu	M	MS, B5	2,4-D, KIN	[22]
Mangifera indica	Nu	M	MS, B5	BAP, 2, 4–D GA$_3$	[110]
Mangifera indica	Nu, ZE	M, J	MS, B5	2, 4-D, GA$_3$	[23]
Mangifera indica	Nu	M	MS, B5	2, 4-D, GA$_3$	[24]
Mangifera indica	Protoplasts isolated from pro embryo-genic masses	J	MS, B5	2, 4-D, NAA, KIN GA$_3$	[111]
Musa (AAA, ABB)	Rh, Basal Sheath	M	SH, MS	Dicamba, Zea	[112]
Myrciaria cauliflora	Nu	M	MS	2,4-D	[113]

C = Cotyledon, Ep = epicotyl, H = hypocotyl, Inf = inflorescence, L = leaf, Nu = nucellus, P = petiole, R = root, Rh = rhizome, S = stem, ST = shoot tip, ZE = zygotic embryo, M = mature, J = juvenile, MS = Murashige and Skoog, B5 = Gamborg et al. [115], MT = Murashige and Tucker [39]; W = White's [116], * = specific formulation, 2,4-D = 2, 4 dichlorophenoxy acetic acid, BA = 6-benzylamine purine, KIN=kinetin, TDZ = thidiazuron, NAA = α-naphthalene acetic acid, IAA = Indole-3-acetic acid, IBA = Indole-3-butyric acid, Zea = zeatin, GA$_3$ = gibberellic acid.

Among the tropical fruit trees, regeneration of viable plantlets from somatic embryos is a more frequently encountered problem than the production of somatic embryos from somatic embryogenesis. The problem may occur at any stage of development like maturation, germination, shoot apex elongation or acclimatization. Although somatic embryogenesis has been reported for several tropical fruit tree species (Table 2), the quality of somatic embryos with regard to their germinability or conversion into plants has been very poor This is because the apparently normal looking somatic embryos are actually incomplete in their development. Unlike seed embryos, the somatic embryos normally do not go through the final phase of embryogenesis called 'embryo maturation' which is characterised by the accumulation of embryo specific reserve food materials and proteins which impart desiccation tolerance to the embryos [41]. Abscisic acid (ABA) which prevents precocious germination and promotes normal development of embryos by suppression of secondary embryogenesis and pluricotyledonary [18] is reported to promote embryo maturation in several species. A number of other factors such as temperature shock, osmotic stress, nutrient deprivation and high density inoculation can substitute for ABA, presumably by inducing the embryos to synthesize the hormone. ABA is known to trigger the expression which is normally expressed during down phase of seeds [41]. Cytokinin can be important for somatic embryo maturation and has been demonstrated to influence development of cotyledon and shoot apex [17].

2.3 Anther Culture

In vitro androgenesis has been described as a process of deviation of development from normal gametophytic to a sporophytic pathway. This deviation generally leads to callus production or embryo formation. The plants can subsequently be obtained either via organogenesis or embryogenesis from the androgenic callus or via direct germination of androgenic embryos. In the tropical fruit trees androgenesis and plantlet regeneration have been reported in *Citrus* [42], sugar apple [43], papaya [44], longan [45] and *Litchi* [46]. The androgenic callus formation from *in vitro* culture of anthers has been reported in guava [47] and *Feijoa* [48]. This meagre progress with anther culture particularly in woody tropical fruit species that are difficult to culture suggests that the technique could be a reproducible method in regeneration of tropical fruit trees, but an extensive research in this area is still needed. Anther culture following pollen storage has potential for conservation [49]. Cryogenic storage of pollen would be space efficient and economical.

3. Somaclonal Variation

The term 'somaclonal variation' refers to the phenotypic and genotypic variation observed in plants regenerated from any form of cell culture [50]. The degree of variation has been shown to depend on tissue being cultured [51] and also on the length of time that the cells or tissues have been maintained *in vitro* [51]. Somaclonal variation may be a viable approach for obtaining horticulturally useful traits in tropical fruit trees. In addition, a long generation time for most fruit species like, seedlessness in *Musa*, etc. make this approach even more appealing for many species. Progress has been made with a few fruit species to use this technique to obtain disease resistance [117, 120–122], salt tolerance [119], thornlessness [99] and toxin resistance [118]. Additional research still needs to be conducted to assess the phenotypic and genotypic stability of these traits.

4. Synthetic Seed

The 'synthetic' or 'artificial' seed technology is an exciting and rapidly growing area of research in plant cell and tissue culture. Production of artificial seeds has unravelled new vistas in plant propagation. It is an excellent technique for propagation of rare hybrids, elite germplasm and genetically engineered plants. Germplasm can be stored effectively in the form of synthetic seeds. They serve as the most efficient delivery system. Synthetic seeds have been produced using either of the two methods: a hydrated system [52] or a desiccated one [53]. In the tropical fruit trees, the artificial seed technology is progressing well. Encapsulation of somatic embryos and plantlet regeneration have been reported in guava [38], mango [54] and papaya [55]. Plants were also regenerated from encapsulated shoot tips of banana [56, 57].

5. Transgenic Plants: Achievements in Tropical Fruit Trees

The development of recombinant DNA technology and efficient systems of controlling morphogenesis from the culture of cells and tissue have opened the opportunity for genetic manipulation of plants at the cellular level. The goal of gene transfer techniques is to produce improved varieties through the incorporation of horticulturally important genes (such as pest and disease resistance, drought and cold tolerance, herbicide resistance, improved fruit quality, reduced juvenility, dwarfism, etc.) into existing cultivars. Methods available for plant transformation are arranged in three main groups: (i) those using biological vectors (virus- or *Agrobacterium*-mediated transformation), (ii) direct DNA transfer techniques (chemical-, electrical-, or microlaser-induced permeability of protoplasts or cells), and (iii) non-biological vector system (microprojectiles, microinjection or liposome fusion). A comprehensive review on transformation methods has been compiled by Potrykus [58]. *Agrobacterium* based transformation shows an advantage over other methods since it targets transgenes to the nucleus and integrates them into the host DNA. Several trasformations have been reported based on *Agrobacterium*-mediated transformation of cells or explants, e.g. in *Citrus* [125], papaya [134] and mango [131]. The recovery of transgenic plant is mainly dependent on the frequency of gene introduction and the ability of the transformed cells to differentiate into plants, i.e. an efficient *in vitro* regeneration protocol is a pre-requisite. Pang and Sanford [133] were the first to demonstrate transformation of papaya by co-cultivating leaf discs, stems and petioles with *A. tumefaciens*. Although transformation was confirmed by nopaline assays, they were not able to regenerate the callus into plant. Fitch et al. [129] first demonstrated papaya with the neomycin phosphotransferase II (NPTII) and β-glucuronidase (GUS) genes using immature embryo explants via microprojectile bombardment. Fitch et al. [130] regenerated papaya plants resistant to papaya ring spot virus by incorporating PRV cp gene. The frequency of transformation in both cases was very low. In *Citrus*, successful transformation is reported using different methods (Table 3), but transformation frequencies were much lower. Transformation has been reported in a few tropical fruit trees, some of which are listed in Table 3.

Although elegant protocols have been worked out using the biological vector *Agrobacterium tumefaciens* as well as direct gene transfer in basic and applied science, there are still many problems which have to be solved in terms of a reproducible method, but these problems are more related to the biological or genetical phenomena than to the delivery of DNA into plant cells. The different methods could deliver DNA into the cells, but the events in the cell and the genetic compartments, organelles and nucleus are not controlled and the genetic integration of

Table 3. **Tropical fruit trees in which stable transformed plants have been obtained**

Species	Explant	Transformation method	Foreign gene	Result	Reference
Citrus reticulata CVS 'Onta' (Ponkan) 'Kara' (mandarin)	Electroporation Embryogenic callus sub-cultured in liquid medium	Direct DNA transfer by Electroporation	gus	Reduced colony formation	[123]
Citrus jambhiri (rough lemon)	Protoplast of nucellar callus	Direct DNA transfer with 20% PEG6000	cat and npt II	Selection of micro-colonies with paramomycin (20–40 µg/ml). Transgenic plants	[124]
Citrus sinensis CVS 'Trovita' 'Washington navel'	Cell suspension culture derived from embryo callus	*Agrobacterium*	npt II and hpt	Transgenic plants, embryoids resisting Kan	[125]
Citrus sinensis CV 'Pineapple'	Internodal stem segments of seedlings	*Agrobacterium*	gus with intron and npt II	Transformed shoots grafted *in vitro* onto seedling rootstocks	[126]
Citrus sinensis × *Poncirus trifoliata* (root stock)	Internodal stem segments of 5 week old seedlings	*Agrobacterium*	gus with intron and npt II	Transformed shoots grafted *in vitro* onto seedling rootstocks	[127]
Poncirus trifoliata (root stock)	Epicotyl segment	*Agrobacterium*	gus, npt II	Transformed plants (Resistance to Kan)	[128]
Carica papaya	Zygotic and somatic embryos and hypocotyl	Microprojectile bombardment	gus, npt II and the coat protein of papaya ring spot virus	Transformed somatic embryos and leafy shoots. Resistance to Kan	[129]
Carica papaya	Immature zygotic embryos	Particle bombard-ment	coat protein of papaya ring-	Transgenic papaya plants having increased resistance to PRV	[130]

(Contd)

Table 3. *(Contd)*

Species	Explant	Transformation method	Foreign gene	Result	Reference
Mangifera indica	Somatic proembryos	*Agrobacterium*	spot virus gus, npt II	Proembryos resistant to Kanamycin	[131]
Musa (AAA group)	Embryogenic cell suspension initiated using immature male flower	Microprojectile bombardment	npt II, Vid A or BBTV	Resistance to Kanamycin	[132]

foreign DNA is random. Targeted transformation is still at its infancy [59]. Gene silencing and interactions between different transgenes result in unexpected expression patterns of foreign genes [59, 60]. Several independent transformants with a specific gene construct are still necessary to find one transgenic plant with the proposed expression pattern [60]. Transgene-mediated suppression of a gene by antisense constructs can be achieved. Up to now, plant biotechnology has mainly focused on a single gene strategy. It is still cumbersome to change physiological traits which are determined by multiple genes and/or quantitatively inherited.

6. Conclusions

Considerable progress has been made in the recent past on *in vitro* plant regeneration via organogenesis and somatic embryogenesis in tropical fruit trees by manipulation of growth media and culture conditions as well as testing a variety of explant sources. To improve the propagation system and to overcome the main bottlenecks, in particular, maturation and low germination frequency, the knowledge of developmental physiology need to be enhanced. Refinements in protocols are also necessary to get good quality embryos to facilitate storage, germination and encapsulation of these embryos. Numerous characteristics in tropical fruit trees which cannot be improved by conventional breeding need biotechnological intervention.

Besides the fundamental aspects, a wide array of practical problems need to be solved such as mechanical handling and automated planting. In addition, it would be necessary to reduce the production cost for commercial application.

References

1. R.H. Zimmerman, Propagation of fruit, nut and vegetable crops—over view, in: (Eds, R.H. Zimmerman, R.J. Griebach, F.A. Hammerschlag, R.H. Lawson) Tissue Culture as a Plant Protection System for Horticultural Crops, Martinus Nijhoff, Dordrecht, 1986, pp. 183–200.
2. P. Maheshwari, N.S. Rangaswamy, Polyembryomy and *in vitro* cultures of embryos of *Citrus* and *Mangifera.*, Indian J. of Horticulture, 15 (1958) 272–282.
3. M. Barlass, K.G.M. Skene, *In vitro* plantlet formation from *Citrus* species and hybrids, Sci. Hort., 17 (1982) 333–341.
4. S. Nair, P.K. Gupta, A.F. Mascarenhas, *In vitro* organogenesis from leaf explants of *Annona squamosa* Linn. Plant Cell Tiss. Org. Cult., 2 (1983) 198–200.
5. H.K.L. Goh, A.N. Rao, C.S. Loh, *In vitro* plantlet formation in mangosteen (*Garcinia mangostana* L.), Ann. Bot., 68 (1990) 113–121.
6. M.N. Normah, A.B. Nor-Azza, R. Aliudin, Factors affecting *in vitro* shoot proliferation and *ex vitro* establishment of mangosteen, Plant Cell Tiss. Org. Cult., 43 (1995) 291–294.
7. M. Jordan, Multiple shoot formation and rhizogenesis from cherimola (*Annona cherimola* L.) hypocotyle and petiole explants, Gartenbauwissenschaft, 53 (1988) 234–237.
8. H.C. Chaturvedi, G.C. Mitra, Clonal propagation of *Citrus* from somatic callus cultures, Hort Science, 9 (1974), 118–120.
9. R. Raj Bhansali, A.C. Arya, Organogenesis in *Citrus limettoides* (sweet lime) callus culture, Phytomorphology, 29 (1979) 97–100.
10. M.N. Amin, V.S. Jaiswal, Rapid clonal propagation of guava through *in vitro* shoot proliferation on nodal explants of mature tree, Plant Cell Tiss. Org. Cult., 9 (1987) 235–243.
11. R.A. Drew, Rapid clonal propagation of papaya *in vitro* from mature field-grown trees, HortScience, 23 (1988) 609–611.

12. S.S. Ma, C.T. Shii, Growing banana plantlets from adventitious buds, J. of the Chinese Society for Hort Science, 20 (1974) 6–12.

13. F. J. Novak, Musa (Bananas and Plantains), in: F.A. Hammerschlag and R.E. Litz (Eds), Biotechnology of Perennial Fruit Crops, Biotechnology in Agriculture No. 8, CAB International, Wallingford, UK, 1992, pp. 449–488.

14. S.S. Cronauer, A.D. Krikorian, Rapid multiplication of bananas and plantains by *in vitro* shoot tip culture, Hort Sci., 19 (1984a) 234–235.

15. S.S. Cronauer, A.D. Krikorian, Multiplication of *Musa* from excised stem tips, Ann. Bot., 53 (1984b) 321–328.

16. T. Murashige, F. Skoog, A revised medium for rapid growth and bioassays with tobacco tissue cultures, Physiologia Plantarum, 15 (1962) 473–497.

17. R.E. Litz, D.J. Gray, Organogenesis and somatic embryogenesis, in: F.A. Hammerschlag, R.E. Litz. (Eds), Biotechnology of Perennial Fruit Crops. CAB International, Wallingford, UK (1992) pp. 3–34.

18. P.V. Ammirato, Embryogenesis, in: D-A. Evans, W.R. Sharp, P.V. Ammirato, Y. Yamada (Eds) Handbook of Plant Cell Culture, Vol. 1. Techniques for Propagation and Breeding. Macmillan, New York, (1983) pp. 82–123.

19. T.A. Thorpe, (1988). *In vitro* somatic embryogenesis, ISI Atlas of Sci.: Animal and Plant Sci. Vol. 1: 81–88.

20. F.F. Stevenson, The behaviour of *Citrus* tissue and embryo *in vitro*. Unpublished Ph.D. thesis, (1956) University of Michigan.

21. T.S. Rangan, T. Murashige, W.P. Bitters, *In vitro* initiation of nucellar embryos in monoembryonic citrus. Hort. Sci. (1968) 3, 226–227.

22. S.G. DeWald, R.E., Litz, G.A. Moore, Optimizing somatic embryo production in mango. J. of the Amm. Soc. HortSci. 114 (1989a) 712–716.

23. S. Pandey, Plant regeneration through somatic embryogenesis in two tropical trees, *Mangifera indica* L. and *Sterculia alata* Roxb. Unpublished Ph.D. Thesis, Banaras Hindu University, 1998.

24. H. Ara, U. Jaiswal, V.S. Jaiswal, Somatic embryogenesis and plantlet regeneration in Amrapali and Chausa cultivars of mango (*Mangifera indica* L.). Curr. Sci., 78 (2000a) 164–169.

25. B. Castillo, M.A.L. Smith, U.L. Yadav, Liquid system scale up of *Carica papaya* L. Somatic embryogenesis, J. Hort. Sci. and Biotech. 73 (1998) 307–311.

26. W.R. Sharp, M.R. Sondahl, L.S. Caldas, S.B. Maraffa, The physiology of *in vitro* asexual embryogenesis. Hort. Rev., 2 (1980), 268–310.

27. R. Raj Bhansali, J. A. Driver, D.J. Durzan, Adventitious embryogenesis and plant regeneration from rescued embryos of peach *Prunus persica* (L.) India J. of Exp. Biol. 29 (1991), 334–337.

28. G.D. March, E. Grenier, N. Minnay, G. Sulmont, H. David, A. David. Potential of somatic embryogenesis in *Prunus avium* immature zygotic embryos. Plant Cell Tiss. Org. Cult., (1993), 34: 209–215.

29. F. De Pasquale, F. Carimi, F.G. Crescimanno, Somatic embryogenesis from styles of different cultivars of *Citrus limon* (L.) Burm. Aust. J. Bot. 42 (1994) 587–594.

30. M.I.S. Gill, Z. Singh, B.S. Dhillon, S.S. Ghoshal, Somatic embryogenesis and plantlet regeneration in mandarin (*Citrus reticulata* Blanco). Sci. Horti. 63 (1995) 167–174.

31. S.A. Merkle, Strategies for dealing with limitations of somatic embryogenesis in hardwood trees. Plant Tiss. Cult. Biotech. 1 (1995) 112–121.

32. M.M.M. Fitch, R.M. Manshardt, Somatic embryogenesis and plant regeneration from immature zygotic embryos of papaya (*Carica papaya* L.) Plant Cell Rep. 9 (1990) 320–324.

33. S.G. DeWald, R.E. Litz, G.A. Moore, Maturation and germination of mango somatic embryos, J. Amer. Soc. HortSci. 114 (1989b), 837–841.

34. R.E. Litz, Somatic embryogenesis from cultured leaf explants of the tropical tree *Euphoria longan* Stend. J. of Pl. Physiol. 132 (1988) 459–466.

35. S.W. Adkin, Y.M. Samosir, A. Ernawati, I.D. Godwin, R.A. Drew, Control of ethylene and use of polyamines can optimize the condition of somatic embryogenesis in coconut (*Cocos nucifera* L.) and papaya (*Carica papaya* L.) Acta. Hortic. 461 (1998) 459–466.

36. R.E. Litz, G.A. Moore, C. Srinivasan, *In vitro* systems for propagation and improvement of tropical fruits and palm, Hort. Rev. 17 (1985) 175–198.

37. F. Carimi, M.C. Tortorici, F. De-Pasquale, F.G. Crescimanno, Somatic embryogenesis and plant regeneration from under developed ovule, stigma/style explant of sweet orange navel group (*Citrus sinensis* L. Osb.), Plant Cell. Tiss. Org. Cult. 54 (1998) 183–189.

38. N. Akhtar, Studies on induction of somatic embryogenesis and production of artificial seeds for micropropagation of a tropical fruit tree guava (*Psidium guajava* L.) Unpublished Ph.D. thesis, Banaras Hindu University, 1997.

39. T. Murashige, D.P.H. Tucker, Growth factor requirements of *Citrus* tissue culture, in: Chapman, H.D. (Ed), Proc. of the first Int. Citrus Symp. Vol. 3, University of California, Riverside, (1969) pp. 1155–1161.

40. R.E. Litz, R.D. Hendrix, P.A. Moon, V.M. Chavez, Induction of embryogenic mango cultures as affected by genotype, explanting, 2,4-D and embryogenic nucellar culture. Plant Cell. Tiss. Org. Cult. 53 (1998) 13–18.

41. S.S. Bhojwani, M.K. Rajdan, Plant tissue culture: Theory and practice, a Revised Edition, Elsevier Science Publishers, Netherlands, (1996).

42. M.A. Germana, Y.Y. Wang, M.G. Barbagallo, G. Iannolina, F.G. Crescimanno, Recovery of haploid and diploid plantlets form anther culture of *Citrus clementina* Hort. ex Tan. and *Citrus reticulata* Blanco. J. Hort. Sci. 69 (1994) 473–480.

43. S. Nair, P.K. Gupta, A.F. Mascarenhas, Haploid plants from *in vitro* anther culture of *Annona squamosa* Linn. Plant Cell Rep. 2 (1983) 198–200.

44. H.S. Tsay, C.Y. Su, Genebanks: Use of modern technology for conservation and use of plant genetic resources. FAO Consultancy Report, Rome, October, 1985.

45. Y.Q. Yang, X.W. Wei, Induction of longan plantlets from pollens cultured in certain proper media. Acta Genetica Sinica, 11 (1984) 288–293.

46. L.F. Fu, D.Y. Tang, Induction fo pollen plants of Litchi trees (*Litchi chinensis* Sonn.) Acta Genetica Sinica, 10 (1983) 369–374.

47. S.B. Babbar, S.C. Gupta, Induction of androgenesis and callus formation in *in vitro* cultured anthers of a myrtaceous fruit tree (*Psidium guajava* L.). The Botanical Magazine Tokyo 99 (1986) 75–85.

48. J.M. Canhoto, G.S. Cruz, Improvement of somatic embryogenesis in *Feijoa sellowiana* Bero (Myrtaceae) by manipulation of culture media composition. *In vitro* cellular and Dev. Bio-Plant, 30 (1994) 21–25.

49. N. Banerjee, E.A.L. Delanghe, A. tissue cultue technique for rapid clonal propagation and storage under minimal growth conditions of *Musa* (banana and plantain). Plant Cell Rep. 4 (1985) 351–354.

50. P.J. Larkin, W.R. Scowcroft, Somaclonal variation-a noval source of variability from cell culture for plant improvement. Theor. App. Genet. 61 (1982) 197–214.

51. R.M. Skirvin, J. Janick, Tissue culture-induced variation in scented *Pelargonium* spp. Journal of the American Society of HortScience, 100 (1976b) 282–290.

52. K. Redenbaugh, B.D. Paasch, J.W. Nichol, M.E. Kossler, P.R. Viss, K.A. Walker, Somatic seeds-encapsulation of asexual plant embryos Bio/Technol. 4 (1986) 797–801.

53. S.L. Kitto, J. Janick, Production of synthetic seeds by encapsulating asexual embryos of carrot. J. Am. Soc. HorticSci. 110 (1985) 277–282.

54. H. Ara, U. Jaiswal, V.S. Jaiswal, Germination and plantlet regeneration from encapsulated somatic embryos of mango (*Mangifera indica* L.). Plant Cell Rep. 19(1999) 166–170.

55. B. Castillo, M.A.L. Smith, U.L. Yadav, Plant regeneration from encapsulated somatic embryos of *Carica papaya* L. Plant Cell Rep. 17 (1997) 172–176.

56. T.R. Ganapathi, P. Suprasanna, V.A. Bapat, P.S. Rao, Propagation of banana through encapsulated shoot tips. Plant Cell Rep. 11 (1992) 571–575.

57. P. Suprasanna, S. Anupama, T.R. Ganapathi, V.A. Bapat, *In vitro* growth and development of encapsulated shoot tips of different banana and plantain cultivar. J. New Seeds 3 (2001) 19–25.

58. I. Potrykus, Gene transfer to plants: Assessment of published approaches and results. Ann. Rev. Plant Physiol. Plant Mol. Biol. 42 (1991) 205–225.

59. J. Siemens, O. Schieder, Transgenic plants: genetic transformation-recent developments and the state of the art. Plant Tiss. Cult. and Biotech. 2 (1996) 66–75.

60. M. Matzke, A.J.M. Matzke, Genomic imprinting in plants: Parental effects and trans-inactivation phenomena. Ann. Rev. Plant Physiol. Plant Mol. Biol. 44 (1993) 53–76.

61. E.E.P. Lemons, J. Blake, Micropropagation of juvenile and adult *Annona squamosa*. Plant Cell Tiss. Org. Cult. 46 (1996) 77–79.

62. M.N. Amin, V.S. Jaiswal, *In vitro* response of apical bud explants from mature trees of jackfruit (*Artocarpus heterophyllus*). Plant Cell Tiss. Org. Cult. 33 (1993) 59–65.

63. S. Yie, S.I. Liaw, Plant regeneration from shoot tips and callus of papaya *In vitro* 13 (1977), 564–567.

64. R.E. Litz, S.K. O'Hair, R.A. Conover, *In vitro* growth of *Carica papaya* L. cotyledons. Scientia Horticulturae 19 (1983), 287–293.

65. O. Reuveni, D.R. Shlesinger, U. Lavi, *In vitro* clonal propagation of dioecious *Carica papaya*. Plant Cell Tiss. Org. Cult. 20 (1990) 41–46.

66. M. Mondal, S. Gupta, B.B. Mukherjee, Callus culture and plant production in *Carica papaya* (var. Honey Dew). Plant Cell Rep. 13 (1994) 390–393.

67. B. Chakravarty, B.C. Goswami, Plantlet regeneration from long-term callus cultures of *Citrus acida* Roxb. and uniformity of regenerated plants. Sci. Hortic. 82 (1999) 159–169.

68. R. Raj Bhansali and A.C. Arya, Shoot formation in stem and root callus of *Citrus aurantifolia* (Christm.) Swinge grown in culture. Current Science 47 (1978a), 775–776.

69. S. Bouzid, Quelques traits du comportement de boutures de *Citrus* en culture in vitro. Comptes Rendus de l'Academie des Sciences Serie D 280 (1975) 1689–1692.

70. K.P. Paudyal, N. Haq, *In vitro* propagation of Pummelo (*Citrus grandis* L. Osbeck). *In Vitro* Cellular Dev. Biol.-Plant 36 (2000) 511–516.

71. M.N. Normah, S. Hamidah, F.D. Ghani, Micropropagation of *Citrus halimii*-an endangered species of south-east Asia. Plant Cell Tiss. Org. Cult. 50 (1997) 225–227.

72. H. Raman, S.S. Gosal, D.S. Brar, Plant regeneration from callus cultures of *Citrus limon* and *C. jambhiri*. Crop improvement 19 (1992) 100–103.

73. U. Grinblatt, Differentiation of Citrus stem *in vitro*. J. Amer. Soc. Hortic. Sci. 97 (1972) 599–603.

74. R. Raj Bhansali, A.C. Arya, Differentiation in explants of *Citrus paradisi* Macf. (grapefruit) grown in culture. Indian J. Expt. Biol. 16 (1978 b) 409–411.

75. S. Singh, B.K. Roy, S. Bhattacharyya, P.C. Deka, *In vitro* propagation of *Citrus reticulata* Blanco and *Citrus limons* Burmf. Hort Sci. 29 (1994) 214–216.

76. A. Sauton, A. Mouras, A. Lutz, Plant regeneration from citrus root meristems. J. Hortic. Sci. 57 (1982) 227–231.

77. C.J. Goh, P. Lakshmanan, C.S. Loh, High frequency direct shoot bud regeneration from excised leaves of mangosteen (*Garcinia mangostana* L.) Plant Sci. 101 (1994) 173–180.

78. H.K.L. Goh, A.N. Rao, C.S. Loh, Direct shoot bud formation from leaf explants of seedlings and mature mangosteen (*Garcinia mangostana*) trees. Plant Sci. 68 (1990) 113–121.

79. S. Te-Chato, M. Lim, Improvement of mangosteen micropropagation through meristematic nodular callus formation from *in vitro* derived leaf explants. Sci. Hortic. 86 (2000) 291–298.

80. D.K. Das, N. Shiva Prakash, N. Bhalla-Sarin, Multiple shoot induction and plant regeneration in litchi (*Litchi chinensis* Sonn.). Plant Cell Rep. 18 (1999) 691–695.

81. S.S. Raghuvanshi, A. Srivastava, Plant regeneration of *Mangifera indica* using liquid shaker culture to reduce phenolic exudation. Plant Cell Tiss. Org. Cult. 41 (1995) 83–85.

82. H. Harada, Y. Murai, Clonal propagation of *Poncirus trifoliata* through culture of shoot primordia. J. Hort Sci. 71 (1969) 887–892.

83. V.S. Jaiswal, M.N. Amin, *In vitro* propagation of guava from shoot cultures of mature tress. J. Pl. Physiol. 130 (1986) 7–12.

84. M.N. Amin, V.S. Jaiswal, Microprogation as an aid to rapid cloning of a guava cultiver. Sci. Hortic., 36 (1988) 89–95.

85. C.S. Loh, A. N. Rao, Clonal propagation of guava (*Psidium guajava* L.) from seedlings and grafted plants and adventitious shoot formation *in vitro*. Sci. Hort. 39 (1989) 31–39.

86. P. Papadatou, C.A. Pontikis, E. Ephtimiadou, M. Lydaki, Rapid multiplication of guava seedlings by *in vitro* shoot tip culture. Scientia Horti. 45 (1990) 99–103.

87. Y.M. Yasseen, S.A. Barringer, R.J. Schnell, W.E. Splittstoesser, *In vitro* shoot proliferation and propagation of guava (*Psidium guajava* L.) from germinated seedlings. Pl. Cell. Rep. 14 (1995) 525–528.

88. M.E.P. Fuenmayor, N.J.M. Montera, *In vitro* clonal propagation fo guava (*Psidium guajava* L.) from stem shoot of cv. Mara-7. Acta Hort. 452 (1997) 47-52.

89. T.R. Ganapathi, J.S.S. Mohan, P. Suprasanna, V.A. Bapat, P.S. Rao, A low cost strategy for *in vitro* propagation of banana. Curr. Sci. 68 (1995) 646–650.

90. U. Yadav, M. Lal, V.S. Jaiswal, *In vitro* micropropagation of the tropical fruit tree *Syzygium cumini* L. Plant Cell Tiss. Org. Cult. 21 (1990) 87–92.

91. G.B. Lloyd, B.H. McCown, Commercially feasible micropropagation of mountain-laurel, *Kalmia latifolia* by use of shoot tip culture. Proc. of the Intl. Plant. Prop. Soc. 30 (1980) 421–427.

92. W. Knop, Quantitative untersuchungen uber die Ernahrungsprecesse der pflanzen. Landwirtschaftlichen versuch Stationes 78 (1865) 93–107.

93. M.H. Chen and C.C. Chen, Plant regeneration from *Carica* protoplasts. Plant Cell Rep. 11(1992) 404–407.

94. J.S. Yang, C.A. Ye, Plant regeneration from petioles of *in vitro* regenerated papaya (*Carica papaya* L.) shoots. Botanical Bulletin of Academia Sinica 33 (1992) 375–381.

95. M.M.M. Fitch, High frequency somatic emryogenesis and plantlet regeneration from papaya hypocotyl callus. Plant Cell Tiss. Org. Cult. 32 (1993) 205–212.

96. M.H. Chen, C.C. Chen, D.N. Wang, F.C. Chen, Somatic embryogenesis and plant regeneration from immature embryos of *Carica papaya* × *Carica cauliflora* cultured *in vitro*, Can. J. Bot. 69 (1991) 1913–1918.

97. G.C. Mitra, H.C. Chaturvedi, Embryoids and complete plants from unpollinated ovaries and from ovules of *in vitro*-grown emasculated flower buds of *Citrus* spp. Bulletin of the Torrey Botanical Club 99, 184–189.

98. G. Ben-Hayyim, H. Neumann, Stimulatory effect of glycerol on growth and somatic embryogenesis in *Citrus* callus cultures. Zeitschrift fur Pflanzenphysiologie 110 (1983), 331–338.

99. L. Navarro, J.M. Ortiz, J. Juarez Aberrant *Citrus* plants obtained by somatic embryogenesis of nucelli cultued *in vitro*. HortSci. 20 (1985) 214–215.

100. S.D. Obukosia, K. Waithaka, Nucellar embryo culture of *Citrus sinensis* L. and *Citrus limon* L. African Crop Sci. Jour. 8 (2000) 109–116.

101. J. Kochaba, P. Spiegel-Roy, H. Neumann, S. Saad, Effect of carbohydrates on somatic embryogenesis in subcultured nucellar callus of *Citrus* cultivars. Zeitschrift fur Pflanzenphysiologie 105 (1982) 359–368.

102. J. Kochba, P. Spiegel-Roy, H. Safran, Adventive plants from ovules and nucelli in *Citrus*. Planta 106 (1972), 237–245.

103. P.S. Sabharwal, *In vitro* culture of ovules, nucelli and emryos of *Citrus reticulata* Blancovar. Nagpuri, In: Maheshwari, P., Rangaswamy, N.S. (Eds) Plant Tissue and Organ Culture A Symposium. University of Delhi, Delhi (1963) pp. 265–274.

104. N. Nito, M. Iwamasa, *In vitro* plantlet formation from juice vesicle callus of satsuma (*Citrus unshiu* Marc.) Plant Cell. Tiss. Org. Cult. 20 (1990) 137–140.

105. R.L. Branton, J. Blake, Development of organized structures in callus derived from explants of *Cocos nucifera* L. Ann. Bot. 52 (1983 a), 673–678.

106. J.L. Verdeil, C. Huet, F. Graode mange, J. Bufford-Morel, Plant regeneration from cultured immature inflorescence of coconut (*Cocos nucifera* L.) evidence for somatic embryogenesis. Plant cell Rep. 13 (1994) 218–221.

107. R.E. Litz, Somatic embryogenesis in tropical fruit trees, in: Henke, R.R., Hughes, K.W., Constantin, M.J., Holleander, A. (Eds), Tissue Culture in Forestry and Agriculture, Plenum Press (1985) pp. 179–193.

108. G.S. Cruz, J.M. Canhoto, M.A.V. Abreu, Somatic embryogenesis and plant regeneration from zygotic embryos of *Feijoa sellowiana* Berg. Plant Science 66 (1990), 263–270.

109. R.E. Litz, R.J. Knight, S. Gazit, *In vitro* somatic embryogenesis from *Mangifera indica* L. Callus Sci. Hort. 22 (1984) 233–240.

110. D.V. Laxmi, H.C. Sharma, P.B. Kirti, M.L. Mohan, Somatic embryogenesis in mango (*Mangifera indica* L.) cv. Amrapali. Curr. Sci. 77 (1999) 1355–1358.

111. H. Ara, U. Jaiswal, V.S. Jaiswal, Plant regeneration from protoplast of mango (*Mangifera indica* L.) through somatic embryogenesis. Plant Cell Rep. 19 (2000b) 622–627.

112. F.J. Novak, R. Afza, M. Van Duren, M. Perea Dallos, B.V. Conger, T. Xiaolang, Somatic embryogenesis and plant regeneration in supension cultures of dessert (AA and AAA) and cooking (ABB) bananas (*Musa* spp.) Biotechnology 7 (1989), 154–159.

113. R.E. Litz, *In vitro* somatic embryogenesis from callus of jaboticaba, *Myrciaria cauliflora*. Hort. Science 19 (1984 a), 62–64.

114. R.E. Litz, *In vitro* responses of adventitious embryos of two polyembryonic *Eugenia* species. Hort. Science 19 (1984 b), 720–722.

115. O.L. Gamborg, R.A. Miller, K. Ojima, Plant cell cultures. I. Nutrient requirements of suspension cultures of soybean root cells. Experimental Cell Res. 50 (1968) 151–158.

116. R.R. White, A handbook of plant and animal tissue culture (1963), Jaques Cattel Press, Lancaster.

117. H. Raman, B.S. Dhillon, Somaclonal variability for canker resistance in *Citrus aurantifolia* cv. Kagzi lime. VII International Congress on Plant Tissue and Cell Culture, Amsterdam, 1990, p. 164 (Abstract).

118. B. Nadel, P. Spiegel-Roy, Selection of *Citrus limon* culture variants resistant to mal secco toxin. Plant Sci. 53 (1987) 177–182.

119. G. Ben-Hayyim, Y. Goffer, Plantlet regeneration from NaCl-selected salt-tolerant callus cultures of Shamouti orange (*Citrus sinensis* L. Osbeck). Plant Cell Rep. 7 (1989) 680–683.

120. R.E. Litz, V.H. Mathews, R.C. Hendrix, C. Yurgalevitch, Mango somatic cell genetics. Acta Hort. 291 (1991) 133–140.

121. S.C. Hwang, W.H. Ko, Somaclonal variation of bananas and screening for resistance to *Fusarium* wilt, in: G.J. Persley, E.A. De Langhe (Eds) Banana and Plantain Breeding Strategies. ACIAR Proceedings, No. 21, Canberra, 1987, pp. 151–156.

122. S.C. Hwang, Somaclonal resistance in Cavendish banana to *Fusarium* wilt, in: R.C. Ploetz (Ed) Fusarium wilt of banana, APS Press, St paul (1990) pp. 121–125.

123. T. Hidaka, M. Omura, Transformation of *Citrus* protoplast by electroporation. J. Japanese Soc. Hortic. Sci., 62 (1990) 371–376.

124. A. Vardi, S. Bleichman, D. Aviv, PEG mediated transformation of *Citrus* protoplasts and regeneration of transgenic plants. Plant Sci. 69 (1990) 199–206.

125. T. Hidaka, M. Omura, M. Ugaki, M. Tomiyama, A. Kato, M. Oshima, F. Motoyoshi, *Agrobacterium*-mediated transformation and regeneration of *Citrus* spp. from suspension cells. Jpn. J. of Breeding, 32 (1990) 247–252.

126. L.Pena, M. Cervera, J. Juarez, A. Navarro, J.A. Pina, N. Duran-Vila, L. Navarro, *Agrobacterium*-mediated transformation of sweet orange and regeneration of transgenic plants. Plant Cell Rep. 14 (1995 a) 616–619.

127. L. Pena, M. Cervera, J. Juarez, C. Ortega, J.A. Pina, N. Duran-Vila, L. Navarro, High frequency *Agrobacterium*-mediated transformation and regeneration of *Citrus*. Plant Sci. 104 (1995b) 183–191.

128. J. Kaneyoshi (Hiramatus), S. Kobayashi, Y. Nakamura, N. Shigemoto, Y. Doi, A simple and efficient gene transfer system of trifoliate orange (*Poncirus trifoliata* Raf.). Plant Cell Rep. 13 (1994) 541–545.

129. M.M.M. Fitch, R.M. Manshardt, D. Gonsalves, J. Slighton, J. Sanford, Stable transformation of papaya via microprojectile bombardment. Plant Cell Rep. 9 (1990) 189–194.

130. M.M.M. Fitch, R.M. Manshardt, D. Gonsalves, J. Slighton, J. Sanford, Virus resistant papaya plants derived from tissue bombarded with the coat protein gene of papaya ringspot virus. Bio/technology 10 (1992) 1466–1472.

131. H. Mathews, R.E. Litz, D.H. Wilde, S. Merkle, H.Y. Wetzstein, Stable transformation and expression of β-glucouronidase and NPT II genes in mango somatic embryos. *In Vitro* Cellular Dev. Biol.-Plant 28 (1992) 172–178.

132. D.K. Bekcer, B. Dugdale, M.K. Smith, R.M. Harding, J.L. Dale, Genetic transformation of Cavendish banana (*Musa* spp. AAA group) cv. Grand Nain via microprojectile bombardment. Plant Cell Rep. 19 (2000) 229–234.
133. S.Z. Pang, J. Sanford, *Agrobacterium*-mediated gene transfer in papaya. J. Amm. Soc. Horticultural Sci. 133 (1988) 287–291.
134. J.S. Yang, T.A. Yu, Y.H. Cheng, S.D. Yeh, Transgenic papaya plants from *Agrobacterium*-mediated transformation of petioles of *in vitro* propagated multishoots. Plant Cell Rep. 15 (1996) 459–464.

Plant Biotechnology and Molecular Markers
P.S. Srivastava, Alka Narula and Sheela Srivastava (Editors)

17. Tissue Culture of Cashewnut

Sumita Jha and Sudripta Das

Centre of Advanced Study, Department of Botany, University of Calcutta,
35, B.C. Road, Kolkata 700 019, India

Abstract: Cashewnut, *Anacardium occidentale* L. (Anacardiaceae), is one of the most recalcitrant species in tissue culture. Even after eighteen years of research since the first report on cashew tissue culture, limited success has been achieved in obtaining a reproducible protocol for induction, development and conversion of somatic embryos. Some success has been achieved in micropropagation using cotyledonary nodal explants but there are no reports on propagation using explants from mature trees. Embryo culture has been successful in raising whole plants from immature embryos to overcome the problem of embryo abortion in cashew. Micrografting has also been employed by a modified side-grafting procedure by which shoot tips from glass-house raised seedlings and field plants were grafted on *in vitro* raised seedling rootstock. Induction of somatic embryos on different explants excised from immature zygotic embryos like excised cotyledons and excised hypocotyl with radicle, as well as on intact zygotic embryos as small as 1–2 mm has been obtained. The establishment of viable aseptic culture of immature zygotic embryos in cashewnut was restricted due to the exudation of phenolics in the media and the oxidation and browning of explants. Somatic embryos were induced directly on intact zygotic embryo explants of A_1–A_3 size in two out of the five genotypes that were studied, after 4 weeks of culture initially in M_1 media and 4 weeks in the M_2 media. Induction of somatic embryos was related to the size of intact zygotic embryo explant, and to the presence or absence of callusing in explants. Embryos were also induced on excised cotyledons and excised hypocotyls with radicle, but the best results of number of somatic embryos induced per explant (23 embryos) and frequencies of maturation (59.5%) and germination (23.2%) were obtained in intact zygotic embryos of A_2 size, belonging to the variety KV-26. A preconditioning or post-maturation period was necessary for germination of somatic embryos and germination was achieved after 4–5 weeks of preconditioning on MS media containing BA (1.0 mg/l) (M_3), however, only in 8.05–23.2% of the cases somatic embryos were induced on different explants. Plants regenerated from somatic embryos were transferred to MS basal medium for proliferation of shoots, which was found to be very slow. Attempts to transfer somatic embryo derived plants to potted soil were not successful. The results obtained have important implications for the further use of *in vitro* culture techniques for this recalcitrant species for the recovery of somaclones and transgenics.

1. Introduction

Plantation crops are high value commercial crops, of great economic importance and play a vital role in the country's export trade. There is an urgent need today to concentrate more on the research aspects of plantation crops, particularly cashew nut, for rapid propagation and qualitative and quantitative improvement of the yield. Conventional breeding techniques have contributed much to the improvement of perennial tree species. However, the resources of useful genetic variation are nearing exhaustion. Reforms in breeding techniques are therefore imperative. In conventional plant breeding, long periods are necessary to develop new varieties of woody species and also to replace varieties. Particularly with cross-pollinated species propagated with seeds, it is difficult to maintain the superior characteristics of a new variety. The application of tissue culture techniques for propagation and improvement of woody plant species, particularly tree species, holds great promise. By using this approach, a large number of individual plantlets with improved characteristics may be propagated in a short time. From a genetic point of view, rapid propagation in tree species has three advantages, production of population with uniformly superior genotypes; maintenance of characteristics and combinations that cannot be maintained by sexual propagation, rapid multiplication and storage of a superior variety or hybrid is possible through artificial seed technology. If traditional breeding techniques can be combined with biotechnology, new and remarkable progress will be achieved in improvement of tree species. As compared with herbaceous plants, perennial crops present some difficulties for using biotechnologies for their improvement. Despite three decades of research, the generation of woody plant species by cell and tissue culture techniques has been elusive [1, 2]. Even though some tree species can be micropropagated from mature trees, many others can presently be propagated only from tissues of juvenile specimens i.e., embryos or young seedlings [3].

Cashewnut, *Anacardium occidentale* L., belonging to family Anacardiaceae, is an evergreen, tall, tropical fruit tree, upto 12 m in height, which forms a thin peripheral canopy, studded with protruding inflorescences. The kernel of the seed, which remains after removal of testa, is the cashewnut of commerce.

Cashewnut is cultivated in many tropical countries, the main producers of the nut being Brazil, India, Mozambique and Tanzania. Although cashew was introduced in India in the 16th century by the Portuguese, the gene pool that was available to breeders was very low. However, some of the research centres in India, namely Bapatla, Vengurla and Ullal (in South India) were instrumental in assembling the germplasms and evaluating them for yield, quality and other agronomic characters. There are a few named cultivars but efforts were made to select superior high yielding types and propagate them by asexual methods [4]. Though there has been considerable improvement in the crop through conventional breeding, progress has been slow because the tree is heterozygous and takes 10–12 years to reach full cropping. It is propagated mainly by seeds often resulting in high degree of variability.

In cashewnut, area has increased from 0.176 million hectares in 1961 to 0.659 million hectares in 1996–97. The production in cashew has gone up from 0.079 million tonnes to 0.430 million tonnes in 1996–97. India exported cashew kernels worth Rs 13,000 million (US $ 362 million during 1996–97). Export of cashew is rising @ 27% per annum. These export earnings are exceeded only by coffee and rice among agricultural exports. To overcome the problem of low production in cashew in·India, areas under high yielding varieties with clonal saplings is being increased and orchards are being replaced by new high yielding varieties [5]. However, conventional methods of propagation are not efficient enough to provide high yielding planting materials [5, 6]. Reforms in breeding techniques are therefore imperative to meet the increasing demand of cashew nut in the international market. Techniques like micropropagation via multiple axillary branching and *in vitro* organogenesis or embryogenesis offer prospects of faster multiplication of elite genotypes. The application of plant biotechnological methodologies for the improvement of cashewnut is mainly limited by the difficulty of regenerating plants in a reproducible and efficient fashion. Cashew, like other members of Anacardiaceae, is strongly recalcitrant in *in vitro* culture and only limited successes have been achieved to date in this cash crop.

1.2   Organogenesis

In India, cashew tissue culture work was initiated for the first time at Calicut University by Philip and Unni [7] and later Philip [8] reported direct shoot and root organogenesis from proximal ends of cotyledon explants. Leva and Falcone [9] used microcuttings, young leaves and cotyledons from mature seed as explants to evaluate micropropagation using buds, and regeneration from callus cultures. Morphogenesis was achieved from globular calli on leaves and from nodular structures on cotyledons on SH medium with a high concentration of NAA and 6-BAP in combination. Sue et al. [10] screened morphogenetic capacities of different explants. Callus formation was obtained from leaf and petiole explants when different levels of NAA/BAP or 2-4, D/BAP were combined on MS modified medium. Adventitious and secondary roots induction occurred only when NAA was present at high levels (6 mg/1) along with BAP (1 mg/1). Hegde et al. [11] reported direct regeneration of plantlets from cotyledonary segments cultured on LS medium supplemented with kinetin and NAA. There are no reports/publications dealing with successful regeneration of plantlets through shoot organogenesis *in vitro* from any type of explants to date.

1.3   Apical and Axillary Node Culture

Progress with application of micropropagation has been achieved using microshoots and cotyledonary nodes using *in vitro* germinated seedling explants. However seedling explants are normally extremely heterozygous due to outbreeding. Shoot tip/node culture from identified elite trees would ensure genetic fidelity of *in vitro* raised plants. However, many problems persist with explant viability when shoot tips/nodes are used from mature trees. Cashewnut was found difficult to propagate *in vitro* from mature plant tissues (nodal segments or shoot apices) due to recalcitrant nature, microbial contaminations and high phenolic exudation. Lievens et al. [12] cultured nodal cuttings from 6- to 15-month old seedlings and showed axillary growth and shoot bud proliferation. Leva and Falcone [9] cultured shoots on MS and SH media and obtained shoot growth and rooting. They reported that the presence of GA_3 in combination with zeatin riboside improved node formation. Our own attempts to raise aseptic cultures from explants taken from mature trees have failed due to browning and non-viability of explants. Boggetti et al. [13] used glass-house raised plants (1 month, 1 year and 5 years old) to develop methods for multiplication of nodal explants. They reported that sprouting of buds decreased strongly with increase in age of mother plants. Shoots developed in presence of cytokinins were short and produced axillary branches while Gibberellins supported bud sprouting but suppressed rooting. Cytokinins never induced the multiple bud formation and only one bud developed at each axil. However, in presence of cytokinins, the number of side branches per microshoots increased and were excised to give new lateral shoots. Microshoots rooted *in vitro* at a frequency of 42% when cultured for five days with 100 μm IBA.

In cashew, attempts were also made to induce multiple shoot formation using seedling explants. Although highly heterozygous, seedling explants can be used to propagate individual seedling genotypes and provides a potential method for propagation of cashew altered by genetic transformation.

D'Silva and D'Souza [14] reported multiple shoot induction from cotyledonary nodes. Sucrose concentration was reported to affect the number of buds developing from cotyledonary nodes. *In vitro* studies on rapid propagation of five cultivars of cashew were undertaken in our laboratory

[15]. Shoot tip, leaf axil and cotyledonary nodes from seedlings could be induced to multiply on MS medium containing BA, kinetin and zeatin in combination (Fig. 1a-d). Factors affecting multiplication rates, included age of explant source, explant type, medium composition, light requirements, and transfer frequencies. Cotyledonary nodes produced more buds than other type of explants. Nodes had a 90% viability when transferred daily to fresh medium containing activated charcoal for 7 days while exposed to continuous dark. Microshoots from the different varieties could be rooted by the use of IBA. In one variety, high frequency rooting could be obtained by treating shoots with *Agrobacterium rhizogenes*. Genotype was found to affect *in vitro* response. The rates of multiplication in three varieties VTH-174 (Andhra Pradesh), Ullal (Karnataka) and VRI-I(M10/4) (Tamil Nadu) was low as compared to KV-26 (West Bengal).

Fig. 1. (a) Swelling and bulging in cotyledonary node explants; (b) and (c) Proliferation and growth of shoots obtained from multiplication of cotyledonary node and (d) Acclimatized plants in pots before being transferred to the field.

Genotypes also differed in the ability of microshoots to root. Boggetti et al. [13] reported difference in response of three genotypes studied (two Brazilian and one Tanzanian elite selection) and axillary branching from explants was achieved only with one genotype (Tanzanian).

1.4   Embryo Culture

One of the problems in cashew breeding is of embryo abortion. Low percentage of fruit set (3-4%) have been reported in cashewnuts [6]. In our study, immature zygotic embryos from five varieties and of various sizes could be cultured to stimulate normal embryological development [16].

1.5   Micrografting

Different grafting techniques and cuttings have been experimented and the best season and climatic conditions determined according to technical informations obtained in Brazil [EC-STD 1999]. *In vitro* micrografting was performed onto cashew and other Anacardiaceae seedlings, mainly, *Rhus typhina* [17]. Terminal apices from cultured shoots of cashew (1–3 mm) were micrografted onto stems and rooted understocks. For cashew/cashew micrografting, mature cashewnuts from the cashew germplasm were scarified, surface sterilized and cultured onto MS medium in agar to germinate. Once the seedlings reached 5–8 cm in height, they were decapitated just below the cotyledons and then grafted with a short apex. Shoot tips from glass house raised seedlings and field plants micrografted by a modified side-grafting procedure on *in vitro* raised seedling rootstocks (cashew or other Anacardiaceae) gave a successful rate of 40–80% [17]. Terminal apices from cultured shoots of cashew were micrografted onto stems and rooted microcuttings of *Rhus typhina*. On cashew/cashew micrografting at different rootstock position, significant differences were reported in the elongation growth rates and hypocotyl grafts grew stronger than epicotyl ones [18]. Rooting of micrografted shoots of mature tree origin was poor (13%) because the shoots were poorly rejuvenated.

1.6   Somatic Embryogenesis

In vitro somatic embryogenesis potentially offers alternative forms of large scale propagation of plants. Somatic embryos can be used for biotechnological applications such as genetic modification of trees to select desired stress tolerance traits and gene transfer. Jha [19] reported morphogenesis in callus cultures derived from zygotic embryos and occurrence of globular protuberances which developed into embryo-like structures. Hegde et al. [20] observed embryogenesis in cotyledonary segments. However, the obtained embryos could not be germinated. Cardoza and D'Souza [21] reported induction of direct somatic embryos from radicular end of zygotic embryos. Secondary embryos developed from the primary embryos. However, conversion of embryos to whole plant was not achieved. Recently, Ananthakrishnan et al. [22] and Cardoza and D'Souza [21] used nucellus tissues from developing seeds for induction of somatic embryogenesis. Ananthakrishnan et al. [22] reported induction of calli from nucellar explants excised from 1-month old developing fruits of cashew on Murashige and Skoog's medium containing 6.78 μM 2,4-D. Differentiation of somatic embryos from calli was noticed when they were transferred to MS liquid medium supplemented with 4.52 μM 2,4–D. Different stages of somatic embryo development were traced but there was no further development of the torpedo stage in the liquid medium containing 2,4-D. Conversion of somatic embryos to whole plants was not obtained. Cardoza and D'Souza

[21] have reported development of globular somatic embryos from nucellar callus in presence of picloram. Somatic embryos maturated in presence of picloram and putrescine and germination was obtained in MS basal medium. In our laboratory we initiated tissue culture studies in cashew nut for induction of somatic embryos from different explants from juvenile and mature trees but failed to raise embryogenic cultures from such explants. We then initiated tissue cultures using immature zygotic embryos to study the induction, development, maturation and conversion of somatic embryos to whole plants.

2. Materials and Methods

2.1 Plant Materials
Plant materials were collected from the Arabari Forest Range, West Bengal and NRCC, Puttur. Immature green nuts of cashew of different improved varieties (viz.VTH-174, M-10/4, Ullal, M-44/3 and KV-26) were collected from mature trees.

2.2 Sterilization of Explants
The immature green nuts were surface disinfected using a sequence which included rinsing for 5 min with 70% alcohol followed by a 0.1% $HgCl_2$ treatment for 20 min. The nuts were washed thoroughly with sterile distilled water, opened aseptically and the intact immature embryos, ranging from 1 to 10 mm, were excised out. The isolated zygotic embryos were inoculated either intact or explants like cotyledons, hypocotyls with radicle or epicotyl with plumule, were excised from the zygotic embryos and inoculated.

2.3 Induction of Somatic Embryos
Different basal media namely, Murashige and Skoog [23], Gamborg [24] and Lloyd and McCown's Woody Plant Medium [25], supplemented with various auxins like NAA, 2,4-D, IAA, IBA and cytokinins like BA, kinetin, either singly or in combinations were used. Growth adjuvants such as yeast extract, casein hydrolysate and proline were supplemented to the media in different experiments. PVP (0.5%) and activated charcoal (0.3%) were added to the media to prevent browning of explants.

2.4 Maturation and Germination
Globular somatic embryos, induced after 6–8 weeks, developed upto cotyledonary stages in MS basal media with various growth regulators, but lacking PVP, charcoal and yeast extract. Maturation of somatic embryos was obtained in this media, after incubation for 6 weeks. However, the cotyledonary embryos did not germinate in the same media. For germination, mature somatic embryos were subjected to different treatments, under dark conditions. Basal salts such as MS, WPM, in combination with different growth regulators such as BA (1-5 mg/l), abscisic acid (ABA, 0.1 mg/l), mannitol (1.0 g/l) and sucrose (30-100 g/l) were used, singly or in combination. Germination of somatic embryos was obtained after 4-5 weeks.

2.5 Cytological Study
For the study of mitotic chromosomes, shoot tips as well as root tips of regenerated plants were pretreated with 0.002 M 8-hydroxyquinoline for 4 hours, fixed in Carnoy's mixture (alcohol :

chloroform : acetic acid $(6:3:1)$ and stained with 2% aceto-orcein : 1(N) HCl $(9 : 1)$. Photomicrographs were taken with Wild-Leitz MPS 52 microscope.

3.   Results

3.1   Problem of Browning and Establishment of Viable Cultures from Zygotic Embryos

In our laboratory, immature intact zygotic embryos of various sizes (1-10 mm, A_1–A_4), excised cotyledons, excised hypocotyls with radicle and epicotyl with plumule were used as primary explants for induction of somatic embryos. The establishment of viable aseptic cultures of cashewnut was restricted due to the exudation of phenolics in the media and the oxidation and browning of explants. Considering intact embryos as explants, it was found that the oxidation and browning of explants, as well as the exudation of phenolics was more pronounced in embryos 5-10 mm (size A_3–A_4) in length than in 1-2 mm (size A_1–A_2) embryos, suggesting that the size of the explant was in some way directly related to the extent of browning of explants and exudation of phenolics (Table 1). Oxidation was maximum in A_4 embryo explants (80%) where exudation of phenolics preceded browning of explants and within 2 weeks of culture initiation, most of the explants turned brown. When excised explants were taken into consideration, maximum exudation of phenolics was observed in excised cotyledons, exudation taking place mostly from the cut-end (Table 3). In A_2 zygotic embryos, the frequency of oxidation was lower in intact embryos, than in excised cotyledons or excised hypocotyl with radicle, as exudation of phenolics takes place profusely from cut-end of explants. The exudation of phenolics and the subsequent browning of explants was reduced to some extent by frequently subculturing the explants (initially after every 2nd day for the 1st week, followed by weekly subcultures during the induction phase), use of activated charcoal (0.3%) and PVP (0.5%) in the culture media, and dark incubation.

The best response, so far as viability and establishment of aseptic viable cultures is concerned, was obtained in MS basal media containing 2 mg/l NAA, 2 mg/l Kn, 500 mg/l YE, 0.3% activated charcoal and 0.5% PVP (M_1). Although it is believed that auxins enhance phenolic oxidation, but NAA was used at an optimum concentration because, in the presence of NAA, the normal germination of zygotic embryos and development of shoot and root was restricted. Cultures were maintained in this media for 4 weeks, under continuous dark conditions.

. Some intact zygotic embryos showed a tendency to initiate callus after 3 weeks of culture on induction media (M_1), under dark conditions. Callus induction from immature embryo explants was variable, depending on the length of incubation, concentration of growth regulators used and size of zygotic embryos. Callus induced, continued to proliferate if maintained on induction media and similar cultural conditions. Callus was induced both with 2,4-D (2-4 mg/l) and NAA (2 mg/l) but not in the presence of other auxins. Callus induced with 2,4–D was brown in colour and leached exudates in media, turning the media completely brown within 10-12 days of culture. Such calli necrosed after 6 months. Callus induced with 2 mg/l (NAA) was creamish-brown in colour, did not show much leaching or exudation, and could be subcultured and maintained after induction. Callus proliferated from all stages of embryos sampled except for A_2 size. Results of preliminary experiments indicated that the maximum frequency of callusing was from embryos of A_4 size when cultures were incubated for 4 weeks (Table 1, Fig. 3a, b). Most of this callus developed at the epicotyl region of the embryo, along the plumule and at the site of explant contact with the media. But these calli were not embryogenic, there was no differentiation

Table 1. Effect of immature zygotic embryo size on frequency of induction of somatic embryos and their maturation and germination in cashewnut

Explant type	Size***	Frequency of explant browning (%)	Frequency of explants forming callus (%)	No. of explants forming embryos (N = 100)	No. of embryos/ explants ± S.E.	Frequency of embryo maturation (%)*	Frequency of embryo germination (%)**
Intact immature zygotic embryos (var. KV-26)	A$_1$	2.0	4.0	80	12.2 ± 0.8	39.5	15.2
	A$_2$	10.6	0	90	23.06 ± 0.04	59.5	23.2
	A$_3$	50.5	17.5	38	4.02 ± 0.23	38.6	11.6
	A$_4$	80.0	55.4	0	–	–	–

Table 2. Difference in response of explants from cashew genotypes during induction, maturation and germination of somatic embryos

Explant type	Genotype	Frequency of explant browning (%)	Frequency of explants forming callus (%)	No. of explants forming embryos (N = 100)	No. of embryos/ explants± S.E.	Frequency of embryo maturation (%)*	Frequency of embryo germination (%)**
Intact zygotic embryos of A$_2$(3-4 mm) size	VTH-174	18.0	4.0	10	8.0 ± 0.4	20.95	10.6
	M-44/3	22.5	17.2	0	–	–	–
	Ullal	36.0	26.5	0	–	–	–
	M-10/4	55.2	21.0	0	–	–	–
	KV-26	10.6	–	90	23.06 ± 0.04	59.5	23.2

*Somatic embryo maturation: transformation from globular to cotyledonary/torpedo stage.
**Somatic embryo germination: development of root and shoot—a complete plantlet.
Media: For Induction—MS+NAA(2 mg/l)+Kn(2 mg/l)+YE(500mg/l)+PVP(0.5%)+act. charcoal (0.3%) (M$_1$);
MS + NAA(0.1 mg/l) + Kn(1 mg/l)(M$_2$), For Maturation—M$_2$; For germination—MS + BA (1 mg/l) (M$_3$).

Table 3. Effect of explant type on frequency of induction of somatic embryos and their maturation and germination in cashewnut

Explant source	Type	Frequency of explant browning (%)	No. of explants forming embryos (N = 100)	No. of embryos/ explant± S.E.	Frequency of embryo maturation (%)*	Frequency of embryo germination (%)**
Intact zygotic embryos of A$_2$ (3–4 mm) size—(var-KV-26)	Intact	10.6	90	23.06 ± 0.4	59.5	23.2
	Excised cotyledons	26.3	35	10.0 ± 0.0	20.2	8.05
	Excised hypocotyl with radicle	20.5	55	8.0 ± 0.5	58.6	20.1
	Excised epicotyl with plumule	8.4	0	–	–	–

*Somatic embryo maturation: transformation from globular to cotyledonary/torpedo stage.
**Somatic embryo germination: development of root and shoot—a complete plantlet
Media : For Induction—MS + NAA(2 mg/l) + Kn(2 mg/l) + YE(500 mg/l) + PVP(0.5%) + act. charcoal (0.3%) (M$_1$); MS +NAA(0.1 mg/l) + Kn (1 mg/l)(M$_2$). For Maturation—M$_2$; For germination—MS + BA(1 mg/l) (M$_3$).

Scheme for direct somatic embryogenesis in cashewnut

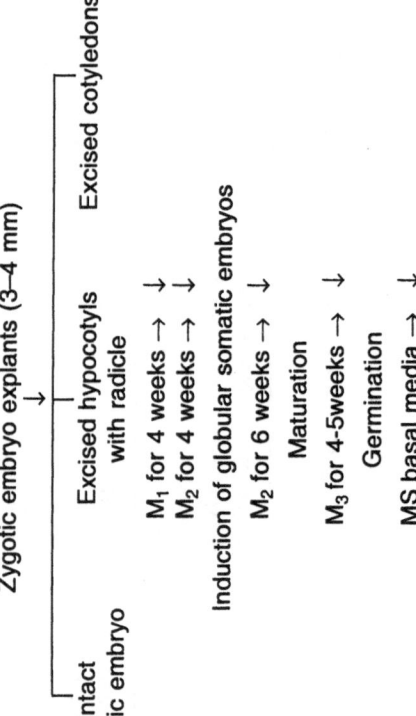

Zygotic embryo explants (3–4 mm)

Intact zygotic embryo   Excised hypocotyls with radicle   Excised cotyledons

Induction of globular somatic embryos
M$_1$ for 4 weeks →
M$_2$ for 4 weeks →

Maturation
M$_2$ for 6 weeks →

Germination
M$_3$ for 4-5weeks →

MS basal media →

Conversion to whole plants

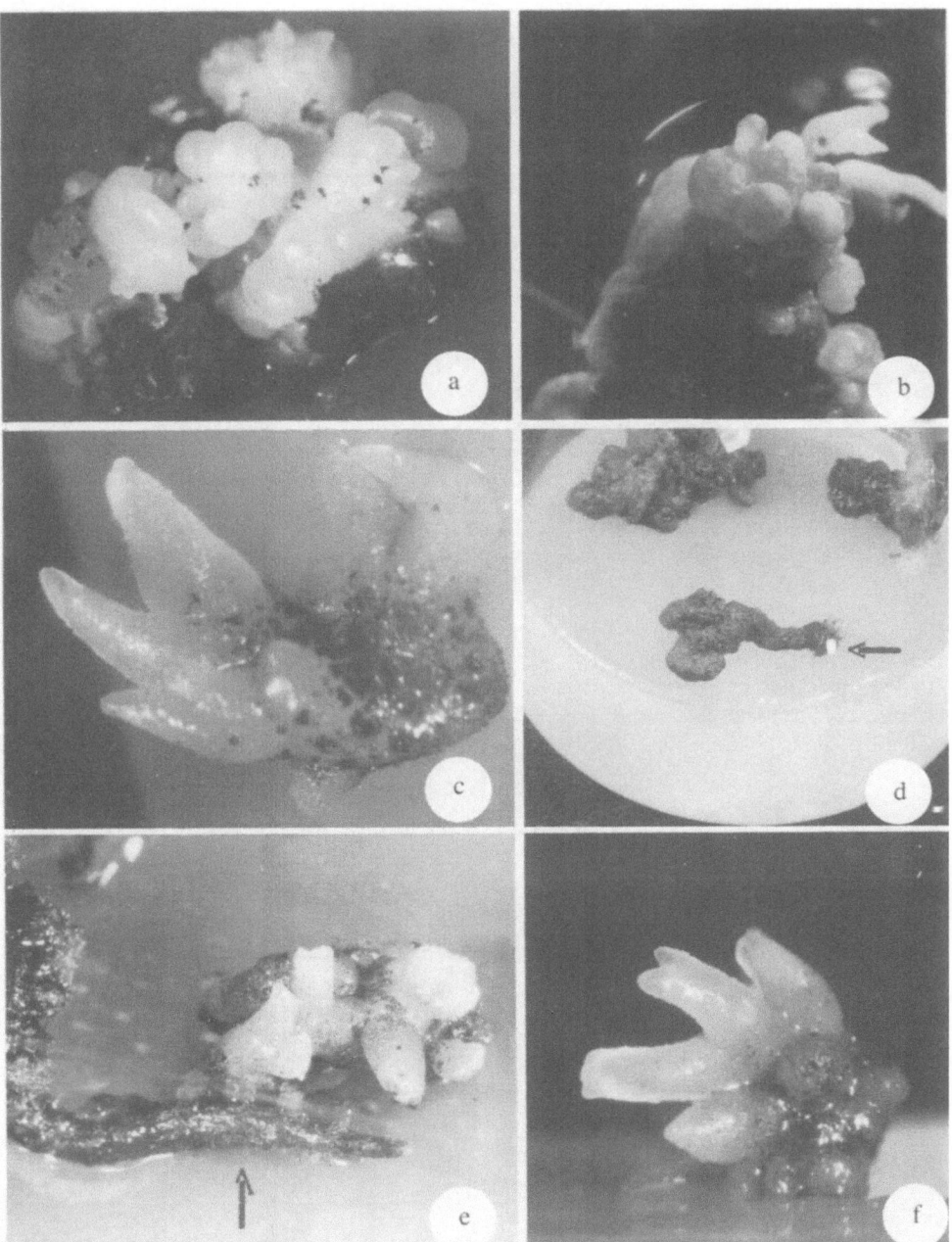

Fig. 2.   **(a–f) Somatic embryogenesis leading to whole plant regeneration in cashewnut, *Anacardium occidentale* L.; (a–c) Cluster of somatic embryos induced on excised cotyledonary node explants. (d–f) Somatic embryos at different stages of development induced on excised radicle explants.**

of somatic embryos from these calli cultured in presence of different combinations and concentrations of NAA and Kn or in auxin unsupplemented media.

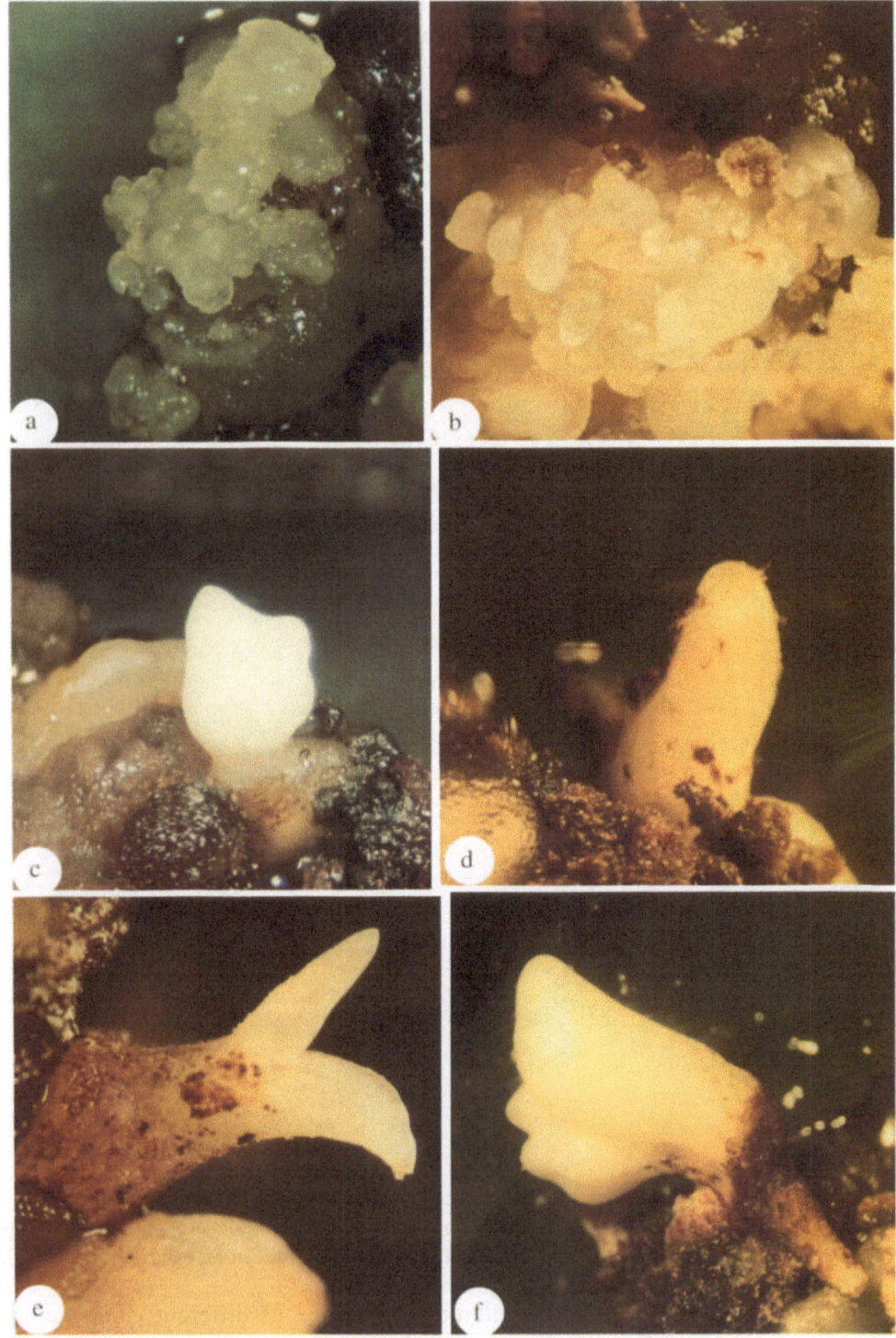

Fig. 3. (a, b) Embryogenic calli induced on intact zygotic embryo explants; (c–f) Different developmental stages of somatic embryos leading to maturation.

3.2 Induction of Somatic Embryos

No zygotic embryo germination or growth of zygotic embryonal axis was observed in intact zygotic embryo explants in the presence of NAA (M_1). Cultures were transferred, after 4 weeks in M_1, to media consisting of MS inorganics, Kn (1 mg/1) and NAA (0.1 mg/1), and lacking PVP, charcoal and yeast extract (M_2), and kept for 4 weeks. The suppression of growth of cashew zygotic embryonal axis in M media was followed by the appearance of white protrusions. These white protrusions subsequently developed into globular somatic embryos. Somatic embryos were thus induced directly on intact immature embryo explants of A_1–A_3 size, after 4 weeks of culture in M_1 media and another 4 weeks in M_2 media. Embryos were observed in culture where there was no callusing, particularly in A_2 explants, and the highest number of embryos induced (23 embryos) per explant were obtained (Table 1). It is interesting that A_3 explants produced few somatic embryos, yet yielded adventitious roots (10-12 roots/explant), in the same media. However, rooting ability was not observed in A_1, A_2 or A_4 explants suggesting cellular competency to differentiate somatic embryos or adventitious roots or calli (since A_2 explants failed to produce any calli) change during cashew ontogeny. No fully developed somatic embryos were observed in A_4 cultures and hardly any (4-5 embryos/explant) in A_3 cultures, indicating that the stage of development of zygotic embryos affects somatic embryo development. Occasionally, we observed secondary embryogenesis, on primary somatic embryos, while still attached to the mother tissue.

3.3 Development of Somatic Embryos

Maturation

Maturation of globular somatic embryos was obtained on the same culture media M_2 after incubation for another 6 weeks. The frequency of somatic embryos maturing i.e. globular embryos ultimately forming torpedo or cotyledonary stages, was maximum in the case of embryos induced on intact zygotic embryos of A_2 size (59.5%), followed by embryos induced on A_1 and A_3 size zygotic embryos (39.5% and 38.6 %, respectively) (Table 1, Fig. 2a-f, 3c-f). But there was no germination of the obtained embryos into complete plantlets in the same media.

Post-Maturation and Germination

It was observed that a preconditioning period or post-maturation treatment of these developing mature embryos was necessary for their germination. Of the various combinations tried out, MS basal media supplemented with BA (1.0 mg/1) (M_3) was found to be the most effective for germination of mature somatic embryos.

Germination was obtained after 4-5 weeks of preconditioning in the abovementioned media M_3, however, in only 11.6-23.2% of the somatic embryos induced on the different intact zygotic embryo explants (A_1–A_3).

Plants regenerated from the somatic embryos were transferred to MS basal media, for proliferation of shoots. However, proliferation of shoots was found to be very slow in the somatic embryo derived plants. Attempts to transfer somatic embryo derived plants to potted soil was not successful. Trials are being carried out to suitably harden the regenerated plants for successful survival in the soil.

3.4 Difference in Response of Different Genotypes During Induction of Somatic Embryos

In a set of experiments, the varying responses of intact zygotic embryo explants of A_2 size of different genotypes were evaluated during somatic embryo induction and further development (Table 2). Globular somatic embryos were successfully differentiated in only two of the five genotypes studied, in var. VTH-174 and KV-26. The number of globular embryos induced per explant was quite low, but the frequency of differentiation of somatic embryos on intact zygotic embryo explant of the 2 genotypes was contrasting, 10.3% explants forming somatic embryos in VTH-174 as against 90.0% in KV-26. The frequencies of maturation of somatic embryos was low to high in the two genotypes (20.95–59.5%) and germination was obtained in only 10.6% and 23.2% of somatic embryos in VTH-174 and KV-26, respectively. Callus was induced in A_2 size zygotic explants of all the genotypes except KV-26, although this callus did not lead to development of embryogenic calli.

3.5   Induction of Somatic Embryos from Explants Excised from Germinated Zygotic Embryos

In a separate set of experiments, explants like excised cotyledons, hypocotyl with radicle and epicotyl with plumule (all excised from A_2 zygotic embryos of variety KV-26) were compared with intact zygotic embryos (A_2) of var. KV-26, in order to ascertain the exact part of the embryonal axis where somatic embryos were induced and the differences in frequencies of induction, maturation and germination of somatic embryos (Table 3). The best results were obtained from intact zygotic embryo explants so far as the number of globular somatic embryos induced per explant and the frequencies of maturation and germination of somatic embryos. There was high frequency of browning of cotyledon explants and exudation of phenolics, we noted an abundance of mucilage in these cultures. No somatic embryos were obtained on epicotyl explants but somatic embryos were induced on the radicle end of hypocotyl explants. The number of somatic embryos obtained from hypocotyl explants was low (8 embryos/explant), as compared to the cotyledon explants (10 embryos/explant), but the maturation and germination frequencies were higher in somatic embryos from hypocotyl explants (58.6% and 20.1%, respectively).

3.6   Cytological Observations

The chromosome number in root tip of zygotic embryo derived plants was noted to be $2n = 40$, in all the 40 regenerated plants derived from somatic embryos that were analysed, with no irregularities in chromosome behaviour. Plants regenerated by somatic embryogenesis were thus euploid, and were free of any noticeable phenotypic variability.

4.   Discussion

The developmental stage of the explant and the explant type determined the type of response obtained *in vitro* in cashew immature zygotic embryo cultures. Although primary cultures from immature embryos followed a pattern of growth somewhat similar to that observed for cultured mature embryos, several aspects were unique to immature embryo explants, particularly of size. The results indicate changes in cellular competency to differentiate somatic embryos or callus or adventitious roots during cashew zygotic embryo ontogeny. Callus initiation was observed from all stages of development of embryos tested, but for A_2 embryos. However, embryos from early stages of development (A_1, A_3) showed a tendency to produce less primary callus than older embryos (A_4). Also, no somatic embryos were observed in explants where callus was induced and vice-versa, suggesting a negative correlation.

Growth and development of zygotic embryo axis in cashewnut was restricted in the presence of high levels of auxin (NAA), irrespective of the size of the embryo. This observation supports earlier findings. Maheswaran and Williams [26], in studies on *Trifolium repens*, proposed that growth suppression of the main embryonal axis is associated with the breakdown in the integrity of the cells as a single embryogenic group, and escape of individual or smaller group of cells to function autonomously. In *Anacardium occidentale*, suppression of growth in the embryonal axis in high auxin media (NAA at 2 mg/l) is followed by appearance of white protrusions in low auxin media (NAA at 0.1 mg/l), which subsequently develop into somatic embryos. While a high level of auxin (NAA at 2 mg/l) was essential for the explants to gain embryogenic competence, prolonged exposure (for more than 4 weeks) in such high auxin supplemented media did not favour somatic embryo differentiation. Hence, it may be assumed that the embryogenic competence of the explant cells gained during 4 weeks of exposure in high auxin supplemented media, leads to the development of globular embryos, when such explant were subcultured onto low auxin and high cytokinin media (NAA 0.1 mg/l and Kn 1.0 mg/l) after 4 weeks. Another noteworthy feature in cashew somatic embryogenesis was while globular embryos developed in the presence of NAA (0.1 mg/l) and Kn (1.0 mg/1), they did not germinate to rooted plants if cultured on the same media. The mature somatic embryos had to be cultured on MS basal media containing BA (1.0 mg/1) for further germination and regeneration of whole plants. However, conversion of mature somatic embryos to whole plants as obtained in the present study, does not occur at a desirable high frequency. Thus, somatic embryogenesis in cashew is not a one step process. The pattern of embryogenesis is somewhat similar to the direct embryogenic pathway in coffee, as reviewed by Sharp et al. [27].

Somatic embryos were induced directly on immature intact zygotic embryo explants in only two out of the five genotypes studied for induction of somatic embryogenesis, at a varying frequency of induction and the number of somatic embryos differentiated per explant. The inability to induce somatic embryos in certain genotypes, as was observed in cashew, is well documented. The ability of closely related plants to produce somatic embryos directly on explants is also under genetic control and differences between varieties are often found [28]. Stamp and Meredith [29] obtained somatic embryos on the zygotic embryos of four cultivars of *Vitis vinifera*, but could not induce them to form on the cultivar 'Pinot noir'. The direct formation of somatic embryos on apple leaf segments was genotype dependent [30]. The frequencies of callus induction in different genotypes varies in cashew. Many reports (e.g. Espinasse et al. [31] in *Helianthus*) illustrate how the capacity of explanted tissues to form callus and the subsequent growth rate of callus cultures, can both be variety dependent. There was variation in the frequencies of explant browning among genotypes in cashew. The extent of blackening or browning and growth inhibition which occurs in cultures is reported to be genotype dependent in species, that naturally contain high levels of tannins or other hydroxyphenols as in cashew, *Juglans*, *Quercus* and *Rhododendron* [28]. Differences were also found between species of the same genus and cultivars within a species. Cultivars of *Sorghum bicolor* released such large quantities of pigmented phenolics that the medium darkened and cultures readily became necrotic [32].

Plant age and the degree of differentiation of tissues are often interrelated and produce interactive effects *in vitro*. Both the size and degree of development of certain organs like cotyledons, hypocotyls and epicotyls depend on age [28]. During seed ontogeny, the physiological changes and their accompanying hormonal control may play a role in the differential responses of explants of varying maturity, as was found in cashew. Zygotic embryos, at the A_3–A_4 size, undergo rapid enlargement within the hardening shell *in situ*. During this period, sugar and protein content may be decreasing. These changes may somehow relate to changes in the morphogenic potential of cashew embryo explants. Competence of zygotic embryos of *Picea* for somatic embryogenesis has been shown to be limited to a specific stage of development prior to the accumulation of proteins [33]. Such relationships between changes in seed physiology, nutrient content and cellular competency to form adventitious structures, remains to be understood.

In *A. occidentale*, regenerated plants from somatic embryos were cytologically stable and normal. Since these somatic embryos can be repetitively embryogenic, they may be used as target tissues for transformation studies in cashewnut. The results obtained in this study have important implications in the further use of *in vitro* culture techniques for this species.

Although diploid normal rooted plants derived from somatic embryos have been successfully regenerated in the present study, procedure for successful survival of such plants on transfer to potted soil have yet to be standardized. The plants obtained following multiplication and proliferation of microshoots from cotyledonary nodes in *Anacardium occidentale* have survived on transfer to potted soil, but somatic embryo derived plants transferred (23 plants) under identical conditions have failed to survive. Further studies are needed for success of propagation through somatic embryogenesis of cashew.

There are two major limitations to the application of somatic embryogenesis for propagation and genetic manipulation—first of these is the low multiplication rates, that is, the low numbers of field plantable clonal plantlets produced per embryogenic culture, and second is the inability to initiate embryogenic cultures from mature trees.

The limitations of low multiplication rates can be further subdivided into problems such as low frequency somatic embryo production, production of malformed embryos, incomplete maturation, low germination and low conversion of germinants to plantlets capable of surviving transfer to *ex vitro* conditions.

Somatic embryogenic cultures in many species fail to demonstrate continued embryo production. Generally, there are also reports of direct embryogenesis, where, for example, individual somatic embryos arise from explanted zygotic embryo cotyledon tissues and in such cases only a single population of embryos is produced, some of which may mature and convert to plantlets. In some cases however, these primary embryos fail to mature and give rise to successive cycles of new embryos. This successive generation of new embryos is known as repetitive, recurrent or secondary embryogenesis. It is this phenomenon that gives somatic embryogenesis its great potential for mass propagation and gene transfer, since a single culture undergoing repetitive embryogenesis is theoretically capable of regenerating an unlimited number of somatic embryos.

A more frequent problem than low embryo production is that of regenerating viable plantlets from somatic embryos. The bottleneck may occur at any stage, including maturation, germination, shoot apex elongation or acclimatization. The standard use of simple two-step media sequences to promote the induction and developmental stages of embryogenesis is proving inadequate to accommodate the multiple and distinct phases that somatic embryos undergo in the course of their ontogeny and subsequent development. Therefore, the more closely the pattern of somatic embryo gene expression matches that of zygotic embryos, the greater are the chance of obtaining highly efficient regeneration systems. Such normalization of gene expression patterns will be achieved through the optimization of media and culture protocols for each individual stage of embryo development.

Proliferation of embryogenic cells takes a number of forms and is apparently influenced by a variety of factors, some of which can be controlled during the culture process, and some of which are yet undefined. Factors such as the effects of plant growth regulators, reduced nitrogen, plant species and genotype of the cultured material have been investigated for induction and proliferation of embryos. By recognizing the critical factors involved at each stage, and those that exert their influence throughout the process, the protocols at each stage can be tailored to more closely simulate the conditions *in planta*.

Besides factors such as explant type and growth regulator regimes, a substantial number of other factors can affect the induction of the embryogenic state. Perhaps one of the most important is plant genotype as suggested from our earlier studies in cashew. Inasmuch as the induction of somatic embryogenesis plausibly involves activation of the same genetic pathways as zygotic embryogenesis, somatic embryogenesis should be a universal phenomenon for all seed-bearing plants. Nevertheless, individual genotypes, within a given species, such as cashew, vary greatly in embryogenic capacity. Such genotypic differences in embryogenic capacity might reflect current differences in the ability to activate key elements in the embryogenic pathway. In addition, individual genotypes may have unique requirements for optimal regeneration capacity, and it would also be expected to have impact on culture proliferation rates, another problem in cashew.

With informed manipulation of these factors, not only will proliferative embryogenic cultures realize their potential for virtually unlimited propagule production, but the somatic embryos produced will come to have the vigor and germination associated with their zygotic counterparts.

References

1. J.M. Bonga, D.J. Durzan, Cell and tissue culture in forestry. Martinus Nijhoff./Dr. W. Junk Publishers, Dordrecht, Boston, Lancaster (1987).
2. Z. Chen, D.A. Evans, W.R. Sharp, P.V. Ammirato, M.R. Sondahl, Handbook of plant cell culture Vol 6(1990). Perennial Crops. McGraw-Hill Publishing Company.
3. N.P. Samson, J. Ponsamuel, S. Ganeshan, *In vitro* studies on callus induction and morphogenesis of cashew (*Anacardium occidentale* L.). XIII Plant Tissue Cult. Conf., Oct. 18–20, Shillong. Abstr. 86(1990).
4. Anonymous, Annual Report, National Research Center for Cashew, ICAR, Puttur, Karnataka (1994).
5. K.L. Chadha, J. Indian, Horticulture 46 (1989): 131.
6. M. Sivantham, L. Pugalendhi, S. Jeeva, D. Somasundram, The Cashew 4 (1990): 17.
7. V.J. Philip, P.N. Unni, *In vitro* propagation of cashew for crop improvement, in: E.V.V. Bhaskara Rao, H.M. Khan (Eds). Cashew Research and Development, CPCRI, Kasargod (1979) pp.78–82.
8. V.J. Philip, *In vitro* organogenesis and plantlet formation in cashew (*Anacardium occidentale* L.) Ann. Bot. 54 (1984): 149–152.
9. A.R. Leva, A.M. Falcone, Propagation and organogenesis in vitro of *Anacardium occidentale* L. Acta Hort. 280 (1990): 143–145.
10. M.O. Sy, L. Martinelli, A. Scienza, *In vitro* organogenesis and regeneration in cashew (*Anacardium occidentale* L.). Acta Hort. 289 (1991): 267–268.
11. M. Hegde, M. Kulasekaran, K.G. Shanmugavelu, S. Jayasankar, *In vitro* culture of cashew seedlings and multiple plantlets from mature cotyledons. Indian Cashew J. 20 (2) (1991): 19–24.

12. C. Lievens, M. Pylyser, P.H. Boxus, First results about micropropagation of *Anacardium occidentale* by tissue culture. Fruits, 44 (1989): 555–557.

13. B. Boggetti, J. Jasik, S. Mantell, *In vitro* multiplication of cashew (*Anacardium occidentale* L.) using shoot node explants of glasshouse-raised plants. Plant Cell Rep. 18 (1999): 456–461.

14. I. D'Silva, L. D'Souza, *In vitro* bud proliferation of *Anacardium occidentale* L. Plant Cell Tiss. Org. Cult., 29 (1992): 1–6.

15. S. Das, T.B. Jha, S. Jha, *In vitro* propagation of cashewnut. Plant Cell Rep., 15 (1996): 615–616.

16. S. Das, T.B. Jha, S. Jha, Factors affecting *in vitro* development of embryonic axes of cashewnut. Sci. Hort. 82 (1999): 135–144.

17. C.J. Rodrigues Jr., Development of Selection and Clonal propagation techniques for multiplication of elite yield and anthracnose tolerant cashew (*Anacardium occidentale*), Summary Reports of European Commission supported STD-3 projects, published by CTA 1999, Tropical and Subtropical Agriculture, Third STD Programme, 1992–95, pp. 112–116.

18. E.E. Mneney, S.H. Mantell, *In vitro* micrografting of cashew. Plant cell, tissue and organ culture 66(1) (2001): 49–58.

19. T.B. Jha, *In vitro* morphogenesis in cashewnut, *Anacardium occidentale* L. Ind. J. Exp. Biol. 26 (1988): 505–507.

20. M. Hegde, M. Kulasekaran, S. Jayasankar, K.G. Shanmugavelu, *In vitro* embryogenesis in cashew. Indian Cashew J. 21 (4) (1992): 17–25.

21. V. Cardoza, L. D'Souza, Somatic embryogenesis from nucellar tissue of cashew (*Anacardium occidentale* L.). *In Vitro* Cellular and Developmental Biology 36 (2000): 3 Part II P-2053.

22. G. Annathakrishnan, R. Ravikumar, R. Prem Anand, G. Vengadesan, A. Ganapathi, Induction of somatic embryogenesis from nucellus derived callus of *Anacardium occidentale* L. Sci. Hort. 79 (1999): 91–99.

23. T. Murashige, F. Skoog, A revised medium for rapid growth and bioassays with tobacco tissue cultures. Physiol. Plant. 15 (1962): 473–497.

24. O.L. Gamborg, R.A. Miller, K. Ojima, Nutrient requirements of suspension cultures of soybean root cells. Exp. Cell Res. 50 (1968): 151–158.

25. G. Lloyd, B. McCown, Commercially-feasible micropropagation of mountain laurel, *Kalmia latifolia*, by use of shoot tip culture. Comb. Proc. Intl. Plant Prop. Soc. 30 (1980): 421–427.

26. G. Maheswaran, E.G. Williams, Direct somatic embryoid formation on immature embryos of *Trifolium repens, T. pretense* and *Medicago sativa*, and rapid clonal propagation of *T. repens*. Ann. Bot. 54 (1984): 201.

27. W.R. Sharp, M.R. Sondahl, L.S. Caldas, S.B. Maraffa, The physiology of *in vitro* asexual embryogenesis. Hort. Rev. 2 (1980): 268–310.

28. E.F. George, Plant propagation by tissue culture. The Technology, Exegetics Limited, London (1993).

29. J.A. Stamp, C.P. Meredith, Proliferative somatic embryogenesis from zygotic embryos of grapevine. J. Am. Soc. Hort. Sci. 113(1988): 941–945.

30. M. Welander, Biochemical, anatomical studies of birch (*Betula pendula* Roth) buds exposed to different climatic conditions in relation to growth *in vitro*, in: Genetic Manipulation of Woody Plants, J.W. Hanover, D.E. Keathley (Eds), Plenum Press, New York, London (1988) pp. 79–99.

31. A. Espinasse, C. Lay, J. Volin, Effects of growth regulator concentrations and explant size on shoot organogenesis from callus derived from zygotic embryos of sunflower (*Helianthus anuus* L.). Plant Cell Tiss Org. Cult. 17 (1989): 171–181.

32. T. Cai, L. Butler, Plant regeneration from embryogenic callus initiated from immature inflorescences of several high-tannin sorghums. Plant Cell Tiss. Org. Cult. 20 (1990): 101–110.

33. D.A. Roberts, B.S. Flinn, D.T. Webb, F.W. Webster, B.C.S. Sutton, Characterization of immature embryos of interior spruce by SDS-PAGE and microscopy in relation to their competence for somatic embryogenesis. Plant Cell Rep. 8(1989): 285–288.

Plant Biotechnology and Molecular Markers
P.S. Srivastava, Alka Narula and Sheela Srivastava (Editors)

18. Changing Scenarios in Indian Horticulture

Sanjay Saxena and Vibha Dhawan

Bioresources and Biotechnology Division, Tata Energy Research Institute, Habitat Place,
Lodhi Road, New Delhi 110 003, India

Abstract: India holds enormous potential for production of horticultural crops. However, despite ranking high in terms of overall production, in most species the yields are far below the world average. Enhancement in productivity levels is not only necessary because land is a finite resource but also to remain cost competitive in the global market. Along with productivity, it is also imperative to improve the quality of the produce. This will not only enable the Indian farmers to compete with the imported products more effectively but also gear them to sell their produce in the international market. This communication highlights some of the strategies that could possibly be adopted to improve production, productivity and the quality of horticulture produce in the country.

1. Introduction

Horticulture covers a wide spectrum of crops such as fruits, vegetables, roots and tubers, medicinal and aromatic plants, plantation crops, ornamentals and spices. It provides nutritional security, offers a remunerative means for diversification of land use for improving productivity and returns, and increases employment opportunities. Horticultural products especially spices are also valuable foreign exchange earners for the country. However, since independence till the beginning of the Eighth Five Year Plan the government policies were focussed towards developing agriculture in the country with virtually no attention been paid to horticulture. Subsequently, the government realized the value of horticulture in Indian economy and gave a major thrust to the same by raising the budgetary allocation from a meager amount of Rs. 24 crores in the Seventh Five Year Plan to Rs. 1000 crores in the Eighth Five Year Plan. Besides this 40 times hike in the allocation, a large number of concessions, subsidies and incentives were given to the growers. One important trend observed in the last few years is that horticulture development has gradually moved out of its rural confines into urban areas and from traditional agricultural enterprise to the corporate sector. The adoption of improved technology, greater commercialization and professionalism in the management of production and marketing has brought about a perceptible change in the concept of horticulture development in the country.

Today India holds a major share in the world trade of spices. It also ranks second in the world both in production of fruits as well as vegetables. However, this increase in production is largely on account of increase in area under cultivation rather than increase in productivity levels. Barring few crops such as grapes, litchi and citrus where the yields have improved during the last decade, in most other important fruit crops the yields have either declined or have shown only a marginal increase (Table 1). It is only recently that there has been an awareness to increase our land productivity and quality of product, and government is taking serious measures for adoption of modern technologies such as use of hybrid seeds, tissue culture for producing clonal plants, molecular techniques, biofertilizers, high-tech agro-techniques etc.

Table 1.   Status of some important fruit crops in India

Fruit	Year	Area (000 ha)	Production (000 tonnes)	Yield (kg/ha)
Apple	1991–92	194.6	1147.7	5900
	1999–2000	238.3	1047.4	4395
Mango	1991–92	1077.6	8751.6	8120
	1999–2000	1486.9	10503.5	7064
Guava	1991–92	94.0	1095.1	11650
	1999–2000	150.9	1710.5	11335
Pineapple	1991–92	57.1	768.5	13470
	1999–2000	75.5	1025.3	13572
Sapota	1991–92	27.2	396.3	14540
	1999–2000	64.4	800.3	12427
Grapes	1991–92	32.4	668.2	20650
	1999–2000	44.3	1137.8	25684

Source: CMIE, 2001.

Except for grapes wherein India has the highest productivity in the world, in most other fruit crops the productivity levels are much lower (Table 2).

Table 2.   Yield (kg/ha) of various fruit crops in the world in 2001

Fruit	World average	India	Leading country	
			Name	Yield
Apple	10767.4	6493.5	Belgium-Luxembourg	52631.6
Apricot	6281.7	4125.0	Switzerland	21515.2
Banana	16338.5	32653.1	Nicaragua	55398.2
Citrus	13709.2	19195.9	USA	33692.5
Grapes	8480.1	26760.6	India	26760.6
Papaya	15983.0	11298.2	Costa Rica	46933.3
Pineapple	17932.3	13750.0	Panama	52681.7

Source: FAO, 2002.

India's position with respect to productivity of vegetables is no better either (Table 3). In most vegetables the yields are marginally to significantly lower than the world average. In fact, the difference between our yields and that of the leading country in the respective crop is so large that it may not be even worthwhile to draw a comparison. It is true that one cannot expect India to be the world leader in each and every crop as climatic conditions that strongly influence the productivity level may not be ideal for all the crops in India. Also, the practices followed by small and marginal farmers in our country are very different from those followed in large mechanized farms in USA and European countries. However, these figures clearly bring out the potential that exists before us in improving the yields of our crops.

It is really hard to believe that a country ranking second in the world in terms of production of fruits and vegetables having below world average per capita consumption of these commodities (Table 4). This is despite the fact that most of our population is vegetarian.

The floriculture sector in India has shown a steady growth in exports over the years, but in

Table 3. Yield (kg/ha) of various vegetable crops in the world in 2001

Fruit	World average	India	Leading country	
			Name	Yield
Green beans	6901.7	2800.0	Kuwait	23130.4
Cabbage	20778.9	17916.7	Korea	61591.6
Carrot	21204.7	14583.3	Austria	57067.7
Cauliflower	17755.0	16250.0	Kuwait	45284.0
Okra	6896.3	10967.7	Cyprus	16438.4
Potatoes	15967.8	18642.8	New Zealand	50000.0
Onion (dry)	17070.2	9800.0	Austria	59777.8
Pumpkins, squash and gourds	12496.0	9750.0	Netherlands	55000.0
Tomatoes	26769.9	17000.0	Netherlands	433333.4

Source: FAO, 2002.

Table 4. Per capita supply of vegetables and fruits in the world in 2000

Country	Vegetables (kg/year)	Fruits (kg/year)
India	62.5	41.5
Asia	116.2	46.2
European Union (EU)	112.5	83.0
USA	125.8	124.8
World average	101.9	59.8

Source: FAO, 2002.

terms of value we stand 25th in the world trade which is an indication of the long road that lies ahead of us.

2. Constraints

Some of the constraints in developing horticulture sector in the country include: (a) inadequacy of good quality seeds and other planting material, (b) low productivity, (c) poor quality of the product, (d) inadequate efforts for product diversification and consumption, (e) lack of awareness, (f) slow pace in adoption of improved technology, (g) lack of infrastructure for post-harvest management and marketing, (h) inadequacy of trained manpower and human resources in horticulture, (i) lack of proper database on demand projection, price realization etc. and (j) poor transportation system, credit facilities and price support.

3. Strategies

To meet the growing demand of horticultural products on account of increase in population and average household income, and to improve per capita consumption there is an urgent need to increase the production of fruits and vegetables in the country. However, it must be emphasized that this increase in production is to be achieved largely through increase in productivity rather than increase in area, as has been the case so far. This is because land being a finite resource will become a limiting factor at some point of time. Improvement in productivity levels is also necessary to remain cost competitive in the global market. This is more so because recently, under WTO agreement India has withdrawn several restrictions on import of agricultural/horticultural items resulting in easy availability of superior quality food items at a reasonable price. This will pose a major challenge to the Indian farming community. To counteract this threat it is absolutely essential for the Indian farmer to enhance productivity and cut down the cost of production without compromising on the quality. High quality standards would not only put them on an even platform with the imported products but will also enhance their prospects of exploiting the foreign market to export their produce for better price realization. Under this changing scenario of Indian horticulture some of the suggested strategies for improving production, productivity and quality of horticultural products in the country are as follows.

3.1   Use of Disease-Free Planting Material

The level of technological and extension support with regard to planting material that is provided by the government in agricultural sector is not available to farmers practicing horticulture. Consequently, even after more than 55 years of independence a majority of the farmers in the country use uncertified seeds or other forms of planting material to grow horticultural crops. This is on account of unawareness, lack of financial resources and even unavailability of authentic planting material in several crops. The problem of disease-infested planting material is even more pronounced in those crops where the propagules are regenerated vegetatively. In many crops such as potato (tubers), sugarcane (ratoons), strawberry (runners), banana (suckers) etc., the propagules derived from the previous crop are used to raise the new crop. Such propagules accumulate diseases on account of perpetual exposure to the field conditions leading to decline in yields. They also contribute towards spread of diseases in virgin areas.

Tissue culture offers rapid and reliable means of large-scale production of disease-free planting material. All the plants raised through tissue culture are free from most of bacterial and fungal diseases. One can also produce virus-free plants by meristem culture followed by micropropagation. Therefore, the use of tissue-cultured material in horticulture should be encouraged.

3.2   Use of High Yielding Superior Quality Planting Material

Ever since the beginning of human civilization, mankind has been selecting superior individuals and improving them further through crossing of different parents with desirable traits. Simultaneously, the technique of vegetative cloning of plants was also perfected. With the advancements in biotechnology, the pace of obtaining plants with desirable traits has gained further momentum because it is now possible to create new and unique genetic combinations that involve distinctly or even totally unrelated parents.

Cloning of elites can be done conventionally through cuttings, grafting or using other vegetative propagules such as roots, suckers, rhizomes etc. However, vegetative propagules are generally available in small numbers. They are often bulky and difficult to transport over long distances. In many plant species particularly in trees, large-scale propagation through cuttings is not possible because by the time the plant is evaluated for its productivity, the cuttings had already lost their capacity to root. Moreover, the vegetative propagules if infested with a disease would result in the spread of the disease even in the virgin areas. In contrast, through micropropagation, one can not only obtain disease-free plants, but all the tissue-cultured plants are clonaly uniform i.e., they are genetically alike and behave just like the mother plant. This way all the desirable traits of the mother plant can be passed on to the progenies unaltered. A desirable genotype that could not be multiplied because of virus infestation could now be freed of known viruses and then mass multiplied. Since the technology has the potential of producing millions of plants starting from a single shoot-tip, it becomes economically and technologically viable to free otherwise elite genotypes of horticultural species of the known viruses and then mass multiply them.

Considerable progress has been made within the country with regard to tissue culture of horticultural species and micropropagation protocols have been developed for several species. These can broadly be classified as follows.

3.2.1 Fruit Crops

Banana
This is an important fruit crop of India grown widely in almost all parts of the country, especially in Southern parts. The plant is propagated through suckers drawn from the previous crop. Consequently, with every passing generation, there is an accumulation of diseases leading to decline in vigour and yields. Also, there remains a major risk of the spread of the disease along with the propagule. Tissue cultured plants of banana have become extremely popular because of higher and consistent yield as compared to conventional propagules. The success met with tissue cultured plants of banana at the field level has generated lot of interest towards the technology among the farmers and they have become more receptive to the idea of trying other plants. There are several varieties of banana for which micropropagation protocols are now available, and depending upon the regional priorities and the end-use that the crop would be put to, suitable varieties are selected for cultivation. This includes Robusta, Dwarf Cavendish, Grandinaine, Williams, Elakki, Basari, Madukar etc. Today, several private companies in India like Khoday Biotek, A.V. Thomas, Cochin; Décor Plant Culture, Mumbai; Godrej Plant Biotech, Mumbai; Growmore Bio-tech (P) Ltd., Hosur; Harrison Malyalam Limited, Hosur etc. together are producing more than a million tissue cultured plants of banana for distribution among the local farmers. Although tissue culture of banana is a success story, there are reports to suggest that after certain passages (usually around 10) the cultures become more prone to somaclonal variations and therefore, as much as possible, subculturing beyond 10 passages should be avoided.

Strawberry
Conventionally, strawberry is propagated through runners. Plants raised through runners give proper yield only up to two generations. Beyond two generations the plant may look healthy but the yields are significantly lower. Moreover, continuous use of runners of the previous crop for the new one results in accumulation of various pathogens resulting in lower yields. Also, there is a potential risk of spread of diseases along with the propagules. In contrast, the plants raised through tissue culture are free of diseases. During the last few years several day-neutral varieties have been developed in USA. Importing material of such varieties on a large-scale will not only be expensive for the grower but would involve outflow of valuable foreign exchange. Also, it would be difficult for a small farmer to procure planting material from abroad. Using tissue culture one can bulk-up the planting material of these new varieties within the country and make the plant available to the growers at a much lower price. Considerable success has been made in this regard at TERI's tissue culture facilities and more than 3 lakh tissue cultured plants have so far been

dispatched to various growers. It is rather interesting that a species of hill is now being cultivated in plains. Overseas, the strawberry mother plants are produced through tissue culture and their subsequent multiplication is done through runners. However, it is important to note that unlike in India, these runners are produced under very hygienic conditions and therefore, give normal yields. Perhaps, it would be worthwhile to adopt similar kind of approach in India as well.

Banana and strawberries are the two fruit crops for which the tissue culture technology has been fully commercialized. In addition, there are few other fruit crops such as apple, pineapple, *Prunus*, raspberry, pomegranate and *Zizyphus* that are produced by various tissue culture companies operating in India on a small-scale. The demand for these species is restricted mainly on account of narrow geographical distribution. Success has also been achieved in regenerating plants of citrus, guava etc. by tissue culture inside the lab. However, these protocols either suffer from certain deficiencies because of which they could not be applied for commercial propagation or there are other technical problems that restrict their usage. For example, those species, which are multiplied through grafting, the self-rooted scions produced through tissue culture may not perform well in the field. In such cases grafting becomes inevitable and there is a possibility that one may not be able to multiply both the scion and the stock by tissue culture. Even where production of both scion and stock could be achieved, the grafts may not be very successful. It is only a matter of time, when such technical problems would be overcome and these species could be multiplied on a commercial scale. There are still few species such as mango and *Litchi* in which there is an urgent need to develop tissue culture procedures but no significant progress has been made so far.

3.2.2   Cash Crops
Potato and sugarcane are two major cash crops in which tissue culture technology has been applied quite extensively.

Potato
Potato is highest consumed single vegetable of the world and accounts for the largest area under cultivation. It serves as an important component of the Indian cuisine and also finds its way in processed food industry as chips and French fries. Conventionally, potato is propagated through tubers that tend to accumulate diseases with repeated cycles of propagation. This accumulation of diseases eventually brings about significant reduction in yields. Through tissue almost disease-free planting material (plantlets, microtubers and minitubers) could be produced on a large-scale as being done by several labs in the country. The secondary farmers, who are planting minitubers, are benefited through higher yields and better price realization for their produce. Potato chip is a major agri-industry, which is highly dependent on the quality of the potato tuber used in processing. Most of the potato varieties grown in India do not yield an even slice resulting in heavy wastage. Also they contain fairly high percentage of sugars that get oxidized and impart brown colour to the chips during baking process. This results in serious losses and low cost realization. Using tissue culture new and exotic varieties of potatoes that are primarily meant for making of potato chips have been successfully multiplied on a large scale within a short period of time. At present only 8% of the plantation is by certified seeds. Multiplication of new varieties (both suitable for processing and table varieties) will not only ensure better price realization to the farmers, but will also benefit the food processing industry.

Sugarcane

Sugarcane is an important cash crop and a major source of raw material for the sugar industry in India. Conventionally, sugarcane is propagated through ratoons and as in the case of potato, tends to accumulate lot of diseases over a period of time adversely affecting the production and the productivity. One of the commonest diseases in sugarcane is 'Red rot'. Till date there is no variety available which is totally resistant to 'Red rot'. Consequently, any variety of sugarcane has a very short life and has to be replaced with a new variety periodically. Since the conventional methods of cuttings are very slow, it takes several years before a newly released variety of sugarcane is available to the farmers on a large-scale. The plants raised through tissue culture are not only free of most bacterial and fungal diseases but also through this method of propagation, the newly released varieties could be made available to the growers within a short period of time. Presently, several varieties of sugarcane obtained from different sources are being multiplied on a large-scale at TERI's production facilities. The plants have survived very well in the field and their performance is very encouraging. The tissue-cultured plants produced more number of tillers as compared to their conventional counter parts. It is desirable to use the first generation tissue-cultured plants as seed stock rather than sending them to mills for recovery of sugar. The ratoon crop from tissue cultured plants give higher yields ($1^1/_2$ times higher than the conventional) and in few varieties up to 10% increase in sugar recovery (depending upon climate and management practices) has been observed. We may have surplus sugar even without the use of tissue cultured plants but in today's era of globalization, it is important to produce food including sugarcane at a competitive price. Also, the current emphasis on ethanol production from molasses for subsequent use in mixing with the petrol has further renewed interest in this crop.

3.2.3 Spices and Aromatic Plants

Micropropagation protocols are now available for several spices and aromatic plants that are found in India (Table 5). However, of all the spices, commercialization is largely confined to cardamom and black pepper and among aromatic plants it is mainly Vanilla and Patchouli. With increased awareness and development of efficient micropropagation protocols demand for other spices and aromatic plants is also catching up.

3.2.4 Medicinal Plants

During the recent past there has been a lot of interest developed towards the tissue culture of medicinal plants. This is mainly due to the fact that there has been a substantial increase in the demand of medicinal plants in the world market that provides ample opportunities to the Indian growers for good economic returns. However, Indian exports have suffered badly on account of inferior and inconsistent quality, and uncertain supplies. Tissue culture can provide solutions to many of these problems as superior quality plants containing high active principle can be produced in very large numbers. In anticipation of a surge in the demand from the growers, several tissue culture companies such as Nandan Agro Farms, Hyderabad; Labland Biotech, Mysore; Unicorn Natural Products (P) Ltd., Hyderabad; PCD Enterprises, Nainital; Growmore Biotech (P) Ltd., Hosur; Whitefield Agrotech, Bangalore; Cipla, Mumbai; Greenearth Biotechnologies Ltd., Bangalore etc. have undertaken mass multiplication of medicinal plants. This includes species such as *Chlorophytum borolivilianum, Withania somnifera, Phyllanthus, Aloe vera, Commiphora mukul, Gymnema sylvestris, Catharanthus roseus* etc. Since tissue culture of medicinal plants has been discussed in great detail in this volume, the same is not being repeated here.

Table 5.   Status of tissue culture work in various spices and aromatic plants in India

S. No.	Species	Explant	Mode of propagation	Remarks
1.	*Piper nigrum* (Black pepper)	Shoot tip and nodal segments	Axillary	Tissue cultured plants were successfully transferred to the field; reports suggest early bearing in some tissue cultured plants
2.	*Piper longum* (Indian long pepper)	Shoot tip and nodal segments	Axillary	Tissue cultured plants transferred to the field flowered early and had more number of axillary shoots
3.	*Piper chaba* (Java long pepper)	Shoot tip and nodal segments	Axillary	Two year old tissue cultured plants transferred to the field flowered early and had more number of axillary shoots
4.	*Piper betle* (Betelvine)	Shoot tip and nodal segments	Axillary	Transplantation success was 80%
5.	*Piper colubrinum*	Shoot tip and nodal segments	Axillary	Field survival was 85%; TC plants flowered earlier than the control
6.	*Piper barberi*	Shoot tip and nodal segments	Axillary	Micropropagation protocol can be used for rapid multiplication of this endangered species
7.	*Elettaria cardamomum* (Cardamom)	Rhizome bits with vegetative buds	Axillary	Technology for micropropagation has been fully commercialized. Up to 40% increase in yield has been reported in the TC plants over the control
8.	*Amomum subulatum* (Large cardamom)	Rhizome bits with vegetative buds	Axillary	The rooted plantlets could be separated and transferred to the soil with 80% success
9.	*Zingiber officinale* (Ginger)	Vegetative buds, immature inflorescences and rhizome bits	Axillary	Field evaluation suggest that it takes three crop seasons for the micropropagated plants to develop rhizomes of normal size, hence they can not be used directly for commercial planting
10.	*Curcuma amada* (Mango ginger)	Vegetative buds, rhizome explants with buds	Axillary	It takes three crop seasons for the micropropagated plants to develop rhizomes of normal size, hence they can not be used directly for commercial planting
11.	*Curcuma longa* (Turmeric)	Vegetative buds and rhizomes	Axillary	Being small, the micropropagated plants can not be used directly for commercial planting
12.	*Curcuma aromatica* (Kasturi turmeric)	Vegetative buds and rhizome bits	Axillary	Transplantation success was over 80%; the rhizomes produced by micropropagated plants are small and not suited for commercial planting

13.	*Kaempferia galanga* (Galangal)	Vegetative buds and rhizome bits	Axillary	As above
14.	*Kaempferia rotunda*	Vegetative buds and rhizome bits	Axillary	As above
15.	*Vanilla fragrans* (Vanilla)	Seeds, shoot tip and nodal segments	Axillary	Process of micropropagation of vanilla has been fully commercialized and there are several companies that are engaged in mass multiplication of this species
16.	*Cinnamomum zeylanicum*	Shoot tip and nodal segments	Axillary	TC plants established in the field with over 90% success. Plants grew to a height of 1-2 ft within one year of planting
17.	*Cinnamomum camphora* (Camphor)	Shoot tip	Axillary	TC plants grew up to a height of 8 ft within one year of field planting; there was lack of expansion of leaves in some cases
18.	*Cinnamomum cassia* (Chinese cassia)	Shoot tip and nodal segments	Axillary	*In vitro* plants hardened up to the nursery stage; field transfer not reported
19.	*Thymus vulgaris* (Thyme)	Seedlings and shoot tips	Axillary	Transplantation success inside the greenhouse was only 60%
20.	*Mentha piperita* (Peppermint)	Shoot tips and nodal segments	Axillary	Hardening survival at the greenhouse stage was 60%
21.	*Mentha spicata* (Spearmint)	Seedlings and shoot tips	Axillary	As above
22.	*Marjorana hortensis* (Marjoram)	Seedlings and shoot segments	Axillary	Field transfer of *in vitro* plants not reported
23.	*Origanum vulgare* (Oregano)	Seeds and shoot segments	Axillary	As above
24.	*Salvia officinalis* (Sage)	Seedlings and shoot tips	Axillary	As above
25.	*Lavendula angustifolia* (Lavender)	Seedlings and shoot tips	Axillary	As above
26.	*Ocimum sanctum* (Sacred basil)	Seedlings and nodal explants	Axillary	As above
27.	*Petroselinum crispum* (Parsley)	Seedlings and nodal explants	Axillary	Greenhouse survival was only 40%
28.	*Apium graveolens* (Celery)	Seedlings and stem cuttings	Axillary	Greenhouse survival was 50%

(Contd)

Table 5.   *(Contd)*

S. No.	Species	Explant	Mode of propagation	Remarks
29.	*Pimpinella anisum* (Anise)	Seedlings and stem cuttings	Axillary	Greenhouse survival was only 60%
30.	*Anethum graveolens* (Dill)	Seedlings and stem cuttings	Adventitious shoots	60% plants survived at the greenhouse stage; field transfer not reported
31.	*Foeniculum vulgare* (Fennel)	Seedlings, shoots and nodal segments	Adventitious shoots	60% plants survived at the greenhouse stage; field transfer not reported
32.	*Bunium persicum* (Kala zira)	Petiolar segments	Somatic embryogenesis	Few plants were transferred to the pots; plants obtained by germinating small tubers do not establish in soil
33.	*Crocus sativus* (Saffron)	Shoot meristem with a pair of leaf primordium	Somatic embryogenesis	50% of the *in vitro* formed corms germinated upon transfer to the field
34.	*Syzygium aromaticum* (Clove)	Shoot-tips and axillary buds	Axillary	Field transplantation not achieved

Sources : IISR, Calicut, 1997 and DBT, 2000.

It is rather sad that it is only those growers who are export oriented are going for tissue cultured plants while those who cater to the domestic demand continue to rely on harvesting from natural forest or adopting conventional means of propagation. However, the growers are not entirely to be blamed for this situation because all said and done, the farmers are looking for higher economic returns and the cost of tissue cultured plants is certainly higher than the conventional propagules. While in exports, the impact of higher cost of the tissue-cultured plants is more than neutralized through higher returns, the domestic market for the medicinal plants continues to be highly disorganized with growers not getting any significantly higher returns on account of the superior quality of their produce. Unless and until the domestic market matures and the farmer is paid not only for the quantity but also for the quality, the present trend of harvesting from natural forest will continue. It is heartening that over the last few years there has been some positive change in this direction and many pharmaceutical companies involved in plants and plant-based products are willing to pay a higher price for a better quality raw material. As a result even the domestic suppliers are now exploring the possibilities of cultivating tissue-cultured plants. However, it is only a beginning and much more is desired. This includes: (a) development of efficient micropropagation protocols so that the cost of the plantlet is really low, because we must realize that when the supplier is simply collecting the plant from the natural forest he is not incurring any cost towards the planting material. Even if he undertakes cultivation, the conventional propagules are much cheaper than the tissue cultured plants, (b) in several cases, although the plants have been raised through tissue culture and are therefore, free of most diseases, no proper studies have been made to estimate the actual active principle in the tissue cultured plants. In such a situation it becomes difficult for the grower to convince his buyer about the superiority of the crop. Therefore, while propagating plants through tissue culture, active principle estimation is must, (c) it is known that the amount of active principle produced by a plant is affected not only by the genotype but also the environment in which it has been grown. Therefore, it is imperative that besides emphasizing on a good genotype, equal attention is paid to work out the most suitable climatic conditions for growing that clone so that the end product is of really high quality, and (d) initiatives are required from the government to organize the market of medicinal plants as much as possible so that the growers are not left to the mercy of the unscrupulous traders who do not pay the growers their due. The multiplication of elite planting material and growing them under properly managed fields will ensure uniform quality of the extract. This will help the pharmaceutical industry in India as well as contribute in capturing international market.

3.2.5 *Ornamentals*

So far as tissue culture of ornamentals is concerned lot of work has been done in the country and almost every tissue culture company is working on some or the other ornamental species. Some of the major species that are being produced commercially include *Ficus* spp., *Syngonium, Spathiphyllum, Dieffenbachia, Philodendron, Cordyline, Calathea,* Orchids, *Gerbera, Zantedeshia, Anthuriums,* etc. There are many reviews appearing in recent past that provide a detailed account of the mode of propagation and the success met with various ornamentals. Therefore, this aspect of tissue culture of ornamentals is being deliberately omitted here. On the whole, ornamentals are the easiest species to deal with in tissue culture. Most of those species, which in general, are not propagated through tissue culture is not on account of non-availability of micropropagation protocols but because there is either no requirement or the tissue-cultured plants are significantly more expensive than the conventional propagules. Somehow the growth of tissue cultured ornamentals has been much slower as compared to the international market. This is because of several reasons as follows:

Grower's Account

- Lack or incomplete scientific knowledge amongst growers is one of the major factors in the popularity of tissue cultured plants including ornamentals.
- Reluctance on part of growers to accept change and adopt new technology.
- Many a times, the traders for business reasons or out of ignorance tend to oversell tissue culture technology raising very high expectations among the minds of the growers. If these expectations fall short, then it adversely affects the demand of the tissue cultured plants.

Seller's Account

- Tissue culture plants are sold through retailers who are more concerned about their profits rather than the quality they are selling. Hence, if a trader makes more profit by selling conventional plants then he will promote only those rather than tissue-cultured plants.
- Although tissue cultured plants can be produced inside the lab round the year, however, they can be taken to the open nursery only in a particular season (this problem is more

pronounced in those regions where there is extremity of both summer and winter, e.g. most of north India). This way the tissue-cultured plants are available to the client only in a specific season. In contrast, the traders are able to bring conventionally raised ready-to-plant plants from other locations at low rates and sell them at a premium during off-season thereby affecting the demand of the tissue-cultured plants in the proper season.

Buyer's Account
- In domestic market, except for few metropolitan towns where the per capita income is high and people are more quality conscious, in other areas the demand is influenced more by the price rather than the quality; since conventionally raised plants are cheaper, they sell more.
- Purchase of ornamental plants is usually not done on the basis of the regeneration process but mainly on looks (a customer prefers to buy which is aesthetically appealing without caring whether the plant has been raised conventionally or through tissue culture).

Technical/Other Reasons
- *Field failures*: There are certain species such as Dahlia, Chrysanthemum etc. in which the tissue cultured plants are very weak, lanky and produce tiny flowers (much smaller than the normal ones). In contrast the cutting-raised plants are healthy and produce normal sized flowers. It is, therefore, necessary to combine both the micropropagation and macropropagation techniques wherein few disease-free mother plants are produced by tissue culture. They are transferred to the field and subsequently, cuttings are derived from them.
- *Imbalances of capacities in production and consumption areas*: To keep the hardening cost as low as possible, most of the tissue culture companies are concentrated in those areas where the climate is relatively moderate such as Bangalore, Pune etc. If the plants produced in this part of the country are to be sold in north India or far off places then the transportation cost becomes prohibitory thereby affecting the market of tissue cultured plants.
- There are certain plants for which there is a genuine demand for tissue cultured material but the micropropagation protocols are not available.

As against the domestic market, tissue culture raised ornamentals find a better international market. In fact, most of the tissue-cultured plants that are exported from India are ornamentals. However, India has failed to make any dent in the international market and its share in the world trade of tissue cultured plants is insignificant. Some of the factors responsible for dismal performance in foreign markets are:

- Non-adherence of delivery schedules by the producer.
- Supply of plants of inferior/inconsistent quality; also, there in no effective mechanism from the government to prevent such dispatch from the country. In fact this factor has contributed most in tarnishing the image of the country in foreign market.
- Inability of the Indian companies to produce plants as per the specifications of the client
- International competition.
- Lack of cooperation and coordination among the Indian companies resulting in under-cutting of prices and thereby lower price realization.

High yielding planting material can also be obtained through breeding and genetic engineering. So far several new varieties have been developed through breeding that are being grown commercially. The only drawback with hybrid seeds is that they are relatively expensive and can be used for raising only a single crop, that is, for every crop new seeds have to be purchased. Recently, genetic engineering technology has gained lot of grounds and several improved varieties carrying several agronomically important traits such as higher yield, disease and insect resistance, drought tolerance, herbicide resistance, enhanced nutritional status etc. have been developed. The first and the only transgenic crop for which the Government of India has granted permission for commercial cultivation is Bt cotton. Although in India research is underway to develop transgenic plants of several other economically important plant species, it will take some more time before these genetically modified plants become available to the Indian farmers for commercial use. Even after the availability of transgenic plants, it will take some more time to popularize them among the masses. This will be more so in food crops for human consumption.

3.3 Use of High-Tech Agro-Techniques

Inadequate technological upgradation by the farmers has been one of the major reasons of low productivity. During the past few decades several technological advancements have taken place in the production technology which requires greater commercialization and professionalism for production of fruits. Some of these include high density planting, drip irrigation, protected cultivation, biofertilizers, use of new and high-yielding varieties etc.

Introduction of high density planting is one of the major advances in the field of fruit production. High-density plantation is a worldwide phenomenon that has been successfully adopted in apple, banana, peach, plum, pear, pineapple and papaya. In India too, substantial increase in yield on account of high density planting have been reported in banana, mango, papaya and guava. In pineapple, a plant density of 63,758 plants per hectare coupled with improved management practices increased the yields from 15-20 tonnes per hectare to 70-80 tonnes per hectare. However, high density planting in India is largely restricted to few demonstration plots only with most of the orchards being still under the traditional low-density system. Much needs to be done in this regard.

Use of drip irrigation not only results in higher productivity (10–50% over conventional methods) but also saves 50–70% water. In addition, using drip, fertilizers, pesticides and other soluble chemicals can be applied along with irrigation water leading to their efficient use, reduced incidence of diseases, less weed growth, better quality products, and low labour and operational costs. Today, 260,000 ha of area in the country is under drip irrigation using which higher yields have been obtained in grapes, banana, mango, guava, pomegranate, sapota, cabbage, coconut, arecanut, roses etc. Although drip irrigation is gaining popularity in the country there is still a long way to go. The major constraints in popularization of this system are high cost of the equipment and maintenance as the equipment is highly sensitive to clogging with the existing water quality and finally, the water-soluble fertilizers are far too expensive and available only at selected outlets. Besides bringing about technological improvement, government initiatives in terms of subsidies are also desired.

The advent of protected cultivation in microclimate regulated/modified greenhouses result in production of high quality vegetables, flowers and other ornamentals. As compared to western world, the concept of protected cultivation in India is rather recent and is largely confined to cultivation of ornamentals. In spite of its late entry, Indian industry has made rapid strides and nearly 250–300 ha of land is now under protected cultivation of floricultural crops. Nearly ninety percent of the area is under roses and the remaining is shared by other ornamentals. High initial and operational costs have deterred the Indian farming community to adopt this system.

Many soil microorganisms enhance nutrient uptake in plants. Those which have a direct beneficial effect on the plants, may have considerable potential as biofertilizers. Two main groups of plant-beneficial micro-organisms are: (i) nitrogen fixing micro-organisms such as blue-green algae (cyanobacteria), soil bacteria of the genera *Azotobacter, Klebsiella, Bradyrhizobium, Rhizobium* and Actinomycetes and (ii) mycorrhizal fungi. Mycorrhiza is the form of a symbiotic relationship between certain fungi, particularly vesicular-arbuscular (VA) mycorrhizae and the roots of vascular plants. Unlike rhizobial associations, VA mycorrhizal fungi are non-specific and can affect a wide range of host plants. In certain circumstances, mycorrhizal infection can significantly increase the rate of uptake of nutrients, particularly phosphorus and nitrogen from deficient soils. In addition, they can mobilize other trace elements such as copper, zinc and iron. Besides enhancing the productivity, mycorrhizae are also very effective in reclamation of wastelands.

Biofertilizers can be an effective substitute for chemical fertilizers. They are not only environmentally benign but also the crops grown without the use of chemical fertilizers command a better price realization especially in the international market. Use of biofertilizers in horticultural crops has not gained much popularity in India as yet. This is largely due to lack of awareness, non-availability and technical constraints associated in the use of biofertilizers. Relatively low cost of chemical fertilizers on account of government subsidies has also deterred the farmers to switch over from chemical fertilizers to biofertilizers on a large-scale.

3.4   Application of Frontier Technologies

As described earlier, breeding, micropropagation, biofertilizers, genetic engineering and other molecular techniques, either singly, or in combination can play a significant role in augmenting horticultural production in the country. In addition to quantitative gains these frontier technologies can also bring about marked improvement in the quality of the horticultural products. Except for genetic engineering where there are still some perceived technological problems that needed to be resolved before the GM crops are cleared for mass consumption in India, all other technologies are well proven and should therefore be promoted for adoption by various end users. More than the cost, it is the lack of awareness that is hindering the wider use of these technologies.

3.5   Post-harvest Management

Although there is no consensus with regard to exact quantum of post-harvest losses of fruits and vegetables in India, however, taking into account the estimates made by various agencies this figure could be anywhere between 25 and 30%. In terms of value, the estimated loss could be over Rs. 23,000 crores per annum. These post-harvest losses in supply chain of horticulture produce are attributed to: (a) mishandling of produce, (b) improper and inadequate facilities for storage, and (c) improper packaging and transportation. If these losses can be minimized, the additional horticultural produce available will help in achieving per capita increase in consumption of fruits and vegetables. Surpluses, if any, may be diverted to the food processing industry in the country. Post-harvest losses is one important aspect of food production that has not received the attention it deserves and still much is desired to be done.

The areas that need to be targeted to reduce post-harvest losses can be broadly classified into following three categories:

- *Storage losses:* The first step towards cutting down on the storage losses is to minimize the damage to the produce during harvest. After harvesting, the fruits and the vegetables should be washed and treated to eliminate or minimize bacterial and fungal infection and attack by insects and other pests. Then only the properly packed produce should be sent for refrigeration or cold storage. The present cold storage capacity in India is about 87 lakh tonnes out of which nearly 80-90% is utilized for storage of potato and potato seed. It is imperative on part of the Central and the State Governments to develop adequate infrastructure for proper storing of fruits, vegetables and flowers at different temperatures depending upon the requirement of the species.

- *Transportation:* Transportation is the link between the farmer and the consumer. The transportation of perishable horticultural produce to the consumer requires the selection of the fastest and most efficient mode of transport to deliver the consignment in the best possible condition at the lowest possible cost. In this regard, infrastructure for both rail and road transport, the mainstay of Indian transport system will have to be strengthened. Packaging, loading and unloading, and containerization are some of the other related aspects of transportation that are as important as the transportation process *per se* but are often neglected. Uniformity in the size of the containers/crates used for different purposes, will not only lead to effective utilization of space during transportation, but will also reduce the time spent in loading and unloading operations.

- *Processing:* Although not an integral part of post-harvest management, facilities for food processing will help in value addition of the product resulting in better cost realization. At present, we do not even cultivate varieties that are better suited for processing purposes. Also, it is usually the substandard produce or surplus, which is diverted for processing.

3.6 Horticulture Informatics

To develop plans and strategies for any developmental activity, a comprehensive database is a must. In India, as compared to agriculture, horticulture sector is much less organized. Consequently, with regard to statistics on area, production and productivity, information on very few horticultural crops is available. In floriculture crops, the situation is even worse as statistics on production-related aspects is either missing or there is huge variation in the data provided by various sources. It is, therefore, desirable to create authentic databases on various aspects of horticulture including area, varieties, total production, yield, post-harvest losses, export, processing, price realization etc. for all the horticultural crops. Some projections on demand of various crops in future should also be made available to the grower so as to enable them to plan their strategy for better price realization. Other databases providing information on soil type, climate, rainfall, pest attack and marketing will also be very useful.

3.7 Marketing

Marketing is an integral component of any production process. However, so far as horticultural crops are concerned very little effort has been made by the government in this regard. Unlike for agricultural crops there are no governmental agencies such as FCI that makes direct purchase of horticultural crops from the farmers. Also, there is no minimum support price offered by the government for the horticultural crops. Apart from local 'Mandis' (wholesale markets) there are no other places where the farmer could go to sell his produce. Very often the poor farmers do not get the real value of their produce and suffer heavily on account of middlemen who control all the business transactions in these 'Mandis'. There are many exotic vegetables such as asparagus, broccoli, celery, Chinese cabbage, kale, leek lettuce, parsley, etc. that can fetch very good price to the farmers. However, the farmers hesitate to grow such crops as their market is highly restricted and not so easily accessible to them. Similarly, on the export front, there are very few international airports in India where facilities for cold storage are available. In addition, there is hardly any information/statistics available on the world markets, price trends, marketing agencies, etc. Much needs to be done for easy and organized marketing of horticultural crops in India.

3.8 Credit Facilities

Credit availability at reasonable interest rate is absolutely essential for the small and marginal farmers to enable them to adopt modern technology and improved horticultural practices. Access to credit facilities is also very critical for sustaining competitiveness. Agricultural/horticultural credit to the needy growers has increased over the years but how far the credit that is made available to the grower is put to use is a point to ponder. There is a serious problem of overdues that has been inhibiting credit expansion and economic viability of the lending institutions. Loan waivers by State Governments for making political gains have caused severe problems of recovery.

3.9 Research and Extension

With the objective of increasing production, new high-yielding varieties with other desirable traits such as disease resistance, better flavour etc. must be developed on a continuous basis. Wherever possible, the conventional breeding must be linked or supplemented with latest molecular techniques to accrue higher gains. GM technology holds lot of potential that must be tapped.

There are several exotic vegetable and fruit crops that could be sold at a premium both in the domestic as well as international markets. In spite of the fact that many of these crops can possibly be grown in India, their cultivation has not been undertaken on the desired scale because either the agroclimatic conditions suitable for their growth are not known or the cultural practices required for raising crop have not been established. In many cases it is mere lack of awareness about the new introductions in the country that has deprived the Indian farmer from growing them. It is, therefore, important to strengthen our extension network in horticulture as we have done in agriculture.

4. Conclusions

India has made significant achievements in the field of horticulture in terms of overall production, however, there is still a lot that remains to be done with regard to productivity and quality. There are several impediments in our way to progress and accordingly a multi-pronged approach is desired to deal with the problem. This is just not a requirement but also a necessity because with the opening of Indian market there would be a large-scale invasion of foreign horticultural products in the country that could seriously effect the very existence of the Indian farmer. However, instead of a threat the integration of the Indian market with the world markets should be taken as an opportunity that every Indian farmer must try to seize effectively. This would be possible only through concerted and integrated efforts on part of the policy makers, researchers, administrators and the growers.

References

1. CMIE (2001) Agriculture. Centre for Monitoring Indian Economy Pvt. Ltd., Mumbai, India (322 p).
2. DBT (2000) Plant Tissue Culture from Research to Commercialization—A Decade of Support. Department of Biotechnology, Ministry of Science & Technology, Govt. of India (224 p).
3. FAO (2002) http://apps.fao.org/.
4. IISR (1997) Protocols for micropropagation of spices and aromatic crops. Nirmal Babu K., Ravindran P.N. and Peter K.V. (Eds). Indian Institute of Spices Research, Calicut, Kerala (35p).

Plant Biotechnology and Molecular Markers
P.S. Srivastava, Alka Narula and Sheela Srivastava (Editors)

19. Cryopreservation: A Potential Tool for Long-term Conservation of Medicinal Plants

Sonali Dixit, Sangeeta Ahuja[1], Alka Narula[2] and P.S. Srivastava[2]

Amity Institute of Biotechnology, Amity Campus, Sector 44, Noida 201303, India

[1]Department of Biochemistry and Molecular Genetics, University of Virginia, Charlottesville, Virginia-22903, USA

[2]Centre for Biotechnology, Jamia Hamdard, Hamdard Nagar, New Delhi 110062, India

Abstract: Medicinal plants are one of the most important groups of plant genetic resources. Their use in biotechnology has assumed considerable significance because of overexploitation of these plants to meet the increasing demand. As cells cultured *in vitro* are prone to spontaneous changes, continuous culture of plant cells is often undesirable. Cryopreservation is a safe and cost-effective technique for preservation of germplasm and management of *in vitro* produced materials for biotechnological applications. The present article is a brief account of cryopreservation techniques and their application for medicinal plant conservation.

1. Introduction

Traditional medicinal systems are part of a time-honoured and time-tested culture, that still intrigues people today. A culture that has successfully used plants to treat primary and complex ailments for over 3,000 years obviously has a contemporary relevance. In an age when toxic drugs are increasingly unwelcome and when people are using viable alternatives, this heritage of medicinal plants must be documented and conserved for effective use in future.

During the past decade, a dramatic increase in exports of medicinal plants attests to worldwide interest in these products. Nevertheless, most of these plants being taken from the wild, hundreds of species are now threatened with extinction because of overharvesting, destructive collection techniques, and conversion of habitats to crop-based agriculture.

Preservation of these genetic resources is currently at the forefront of conservation activities and biotechnology has played an important role in international conservation programs [1]. Traditionally, plant genetic resource management involves conserving germplasm as seeds at low temperature, or as field plantings (field genebanks) for vegetatively propagated plant species. These approaches are now complemented by *in vitro* conservation methods that can be used in combination with traditional practices and offer added security for field genebank conservation [2]. The ideal genetic resource conservation program consists of active collections that are available for distribution or characterization and base collections held for the sole purpose of long-term preservation. Base collections of vegetatively propagated plants are more difficult to achieve and recently, cryopreservation has been identified as the best option for long-term conservation of germplasm of these species [3, 4, 5]. Cryopreservation, i.e., non-lethal storage of plant tissues at ultra-low temperature usually that of liquid nitrogen (–196°C) is the only available method for the long-term conservation of germplasm of these problem species. Cryopreservation has manifold applications in conservation and biotechnology. A number of medicinal plant species have been subjected to cryopreservation (Table 1). The major advantage of storage of biological material at such a low temperatue is that both metabolic processes and biological deterioration are considerably slowed or even halted [6, 7]. Additionally, continued maintenance of plants in tissue culture can lead to loss of morphogenic, genetic and biosynthetic capacity, which may confound successful exploitation [8]. It scores advantages over other conservation strategies as it minimizes the risk of contamination, cost of maintenance and cost of labor [9]. Through cryopreservation, it may be possible to establish a reserve of freshly initiated competent cultures which after thawing and recovery can be reintroduced into culture.

2. International and National Programmes for Cryopreservation of Medicinal Plants

Interests and concern of the international scientific community in this area has lead to formulations of several national and international level programmes, which are devoted to cryopreservation of medicinal plants. For example, G-15 Genebanks for medicinal and aromatic plants were initiated from the Summit Level Meeting of Group on South-South Consulting and Cooperation of the G-15 countries held in Kuala Lumpur (January 1990). Malaysia together with Indonesia and India represent the Asian region where India is a Regional Coordinator. India has also been given the overall responsibility for coordinating the activities of the G-15 nations for the establishment of gene banks for medicinal and aromatic plants. Under the aegis of this programme, Department of Biotechnology, Government of India, has constituted a network of three national gene banks at Tropical Botanical Garden and Research Institute (TBGRI), Thiruvananthapuram; Central Institute of Medicinal and Aromatic Plants (CIMAP), Lucknow; and National Bureau of Plant Genetic Resources (NBPGR), New Delhi. One of the important mandates of this group is to develop cryopreservation protocols for long-term conservation of medicinal plants.

3. Cryopreservation Techniques

Some plant organs such as orthodox seeds and frost-hardy dormant buds contain very low amounts of water and can thus be cryopreserved directly, without any pretreatment. However, most of the experimental systems employed in cryopreservation (cell suspensions, calli, shoot tips, embryos) contain high amounts of cellular water and are thus extremely sensitive to freezing injury since most of them are not inherently freezing-tolerant. Cells have thus to be dehydrated artificially to protect them from the damages caused by the crystallization of intracellular water into ice. The techniques employed and the physical mechanisms upon which they are based are different in classical and new cryopreservation techniques [10]. Classical techniques involve freeze-induced dehydration, whereas new techniques are based on vitrification, i.e. the transition of water directly from the liquid phase into an amorphous phase or glass, whilst avoiding the formation of crystalline ice.

Classical cryopreservation techniques involve slow cooling down to a defined prefreezing temperature followed by rapid immersion in liquid nitrogen. They are generally operationally complex since they require the use of sophisticated and expensive programmable freezers. In some cases, their use can be avoided by performing the freezing step with a domestic or laboratory freezer [11].

Table 1.   Summary of different techniques used for cryopreservation

Technique	Explants	Protocol	Reference
1. Vitrification	Shoot tips/embryogenic tissues/cell cultures	Explant is treated with LS (1 M glycerol) for 20 min at 25°C followed by dehydration with PVS$_2$ (30% glycerol, 15% EG, 15% DMSO) at 0°C for 90 min, rapid freezing in LN, rapid thawing at 40°C for 1–2 min, UL (1.2 M sucrose) and culture for recovery growth	[27]
2. Encapsulation dehydration	Shoot tips/embryogenic tissues	Explant is encapsulated in calcium alginate and precultured in high sucrose solution (0.5–0.75 M), followed by dehydration in laminar airflow for 4–5 h, rapid freezing in liquid nitrogen, rapid thawing at 40°C for 1–2 min and culture for recovery growth	[28]
3. Encapsulation-vitrification	Shoot tips	Excised encapsulated meristems containing 2 M glycerol +0.4 M sucrose were dehydrated with PVS$_2$ for 2h at 0°C and subsequently plunged in LN	[29]
4. Pregrowth	Zygotic and somatic embryos	Pre-growth technique consists of cultivating samples in the presence of cryoprotectants, then freezing them rapidly by direct immersion in liquid nitrogen	[30]
5. Pregrowth desiccation	Stem segments	Pre-growth desiccation refers to the pre-culture of the explant on a medium with high concentration of sucrose or ABA or Proline and desiccation/drying followed by freezing in liquid nitrogen	[31]
6. Desiccation	Large number of recalcitant and intermediate seeds	Desiccation is usually performed in the air current of a laminar flow cabinet, but more precise and reproducible dehydration conditions are achieved by using a flow of sterile compressed air or silica gel	[10]
7. Droplet freezing	Shoot tips	Apices are pretreated with liquid cryoprotectant in medium then placed on aluminum foil in minute droplets of cryoprotectant and frozen directly by rapid immersion in LN	[32]

In the new vitrification-based procedures, cell dehydration is performed prior to freezing by exposure of samples to concentrated cryoprotective media and/or air desiccation. This is followed by rapid cooling. As a result, all factors, which affect intracellular ice formation, are avoided. Glass transitions (changes in the structural conformation of the glass) during cooling and rewarming have been recorded with various materials using thermal analysis. Vitrification-based procedures offer practical advantages in comparison to classical freezing techniques. Like ultrarapid freezing (above), they are more appropriate for complex organs (shoot tips, embryos) which contain a variety of cell types, each with unique requirements under conditions of freeze-induced dehydration. By precluding ice formation in the system, vitrification-based procedures are operationally less complex than classical ones (e.g., they do not require the use of controlled freezers) and have greater potential for broad applicability, requiring only minor modifications for different cell types [10]. A common feature to all these new protocols is that the critical step to achieve survival is the dehydration step, and not the freezing step, as in classical protocols. Seven different vitrification-based procedures can be identified: (1) encapsulation-dehydration; (2) a procedure actually termed vitrification; (3) encapsulation-vitrification; (4) desiccation; (5) pregrowth; (6) pregrowth-desiccation; and (7) droplet freezing (Tables 2 and 3).

Table 2. Advantages and disadvantages of commonly used cryopreservation techniques

Technique	Advantages	Disadvantages
Slow freezing	Stability from relatively nontoxic cryoprotectants	Requires expensive equipment, slow recovery, low applicability to tropical species
Vitrification	No special equipment needed, fast procedure, fast recovery	Vitrification solutions are toxic to many plants, cracking is possible, requires careful timing of solution changes
Encapsulation-dehydration	No special equipment needed, non toxic cryoprotectants, simple thawing procedures	Requires handling each bead several times, some plants do not tolerate high sucrose concentrations
Dormant bud desiccation	Easy, useful for many temperate tree species	Requires freezing equipment, larger storage space, recovery requires grafting or budding, works best in cold temperate regions

4. Some Important Case Studies

4.1 Cryopreservation of Shoot Tips of *Dioscorea* spp.

In vitro grown shoot tips of two medicinally important species of *Dioscorea*, *D. floribunda* and *D. deltoidea*, were successfully cryopreserved using vitrification and encapsulation-dehydration techniques. For vitrification the excised shoot tips were precultured for 16h on MS medium containing 0.3M sucrose followed by loading for 20 min at 25°C (2 M glycerol + 0.4 M sucrose), dehydration with plant vitrification solution (PVS$_2$) (30% glycerol, 15% Ethylene Glycol (EG), 15% DMSO and 0.4 M sucrose) for 90 minutes at 0°C prior to plunging in liquid nitrogen (LN). After storage in LN for atleast 1 h the shoot tips were unloaded for 20 min at 25°C (1.2 M sucrose) and transferred to medium for recovery growth. During recovery growth, apices of *D. floribunda* and *D. deltoidea* regenerated directly with a frequency of 30 and 75%, respectively (Fig. 1 A, C, D).

Table 3.   Summary of cryopreservation studies on some important medicinal plant species

Plant	Explant	Reference
Atropa belladonna	Protoplasts, cells	[33]
Anisodus acuntangulus	Suspension cultures, cells	[34]
Catharanthus roseus	Cells	[25]
Coleus blumei	Cells	[25]
Chicory	Shoot tips	[35]
Cinchona ledgeriana	Protoplasts	[36]
Datura innoxia	Protoplasts	[37, 38]
D. stramonium	Cell suspension	[38]
Dioscorea caucasia	Organogenic callus	[39]
D. balanica	Organogenic callus	[39]
D. bulbifera	Somatic embryos	[13]
D. floribunda	Shoot tips	[40, 41]
D. deltoidea	Cell cultures	[42]
	Shoot tips	[13]
Digitalis lanata	Cell cultures	[43]
D. thapsi	Cell cultures	[17]
Eucalyptus	Leaf	[44]
Gentiana scabra	Axillary buds	[45]
Holostemma annulare	Shoot tips	[46]
Nicotiana tabacum	Suspension cultures	[47]
N. sylvestris	Suspension cultures	[48]
N. plumbaginifolia	Suspension Cultures	[48]
Olea europe	Shoot tips	[49]
Panax ginseng	Hairy roots	[15]
	Cells	
P. quinquefolium	Cell cultures	[50]
Papaver somniferum	Transformed Cells	[51]
Polygonum avuculare	Suspension cells	[52]
Trifolium repens	Shoot tips	[53]

For encapsulation-dehydration the shoot tips pregrown in 0.3 M sucrose were encapsulated in calcium alginate followed by preculture in 0.75 M sucrose, dehydration for 51/2 and 5 h respectively, rapid freezing and rapid thawing. The encapsulated shoot tips were recovered with high frequency direct regeneration in both the species (Fig. 1B) [12, 13]. Interestingly, the diosgenin content in the plants recovered after cryopreservation was found to be stable using HPLC analysis. Molecular studies using RAPD analysis proved that the plants were genetically stable [14].

4.2   Cryopreservation of Somatic Embryos of *Dioscorea bulbifera*

Somatic embryos/embryogenic tissues of *Dioscorea bulbifera* were cryopreserved using encapsulation-dehydration technique. The embryogenic tissues of about 1–2 mm in diameter, with a group of embryoids were encapsulated into calcium alginate beads. These were then precultured in 0.5 M sucrose for 7 d followed by dehydration under laminar airflow for 4 h. High frequency (75%) of embryogenic survival was recorded after storage in LN (Fig. 2 A-D). The plants hence produced and transferred to field have been found to be morphologically similar to the non-treated controls. The diosgenin content in the plants recovered after cryopreservation

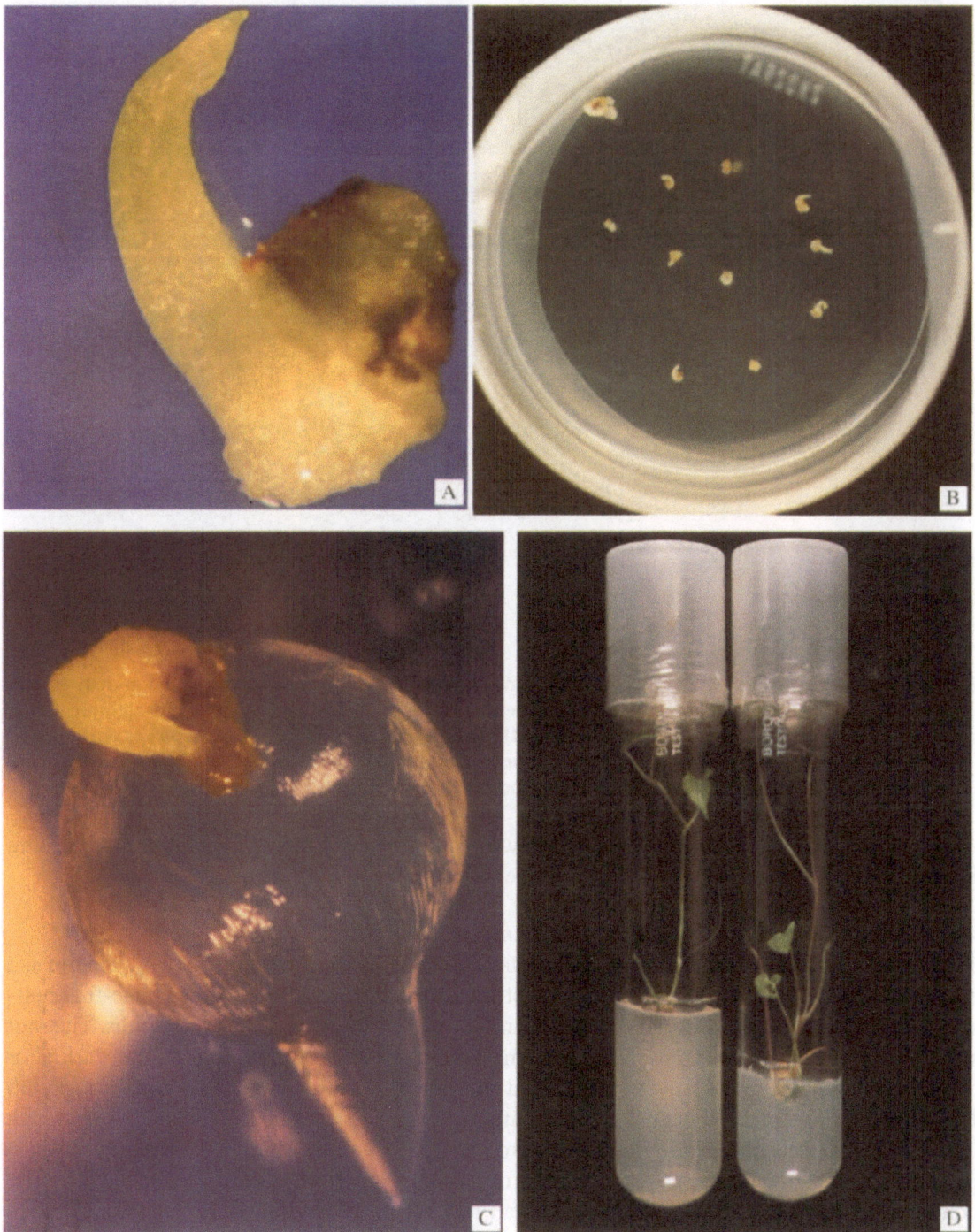

Fig. 1. Recovery growth of shoot tips of *D. deltoidea* after cryopreservation. (A) Close-up cryopreserved shoot tip showing growth without any intermediary callus phase; (B) High frequency direct regeneration from cryopreserved shoot tips; (C) Close-up of shoot tip showing recovery growth cryopreserved using encapsulation-dehydration and (D) Well developed shoots.

Fig. 2.   Recovery growth of encapsulated embryogenic tissue of *D. bulbifera* using encapsulation-dehydration technique. **(A)** Development of somatic embryos directly emerging out of an alginate bead; **(B)** Maturation of somatic embryos. Note the numerous cotyledonary stage embryos on the cryo-preserved tissue; **(C)** Single somatic embryo growing to give rise to a complete plantlet after freezing; **(D)** Plants established in small pots transferred from *in vitro* cultures.

was analyzed using HPLC and the content was found to be stable. Molecular studies using RAPD analysis proved that the plants were genetically stable [13].

4.3   Cryopreservation of Hairy Roots of *Panax ginseng*

The protocol for cryopreservation of hairy roots of *Panax ginseng* was developed by Yoshimatsu et al. [15]. Hairy root segments including root tips were placed on to phytohormone-free half-strength Murashige and Skoog solid medium and stored at 4°C in the dark for 4 months. The root segments resumed elongation when the temperature was raised to 25°C in the dark. For cryopreservation, the root tips were precultured with 0.1 mg l^{-1} 2,4-D for 3 d and dehydrated with PVS_2 for 8 min before immersion in liquid nitrogen. Sixty percent survival could be obtained. The hairy roots regenerated from cryopreserved root tips grew well and showed the same ginsenoside productivity and patterns as those of the control hairy roots cultured continuously at 25°C. The conservation of T-DNAs in the regenerated hairy roots was proved by PCR analysis.

4.4   Cryopreservation of Transformed Calli of *Papaver somniferum*

The transformed *P. somniferum* cells maintained on MS solid medium at 22°C in the dark were precultured in 50% loading solution (1 M glycerol + 0.2 M sucrose) at 20°C in the dark for 1 d, dehydrated with PVS_2 at 25°C for 35 min without loading, and then cryopreserved in liquid nitrogen. After rapid thawing and washing, the cells were precultured on MS solid medium at 22°C in the dark. All the four clones used for cryopreservation regenerated successfully showing the same morphological characteristics as the untreated cultures. To confirm the conservation of T-DNA derived from *Agrobacterium rhizogenes*, the existence of T-DNA in the regenerants was examined by PCR analysis. Amplification of T-DNA bands was clearly observed in the regenerated cells as well as in the untreated ones. Preliminary evaluation of genetic stability using RAPD analysis was performed and no significant difference was observed between the untreated and cryopreserved cells [16].

4.5   Cryopreservation of Cell Cultures of *Digitalis thapsi*

Cell cultures of *Digitalis thapsi* were treated for 3 d with 0.15 M mannitol followed by treatment with a cryoprotectant solution composed of 0.5 M DMSO, 0.5 M glycerol and 1 M sucrose, slow cooled for 30 min at –20°C and frozen by rapid immersion in LN, rapid thawing and transfer of cells without washing to a standard semi-solid medium. High viability (60%) was recorded and the cultures originating from cryopreserved cells retained their capacity to accumulate carotenoids [17].

5.   Monitoring Genetic Stability of Regenerants from Cryopreserved Germplasm

It is important to consider that the plants regenerated from cryopreserved germplasm have been exposed to a range of different experimental conditions including tissue culture, pre-growth, cryoprotection, freezing-thawing, recovery (re-growth) and manipulations to enhance regeneration. All these stages have the potential to influence genetic stability [18]. Successful post-thaw storage recovery must not only be assessed in terms of survival (viability) of plant tissues, but also the ability to regenerate and produce complete plants. This necessitates a tissue culture regeneration system. It is therefore essential to consider the effects that *in vitro* regeneration will have on the genetic stability of the surviving plants as these may show somaclonal variation [19]. For instance, the time taken to regenerate plants and the quality of germplasm recovered after cryopreservation are likely to be important features in maintenance of genetic stability and operation of a functional gene bank [20-22]. It is thus imperative to assess the genetic stability of plant material regenerated from cryopreserved germplasm and to determine if it is genetically identical to the mother stock (Germplasm prior to storage in LN).

Studies on ginsenoside production from cryopreserved cells of *Panax ginseng* revealed that total amount of ginsenosides together with product remain unchanged [15]. Furthermore, it was demonstrated that alternative conservation method such as preservation under mineral oil for six months as well as continuous subculturing for 14 months failed to preserve biosynthetic capacity of the cells [23]. Benson and Hamil [24] reported stability in the biosynthetic capacity in transformed roots of *Beta vulgaris* after cryopreservation. All recovered cultures of *Coleus blumei* showed the same growth and production characteristics of rosmaric acid as controls [25]. Stability was also demonstrated for recovery after different storage periods in liquid nitrogen (from 1 day to 15 months). These works clearly show that cultures were stable even after successive cryopreservation cycles. Similar conclusions of successful application of cryopreservation can be made from freezing experiments with biotin producing callus cultures [26].

6.   Conclusions

Cryopreservation of medicinal plants has multifacet advantages. The technology of cryopreservation has been refined and it enables the storage of *in vitro* cultures for the long-term conservation of medicinal plants. The retention of biosynthetic potential of the retrieved cultures amply demonstrates the use of this technology for the storage of rare, high alkaloid/secondary metabolites/medicines producing cell cultures for pharmaceutical purposes. Medicinal plants are potential candidates for transformation as well and cryopreservation of transgenic lines is an important line of research. The successful cryopreservation of cell cultures of tobacco, *Datura, Panax, Dioscorea*, *Catharanthus*, *Anisoidus*, *Atropa,* etc., without any evidence of deterioration coupled with potential expected benefits warrant the extension of this technique to other species. As far as genetic stability is concerned the chances of an undesired cell selection after cryogenic storage seem to be less important than expected. It is even more remarkable that in most of the cases the important characters of the preserved cell lines did not change. Finally, it may be concluded that routine cryopreservation procedures developed and applied on a large scale in medicinal plants will help in conservation of this important group of genetic resource.

References

1.   E.E. Benson, An introduction to plant conservation biotechnology, in: E.E. Benson (Ed), Plant Conservation Biotechnology, Taylor and Francis, London, 1999, pp. 3–10.
2.   S.E. Ashmore, in: Status report on development and application of *in vitro* techniques for the conservation and use of plat genetic resources, IPGRI, 1997, pp. 27–38.
3.   D.H. Touchell, Conservation of threatened flora by cryopreservation of shoot apices, in: F. Engelmann, H. Takagi, (Eds), Cryopreservation of Tropical Plant Germplasm: Current Research Progress and Application, JIRCAS/IPGRI, 2000, pp. 269–272.
4.   F. Engelmann, Importance of cryopreservation for the conservation of plant genetic resources, in: F. Engelmann, H. Takagi, (Ed.), Cryopreservation of Tropical Plant Germplasm—Current Research Progress and Applications, IPGRI, Rome & JIRCAS, Tsukuba, 2000, pp. 8–20.
5.   B.M. Reed, Implementing cryogenic storage of clonaly propagated plants, CryoLett. 22 (2001) 97–104.
6.   K.K. Kartha, Meristem culture and germplasm preservation, in: K.K. Kartha, (Ed), Cryopreservation of Plant Cells and Organs, CRC Press, Boca Raton, 1985, pp. 115–134.
7.   K.K. Kartha, Cryopreservation of secodary metabolite producing plant cell cultures, in: F. Constable, I.K. Vasil (Eds), Cell Culture in Phytochemistry, Academic Press, New York, 1987, pp. 217–227.
8.   K. Harding, Stability assessments of conserved plant germplasm, in: E.E. Benson (Ed), Plant conservation biotechnology, Taylor and Francis, London, 1999, pp. 97–108.
9.   M. Taylor, Field conservation of root and tuber crops in South Pacific. Presentation at the Consultation meeting on management of field and *in vitro* gene bank, Columbia, January, 1996.
10.   F. Engelmann, *In vitro* conservation methods, in: B.V. Ford-Lloyd, J.H. Newburry, J.A. Callow, (Eds), Biotechnology and Plant Genetic Resources: Conservation and Use, CABI, Wellingford, 1998, pp. 119–162.
11.   K. Kartha, F. Engelmann, Cryopreservation and germplasm storage, in: I.K. Vasil, T.A. Thorpe, (Eds), Plant Cell and Tissue Culture, Kluwer, Dordrecht, 1994, pp. 195–230.
12.   S. Ahuja, Cryopreservation of *Dioscorea* spp. using vitrification encapsulation-dehydration and techniques. Ph.D. Thesis. Jamia Hamdard, New Delhi, 2001.

13. S. Dixit, *In vitro* Regeneration and Cryopreservation of Asian Yams and Sweet Potato. Ph.D. Thesis. Jamia Hamdard, New Delhi. 2001.
14. S. Ahuja, B.B. Mandal, S. Dixit, P.S. Srivastava, Molecular phenotypic and biosynthetic stability of plants recovered from cryopreserved shoot-tips of *Dioscorea floribunda*. Plant Sci. 3 (2002) 971–977.
15. K. Yoshimatsu, H. Yamaguchi, K. Shimomura, Traits of *Panax ginseng* hairy roots after cold storage and cryopreservation, Plant Cell Rep. 15 (1996) 555–560.
16. K. Yoshimatsu, K. Touno, K. Shimomura, Cryopreservation of medicinal plant resources: retention of biosynthetic capabilities in transformed cultures, in: F. Engelmann, H. Takagi, (Eds), Cryopreservation of Tropical Plant Germplasm—Current Research Progress and Applications, IPGRI, Rome & JIRCAS, Tsukuba, 2000, pp. 77–88.
17. M. Moran, M. Cacho, J. Fernandez-Tarrago, P. Corchete, A protocol for cryopreservation of *Digitalis thapsi* L. cell cultures, Cryo Lett. 20 (1999) 193–198.
18. K. Harding, Stability of ribosomal RNA genes in *Solanum tuberosum* L. plants recovered from cryopreservation, Cryo Lett. 18 (1997) 217–230.
19. W.R. Scowcraft, Genetic variability in tissue culture: impact on germplasm conservation and utilization, IBPGR/Rome, 1984, pp. 41.
20. K. Harding, E.E. Benson, A study of growth, flowering and tuberization in plants derived from cryopreserved potato shoot tips: implications for *in vitro* germplasm collections, Cryo Lett. 15 (1994) 59–66.
21. K. Harding, Approaches to access the genetic stability of plants recovered from *in vitro* culture, in: M.N. Norman, (Ed), Proceedings of the International Workshop on *In Vitro* Conservation of Plant Genetic Resources Kuala Lumpur, Malaysia, 1996, pp. 137–170.
22. K. Harding, Stability assessments of conserved plant germplasm, in: E.E. Benson (Ed.), Plant Conservation Biotechnology, Taylor and Francis, London, 1999, pp. 97–108.
23. L. Mannoven, L. Toivonen, V. Kauppinen, Effects of long term preservation on growth and productivity of *Panax ginseng* and *Catharanthus roseus* cell cultures, Plant Cell Rep. 9 (1990) 173–177.
24. E.E. Benson, J.D. Hamill, Cryopreservation and post freeze molecular and biosynthetic stability in tansformed roots of *Beta vulgaris* and *Nicotiana rustica*, Plant Cell Tiss. Org. Cult. 24 (1991) 163–172.
25. I. Reuff, Untersuchungen zur Kryokonservierung pflanzlicher Zellkulturen am Biespiel von *Coleus blumei* and *Berberis wilsoniae*, Ph.D. Thesis, Tubingen, 1987.
26. A. Kuriyama, K. Watanabe, S. Ueno, H. Mitsuda, Effect of post thaw treatment on the viability of cryopreserved *Lavanula vera* cell, Cryo Lett. 11 (1990) 171–178.
27. A. Sakai, S. Kabayashi, I. Oiyama, Cryopreservation of nucellar cells of navel orange (*Citrus sinensis* osb. Var. Brasieleinsis Tanaka) by vitrification, Plant Cell Rep. 9 (1990) 30–33.
28. J. Fabre, J. Dereuddre, Encapsulation/dehydration: A new approach to cryopreservation of *Solanum* shoot tips, Cryo Lett. 11 (1990) 413–426.
29. D. Hirai, K. Shirai, S. Shirai, A. Sakai, Cryopreservation of *in vitro* grown meristems of strawberry (*Fragaria* × *ananasa* Duch.) by encapsulation-vitrification, Euphytica 101 (1998) 109–115.
30. B. Panis, Cryopreservation of banana (*Musa* spp.) germplasm. Dissertiones de Agricultura, Katholieke University Leuven, Belgium, 1995.
31. A. Uragami, A. Sakai, M. Nagai, Cryopreservation of dried axillary buds from plantlets of *Asparagus officinalis* L. grown *in vitro*, Plant Cell Rep. 9 (1990) 328–331.
32. A schäfer-Menuhr, G. Mix-Wagner, H.M. Schumacher, Cryopreservation of potato cultivars—design of a method for routine application in genebanks, Acta Hort. 447 (1997) 477–482.
33. Y.P.S. Bajaj, Regeneration of plants from forzen (−196°C) protoplasts of *Atropa belladonna, Datura innoxia* and *Nicotiana tabacum*, J. Exp. Bot. 16 (1988) 947–953.
34. Z. Zheng Guang, M. Jingbo, W. Shiling, Cryopreservation of calli and their suspension culture cells of *Anisodus acutangulus*, Acta Bot Sci, 25 (1983) 512 – 517.
35. M.A.C. Demeulemeester, B. Vandenbussche, M.P. Proft, Regeneration of Chicory plants from cryopreserved *in vitro* shoot tips, Cryo Lett. 14 (1993) 57–64.
36. C.S. Hunter, *In vitro* propagation and germplasm storage of *Cinchona*, in: P. Alderson, L.A. Withers, (Eds), Plant Tissue Culture and its Agricultural Applications, Butterworth, London, pp. 291–301.

37. G. Weber, E.J. Roth, H.F. Schmeiger, Storage of cell suspensions and protoplasts of *Glycine max* (L.) Merri., *Brassica napus* (L.), *Datura innoxia* (Mill.) and *Daucus carota* (L.), Z. pflanzenphysiol. 109 (1983) 29–39.

38. Y.P.S. Bajaj, Regeneration of plants from cell suspensions frozen at –20, –70, –196°C, Physiol. Plant. 37 (1976) 263–278.

39. L. Chaulafich, D. Grabishich, R. Vriichich, L.A. Volkova, A.S. Popov, Somatic embryo production *in vitro* in *Dioscorea causia* lipsky. and *D. balcquice*. Kosanin and cryopreservation of their organogenic callus tissue, Russ. J. Plant Physiol. 41 (1994) 821–826.

40. B.B. Mandal, K.P.S. Chandel, S. Dwivedi, Cryopreservation of yam (*Dioscorea* spp.) shoot apices by encapsulation-dehydration, Cryo Lett. 17 (1996) 165–174.

41. B.B. Mandal, Development of cryopreservation for long-term conservation of tropical plant germplasm at NBPGR, in: F. Engelmann, H. Takagi, (Eds), Cryopreservation of Tropical Plant Germplasm: Current Research Progress and Application. JIRCAS/IPGRI, 2000, pp. 282–286.

42. R.G. Butenko, A.S. Popov, L.A. Volvoka, N.D. Chernyak, M. Nosov, Recovery of cell cultures and their cultures and their biosythetic capacity after storage of *Dioscorea deltoidia* and *Panax ginseng* in liquid nitrogen, Plant Sci. Lett. 33 (1984) 285–292.

43. E.M. Goldner, U. Seitz, E. Reinhard, Cryopreservation of *Papaver somniferum* cells, Cryo Lett. 19 (1991) 147–159.

44. M. Pacques, M. Poissonnier, E. Dumas, V. Monod, Cryopreservation of dormant and non dormant broad leaved trees, in: Proceedings of Third International ISHS Symposium on *in vitro* Culture and Horticulture Breeding, Jerusalem, Israel, 1996.

45. M. Suzuki, M. Ishikawa, T. Akihama, A Novel preculture method for the iduction of dessication tolerance in *Gentiana* axillary buds for cryopreservation, Plant Sci. 135 (1998) 69–76.

46. S.W. Decruse S. Seeni, P. Pushpagandhan, Cryopreservation of alginate coated shoot tips of *in vitro* grown *Holostemma annulare* (Roxb.) K. Schum, and endangered medicinal plant: Influence of pre-culture and DMSO treatment on survival and regeneration, Cryo Lett. 20 (1999) 243–250.

47. K. Sami, S. Ethington, L. Ximing, J. Widholm, Recovery and characterization of amino acid analog resistant carrot and tobacco suspension-cultured cells after cryostorage for over 10 years, Physiol. Plant. 103 (1998) 30–33.

48. A.D. Maddox, M.F. Gonsalves, R. Sheilds, Successful preservation of suspension cultures of three *Nicotiana* spp. at the temperature of liquid nitrogen, Plant Sci. Lett. 28 (1994) 157–162.

49. Z.D. Martinez, R. Arraya Gracia, M.A. Revilla, Cryopreservation of *in vitro* grown shoot tips of *Olea europea* L. var. Arbequina, Cryo Lett. 20 (1999) 29–36.

50. B. Dietrich, A.S. Papov, B. Pfeiffer, D. Newmann, R.G. Butenko, M. Luckner, Cryopreservation of *Digitalis lanata* cell cultures, Planta Med. 46 (1982) 82–87.

51. C. Gazeau, H. Elleuch, A. David, C. Morisset, Cryopreservation of transformed *Papaver somniferum* cells, Cryo Lett. 19 (1998) 147–159.

52. T.W. Swan, E.A. Deakin, G. Junjan, G.R. Souch, M.E. Spencer, A.M. Lynch, Cryopreservation of cells suspensions of *Polygonum avicular* using traditional controlled rate freezing and encapsulation-dehydration protocol, a comparison of post thaw cell recovery, Cryo Lett. 19 (1998) 237–248.

53. T. Yamada, A. Sakai, T. Matsumura, S. Hignichi, Cryopreservation of apical meristems of white clover (*Trifolium repens* L.) by vitrification, Plant Sci. 78 (1991) 81–87.

Plant Biotechnology and Molecular Markers
P.S. Srivastava, Alka Narula and Sheela Srivastava (Editors)

20. Molecular Mapping and Marker Assisted Selection of Traits for Crop Improvement

Anushri Varshney, T. Mohapatra and R.P. Sharma

National Research Centre on Plant Biotechnology, Indian Agricultural Research Institute,
New Delhi-110 012, India

Abstract: Genetic markers, the heritable entities that are associated with economically important traits can be used by plant breeders as selection tools. By using molecular markers, breeders can by-pass traditional phenotype-based selection methods, which involve growing plants to maturity and closely observing their physical characteristics in order to infer underlying genetic make up. The purpose of this chapter is to describe the available genetic marker types and provide the status of gene mapping and marker-assisted selection in important crop species. This review highlights how genetic markers have been used in mapping genes both for qualitative and quantitative traits and defines the potential use of genetic maps for marker-assisted selection.

1. Introduction

Crop plants have evolved initially by incidental consequences of human gatherers, and more recently through sophisticated plant breeding programmes. While changes in cultural practices and mechanization have had significant impact on agricultural productivity, yield gains in most crops have been due to genetic improvement. Although the gains have already achieved, further improvement of agricultural productivity and quality are demanded continuously mainly due to population growth, the increasing cost of inputs such as water, fertilizer and energy, concerns about the effects of agrochemicals on the ecosystem, and rapidly changing consumer preferences. Plant breeding, a process being used for centuries is largely depending on selection for desirable traits. These selections often take many cycles of breeding in order to place desirable agronomic and quality characteristics from different parents into a single genotype. Recent advances in biotechnology have led to the development of a number of novel tools that offer the promise of making plant breeding more precise and faster. Among the most promising are molecular markers, which are segments of plant DNA that breeders use to detect the presence or absence in experimental plants of specific alleles of interest and thus use them as selection tools [1, 2]. Such a selection of desirable plants based on linked markers is termed as Marker-Assisted Selection (MAS). By using molecular markers, breeders can by-pass traditional phenotype-based selection methods, which involve growing plants to maturity and closely observing their physical characteristics in order to infer underlying genetic make up. Several molecular marker systems have been developed and put to use. The more frequently used ones are discussed as follows.

2.   Molecular Markers

2.1   RFLP
Restriction fragment length polymorphisms, the first molecular marker developed by Botstein et al. [3] are detected by the use of restriction enzymes that cut genomic DNA molecules at specific nucleotide sequences (restriction sites), thereby yielding variable size DNA fragments. Identification of genomic DNA fragments is done by Southern blotting, a procedure whereby DNA fragments, separated by electrophoresis, are transferred to nitrocellulose or nylon filter [4]. In this, filter-immobilized DNA is allowed to hybridize to radioactively labeled probe DNA. RFLP is a co-dominant marker in which the probes are usually small (500 to 3000 bp), cloned DNA fragments (e.g. genomic or cDNA). The filter is placed against photographic film, where radioisotope disintegration from the probe results in visible bands.

2.2   RAPD
Random amplified polymorphic DNA is a dominant marker based on polymerase chain reaction (PCR). It employs a single decamer primer of arbitrary sequence, which is annealed to the template DNA typically at 37°C [5]. The variation in RAPD profile is in the form of presence or absence of a band resulting from variation in primer binding sites. A major limitation of this marker system is non-reproducibility due to low annealing temperature. However, utility of a desired RAPD marker can be increased by sequencing its termini and designing longer primers (e.g. 24 nucleotides) for specific amplification of markers [6]. Such sequenced characterized amplified regions (SCARs) are similar to sequence-tagged-sites (STS, [7]) in construction and application.

2.3   CAPS
Cleaved amplified polymorphic sequences are based on the restriction enzyme site variation in the DNA fragments generated by PCR [8]. The source of the sequence information for the primers can come from a gene bank, genomic or cDNA clones, or cloned RAPD bands. This marker is a co-dominant marker.

2.4   SSRs
Simple sequence repeats or microsatellites are ubiquitous in eukaryotes. SSR polymorphism reflects variation in the number of repeat units in a defined region of the genome. The frequency of repeats longer than 20 bp has been estimated to occur every 33 kb in plants. Nucleotide sequence flanking the repeat is used to design primers to amplify different number of repeat units in different varieties. These primers are very useful for rapid and accurate detection of polymorphic loci and the information could be used for developing a physical map based on these sequence tags. This type of polymorphism is highly reproducible.

2.5   AFLP
The amplified fragment length polymorphism markers are generated by selective amplification of DNA fragments obtained by restriction enzyme digestion [9]. High molecular weight DNA is digested by two restriction enzymes: one hexacutter (e.g. EcoRI) and one tetracutter (e.g. Mse I). Adapter molecules are ligated to the ends of DNA fragments. Two primers possessing sequence

complementarity to the adapter as well as few extra random nucleotides at their 3' ends are used for selective amplification of fragments employing PCR. The amplified products are separated on sequencing gels or even ordinary PAGE and visualized by silver staining. Alternatively, the primers are labeled either by radioisotope or fluorescent dye so that the AFLP profile can be obtained by autoradiography or by using image analysis. The highest number of amplified products (50-100) is produced in AFLP among all the DNA profiling systems. This increases the probability of detecting polymorphism many folds. The technique is, at present, lengthier and costlier than other PCR based techniques. It requires good quality DNA for ensuring complete digestion by enzymes. Partial digestion of DNA results in non-reproducible variation in DNA profiles.

2.6 SNP

Molecular markers are polymorphic when there is DNA sequence variation between the individuals under study. Molecular markers are, therefore, simply an indicator of sequence polymorphism. Sequence polymorphism between individuals can take many forms, for instance, it can be due to the insertion or deletion of multiple bases, or it can be due to single nucleotide polymorphisms (SNPs; [10]). Insertions, deletions and SNPs are important in determining sequence variation between individuals. SNPs are abundant in plant genomes. They are being used for genotyping human populations for certain genetic diseases. The cost of developing SNPs is very high, since for each locus DNA has to be sequenced and suitable PCR primers designed. The primers must then be used to amplify the corresponding fragment from all other possible genotypes. These fragments must then be sequenced and the sequences compared with one another to determine the SNPs for each haplotype [11]. The term 'haplotype' is used in the context of SNPs instead of the term 'allele'. There are number of methods for identifying SNPs within a genetic locus namely direct sequencing, single-strand conformation polymorphism (SSCP), chemical cleavage of mismatches (CCM) and enzyme mismatch cleavage (EMC).

3. Molecular Mapping of Genes of Agricultural Importance

Earlier, construction of genetic linkage maps using morphological markers could not be initiated in most crop plants due to lack of sufficient number of molecular markers. Map construction was highly laborious, took many years and required several mapping populations since all the morphological markers could not be obtained in a single cross. These maps contained limited number of markers and, therefore, could not be used for efficient mapping of target genes. With the availability of a large number of molecular markers, like RFLP, RAPD, AFLP, microsatellites, etc. saturation mapping of plant genomes has become a reality. Molecular genome maps have been constructed in almost all important crop plants. The number of markers employed to construct these maps and marker density varies greatly. Most of these maps are based on RFLP markers. The recent mapping efforts have included mostly the PCR based markers such as AFLP, STMS, RAPD, CAPS, SCAR and STS. Among the crop plants, the rice genome map is considered most saturated. The map reported by Harushima et al. [12] contained the maximum number of markers (2275). Significantly, this map was made using a single F_2 population. Recently, this map has been further saturated by combining additional STS and STMS markers [13]. In most of the crop plants F_2 population has been used since it could be generated in the shortest possible time with the least effort. However, for mapping genes, particularly those for

Table 1.   Molecular mapping of agriculturally important genes in crop plants

Crop	Pathogen/Trait	Gene	Marker(s)	Reference
1. Disease resistance				
Rice	*Pyricularia oryzae*	*Pi-2,4*	RFLP	[18]
		Pi11	RFLP	[139]
		Pi-5(t), Pi–7(t)	RFLP	[140]
		Pi-Z^{-6}	RFLP	[141]
		Pi-10	RFLP	[142]
		Pi-12(t)	RFLP	[143]
		Pi-18(t)	RFLP	[144]
		Pib	RFLP	[145]
		Pik^m	RFLP	[146]
		Pita-2, Pita	RFLP	[147]
		Pi-5 (t)	AFLP	[148]
		Pi20	RFLP	[149]
		Pi44	RFLP	[150]
		Pb1	RFLP	[151]
	Pyricularia grisea	*Pi-1(t)*	RFLP	[152]
		QTL (1)	RFLP	[153]
	Xanthomonas oryzae pv. *oryzae* (Bacterial blight)	*Xa-1, Xa-3, Xa-4*	RFLP	[20]
		Xa-5	RFLP	[19]
		Xa-13	RFLP	[154]
		Xa-21	RFLP	[155]
		Xa-1	RAPD	[156]
		Xa3, Xa4, Xa5, Xa10	RFLP	[156]
		Xa13	RFLP	[157]
		Xa22(t)	RFLP	[158]
		Xa-1	RAPD	[159]
		Xa 23 (t)	SSR	[160]
	Rice yellow mottle virus	*RYMV (QTL)*	RFLP	[161]
		RYMV	RFLP & STS	[162]
	Rice stripe	*Stv-b^i*	RFLP	[163]
	Tungro	*RTSV*	RFLP	[164]
	Puccinia striiformis f. sp. *tritici*	*Yr5*	RGA	[165]
	Rhizoctonia solani Kuhn	*Rsb1*	RFLP, RAPD, AFLP, SSR	[166]
Wheat	*Erysiphe graminis* p.v. *tritici*	*Pm1, Pm2, Pm3b, Pm4a*	RFLP	[167]
		Pm1, Pm2	RFLP	[168]
		Pm2	RFLP	[169]
		Pm3b, Pm4a	RFLP	[170]
		Pm2	RFLP	[171]
		Pm1	RFLP	[172]
		Pm12	RFLP	[173]
		Pm21	RAPD	[174]
		Pm	RFLP	[175]
		Pm4b	AFLP	[176]

Crop	Pathogen/Trait	Gene	Marker(s)	Reference
		Pm4a & Pm4b, Pm6	STS, RFLP	[177]
		Pm13	RFLP, RAPD, STS, DDRT-PCR	[178]
		MlG	SSR	[179]
	Adult plant resistance to powdery mildew	*APR*	STMS, RFLP	[180]
	Common bunt	*Bt-10*	RAPD	[181]
		Bt-11	RAPD	[182]
	Karnal bunt	*KB*	RFLP	[183]
	Durable stem rust	*Sr2*	RFLP	[169]
		Sr2	STS	[184]
		Sr2	RFLP	[185]
		Sr22	RFLP	[186]
	Puccinia recondite	*Lr9*	RFLP, RAPD	[187]
		Lr18	N-band	[188]
		Lr1	RFLP	[189]
		Lr9	RFLP	[190]
		Lr 19	RFLP	[190]
		Lr24	RFLP	[190]
		Lr24	RAPD	[191]
		Lr 29	RAPD	[192]
		Lr32	RFLP	[190]
		Lr34	RFLP	[169]
		Lr24	RAPD, SCAR	[193]
		Lr10	STS	[194]
		Lr10	RFLP	[195]
		Lr23	RFLP	[195]
		Lr27	RFLP	[195]
		Lr31	RFLP	[195]
		Lr34	RFLP, RAPD	[195]
		Lr34	RFLP	[196]
		Lr13	RFLP, STMS	[197]
		Lr35	PCR	[197]
		Lr28	RAPD, STS	[37]
		Lr3	RFLP	[198]
		Lr3	mRNA fingerprinting, cDNA cloning	[199]
	Stripe rust	*Yr15*	RFLP	[200]
		Yr15	RAPD, STMS	[201]
		YrH52	STMS, RFLP	[40]
	Loose smut	*T19*	Monoclonal antibody	[202]
		T10	RAPD, RFLP	[203]
	Septoria nodorum	–	RAPD	[204]
	Septoria tritici	–	AFLP	[205]
	Fusarium head blight	–	AFLP, RFLP	[206]
		–	AFLP	[207]
	Yellow rust	*YrMoro*	AFLP, STS	[208]

(Contd)

Table 1. *(Contd)*

Crop	Pathogen/Trait	Gene	Marker(s)	Reference
	Tilletia indica	QTL *(1)*	SSR, AFLP	[209]
	Wheat streak mosaic virus	*Wsm1*	STS, RAPD	[210]
Maize	*Heliminthosporium turcicum*	*Ht1*	RFLP	[211]
	Maize dwarf mosaic virus	*mdm1*	RFLP	[212]
	Cercospora zeamaydis	QTL *(>10)*	–	[213]
	Maize streak virus	QTL *(1)*	RFLP	[214]
	Maize mosaic virus	QTL *(1)*	RFLP	[215]
	Maize stripe virus	QTL *(1)*	RFLP	[216]
	Sugarcane mosaic virus	*Scm1*	RFLP, SSR	[217]
		Scmv1, Scmv2	RGA-CAPs	[218]
		Scmv1, Scmv2	AFLP, SSR	[219]
Barley	*Erysiphe graminis*	QTL *(2)*	RFLP	[220]
		Rar1	AFLP	[221]
	Rynchosporium secalis	*Rrs 13*	RFLP	[222]
	Puccinia hordei	*Rph Q*	RAPD	[223]
		QTL-Rphq (6)	AFLP	[224]
		Rph7.g	RFLP	[225]
	Xanthomonas campestris pv. *hordei*	QTL *(3)*	–	[226]
	Puccinia striiformis f.sp. *hordei*	QTL *(1)*	–	[227]
	Barley yellow dwarf Luteovirus	*Yd2*	AFLP	[228]
	Barley yellow mosaic virus	*rym5*	CAPs, SSR	[229]
	Cochliobolus sativus	*Vhv1*	AFLP	[230]
	Pyrenophora graminea	QTL *(2)*	–	[231]
Sorghum	*Sporisorium reilianum*	*Shs*	RFLP/RAPD	[232]
Tomato	*Stemphylium vesicarum*	*Sm*	RFLP	[233]
	Cladosporium fulvum	*cfa*	RFLP	[234]
	Fusarium oxysporum	I_2	RFLP	[235]
	Pseudomonas syringae	*Pto*	RFLP	[236]
	Leveillula tourica	*Lv*	RAPD/RFLP	[237]
	Verticillium dahliae	*Ve*	RAPD	[238]
		Ve	RFLP	[239]
	Medoidogyne sp.	*Mi*	RFLP	[240]
	Psuedomonas solanacearum	QTL *(3)*		[241]
	Oidium lycopersicum	*Ol-1*	RFLP, RAPD, SCAR	[242]
	Alternaria solani	QTL *(1)*	RFLP, RGA	[89]
Potato	Potato virus X	Rx_1, Rx_2	RFLP	[243]
		Nb	AFLP	[244]
	Phytophthora infestans	QTL *(11)*	RFLP	[97]
		R2	AFLP	[99]
	Potato X potexvirus	Nx_{phu}	RFLP	[245]
	Potato Y potyvirus	Ry_{adg}	RFLP	[246]

Crop	Pathogen/Trait	Gene	Marker(s)	Reference
Soybean	*Phytophthora sojai*	*Rps 1*	RFLP	[247]
	Soybean mosaic virus	*Rsv*	RFLP/SSR	[248]
	Pseudomonas syringae pv. *glycinea*	*Rpg 1*	RFLP	[249]
Common bean	*Uromyces appendiculatus*	PI 181996	RAPD	[250]
		Ur-9, Fin	RAPD	[251]
	Potyvirus	*I*	RAPD	[252]
	Xanthomonas campestris	*QTL (7)*	RFLP	[253]
	Colletotrichum lindemut hianum	*Co-4², Co-7*	RAPD, SCAR	[254]
Pea	Pea seed borne mosaic virus	*sbm-1*	RFLP	[255]
	Ascochyta pisi	*QTL (3)*	RFLP	[256]
Tobacco	*Chalara elegans*	*Brr*	RAPD	[257]
Apple	*Venturia inaequalis*	*Vf*	RAPD	[258]
		Vf	AFLP & SCAR	[259]
Melon	*Fusarium* sp.	*Form 2*	RAPD	[260]
Mungbean	*Erysiphe polygoni*	*QTL (3)*	RFLP	[261]
Cocoa	*Phytophthora palmivora*	*QTL (5)*	AFLP	[262]
Oil palm	*Fusarium* sp.	*QTL (1)*	SSR & AFLP	[263]
Rubber	*Microcyclus ulei*	*QTL (8)*	–	[264]
	Phyllochora herberi	*Phr*	Isozyme	[265]
Sugarcane	*Puccinia melanocephala*	–	RFLP	[266]
Brassica	*Leptosphaeria maculans* (Desm.) Ces.et de Not	*QTL (10)*	–	[267]
	Plasmodiophora brassicae	*Pb-Bn1, QTL (2)*	–	[268]
	Sclerotinia sclerotiorum	*QTL (1)*	RFLP, AFLP, SSR, RAPD	[269]
Chick pea	*Fusarium* sp.	*Race 4*	ISSR (Inter-Simple Sequence repeat)	[270]
Pearl millet	*Puccinia substriata* var. *indica*	*Rr₁*	RAPD,RFLP	[271]
Rose	*Diplocarpon*	*Rdr1*	RAPD, AFLP	[272]
Grape	Powdery mildew	*Run1*	AFLP	[273]
Pepper	Potato virus Y	*Pvr4*	RAPD, SCAR	[274]
Cassava	Cassava Mosaic Virus	*CMD2*	SSR, RFLP	[275]
Rye	Rust	*Lr26* *Sr31* *Yr9* *SrR*	AFLP, RGA, STS	[276]
Banana	Banana Streak Virus	–	AFLP	[277]
Tobacco	*Ralstonia solanacearum*	*QTL (1)*	AFLP	[278]
Lentil	*Colletotrichum truncatum*	*LCt-2*	RAPD, AFLP	[279]

(Contd)

Table 1. *(Contd)*

Crop	Pathogen/Trait	Gene	Marker(s)	Reference
2. Nematode and insect resistance				
Potato	*Globodera rostochiensis*	*QTL (2)*	–	[280]
	Globodera rostochiensis, *G. pallida*	*Grp1*	AFLP, CAPs & RFLP	[281]
	G. pallida	*QTL (1)*	AFLP, SSR	[282]
		QTL (1)	AFLP	[283]
Tomato	*Globodera rostochiensis*	*Hero*	SSR	[284]
	Meloidogyne spp.	*Mi-1*	RFLP	[285]
Sorghum	Head bug	*B2/b2*	RFLP, SSR	[286]
	Schizaphids graminum	*QTL (1)*	RAPD, SSR	[287]
Soybean	*Helicoverpa zea* Boddie	*QTL (1)*	RFLP	[77]
	Heterodera glycines Ichinohe	*QTL (1)*	RFLP	[288]
Wheat	*Diuraphis noxia* Mordvilko	*Dn2*	RAPD, SCAR	[289]
		Dn2, Dn4	RFLP	[290]
		Dn4	SSR	[291]
		Dn6		
	Hessian fly	*H23, H24*	RFLP	[292]
		H3, H5, H6, H9- H17	RAPD	[293, 294]
		H21	RAPD	[295]
		H6	RAPD, STS	[296]
	Pratylenchus neglectus	*Rlnn1*	AFLP, RFLP	[297]
	Cereal cyst nematode resistance	*Cre1*	RFLP	[298]
		Cre1	RFLP	[127]
		Ccn-D1	RAPD, RFLP	[299]
Maize	*Ostrinia nubilalis*	*QTL (1)*	RFLP, SSR	[300]
Rice	*Orseolia oryzae* (Gall midge)	*Gm2*	RFLP	[14]
		Gm4(t)	RFLP	[118]
		Gm7	AFLP, SCAR	[301]
	Brown planthopper	*Bph1*	RFLP	[302]
		Bph10	RFLP	[303]
		Bph(t)	RFLP	[23]
	Green leafhopper	*GLH*	RFLP	[17]
		Grlp3	RFLP	[304]
		Grlp11	RFLP	[304]
		Grh1	RFLP	[305]
	Whitebacked planthopper	*WBPH*	RFLP	[306]
		WBPH	RFLP	[307]
Apple	*Dysaphis devecta* Wlk.	*Sd-1*	AFLP, SSR, RFLP	[308]
3. Abiotic stresses				
Rice	Submergence tolerance	*Sub1*	RFLP	[24]
	Salt tolerance	*Salt*	RFLP	[25]
		OSA3	RFLP	[26]

Crop	Pathogen/Trait	Gene	Marker(s)	Reference
	Phosphorus uptake	*QTL (1) (Pup1)*	RFLP	[27]
	Aluminium tolerance	*QTL (1)*	–	[28]
Wheat	Thermosensitive earliness *per se*	*Eps-A^m1 (QTL)*	RFLP	[309]
	Aluminium tolerance	*Alt2*	RFLP	[310]
		Alt_{BH}	RFLP	[311]
	Tolerance to salt stress	*Kna1*	Protein poly morphism	[312]
Barley	Aluminium tolerance	*Alt (QTL)*	AFLP, SSR	[313]

4. Male sterility, wide compatibility and fertility restoration

Crop	Pathogen/Trait	Gene	Marker(s)	Reference
Petunia	Restorer of fertility	*Rf*	RAPD, AFLP	[314]
Rice	Male sterility and fertility restoration	*tgms1.2*	RFLP	[29]
		tms2	RFLP	[30]
		tms3	RFLP	[31]
		tgms	RFLP	[32]
		tgms-vn1 (tms4)	RFLP	[33]
		pms1	RFLP	[315]
		pms2	RFLP	[315]
		pms3	RFLP	[316]
		ms-h(t)	RFLP	[317]
		Rf-1	RFLP	[318]
		Rf?	RFLP	[319]
		Rf2	RFLP	[320]
		Rf3	RFLP	[321]
		Rf5	RFLP	[322]
		Rfu	RFLP	[323]
		Rf?	RFLP	[114]
	Hybrid breakdown	*Hwd1, hwd2*	RFLP	[324]
	Wide compatibility	*S5*	RFLP	[325]
Wheat	Fertility restoration	*Rf4, Rf3*	RFLP	[326]
Rye	CMS	*Rfg1*	RFLP, RAPD	[227]
	Self-fertility	*S1Z1S5*	Isozyme, RFLP	[228]
Brassica	CMS restorer	*Rfp1*	RFLP, RAPD	[229]
Sorghum	Fertility restorer	*rf4 (QTL)*	AFLP	[330]
Sunflower	Fertility restoration	*Rf1*	RAPD, AFLP, SCAR	[331]
Cotton	CMS fertility restoration	*Rf1*	RAPD, SSR	[332]
Coffee	Pollen viability restoration	*QTL (3)*	AFLP	[333]

5. Grain quality

Crop	Pathogen/Trait	Gene	Marker(s)	Reference
Sorghum	Grain quality and yield components	*QTL (6)*	RFLP, AFLP & SSR	[334]
Rice	Grain aroma	*Fgr*	RFLP	[34]

(Contd)

Table 1. *(Contd)*

Crop	Pathogen/Trait	Gene	Marker(s)	Reference
	Cooked-kernel elongation	*KNE*	RFLP	[35]
	Amylose	*Wx*	RFLP	[335]
Wheat	Flour colour	*QTL (1)*	RFLP, AFLP	[336]
	Grain yield	*QTL (1)*	RFLP	[337]
	Red grain colour	*R3, R1*	RFLP	[169]
	High molecular weight glutanin	*Glu-D1*	PCR-based	[338]
	Grain protein content	*QTL (1)*	STMS, RFLP	[44]
		QTL (1)	SSR	[339]
	Kernel hardness	*ha*	RFLP	[169]
		ha	RFLP	[340]
	Bread making quality	*Glu-D1(1Dx5)*	PCR	[341]
	Amylose content	*Wx-B1*	RFLP	[342]
Sunflower	Grain oil content	*QTL(2)*	AFLP, SSR	[343]

6. Yield, its components and other traits

Crop	Pathogen/Trait	Gene	Marker(s)	Reference
Eucalyptus	Wood density, stem growth and stem form	*QTL (1)*	RAPD	[110]
	Lignification genes	*EgHypar* and *EgTub A1*	SSCP (Single Strand Confirnation Polymorphism)	[344]
Oil palm	Fruit morphology and fertility	*Sh*	AFLP	[263]
Carnation	Flower type	*QTL (1)*	RAPD, SCAR & RFLP	[345]
Soybean	Specific leaf weight and leaf size	*QTL (1)*	RFLP	[346]
	Stearic acid content	*Fas*	SSR	[80]
Wheat	Plant height	*Rht-B1*		
		Rht-D1	RFLP	[347]
	Dwarfing genes	*Rht12*	STMS, RFLP	[43]
		Rht8	SSR	[41]
		Rht-B1, Rht-D1	RFLP	[348]
		Rht-B1, Rht-D1	RFLP	[347]
	Haploid formation	*QTL (1)*	AFLP	[349]
	Green plant formation	*QTL (1)*	AFLP	[349]
	Semi-dwarfing genes Rht-D1b (Rht2)	*Rht-B1b (Rht-1)–*	PCR-based	[350]
	Ear emergence time, plant height	*QTL (1)*	RFLP	[351]
	Preharvest sprouting tolerance	*QTL (1)*	RFLP	[352]
		Major gene	STMS, STS	[45]
	Vernalization response	*Vrn1*	RFLP	[353]
		Vrn1	RFLP	[169]
		Vrn1	RFLP	[43]
		Vrn-A^m1, Vrn-A^m2	RFLP	[354]
		Vrn-D1	STMS	[355]

Crop	Pathogen/Trait	Gene	Marker(s)	Reference
	Cadmium uptake	*Cdu1*	RAPD	[356]
	ABA production and response	–	RFLP	[357]
	Coleoptile pigmentation	*Rc1*	RFLP	[169]
	Milling yield	–	RFLP, STMS	[358]
	Eyespot	*Pch2*	RFLP	[359, 360]
	Tan spot	–	RFLP	[361]
	Na$^+$/K$^+$ discrimination	–	RFLP	[362]
Apple	Growth and development in juvenile apple trees	*QTL (1)*	RAPD	[363]
Potato	For foliar glycoalkaloid and aglycones	*QTL (1)*	RFLP	[364]
Peach	Fruit quality	*QTL (1)*	Isozymes, RAPD, RFLP, AFLP	[365]
Rose	Recurrent blooming, double corolla, thorn density of the shoots	*QTL (1)*	AFLP	[366]
Grape	Seedlessness, berry weight	*QTL (1)*	AFLP, SSR, isozyme, RAPD, SCAR	[367]
Barley	Intermedium spike-C and non-brittle rachis1	*int-c* *btr-1 (QTL)*	AFLP	[368]
Pea	Rhizobium nodulation	*sym9, sym 10*	AFLP, RFLP	[369]
Sugarbeet	Sucrose content, yield and quality	*QTL (1)*	RFLP, AFLP	[370]
Rice	Photoperiod sensitivity	*Se1*	RFLP	[371]
	Semidwarf gene	*sd1*	RFLP	[122]
		Sdg	RFLP	[372]
	Shattering-resistance gene	*Sh2*	RFLP	[122]
		Sh4	RFLP	[373]
		Sht	RFLP	[374]
	Seed dormancy, heading date	*QTL (5)*	RFLP	[375]
	Heading date	*QTL-Hd-1, Hd-2 & Hd-3*	RFLP	[376]
	Yield	*QTL (1)*	SSR, STS	[36]
	Root morphology	*QTL (1)*	–	[377]
Pinus palustris Mill. × *P. elliottii* Engl.	Early height growth	*QTL (1)*	RAPD	[105]
Maize	Popping explosion volume	*QTL (4)*	SSR	[378]
Cotton	Fibre strength	*QTL (2)*	SSR, RAPD	[379]
Sunflower	Agronomic traits [grain weight by plant (GWP), 100-grain weight (TGW), percentage of oil in grain (POG), sowing to flowering date (STF)]	*QTL (1) for TWP, QTL (6) for POG, QTL (2) for STF*	AFLP, SSR	[341]

quantitative traits, permanent mapping populations such as recombinant inbred lines (RILs) and doubled haploids are preferred since they can be maintained over years by selfing and replicated over locations and seasons.

Availability of molecular markers and saturated linkage maps has enabled mapping of genes for qualitative as well as for quantitative traits. The qualitative traits that are controlled by one gene show simple Mendelian pattern of monogenic inheritance such as genes controlling biotic stresses, fruit flesh colour in peach, kernel colour in corn, flower colour in Petunia, etc. The mapping of such genes with different molecular markers is listed in Table1. Quantitative traits such as yield, drought and cold tolerance, wood density etc. that show continuous variation are controlled by many genes. The individual genes controlling the expression of quantitative traits are now called Quantitative Trait Loci (QTL). QTL with relatively strong effects are good targets for marker-assisted selection especially if the trait is difficult to measure. Many QTL have relatively small effects. These QTL are difficult to accurately map, especially with population sizes typical of most mapping studies. So, now the population size for QTL mapping has increased to about 300-500 individuals. However, in order to have an understanding of what the gene does and how it interacts with other genes, it is useful to know where the genes are within the genome and in relation to other genes of interest. The number of QTL controlling a trait varies from 1 to more than 10 for different crop plants (Table 1). The progress in mapping of genes of agricultural importance in some major crops is described here.

3.1   Rice

A large number of genes for qualitative and quantitative traits including disease resistance, insect resistance, cooking quality, drought and flooding tolerance etc. have been mapped using DNA markers as listed in Table1. Mapping of disease resistance genes are of major concern for imparting stability to rice production. Few examples of mapping such genes using molecular markers is described here. For instance, RFLP markers have been used to map gall midge resistance gene $Gm2$ using recombinant inbred lines derived from a cross between 'Phalguna' (resistant variety) and 'ARC6650' (a susceptible land race) [14]. Another gall midge resistance gene, $Gm4t$, which is non-allelic to $Gm2$ and is known to confer resistance against insect biotypes 1, 2, 3 and 4 has also been tagged using RAPD in combination with bulk segregant analysis of a F_3 population [15]. Several of the putative resistance gene analogues (RGAs) have been cloned, sequenced and found to be tightly linked to known disease resistance genes [16]. Genetic mapping of resistance to rice tungro spherical virus (RTSV) and green leaf hopper (GLH) in ARC11554 was achieved using RAPD and RFLP markers [17]. Yu et al. [18] mapped a major locus $Pi-2(t)$ for resistance to blast caused by the fungus *Magnaportha grisea* using RFLP markers. Several of the major genes to the bacterial leaf blight (BLB) pathogen, *Xanthomonas oryzae* pv. *oryzae*, have been tagged with RFLP or RAPD markers [19–21]. Two microsatellite markers tightly linked to BLB were located at approximately 2 and 18 cM from the $xa5$ locus [22]. RFLP tagging of a gene for resistance to brown plant hopper (BPH) was reported by Mei et al. [23].

Genes have also been mapped using RFLP markers for submergence tolerance [24], salt tolerance [25 and 26], phosphorus uptake [27] and Al tolerance [28]. As far as male sterility and fertility restoration is concerned, several reports have been available on mapping genes using RFLP markers [29, 30–33]. Similarly, for traits like grain aroma, cooked kernel elongation, genes have been mapped using RFLP markers [34, 35]. Recently, quantitative trait loci for yield have been mapped using SSR and STS markers [36].

3.2 Wheat

Several reports on mapping of genes for various disease and insect resistance, abiotic stresses, grain quality and other traits are listed in Table 1. Wheat rust disease is a major concern and many successful results have been reported on gene mapping. A sequence-tagged-site (STS) marker linked to *Lr28*, a wheat leaf rust resistance gene has been identified by Randomly Amplified Polymorphic DNA (RAPD) analysis of near isogenic lines (NILs) of *Lr28* in eight varietal backgrounds. Of the 80 primers tested, one RAPD marker distinguished the NILs and the donor parent from susceptible recurrent parent [37]. Comparisons between near isogenic lines (NILs) and their recurrent parents have been useful for identifying molecular markers linked to host genes showing resistance to pathogens. Inter-simple sequence repeat (ISSR primers) markers for stem rust (*SR39*) and leaf rust (*Lr 35*) resistance genes have been developed by Gold et al. [38], which would facilitate the transfer of these genes to elite wheat lines. Microsatellite markers have been used for detecting DNA polymorphism in yellow rust-resistance accessions of *Triticum dicoccoides* [39]. Nine microsatellite markers were identified to be linked to stripe-rust resistance gene YrH52 [40]. Microsatellite markers have also been used to tag several genes or QTL, including the genes *Rht8* [41, 42], *Rht12* and *Vrn1* [43], and QGpc.ccsu.2D.1, a QTL for grain protein content [44]. The problem of pre-harvest sprouting, particularly in amber kernels, is quite common in major wheat growing regions of the world, including India. The improvement in grain protein content and its composition in bread wheat is also a difficult task and remains a major concern to plant breeders. The QTL for pre-harvest sprouting tolerance [45] and grain protein content [44, 46] have been tagged using sequence tagged multiple sites (STMS) and sequence tagged sites (STS) markers.

3.3 Brassica

Brassica juncea (Indian mustard), *B. rapa* (turnip rape) and *B. napus* (rapeseed) are the major oilseed Brassicas. In this group of crops, molecular markers have been employed for mapping of genes primarily for disease resistance and oil and meal quality. Several efforts have been made to identify markers for resistance to white rust caused by the fungus *Albugo candida* (Pers.) Kuntze, which is a widespread and destructive disease in these crops with yield reductions of 30-60% in severely infested fields [47]. A locus (*ACA1*) controlling resistance to *A. candida* has been mapped in *B. napus* using RFLP markers [48]. A single locus controlling resistance to *AC2* in *B. rapa* was mapped using RFLP markers and a segregating population from *Per* (resistant to both *AC2* and *AC7*) × 'R500' (susceptible) [49]. A co-segregating RFLP marker (X140a) and two closely linked RFLP markers (X42 and X83) were identified which were useful for MAS and map based cloning of a single gene (*Acr*) responsible for conferring resistance to *A. candida* in *B. juncea* [50]. Prabhu et al. [51] mapped a resistance gene (*Ac2$_t$*) in *B. juncea* from a Russian source imparting resistance to a predominant Canadian isolate of *A. candida*. *B. juncea* accession BEC-144 from Poland shows resistance to the Indian isolates of the white rust pathogen. Identification of two markers linked in coupling and repulsion phases flanking the gene controlling resistance to *A. candida* in BEC-144 was reported by Mukherjee et al. [52]. This work has been further extended to develop AFLP and CAPS markers for this gene. Moreover the CAPS marker has been validated in different populations revealing thereby its utility in marker assisted selection. Kole et al. [53] mapped genes for resistance to white rust in *B. rapa* using a recombination inbred population and a genetic linkage map consisting of 144 RFLP markers and 3 phenotypic markers. Molecular markers have been generated for the genes conferring resistance to *Leptosphaeria maculans* in *B. napus* by various workers. The resistance locus *LmFr$_1$* was linked to markers cDNA 011 and cDNA 110 [54], and localized onto the linkage group 6 (LG6); [48]. Loci pb-3 and pb-4 conferring resistance to *Plasmodiophora brassicae* in *B. oleracea* were identified and linked to RFLP and AFLP markers [55]. Similarly, Figdore et al. [56] also identified markers 14a on LG1, marker 48 on LG4 and 177b on LG9 linked to clubroot resistance (resistance to *Plasmodiophora brassicae* wor. Race 7) in *B. oleracea*.

A number of studies have been undertaken to generate markers for fatty acids such as linolenic acid, linoleic acid, oleic acid, palmitic acid and erucic acid. Two RAPD markers, K-01$_{1100}$ and 25a were generated and linked to the linolenic acid concentration [57, 58]. RAPD markers linked to oleic, linolenic and linoleic acids were identified in *B. napus* [59]. RAPD marker linked to the linolenic acid content was converted to a co-dominant SCAR marker [59]. Markers linked to genomic regions controlling linolenic acid concentration in *B. napus* corresponding to *fad3* (omega-3-desaturase) gene in *A. thaliana* [60] were also identified [61-63]. In another study, a single QTL containing 6 markers associated with oleic, palmitic and linoleic acid content was detected in *B. rapa* [64]. Sharma et al. [65] recently mapped two major QTLs influencing oleic acid level in *B. juncea* using both single factor analysis of variance and interval mapping. Erucic acid loci have been linked to molecular markers by [66-68] using BSA or RFLP analysis in *B. napus*. In each of the studies, two QTL were detected. These QTL have been positioned on LG 6 and LG12 [66] or on LG7 and LG15 [63]. In an independent study, two QTL associated with the erucic acid level in *B. napus* were detected [67] and mapped onto two different loci termed as E1 and E2. QTL E1 and E2 correspond to the two alleles of the β-ketoacyl-synthase (KCS) derived from *B. campestris* and *B. oleracea*, the two parental species of *B. napus* and encode the *Fatty acid elongation 1 (Fae1)* protein [69]. In *B. rapa* (Syn *campestris*) erucic acid loci were linked to RFLP markers [70].

The seed coat colour gene has been tagged to various RFLP and RAPD markers. The RFLP markers linked to seed coat colour in *B. napus* were identified using the Bulked Segregant Analysis (BSA) approach [71]. Seed coat colour trait in *B. campestris* was tagged with RAPD markers using *B. campestris-oleracea* additional lines [72]. A 3:1 ratio of segregation of brown : yellow seed in *B. rapa* indicated a monogenic control of this trait and was mapped to LG5 [70]. Upadhyay et al. [73] studied segregation of the trait in an F$_2$ population of *B. juncea* and reported duplicate dominant gene action giving a phenotypic ratio 15:1. Two RFLP markers flanking one of the interacting loci were identified. In a recent report, the seed coat colour trait was tagged using a combined approach of BSA and AFLP in *B. juncea* [74].

3.4   Soybean

In soybean, emphasis is laid on genetic mapping of pest and disease resistance genes. Apart from that quantitative traits such as oil quality, plant height, sprout yield etc. have been characterized using molecular maps. The soybean cyst nematode (SCN) (*Heterodera glycines* Inchinoe) is the most economically significant soybean pest. Two SSR markers BARC-Satt 309 and BARC-Satt 168 have been reported that segregate and map 0.4 cM from *rhg1* [75]. When these markers were used to assay lines from SCN-susceptible × SCN - resistant crosses, they proved to be highly effective in identifying lines carrying *rhg1* resistance from those carrying the allele for SCN susceptibility at the *rhg1* locus. In another study, field resistance to SCN race 3 in soybean cv. Forrest was found conditioned by two QTLs. The underlying genes are presumed to include *rhg1* on linkage group G and *rgh4* on linkage group A2. A high density map for the intervals carrying *rhg1* and *rhg4* have been developed using AFLP markers. A12-way analysis of variance showed two loci controlling SCN resistance in Essex × Forrest RILs [76]. Using 139 RFLPs QTLs associated with resistance to corn earworm (*Helicoverpa zea* Boddie) were identified [77]. With the help of AFLP, four markers closely linked to soybean mosaic virus resistance gene, *Rsv1* was mapped, thus demonstrating the utility of genetic mapping for generating markers tightly linked to important plant disease resistance genes [78]. Soybean death syndrome (SDS) caused by *Fusarium solani* f. sp. *glycines* results severe yield losses. Two QTLs for resistance to SDS were mapped in cv. Pyramid using SSR markers namely, BARC-Satt 163 and BARC-Satt 080. Similarly, a QTL was identified from cv. Douglas using SSR marker BARC-Satt 307. Njiti et al. [79] suggested that gene pyramiding would be an effective method for developing cultivars with stable resistance to SDS.

Increasing the stearic acid content to improve soybean oil quality is a desirable breeding objective for food processing applications. Three SSR markers, Satt 070, Satt 474 and Satt 556 were identified to be associated with stearic acid content by Spencer et al. [80]. Identification of these markers may be useful in molecular marker-assisted breeding programmes targeting modifications in soybean fatty acids. RFLP markers have also been used to identify QTLs associated with plant height, lodging and maturity. The major locus associated with plant height was identified as *Dt1* on LG L. *Dt1* was also associated with lodging. In addition, with the help of RFLP markers, two QTLs for plant height (K007 on LG H and A516b on LG N) and one QTL for lodging (cr517 on LG J) were identified. For maturity, independent QTLs were identified in intervals between R051 and N100, and between B032 and CpTI on LG K [81]. RFLP markers were also used for identifying QTLs associated with soybean sprout-related traits. Four QTLs were associated with sprout-yield in the combined analysis done for two years by Lee et al. [82]. They also found that the QTLs conditioning sprout yield were in the same genomic locations as the QTLs for seed weight. These data demonstrates MAS may be feasible for enhancing sprout-yield in soybean.

3.5 Pea

Aphanomyces root rot, caused by *Aphanomyces euteiches* Drechs, is the most important disease of pea worldwide. No efficient chemicals are available to control the pathogen. Thus, to facilitate breeding for Aphanomyces root rot resistance and to better understand the inheritance of partial resistance [83], identified QTLs associated with the disease using DNA markers AFLPs, RFLPs, SSRs, ISSRs and STS. The resulting genetic map consisted of 324 linked markers distributed over 13 linkage groups covering 1,094 cM. A total of seven genomic regions were associated with Aphanomyces root rot resistance. The first one was named as *Aph1*, which was considered a major QTL. Two other specific QTLs, namely *Aph2* and *Aph3* were identified, which were mapped near the *r* (wrinkled/round seeds) and *af* (normal afila leaves) genes. Four other minor QTLs were identified. The resistant alleles of *Aph3* and the two minor QTLs were derived from the susceptible parent. RAPD and SACR markers linked to genes affecting plant architecture of pea, namely, three ramosus genes (*rms2*, *rms3* and *rms4*) and two genes conferring flowering response to photoperiod (*sn and dne*) have been reported by Rameau et al. [84]. In another study, QTLs affecting seed weight in pea were mapped using RFLP markers. Four QTLs were identified in marker intervals on three different linkage groups [85].

3.6   Chickpea

Ascochyta blight is an economically important disease of chickpea caused by the fungus *Ascochyta rabiei*. Udupa and Baum [86] identified and mapped a major locus (*ar1*) using SSRs, which confers resistance to pathotype I, and two independent recessive major loci (*ar2a*), with complementary gene action conferring resistance to pathotype II. In another study, integration of co-dominant STMS markers improved the mapping of ascochyta resistance in chickpea [87]. Resistance gene analogs (RGAs) of *Cicer* were isolated by different PCR approaches and mapped in an inter-specific cross, segregating for *Fusarium* wilt by RFLP and CAPS markers. A total of 13 different RGAs were isolated and classified into nine distinct classes. This study by Huettel et al. [88] provides a starting point for the characterization and genetic mapping of candidate resistance genes in *Cicer* that is useful for MAS and as a pool for resistance genes of *Cicer*.

3.7   Tomato

Tomato is an important vegetable crop. Extensive work has been done on genetic mapping of agriculturally important genes in this crop. Early blight (EB) caused by a devastating fungus, *Alternaria solani* Sorauer causes plant defoliation, reduces yield and fruit quality, and contributes to significant crop loss. QTL mapping using 14 RFLP markers and 23 RGAs identified 10 significant QTLs for EB by Foolad et al. [89]. Potato virus Y (PVY) is also an important disease causing organism which affects the yield of tomato. Resistance against PVY was identified in the wild tomato relative *Lycopersicon hirsutum* PI247087. The locus *pot-1* was mapped using AFLP markers to the short arm of tomato chromosome 3, in the vicinity of the recessive *py-1* locus for resistance to corky root rot [90]. Another important pathogen, cucumber mosaic virus (CMV) gene, *Cmr* was mapped using RFLP and isozyme markers in *L. chilense* and was located on chromosome 12. The chromosome 12 markers were found to be significantly associated with CMV resistance in both qualitative and quantitative models of inheritance. This knowledge of the map location of *Cmr* should accelerate introgression by marker-assisted selection [91].

Identification of tightly linked markers for the genes of importance has facilitated isolation of genes from tomato. For instance, map based cloning strategy was designed to isolate the root-knot nematode resistance gene *Mi* in tomato using PCR-based flanking markers. Fine structure mapping of recombinants with newly developed AFLP and RFLP markers from physically mapped cosmid subclones localized *Mi* to a genomic region of about 550 kb [92]. Two recessive mutations have been discovered in tomato that completely suppress the formation of flower and fruit pedicel abscission zones, i.e. jointless (*j*) and jointless-2 (*j-2*). Both the genes were tentatively localized to chromosome 11 about 30 cM apart. However, RFLP and RAPD markers helped in correctly identifying and mapping the *j-2* locus on chromosome 12 instead of chromosome 11 [93] that enabled map-based cloning of this gene.

Improving organoleptic quality is an important but complex goal for fresh market tomato breeders. A total of 26 traits involved in organoleptic quality variation were evaluated. Physical traits included fruit weight, diameter, colour, firmness and elasticity. Chemical traits were dry matter weight, titratable acidity, pH, and the contents of soluble solids, sugars, lycopene, carotene and 12 aroma volatiles. A total of 81 significant QTLs were detected for the 26 traits using DNA markers [94]. RFLP mapping of 32 independent tomato loci corresponding to genes known to influence fruit ripening and/or ethylene response was reported by Giovannoni et al. [95]. The placement of ripening and ethylene-response loci on the tomato RFLP map would facilitate both the identification of candidate gene sequences corresponding to identified single gene and QTL contributing to fruit development and ethylene response.

3.8 Potato
In vegetable crops like potato, genetic mapping has been done primarily on disease resistance. *Phytophthora infestans* is a very devastating fungus causing late blight of potato. Eleven resistance alleles (R1-R11) are known which confer race-specific resistance to this fungus. In two of the reports, R6 and R7 alleles were mapped by RFLP markers on chromosome XI similar to R3 allele [96] and R2 allele was mapped using AFLP marker [97]. A study on mapping of the resistance gene of root knot nematode (*Meloidogyne chitwoodi*) derived from *Solanum bulbocastanum* in a BC_2 population using RFLP markers have been reported by Brown et al. [98]. RFLP mapping has also been carried out for the potato virus X controlled by a single gene, Nx_{phu}. Four RFLP markers CT220, TG328, CT112 and TG424 from the long arm of chromosome IX that were linked to the hypersensitive phenotype have been reported by Tommiska et al. [99]. Mapping of QTL for resistance to potato cyst nematode (*Globodera rostochiensis*) has been reported by several researchers. In one of the studies, the nematode resistance locus *Gpa2* was mapped on chromosome 12 of potato using 733 AFLP markers. This study also showed that *Gpa2* is linked to the *Rx1* locus conferring resistance to potato virus X [100]. Linkage maps using AFLP and RFLP markers were constructed and used to identify three QTLs on chromosomes V, VI and XII, respectively, for resistance against the potato cyst nematode [101]. In a recent study by Baker at al. [102], nine resistance gene homologues (RGHs) were identified in two diploid clones of potato with a specific primer pair based on conserved motifs in the LRR domain of the potato cyst nematode resistance gene *Gpa2* and the potato virus X resistance gene *Rx1*. AFLP marker was used to facilitate the genetic mapping of the RGHs in the four haplotypes under investigation.

3.9 Sugarcane
Sugarcane is an important cash crop and in order to analyse the inheritance of quantitative traits, extensive study on Quantitative Trait Allele (QTA) mapping was done. The first extensive QTL mapping study performed in cultivated sugarcane was reported by Hoarau et al. [103] based on a population of 295 progenies derived from the selfing of cultivar R570, using about 1,000 AFLP markers. The population was evaluated in a replicated trail for four basic yield components, plant height, stalk number, stalk diameter and brix, in two successive crop-cycles. Forty putative QTAs were found for the four traits of which five appeared in both years. In another study, mapping of QTLs for sugar yield and related tarits, namely pol, stalk weight, stalk number, fiber content and ash content were done using 735 DNA markers. Fifty of the 61 mapped QTLs were clustered in 12 genomic regions of seven sugarcane homologous groups [104].

3.10 Forest Trees
Molecular markers have been successfully applied in tree species also which may be incorporated

into existing improvement programmes in an efficient and cost efficient manner. In loblolly pine, marker-trait associations for components of radial wood density profiles had been found and verification populations have been established to confirm these associations. RAPD markers were employed to map the genome and quantitative trait loci controlling the early growth of a pine hybrid F_1 tree (*Pinus palustris* Mill. × *P. elliottii* Engl.) and a recurrent slash pine tree (*P. elliottii* Engl.) in a (long leaf pine × slash pine) × slash pine BC1 family consisting of 258 progeny [105]. With the help of RFLP markers 13 different height increment and eight different diameter-increment QTLs were detected in loblolly pine by Kaya et al. [106]. Similarly, chemical wood property traits were analysed for the presence of QTLs in a three-generation outbred pedigree of loblolly pine (*Pinus taeda* L.) using DNA markers [107]. Kumar et al. [108] reported multiple-marker mapping of wood density loci in an outbred pedigree of radiata pine using DNA markers. The effect of locations of QTL was found to be significantly associated with the expression of wood density at different ages. These results are encouraging for the application of marker information to early selection in order to increase juvenile wood density.

The single dominant gene (*R*) that confers resistance to the white pine blister rust fungus (*Cronartium ribicola* Fisch.) in *Pinus lambertiana* Dougl. has been mapped using RAPD markers. Thirteen RAPD loci were identified by Harkins et al. [109] that were linked to *R*. This would help in subsequent high-resolution mapping experiments to identify very tightly linked markers to facilitate the eventual cloning of *R*.

RAPD markers have been used to determine the genetic location and effects of genomic regions controlling wood density, stem growth and stem formation in *Eucalyptus* [110]. A total of 86 and 92 markers distributed among 11 linkage groups covered 1295 cM and 1312 cM for *E. urophylla* and *E. grandis*, respectively. This application of marker information will help in early selection of hybrid trees to be vegetatively propagated for the production of clonal varieties.

4. Marker-Assisted Selection (MAS)

Molecular marker-assisted selection involves scoring for the presence or absence of a desired plant phenotype indirectly based on DNA banding pattern of linked markers on a gel or on autoradiogram depending on the marker system. The rationale is that the banding pattern revealing parental origin of the bands in segregants at a given marker locus indicates presence or absence of a specific chromosomal segment which carries the desired allele. This increases the screening efficiency in breeding programmes in a number of ways such that:

(a) the segregants can be scored at the seedling stage for traits that are expressed late in plant development. This includes traits such as grain quality, male sterility and photoperiod sensitivity.

(b) it is possible to screen for traits that are extremely difficult, expensive or time consuming to score and measure such as tolerance to drought, salt, mineral deficiencies and toxicity, root morphology, resistance to nematodes or to specific races or biotypes of diseases or insects.

(c) selection can be practiced for several traits simultaneously, which is difficult or even impossible by conventional means.

(d) heterozygotes are easily identified and distinguished from either homozygotes without resorting to progeny testing. This saves time and effort.

MAS is an attractive option for improvement of certain traits of interest for which phenotypic evaluation is often expensive or unreliable. MAS provides a potential for increasing selection efficiency by allowing earlier selection and reducing plant population size during selection. Breeders can rapidly determine inheritance patterns at the genomic level by directly examining the genetic make up of experimental plants when they are still seedlings. This is especially useful for traits that cannot be identified until the plant is mature such as fruit characteristics and for traits that are difficult to test such as disease resistance. Resistant plants are selected based on DNA markers that are linked to the gene(s) controlling the trait, instead of actually evaluating the disease resistance of the plants. Incorporating natural resistance genes into varieties is the most effective, economic and environmentally safe means of controlling the disease. This is the response to the demand for cost-effective, "green" solutions since it eliminates the need for expensive chemicals to control diseases. It is a uniform method of scoring, tells percentage of genome from each parent, and tells which parts of each chromosome come from each parent. In addition to that as the precision in selection is increased, less unwanted side effects appear in the following generation of plants. MAS can also be used to pyramid two or more desirable genes in a new plant variety.

4.1 Some MAS Advantages in Backcrossing Breeding

There are cases where many conventional backcross programmes fail. For example, despite carefully made backcrosses to the recurrent parent, progeny derived exclusively from self-pollination due to failure of crossing, have been found in backcross programmes. Hence, in conventional breeding, breeders are often not working on the genetic material that they assume they are, because crossing fails more often than expected. A reason why backcrossing fails is the misclassification of a plant for the presence of the donor gene (disease escape instead of disease resistance) and is used as a parent in further backcrossing. All these problems and others like need to make time and resource consuming selfed generations to identify a recessively controlled character are avoided in marker assisted selection.

4.2 MAS Status in Different Crops

Enormous work is being carried out on marker-assisted selection in India and abroad. The progress made in some major crops is presented here.

Rice

The ongoing significant efforts on marker-assisted breeding and gene pyramiding in rice include resistance to blast, blight, gall midge, dwarfing and also drought. Bacterial blight (BB) caused by *Xanthomonas oryzae* pv. *oryzae* (Xoo) is one of the most destructive diseases of rice throughout the world and in some areas of Asia it can reduce crop yield by upto 50%. The most effective approach to combat BB is the use of resistant varieties [111]. So far, 19 resistant genes have been identified [112] and some of these have been incorporated into modern rice varieties. However, the large-scale and long-term cultivation of varieties carrying one of the most important resistant gene *Xa-4* has resulted in significant shifts in the rice frequency of Xoo [113]. In many areas of Indonesia, India, China and Philippines, rice varieties with only *Xa-4* for defense against Xoo have become susceptible to the pathogen. Thus, DNA marker- assisted selection was used to pyramid four bacterial resistance genes, *Xa-4*, *xa-5*, *xa-13* and *Xa-21*. Breeding lines with two, three and four resistance genes were developed and tested for resistance to bacterial blight pathogen. The pyramid lines showed a wider spectrum and a higher level of resistance than lines with only a single gene. To speed up the gene pyramiding process and to facilitate future marker-aided selection, Huang et al. [114] developed PCR markers for two recessive genes *xa-5* and *xa-13*, and used these to survey a range of rice germplasm. The results of the germplasm survey will be useful for the selection of parents in breeding programmes aimed at transferring these bacterial blight resistance genes from one varietal background to another. In India, at Punjab Agricultural University (PAU), Ludhiana, three BB resistance genes *xa-5*, *xa-13* and *Xa-21* were pyramided in PR106 and Pusa 44 background. After multi-location and replicated testing, two PR106 pyramid lines were identified and these have been included in All India Coordinated Testing during 2002. This is the first ever marker- assisted product reaching testing at national level. Pusa 44 pyramid lines were tested in multi-location replicated trials during 2002 [115].

Blast caused by the fungus *Magnaportha grisea* is another devastating disease of rice. The most economical and effective approach to reduce the yield loss is to breed varieties that are resistant to the disease. However, the resistance often breaks down within a few years of cultivar release. Many genes for qualitative blast resistance have been mapped using molecular markers (Table 1) and some of those markers have also been tried in MAS for blast resistance. For example, RG64, a RFLP marker on chromosome 6 is tightly linked (2.8 cM) to *Pi-2(t)*, a major gene for blast resistance [18]. The RG64 rice genome clone was sequenced and primers based on the DNA sequences were found useful in producing polymorphism between the susceptible and resistant varieties after the monomorphic PCR product was digested with restriction enzymes [116]. The CAPS marker was then used to identify rice plants carrying *Pi-2(t)* from an F_2 population derived from the cross between CO39 and CO10151. The effectiveness of the selection for resistant plants based on linked DNA markers was then compared with phenotyping for blast resistance through progeny testing in the F_3 families by blast inoculation. Results indicated that identification of plants carrying *Pi-2(t)* in a large segregating population is possible using one linked marker as well as flanking markers. The accuracy of identification of homozygous resistant genotypes was 96% when RG64 marker was used. The accuracy of selection increased to 100% when two markers flanking the *Pi-2(t)* were scored simultaneously. These results illustrate that marker-assisted identification of linked target gene in a segregating population is efficient in identifying resistant genotypes [117]. This work has been extended futher to pyramid three major genes for blast resistance.

Another objective of marker-assisted breeding in rice is to transfer resistance against gall midge (*Orseolia oryzae*), a major insect pest of rice. The resistance to gall midge biotypes is governed by single dominant genes. At national level, effort has been made to screen rice germplasm for new sources for resistance genes effective against one or more biotypes of the pest [118]. PCR based markers have been designed and currently are being used in marker-assisted selection at different research institutions in the country.

The semi-dwarf gene (*sd-1*) in rice is one of the most important single genes in the history of rice improvement. This single, recessive gene causes reduced culm length and has been widely used to confer lodging resistance, high harvest index, responsiveness to nitrogen fertilizer, and favourable plant type, in the breeding of high-yielding rice varieties [119]. In rice, molecular mapping of *sd-1* has been reported by several workers [120–122]. Chao et al. [123] used 20 mapped clones as probes, based on an existing rice RFLP map [124], and conducted experiments to establish the location of *sd-1* gene. They evaluated the efficacy of marker-assisted selection in F_2 and F_6 plants derived from the cross Milyang 23/Gihobyeo. The application of MAS for *sd-1* gene has potential to greatly improve the efficiency of the Australian rice-breeding programme. In their breeding programme, the semi-dwarf character when detected before maturity even as a heterozygote eliminates the need for progeny testing in a backcrossing programme [125].

Wheat
The wheat stem and leaf rusts are two major pathogens, which can potentially devastate wheat crops. Resistant cultivars have long been depended upon to control disease epidemics. Most of the genes for resistance have been mapped using molecular markers and currently being used in MAS at different national as well as international centres. In India, RAPD markers linked to *Lr19* (leaf rust) gene has been converted into SCAR markers. *Lr 28* gene was also tagged by two flanking RAPD marker $S464_{700}$ and $S326_{350}$. These linked molecular markers are further in use in pyramiding of rust resistance genes, which is difficult, and time consuming by conventional breeding procedures [126]. A population of 220 BC_1F_2 plants segregating for two genes *Cre1* and *Cre3* was evaluated with three molecular markers Xglk 605, Xcdo 588 and Cd 2.2 and the markers were found to provide a reliable means of gene pyramiding and selecting plants carrying the genes in wheat breeding programmes [127].

With overall goal of transferring new developments in genomics to wheat breeding and production, investigators at 12 public wheat-breeding and research programmes across the US including University of California, Colorado State University, Cornell University, Kansas State University, Montana State University, University of Idaho, University of Minnesota, Purdue University, University of Nebraska, USDA and Washington State University have constituted a national wheat Marker Assisted Selection (MAS) consortium that aims to use molecular markers as chromosome landmarks in MAS programme to facilitate introgression of small chromosome segments carrying the genes of interest. Available molecular markers will be used to transfer genes for resistance to fungi, viruses and insects as well as gene variants related to improved bread, pasta and noodle quality. These genes will be incorporated into a minimum of 240 adapted cultivars or breeding lines belonging to all major market classes of US wheat, and since they are transferred by normal recombination, the resulting lines will not be classified as transgenics. These improved cultivars will transfer the value of genomic research to the wheat growers' fields [128].

4.3 Other Crops
Use of molecular markers in marker-assisted selection has been carried out for improvement of several other crops such as sunflower, tomato, sugar beet, barley, soybean, apple etc. throughout the world. These are briefly described crop-wise as follows.

Sunflower
MAS for two rust resistance genes in sunflower was reported using RAPD markers $Ox20_{600}$ and $OO04_{950}$ linked to the gene R_Adv responsible for rust resistance in the proprietary inbred line P_2. This gene confers resistance to most of the pathotypes of *Puccinia helianthi* identified in Australia. These RAPD markers were converted into SCAR markers and the robustness of these markers were demonstrated through the amplification in a diverse range of sunflower germplasm. This

will be useful in further attempts for molecular-assisted breeding to produce durable resistance in sunflower to *P. helianthi* [129].

Tomato

MAS has been carried out for several traits including fruit characteristics in tomato. More recently, MAS was used to transfer the ability to accumulate acylsugars to cultivated tomato. RFLP and PCR-based markers were used through three backcross generations to select plants containing five target regions associated with acylsugar accumulation [130]. In another example, MAS has been demonstrated for QTL influencing blackmold resistance. Blackmold, caused by the fungus *Alternaria alternata*, is a major ripe fruit disease of processing tomatoes. Five QTLs were selected for introgression from *Lycopersicon cheesmanii* into cultivated tomato using marker-assisted selection. RFLP and PCR-based markers flanking and within the chromosomal regions containing QTLs were used for MAS during backcross and selfing generations [131].

Barley

The effectiveness of molecular marker-assisted selection for malting quality trait in barley was reported by Han et al. [132]. In this study, the flanking markers, Brz and Amy2, and WG622 and BCD402B, for two major QTL regions present on chromosomes 1 and 4 were used for MAS. The MAS for QTL1 was more effective than phenotypic selection. It could substantially eliminate undesirable genotypes by early genotyping and keeping only desirable genotypes for later phenotypic selection. The MAS was also used for verification of yield QTL in a barley cross. The objectives of this study were to verify the value of four QTLs for selection and to compare the efficiency of alternative MAS strategies using these QTL vs. conventional phenotypic selection for grain yield. It was shown that MAS was as good as phenotypic selection [133].

Soybean

MAS offers the potential to reduce linkage drag and to pyramid genes with similar phenotypic effects into elite genotypes. One such example was seen in soybean breeding programme where a QTL conditioning corn earworm resistance in the accession PI229358 and a synthetic *Bacillus thuringiensis cry1Ac* transgene from the recurrent parent 'Jack-Bt' were pyramided into BC_2F_3 plants by marker-assisted selection. Segregating individuals were genotyped at SSR markers linked to an antibiosis/antixenosis QTL on linkage group M, and were tested for the presence of *cryAc1*. MAS was used during and after the two backcrosses to develop a series of BC_2F_3 plants with or without *cryAc1* transgene and the QTL conditioning for resistance in BC_2F_3 plants that were homozygous for parental alleles at markers. This work by Walker et al. [134] demonstrated the usefulness of SSR for MAS in soybean, and showed that combining transgene and QTL-mediated resistance to lepidopteran pests might be a viable strategy for insect control.

Apple

MAS is also a promising method to select resistant individuals in horticultural crops like apple. Molecular tools have the potential to give very early information on the genetics of apple seedlings. The aims of apple breeding such as high fruit quality, consistently high yields and durable disease and pest resistance can be achieved more efficiently. Progress in MAS for apple breeding is being achieved mainly in the area of disease resistance especially in the durable incorporation of scab and mildew resistance. The AL07-SCAR and M18-CAPS molecular analysis in progenies, in which both parents are heterozygous for the resistance gene, made it possible to identify clearly the homozygous plants for the *Vf* gene (resistance for apple scab) and these plants showed higher level of resistance than the heterozygous one. Progenies were developed from crosses with parents that carry different resistance genes such as *Vf, Vm, Vb, Pl1* and QTLs in different combinations [135–137].

Kentucky Bluegrass

The MAS has wide range of utility in this grass species. It was reported by Albertini et al. [138] that MAS has helped in avoiding costly and time-consuming phenotypic progeny tests in *Poa pratensis* to study mode of reproduction. Genotypic apomixis in Kentucky bluegrass involves the pathenogenetic development of unreduced eggs from aposporic embryo sacs. Two SCAR primer pairs were tested and identified the apomictic and sexual genotypes among progenies of sexual × apomictic crosses with low bias. Furthermore, when tested on a wide range of Italian and exotic *P. pratensis* germplasm, they were able to unequivocally distinguish sexual from apomictic genotypes. This system should, therefore, allow new selection models to be set up in this species.

5. Future Prospects of MAS

The above review of the available literature reveals that during the last 17 years since the publication of the first paper on the use of RFLP markers for construction of linkage maps in tomato and maize in 1986, molecular markers have been extensively used for mapping and tagging of hundreds of different agriculturally important genes/QTL in various crop species. With the availability of linked markers, the first requirement for successful MAS has been fulfilled. Besides, the feasibility of MAS based on these linked markers has been demonstrated in several crops both for qualitative as well as quantitative traits as evident from the above description. However, MAS is yet to be used routinely in plant breeding programmes.

Utility of the MAS in crop plants is currently limited by factors such as recombination between the marker and the target gene, low level of polymorphism between parents with contrasting traits and lower resolution of QTLs due to interaction with the environment. With the recent developments in both structural and functional genomics, it would not be difficult to find solutions to these problems. Availablity of high-density genetic and physical maps will enable finding markers physically closer to the target gene that would not allow failure of MAS due to genetic recombination. Moreover, cloning and characterization of the target genes, which are possible based of their position on the linkage map, would allow development of allele-specific markers. Use of such markers would completely eliminate the possibility of breakdown of the marker-trait linkage. Besides, markers based on the sequences of the genes would facilitate allele mining in the germplasm resources, thereby leading to identification and utilization of newer alleles in crop improvement. Different alleles of a gene would differ for a number of nucleotides at different positions in their sequence that would be the basis of developing highly polymorphic single nucleotide polymorphism (SNP) markers. The problem of low level of polymorphism in narrow crosses can thus be circumvented. Use of MAS for QTL, particularly those having little effect on trait expression and highly interacting with environment would require greater amount of research effort and newer experimental strategies.

Complete integration of MAS with the conventional plant breeding programmes demands consideration of two important factors: a) size of population and b) cost. Plant breeding experiment requires screening of large segregating populations routinely over generations. Genotyping of large number of samples manually is an extremely difficult task. MAS to be practicable should be amenable to automation that would allow handling of large number of samples. Development and use of PCR based markers such as STS and SCAR will be a key to success of MAS in crop improvement. As the technology develops and gets modified to analyze large number of samples, the cost will automatically go down. The investment in gene tagging, and selection based on molecular markers should be weighted against the overall cost and time involved in traditional breeding program. Even though the cost of MAS is higher at present level of estimation, its integration with traditional plant breeding is desirable because of immense possibilities it offers.

References

1. J.S. Beckmann, M. Soller, Restriction fragment length polymorphisms in genetic improvement: Methodologies, mapping and costs. Theor. Appl. Genet. 67 (1983) 35–43.
2. A. Darvasi, M. Soller, Optimum spacing of genetic markers for determining linkage between marker loci and quantitative trait loci. Theor. Appl. Genet. 89 (1994) 351–357.
3. D. Botstein, R.L. White, M. Skolnick, R.W. Davis, Construction of a genetic linkage map in man using restriction fragment length polymorphisms. Am. J. Hum. Genet. 32 (1980) 314–331.
4. E.M. Southern, Detection of specific sequences of DNA fragments separated by gel electrophoresis. J. Mol. Bio. 98 (1975) 503–517.
5. J.G.K. Williams, A.R. Kubelik, K.J. Livak, J.A. Rafalski, S.V. Tingey, DNA polymorphisms amplified by arbitrary primers are useful as genetic markers. Nucleic Acids Res. 18 (1990) 6531–6535.
6. I. Paran, R.W. Michelmore, Development of reliable PCR based markers linked to downy mildew resistance genes in lettuce. Theor. Appl. Genet. 85 (1993) 985–993.
7. M. Olson, L. Hood, C. Cantor, D. Botstein, A common language for physical mapping of the human genome. Science 245 (1989) 1434–1435.
8. A. Konieczyn, F.M. Ausubel, A procedure of mapping *Arabidopsis* mutations using co-dominant ecotype specific PCR-based markers. Plant J. 4 (1993) 403–410.
9. P. Vos, R. Hogers, M. Bleeker, M. Reijans, T. Lee, M. Hornes, A. Frijters, J. Pot, J. Peleman, M. Kuiper, M. Zabeau, AFLP: a new technique for DNA fingerprinting. Nucleic Acids Res. 23 (1995) 4407–4414.
10. A.J. Brookes, The essence of SNPs. Gene 234 (1999) 177–186.
11. T.A. Brown, Genomics, 1st edn. BIOS Scientific Publishers, Oxford, 1999.
12. Y. Harushima, M. Yano, A. Shomura, M. Sato, T. Shimano, Y. Kuboki, T. Yamamoto, S.Y. Lin, B.A. Antonio, A. Parco, H. Kajiya, N. Huang, K. Yamamoto, Y. Nagamura, N. Kurata, G.S. Khush, T. Sasaki, A high-density rice genetic map with 2275 markers using a single F_2 population. Genetics 148 (1998) 479–494.
13. S.R. McCouch, L. Teytelman, Y. Xu, K.B. Kobas, K. Clare, M. Walton, B. Fu, R. Maghirang, Z. Li, Y. Xing, Q. Zhang, I. Kano, M. Yano, F. Jellstrom, G. DeClerck, D. Schneider, S. Cartinhour, D. Ware, L. Stein, Development and mapping of 2240 new SSR markers for rice (*Oryza sativa* L.). DNA Res. 9 (2002) 199–207.
14. M. Mohan, S. Nair, J.S. Bentur, U.P. Rao, J. Bennett, RFLP and RAPD mapping of the rice *Gm2* gene that confers resistance to biotype 1 of gall midge (*Orseolia oryzae*). Theor. Appl. Genet. 87 (1994) 782–788.
15. S. Nair, A. Kumar, M.N. Srivastava, M. Mohan, PCR-based DNA markers linked to a gall midge resistance gene, *Gm4t*, has potential far marker-assisted selection in rice. Theor. Appl. Genet. 92 (1996) 660–665.

16. R. Mago, S. Nair, M. Mohan, Resistance gene analogues from rice: cloning, sequencing and mapping. Theor. Appl. Genet. 99 (1999) 50–57.

17. L.S. Sebastian, R. Ikeda, N. Huang, T. Imbe, W. R. Coffman, M. Yano, T. Sasaki, S.R. McCouch, Genetic mapping of resistance to rice tungro spherical virus (RTSV) and green leafhopper (GLH) in ARC11554, in: G.S. Khush (Ed), Rice Genetics III, Proceedings of the Third International Rice Genetics Symposium, IRRI, Manila, Philippines, 1996, pp. 560–564.

18. Z.H. Yu, D.J. Mackill, J.M. Bonman, S.D. Tanksley, Tagging genes for balst resistance in rice via linkage to RFLP markers. Theor. Appl. Genet. 81 (1991) 471–476.

19. S.R. McCouch, M.L. Abenes, R. Angeles, G.S. Khush, S.D. Tanksley, Molecular tagging of a recessive gene, *xa-5* for resistance to bacterial blight of rice. Rice Genet. Newsl. 8 (1992) 143–145.

20. S. Yoshimura, A. Yoshimura, A. Saito, N. Kishimoto, M. Kawase, M. Yano, M. Nakagahara, T. Ogawa, N. Iwata, RFLP analysis of introgressed chromosomal segments in three near-isogenic lines of rice for bacterial blight resistance genes *Xa-1*, *Xa-3*, and *Xa-4*. Jpn. J. Genet. 67 (1992) 29–32.

21. S. Yoshimura, A. Yoshimura, N. Iwata, S.R. McCouch, M.L. Abenes, M.R. Baraoidan, T.W. Mew, R.J. Nelson, Tagging and combining bacterial blight resistance genes in rice using RAPD and RFLP markers. Mol. Breed. 1 (1995) 375–387.

22. M.W. Blair and S.R. McCouch, Development of diagnostic markers for the bacterial blight resistance gene, *xa5*, in: G.S. Khush (Ed), Rice Genetics III, Proceedings of the Third International Rice Genetics Symposium, IRRI, Manila, Philippines, 1996, pp. 582–589.

23. M. Mei, C. Zhuang, R. Wan, J. Wu, W. Hu, G. Kochert, Genetic analysis and tagging of genes for brown planthopper resistance in indica rice, in: G.S. Khush (Ed), *Rice Genetics III, Proceedings of the* Third International Rice Genetics Symposium, IRRI, Manila, Philippines, 1996, pp. 590–595.

24. Xu, K., Mackill, D. J., A major locus for submergence tolerance mapped on rice chromosome 9. *Mol. Breed.* 2 (1996) 219–224.

25. G.Y. Zhang, Y. Guo, S.L. Chen, S.Y. Chen, RFLP tagging of a salt tolerance gene in rice. Plant Sci. 110 (1995) 227: 234.

26. J.S. Zhang, C. Xie, Z.Y. Li, S.Y. Chen, Expression of the plasma membrane H^+-ATPase gene in response to salt stress in a rice salt-tolerant mutant and its original variety. Theor. Appl. Genet. 99 (1999) 1006–1011.

27. M. Wissuwa, J. Wegner, N. Ae, M. Yano, Substitution mapping of Pup 1: a major QTL increasing phosphorus uptake of rice from a phosphorus-deficient soil. Theor. Appl. Genet. 105 (2002) 890–897.

28. B.D. Nguyen, D.S. Brar, B.C. Bui, T.V. Nguyen, L.N. Pham, H.T. Nguyen, Identification and mapping of the QTL for aluminium tolerance introgressed from the new source, *Oryza rufipogon* Girff. into indica rice (*Oryza sativa* L.). Theor. Appl. Genet. 106 (2003) 583–593.

29. B. Wang, W.W. Xu, J.Z. Wang, W. Wu, H.G. Zheng, Z.Y. Yang, J.D. Ray, H.T. Nguyen, Tagging and mapping the thermo-sensitive genic male-sterile gene in rice (*Oryza sativa* L.) with molecular markers. Theor. Appl. Genet. 91 (1995) 1111–1114.

30. Y. Yamaguchi, R. Ikeda, H. Hirasawa, M. Minami, A. Ujihara, Linkage analysis of thermosensitive genic male sterility gene, *tms-2*, in rice (*Oryza sativa* L.). Breed. Sci., 47 (1997) 371–373.

31. P.K. Subudhi, R.P. Borkakati, S.S. Virmani, N. Huang, Molecular mapping of a thermosensitive genetic male sterility gene in rice using bulk segregant analaysis. Genome 40 (1997) 188–194.

32. O.U.K. Reddy, E.A. Siddiq, N.P. Sarma, J. Ali, A.J. Hussain, P. Nimmakayala, P. Ramasamy, S. Pammi, A.S. Reddy, Genetic analysis of temperature-sensitive male sterility in rice. Theor. Appl. Genet. 100 (2000) 794–801.

33. N.V. Dong, P.K. Subudhi, P.N. Luong, V.D. Quang, T.D. Quy, H.G. Zheng, B. Wang, H.T. Nguyen, Molecular mapping of a rice gene conditioning thermosensitive genic male sterility using AFLP, RFLP and SSR techniques. Theor. Appl. Genet. 100 (2000) 727–734.

34. S.N. Ahn, C.N. Bollich, S.D. Tanksley, RFLP tagging of a gene for aroma in rice. Theor. Appl. Genet. 84 (1992) 825–828.

35. S.N. Ahn, C.N. Bollich, A.M. McClung, S.D. Tanksley, RFLP analysis of genomic regions associated with cooked-kernel elongation in rice. Theor. Appl. Genet. 87 (1993) 27–32.

36. C. Brondani, P.H.N. Rangel, R.P.V. Brondani, M.E. Ferreira, QTL mapping and introgression of yield related traits from *Oryza glumaepatula* to cultivated rice (*Oryza sativa*) using microsatellite markers. Theor. Appl. Genet. 104 (2002) 1192-1203.

37. S. Naik, K.S. Gill, V.S.P. Rao, V.S. Gupta, S.A. Tamhankar, S. Pujar, B.S. Gill, P.K. Ranjekar, Identification of a STS marker linked to the *Aegilops speltoides*-derived leaf rust resistance of gene *Lr28* in wheat. Theor. Appl. Genet. 97 (1998) 535–540.

38. J. Gold, D. Harder, F.T. Smith, T. Aung, J. Procunier, Development of a molecular marker for rust resistance genes *SR39* and *LR35* in wheat breeding lines. Electronic J. of Biotech. ISSN (1999) 2717–3458.

39. T. Fahima, M.S. Roder, A. Grama, E. Nevo, Microsatellite DNA polymorphism divergence in *Triticum dicoccoides* accessions highly resistant to yellow rust. Theor. Appl. Genet. 94 (1998) 98–103.

40. J.H. Peng, T. Fahima, M.S. Roder, Y.C. Li, A. Dahan, A. Grama, Y. I. Ronin, A. B. Karol, E. Nevo, Microsatellite tagging of the stripe rust resistance gene *YrH52* derived from wild emmer wheat, *Triticum dicoccoides*, and suggestive negative crossover interfernce on chromosome 1B. Theor. Appl. Genet. 98 (1999) 862–872.

41. V. Korzun, M.S. Roder, M.W. Ganal, A.J. Worland, C.N. Law, Genetic analysis of the dwarfing gene (*Rht8*) in wheat. Part I. Molecular mapping of Rht8 on the short arm of chromosome 2D of bread wheat (*Triticum aestivum* L.). Theor. Appl. Genet. 96 (1998) 1104–1109.

42. A.J. Worland, V. Korzun, M.S. Roder, M.W. Ganal, C.N. Law, Genetic analysis of the dwarfing gene *Rht8* in wheat. Part II. The distribution and adaptive significance of allelic variants at the *Rht8* locus of wheat as revealed by microsatellite screening. Theor. Appl. Genet. 96 (1998) 1110–1120.

43. V. Korzun, M.S. Roder, A.J. Worland, A. Borner, Intrachromosomal mapping of genes for dwarfing (*Rht12*) and vernalization response (*Vrn1*) in wheat by using RFLP and microsatellite markers. Plant Breed. 116 (1997) 227–232.

44. M. Prasad, R.K. Varshney, A. Kumar, H.S. Balyan, P.C. Sharma, K.J. Edwards, H. Singh, H.S. Dhaliwal, J.K. Roy, P.K. Gupta, A microsatellite marker associated with a QTL for grain protein content on chromosome arm 2DL of bread wheat. Theor. Appl. Genet. 99 (1999) 341–345.

45. J.K. Roy, M. Prasad, R.K. Varshney, H.S. Balyan, T.K. Blake, H.S. Dhaliwal, H. Singh, K.J. Edwards, P.K. Gupta,. Identification of a microsatellite on chromosome 6B and a STS on 7D of bread wheat showing an association with preharvest sprouting tolerance. Theor. Appl. Genet. 99 (1999) 336–340.

46. H. Singh, M. Prasad, R.K. Varshney, J.K. Roy, H.S. Balyan, H.S. Dhaliwal, P.K. Gupta, STMS markers for grain protein content and their validation using near-isogenic lines in bread wheat. Plant Breed. 120 (2001) 273–278.

47. C.C. Bernier, Diseases of rapeseed in Manitoba in 1971. Can. Plant Dis. Surv. 52 (1972) 108.

48. M.E. Ferreira, S.R. Rimmer, P.H. Williams, T. C. Osborn, Mapping loci controlling *Brassica napus* resistance to *Leptosphaeria maculans* under different screening conditions. Genetics 85 (1995) 213–217.

49. C. Kole, R. Teutonico, A. Mengistu, P.H. Williams, T.C. Osborn, Molecular mapping of a locus controlling resistance to *Albugo candida* in *Brassica rapa*. Phytopath. 86 (1996) 367–369.

50. W.Y. Cheung, R.K. Gugel, B.S. Landry, Identification of RFLP markers linked to the white rust resistance gene (*Acr*) in mustard (*Brassica juncea* (L.) Czern. and Coss). Genome 41 (1998) 626–628.

51. K.V. Prabhu, D.J. Somers, G. Rakow and R.K. Gugel, Molecular markers linked to white rust resistance in mustard *Brassica juncea*. Theor. Appl. Genet. 97 (1998) 865–870.

52. A.K. Mukherjee, T. Mohapatra, A. Varshney, R. Sharma, R.P. Sharma, Molecular mapping of a locus controlling resistance to *Albugo candida* in Indian mustard. Plant Breed. 120 (2001) 483–487.

53. C. Kole, P.H. Williams, S.R. Rimmer, T.C. Osborn, Linkage mapping of genes controlling resistance to white rust (*Albugo candida*) in *Brassica rapa* (syn. *campestris*) and comparative mapping to *Brassica napus* and *Arabidopsis thaliana*. Genome 45 (2002) 22–27.

54. Y. Dion, R.K. Gugel, G.F.W. Rakow, G. Seguin-Swartz, B.S. Landry, RFLP mapping of resistance to the blackleg disease [causal agent, *Leptosphaeria maculans* (Desm) Ces et de Not.] in canola (*Brassica napus* L.). Theor. Appl. Genet. 91 (1995) 1190–1194.

55. R.E. Voorrips, M.C. Jongerius, H.J. Kanne, Mapping of two genes for resistance to clubroot (*Plasmodiophora brassicae*) in a population of doubled haploid lines of *Brassica oleracea* by means of RFLP and AFLP markers. Theor. Appl. Genet. 94 (1997) 75–82.

56. S.S. Figdore, M.E. Ferreria, M.K. Slocum and P.H. Williams, Association of RFLP markers with trait loci affecting clubroot resistance and morphological characters in *Brassica oleracea* L. Euphytica 69 (1993) 33–44.

57. J. Hu, C.F. Quiros, P. Arus, D. Struss, Robbelen, Mapping of a gene determining linolenic acid concentration in rapeseed with DNA-based markers. Theor. Appl. Genet. 90 (1995) 258–262.

58. P.K. Tanhuanpaa, J.P. Vilkki, H.J. Vilkki, Association of a RAPD marker with linolenic acid concentration in seed oil of rapeseed (*Brassica napus* L.). Genome 38 (1995) 414-416.

59. J. Hu, G. Li, D. Struss, C.F. Quiros, SCAR and RAPD markers associated with 18-carbon fatty acids in rapeseed, Brassica napus. Plant Breed. 118 (1999) 145–150.

60. V. Arondel, B. Lemieux, T. Hwang, S. Gibson, H.M. Goodman, C.R. Somerville, 1992. Map-based cloning of a gene controlling omega-3 fatty acid desaturation in *Arabidopsis*. Science 258: 1352–1355.

61. C. Jourden, P. Barret, D. Brunel, R. Delourme, M. Renard, Specific molecular marker of genes controlling linolenic acid content in rapeseed. Theor. Appl. Genet. 93 (1996a) 512–518.

62. C. Jourden, P. Barret, R. Horvais, R. Delourme and M. Renard, Identification of RAPD marker linked to linolenic acid genes in rapeseed. Euphytica 90 (1996b) 351–357.

63. C.E. Thormann, J. Romero, J. Mantet, T.C. Osborn, Mapping loci controlling concentration of erucic and linolenic acids in seed oil of *Brassica napus* L. Theor. Appl. Genet. 93 (1996) 282–286.

64. P.K. Tanhuanpaa, J.P. Vilkki, H.J. Vilkki, Mapping a QTL for oleic acid concentration in spring turnip rape (*Brassica rapa* ssp. *oleifera*). Theor. Appl. Genet. 92 (1996) 952–956.

65. R. Sharma, R.A.K. Aggrawal, R. Kumar, T. Mohapatra, R.P. Sharma, Construction of RAPD linkage map and localization of QTLs for oleic acid level using recombinant inbreds in mustard. Genome 45 (2002) 467-472.

66. W. Ecke, M. Uzunova, K. Weissleder, Mapping the genome of rapeseed (*Brassica napus* L.). II. Localization of genes controlling erucic acid synthesis and oil content. Theor. Appl. Genet. 91 (1995) 972–977.

67. C. Jourden, P. Barret, R. Horvais, N. Foisset, R. Delourme, M. Renard, Identification of RAPD markers linked to loci controlling erucic acid level in rapeseed. Mol. Breed. 2 (1996c) 61–71.

68. P. Barret, R. Delourme, M. Renard, F. Domergue, L. Lessire, M. Delseny, T.J. Roscoe, A rapeseed *FAE1* gene is linked to the E1 locus associated with the variation in the content of erucic acid. Theor. Appl. Genet. 96 (1998) 177–186.

69. M. Fourmann, P. Barret, M. Renard, G. Pelletier, R. delourme, D. Brunel, The two genes homologus to *Arabidopsis FAE1* co-segregate with the two loci governing erucic acid content in *Brassica napus*. Theor. Appl. Genet. 96 (1998) 852–858.

70. R.A. Teutonico, T.C. Osborn, Mapping of RFLP and quantitative trait loci in *Brassica rapa* and comparison to the linkage maps of *B. napus*, *B. oleracea* and *Arabidopsis thaliana*. Theor. Appl. Genet. 89 (1994) 885–894.

71. A.E. Van Deynze, B.S. Landry, K.P. Pauls, The identification of restriction fragment length polymorphisms linked to seed colour genes in *Brassica napus*. Genome 38 (1995) 534–542.

72. B.Y. Chen, R.B. Jorgensen, B.F. Cheng, W.K. Heneen, Identification and chromosomal assignment of RAPD marker linked with a gene for seed coat colour in a *Brassica campestris-alboglabra* addition line. Hereditas 126 (1997) 133–138.

73. A. Upadhyay, T. Mohapatra, R.A. Pai, R.P. Sharma, Molecular tagging and character tagging in Indian mustard (*Brassica juncea*). II. RFLP marker association with seed coat colour and quantitative traits. J. Plant Biochem. Biotech. 5 (1996) 17–22.

74. M.S. Negi, M. Devic, M. Delseny, M. Lakshmikumaran, Identification of AFLP fragments linked to seed coat colour in *Brassica juncea* and conversion to SCAR marker for rapid selection. Theor. Appl. Genet. 101 (2000) 146–152.

75. P.B. Cregan, J. Mudge, E.W. Fickus, D. Danesh, R. Denny, N.D. Young, Two simple sequence repeat

markers to select for soybean cyst nematode resistance conditioned by the *rhg1* locus. Theor. Appl. Genet. 99 (1999) 811–818.

76.   K. Meksem, P. Pantazopoulos, V.N. Njiti, L.D. Hyten, P.R. Arelli, D.A. Lightfoot, 'Forrest' resistance to the soybean cyst nematode is bigenic: saturation mapping of the *Rhg1* and *Rhg4* loci. Theor. Appl. Genet. 103 (2001) 710–717.

77.   B.G. Rector, J.N. All, W.A. Parrott, H.R. Boerma, Identification of molecular markers linked to quantitative trait loci for soybean resistance to corn earworm. Theor. Appl. Genet. 96 (1998) 786–790.

78.   A.J. Hayes and M.A. Saghai, Maroof, Targeted resistance gene mapping in soybean using modified AFLPs. Theor. Appl. Genet. 100 (2000) 1279–1283.

79.   V.N. Njiti, K. Meksem, M.J. Iqbal, J.E. Johnson, M.A. Kassem, K.F. Zobrist, V.Y. Kilo, D.A. Lightfoot, Common loci underlie field resistance to soybean sudden death syndrome in Forrest, Pyramid, Essex, and Douglas. Theor. Appl. Genet. 104 (2002) 294–300.

80.   M.M. Spencer, V.R. Pantallone, E.J. Meyer, D. Ellis-Landau, D.L. Jr Hyten, Mapping of *Fas* locus controlling stearic acid content in soybean. Theor. Appl. Genet. 106 (2003) 615–619.

81.   S.H. Lee, M.A. Bailey, M.A.R. Mian, E.R. Shipe, D.A. Ashley, W.A. Parrott, R.S. Hussey and H.R. Boerma, Identification of quantitative trait loci for plant height, lodging and maturity in a soybean population segregating for growth habit. Theor. Appl. Genet. 92 (1996) 516–523.

82.   S.H. Lee, K.Y. Park, H.S. Lee, E.H. Park and H.R. Boerma, Genetic mapping of QTLs conditioning soybean sprout yield and quality. Theor. Appl. Genet. 103 (2001) 702–709.

83.   M.L. Pilet-Nayel, F.J. Muehlbauer, R.J. McGee, J.M. Kraft, A. Baranger, C.J. Coyne, Quantitative trait loci for partial resistance to Aphanomyces root rot in pea. Theor. Appl. Genet. 106 (2002) 28–39.

84.   C. Rameau, D. Denoue, F. Fraval, K. Haurogne, J. Josserand, U. Laucou, S. Batge, I.C. Murfet, Genetic mapping in pea.2. Identification of RAPD and SCAR markers linked to genes affecting plant architecture. Theor. Appl. Genet. 97 (1998) 916–928.

85.   G.M. Timmerman-Vaughan, J.A. McCallum, T.J. Frew, N.F. Weeden, A.C. Russell, Linkage mapping of quantitative trait loci controlling seed weight in pea (*Pisum sativum* L.). Theor. Appl. Genet. 93 (1996) 431–439.

86.   S.M. Udupa, M. Baum, Genetic dissection of pathotype-specific resistance to ascochyta blight disease in chickpea (*Cicer arietinum* L.) using microsatellite markers. Theor. Appl. Genet. 105 (2002) 470–479.

87.   M. Tekeoglu, P.N. Rajesh, F.J. Muehlbauer, Integration of sequence tagged microsatellite sites of the chickpea genetic map. Theor. Appl. Genet. 105 (2002) 847–854.

88.   B. Huettel, D. Santra, F.J. Muehlbauer, G. Kahl, Resistance gene analogues of chickpea (*Cicer arietinum* L.): isolation, gentic mapping and association with a *Fusarium* resistance gene cluster. Theor. Appl. Genet. 105 (2002) 479–590.

89.   M.R. Foolad, L.P. Zhang, A.A. Khan, D. Liu-Nino, G.Y. Liu, Identification of QTLs for early blight (*Alternaria solani*) resistance in tomato using backcross populations of a *Lycopersicon esculentum* × *L. hirsutum* cross. Theor. Appl. Genet. 104 (2002) 945–958.

90.   G. Parrella, S. Ruffel, A. Moretti, C. Morel, A. Palloix, C. Caranta, Recessive resistance genes against potyviruses are localized in collinear genomic regions of the tomato (*Lycopersicon* spp.) and pepper (*Capsicum* spp.) genomes. Theor. Appl. Genet. 105 (2002) 855–861.

91.   B.S. Stamova, R.T. Chetelat, Inheritance and genetic mapping of cucumber mosaic virus resistance introgressed from *Lycopersicon chilense* into tomato. Theor. Appl. Genet. 101 (2000) 527–537.

92.   I. Kaloshian, J. Yaghoobi, T. Liharska, J. Hontelez, D. Hauson, P. Hogan, T. Jesse, J. Wijbrandi, G. Simons, P. Vos, P. Zabel, V.M. Williamson, Genetic and physical localization of the root-knot nematode resistance locus *Mi* in tomato. Mol. Gen. Genet. 257 (1998) 376–385.

93.   H.B. Zhang, M.A. Budiman, R.A. Wing, Genetic mapping of jointless-2 to tomato chromosome 12 using RFLP and RAPD markers. Theor. Appl. Genet. 100 (2000) 1183–1189.

94.   V. Saliba-Colombani, M. Causse, D. Langlois, J. Philouze, M. Buret, Genetic analysis of organoleptic quality in fresh market tomato. 1. Mapping QTLs for physical and chemical traits. Theor. Appl. Genet. 102 (2001) 259–272.

95. J. Giovannoni, H. Yen, B. Shelton, S. Miller, J. Vrebalov, P. Kannan, D. Tieman, R. Hackett, D. Grierson and H. Klee, Genetic mapping of ripening and ethylene-related loci in tomato. Theor. Appl. Genet. 98 (1999) 1005–1013.

96. A. El Kharbotly, C. Palomino-Sanchez, F. Salamini, E. Kacobsen, C. Gebhardt, *R6* and *R7* alleles of potato conferring race-specific resistance to *Phytophthora infestans* (Mont.) de Bary identified genetic loci clustering with the *R3* locus on chromosome XI. Theor. Appl. Genet. 92 (1996) 880–884.

97. X. Li, H.J. vanEck, J.N.A.M. Rouppe vander Voort, D.J. Huigen, P. Stam, E. Jacobsen, Aitotetraploids and genetic mapping using common AFLP markers: the *R2* allele conferring resistance to *Phytophthora infestans* mapped on potato chromosome 4. Theor. Appl. Genet. 96 (1998) 1121–1128.

98. C.R. Brown, C.P. Yang, H. Mojtahedi, G.S. Santo, R. Masuelli, RFLP anlysis of resistance to Columbia root-knot nematode derived from *Solanum bulbocastanum* in a BC$_2$ population. Theor. Appl. Genet. 92 (1996) 572–576.

99. T.J. Tommiska, J.H. Hamalianen, K.N. Watanabe, J.P.T. Valkonen, Mapping of the gene Nx_{phu} that controls hypersensitive resistance to potato virus X in *Solanum phureja* IvP35. Theor. Appl. Genet. 96 (1998) 840–843.

100. J. Rouppe van der Voort, P. Wolters, R. Folkertsma, R. Hutten, P. vanZandvoort, H. Vinke, K. Kanyuka, A. Bendahmane, E. Jacobsen, R. Janssen, J. Bakker, Mapping of the cyst nematode resistance locus *Gpa2* in potato using a strategy based on comigrating AFLP markers. Theor. Appl. Genet. 95 (1997) 874–880.

101. B. Caromel, D. Mugniery, V. Lefebvre, S. Andrzejewski, D. Ellisseche, M.C. Kerlan, P. Roueselle, F. Rousselle-Bourgeois, Mapping QTLs for resistance against *Globodera pallida* (Stine) Pa2/3 in a diploid potato progeny originating from *Solanum spegazzinii*. Theor. Appl. Genet. (online) 2003.

102. E. Bakker, P. Butterbach, J. Rouppe vander Voort, E. van der Vossen, J. van Vliet, J. Bakker, A. Goverse, Genetic and physical mapping of homologues of the virus resistance gene *Rx1* and the cyst nematode resistance gene *Gpa2* in potato. Theor. Appl. Genet. (online) 2003.

103. J.Y. Hoarau, L. Grivet, B. Offmann, L.M. Raboin, J.P. Diorflar, J. Payet, M. Hellmann, A.D. Hont, J.C. Glaszmann, Genetic dissection of a modern sugarcane cultivar (*Saccharum* spp.). II. Detection of QTLs for yield components. Theor. Appl. Genet. 105 (2002) 1027–1037.

104. R. Ming, Y.W. Wang, X. Draye, P.H. Moore, J.E. Irvine, A.H. Paterson, Molecular dissection of complex traits in autopolyploids: mapping QTLs affecting sugar yield and related traits in sugarcane. Theor. Appl. Genet. 105 (2002) 332–345.

105. C. Weng, T.L. Kubisiak, C.D. Nelson, M. Stine, Mapping quantitative trait loci controlling early growth in a (long leaf pine × slash pine) × slash pine BC, family. Theor. Appl. Genet. 104 (2002) 852–859.

106. Z. Kaya, M.M. Sewell, D.B. Neale, Identification of quantitative trait loci influencing annual height- and diameter-increment growth in loblolly pine (*Pinus taeda* L.). Theor. Appl. Genet. 98 (1999) 586–592.

107. M.M. Sewell, M.F. Davis, G.A. Tuskan, N.C. Wheeler, C.C. Elam, D.L. Bassoni, D.B. Neale, Identification of QTLs influencing wood property traits in loblolly pine (*Pinus taeda* L.). II. Chemical wood properties. Theor. Appl. Genet. 104 (2002) 214–222.

108. S. Kumar, R.J. Spelman, D.J. Garrick, T.E. Richardson, M. Lausberg, P.L. Wilcox, Multiple-maker mapping of wood density loci in an outbred pedigree of radiata pine. Theor. Appl. Genet. 100 (2000) 926–933.

109. D.M. Harkins, G.N. Johnson, P.A. Skaggs, A.D. Mix, G.E. Dupper, M.E. Devey, B.B. Jr. Kinloch, D.B. Neale, G.N. Johnson, Saturation mapping of a major gene for resistance to white pine blister rust in sugar pine. Theor. Appl. Genet. 97 (1998) 1355–1360.

110. D. Verhaegen, C. Plomion, J. M. Gion, M. Poitel, P. Costa, A. Kremer, Quantitative trait dissection analysis in *Eucalyptus* using RAPD markers. 1. Detection of QTL in interspecific hybrid progeny, stability of QTL expression across different ages. Theor. Appl. Genet. 95 (1997) 597–608.

111. G.S. Khush, D.J. Mackill, G.S. Sidhu, Breeding rice for resistance to bacterial blight, in: Bacterial blight of rice, International Rice Research Institute, Los, Banos, Manila, Philippines, 1989, pp. 207–217.

112. T. Kinoshita, Report of committee on gene symbolization, nomenclature and linkage groups. Rice Genet. Newsl. 12 (1995) 9–153.

113. T.W. Mew, C.M. Vera Cruz, E.S. Medalla, Changes in race frequency of *Xanthomonas oryzae* pv. *oryzae* in response to the planting of rice cultivars in the Philippines. Plant Dis. 76 (1992) 1029–1032.

114. Q. Huang, Y. He, R. Jin, H. Huang and Y. Zhu, Tagging of the restorer gene for rice HL-type CMS using microsatellite markers. Rice Genet. Newsl. 16 (1999) 75–77.

115. H.S. Dhaliwal, J.S. Sidhu, K. Singh, Y. Vikla, A. Dass, R.K. Goyal, T. Jhang, Gene mapping, marker-assisted pyramiding and introgression of desirable genes in rice at PAU, Ludhiana. Rice Functional Genomics Workshop, May 20-21, 2002, National Research Centre on Plant Biotechnology, IARI, New Delhi, pp. 19 (abstr.).

116. M.N.V. Williams, N. Pande, S. Nair, M. Mohan, J. Bennett, Restriction length polymorphism analysis of polymerase chain reaction products and amplified from mapped loci of rice (*Oryza sativa* L.) genomic DNA. Theor. Appl. Genet. 82 (1991) 489–498.

117. S. Hittalmani, M. Foolad, T. Mew, R. Rodrigues, H. Huang, Identification of blast resistance gene, *Pi-2(t)* in rice plants by flanking DNA markers, in: H.I. Oka and G.S. Khush, (Eds), Rice Genetics Newsletter, a publication of Rice Genetics Cooperative Genetic Resources Section, National Institute of Genetics, Misima, Japan. Vol. 11, 1994, pp. 144.

118. M. Mohan, P.V. Sathyanarayanan, A. Kumar, M.N. Srivastava, S. Nair, Molecular mapping of a resistance-specific PCR-based marker linked to a gall midge resistance gene (*Gm4t*) in rice. Theor. Appl. Genet. 95 (1997) 777–782.

119. R.C. Aquino, P.R. Jennings, Inheritance and significance of dwarfism in an indica rice variety. Crop Sci. 6 (1966) 551–554.

120. M.Y. Eun, Y.G. Cho, T.Y. Chung, Molecular markers linked tightly with semidwarf (*sd-1*) character and shattering habits in rice. Korean Soc. Mol. Biol. SE 4 (1991) 182–185.

121. Z.H. Yu, Molecular mapping of rice (*Oryza sativa* L.) genes via linkage to restriction fragment length polymorphism (RFLP) markers. Ph.D. thesis. Cornell University, Ithaca, N. Y, 1991.

122. Y. Ogi, H. Kato, K. Maruyama, A. Saito, F. Kikuchi, Identification of RFLP markers closely linked to the semidwarfing gene at the *sd-1* locus in rice. Japan J. Breed. 43 (1993) 141–146.

123. Y.G. Cho, M.Y. Eun, S.R. McCouch, Y.A. Chae, The semidwarf gene, *sd-1*, of rice (*Oryza sativa* L.). II. Molecular mapping and marker-assisted selection. Theor. Appl. Genet. 89 (1994) 54–59.

124. S.R. McCouch, G. Kochert, Z.H. Yu, Z.Y. Wang, G.S. Khush, W.R. Coffman, S.D. Tanksley, Molecular mapping of rice chromosomes. Theor. Appl. Genet. 76 (1988) 815–829.

125. S. Garland, R. Henry, Application of molecular markers to rice breeding in Australia. RIRDC Publication No. 01/38, RIRDC Project No., USC-2A, 2001.

126. K.V. Parbhu and co-workers, Personal communication (2003).

127. F.C. Ogbonnaya, O. Moullet, R.F. Eastwood, J. Kollmorgen, H. Eagles, R. Appels, E.S. Lagudah, The use of molecular markers to pyramid cereal cyst nematode resistance genes in wheat, in: A.E. Slinkard (Ed), Proc. 9th Int. Wheat Genet. Symp., Vol. 3, Univ. Extension Press, Univ. of Saskatchewan, Saskatoon, 1998, pp. 138–139.

128. NPGI Progress Report, USA (2001).

129. W.R. Lawson, K.C. Goutler, R.J. Henry, G.A. Kong, J.K. Kochman, Marker-assisted selection for two rust resistance genes in sunflower. Mol. Breed. 4 (1998) 227-234.

130. D.M. Lawson, M.A. Lunde, M.A. Mutschler, Marker-assisted transfer of acylsugar-mediated pest resistance from the wild tomato, *Lycopersicon pennellii*, to the cultivated tomato, *Lycopersicon esculentum*. Mol. Breed. 3 (1997) 307–317.

131. V.J.M. Robert, M.A.L. West, S. Inai, A. Caines, L. Arntzen, J.K., Smith, D.A. St Clair, Marker-assisted introgression of blackmold resistance QTL alleles from wild *Lycopersicon cheesmanii* to cultivated tomato (*L. esculentum*) and evaluation of QTL phenotypic effects. Mol. Breed. 8 (2001) 217–233.

132. Han, F., Romagosa, I., Ullrich S. E., Jones, B. L., Hayes, P. M., Wesenberg, D. M., (1997). Molecular marker-assisted selection for malting quality traits in barley. Mol. Breed. 3: 427-437.

133. Romagosa, I., Han, F., Ullrich, S. E., Hayes, P. M., Wesenberg, D. M., 1999. Verification of yield QTL through realized molecular marker-assisted selection responses in a barley cross. Mol. Breed. 5: 143–152.

134. D. Walker, H.R. Boerma, J. All and W. Parrott, Combining *cryIAc* with QTL alleles from PI 229358 to improve soybean resistance to lepidopteran pests. Mol. Breed. 9 (2002) 43-51.

135. L. Gianfranceschi, B. Koller, N. Seglias, M. Kellerhals, C. Gessler Molecular markers in apple for resistance to scab caused by *Venturia inaequalis*. Theor. Appl. Genet. 93 (1996) 199–204.

136. S. Tartarini, S. Sansavini, B. A. Vinatzer, F. Gennari, Development of reliable PCR markers for the selection of the *Vf* gene conferring scab resistance in apple. Plant Breed. 118 (2000) 183–186.

137. T. Markussen, J, Kruger, H. Schmidt, F. Dunemann, Identification of PCR-based markers linked to the powdery mildew gene *Pl1* from *Malus robusta* in cultivated apple. Plant Breed. 114 (1995) 530–534.

138. E. Albertini, G. Barcaccia, A. Porceddu, S. Sorbolini, M. Falcinelli, Mode of reproduction is detected by *Parth1* and *Sex1* SCAR markers in a wide range of facultative apomictic Kentucky bluegrass varieties. Mol. Breed. 7 (2001) 293–300.

139. L.H. Zhu, Y. Chen, Y.B. Xu, J.C. Xu, H.M. Cai, Z.Z. Ling, Construction of a molecular map of rice and gene mapping using a double haploid population of a cross between Indica and Japonica varieties. Rice Genet. Newsl. 10 (1993) 132–134.

140. G.L. Wang, D.J. Mackill, M. Bonman, S.K. McCouch, M. Champoux, R. Nelson, RFLP mapping of genes conferring complete and partial resistance to blast in a durably resistance rice cultivar. Genetics 136 (1994) 1421–1434.

141. T.N. Mew, A.R. Parco, S. Hittalmani, T. Inukai, R. Nelson, R.S. Zeighar, N. Huang, Fine mapping of major genes for blast resistance in rice. Rice Genet. Newslett. 11 (1994) 126–128.

142. N.I. Naqvi, J.M. Bonman, D.J. Mackill, R.J. Nelson, B.B. Chattoo, Identification of RAPD markers linked to a major blast resistance gene in rice. Mol. Breed. 1 (1995) 341–348.

143. K.L. Zheng, J.Y. Zhuang, J. Lu, H.R. Qian, H.X. Lin, Identification of DNA marker linked to blast resistance genes in rice, paper presented at the FAO/IAEA Int. Symp. on the use of induced mutations and molecular techniques for crop improvement, Vienna (1995).

144. S.N. Ahn, Y.K. Kim, S.S. Han, H.C. Choi, H.P. Moon, S.R. McCouch, Molecular mapping of a gene for resistance to a Korean isolate of rice balst. Rice Genet. Newsl. 13 (1996) 74–75.

145. M. Miyamoto, I. Ando, K. Rybka, O. Kodama, S. Kawasaki, High resolution mapping of the indica-derived rice blast resistance genes. 1. *Pi-b*. Mol. Plant-Microbe Interact. 9 (1996) 6–13.

146. R. Kaji, T. Ogawa, RFLP mapping of blast resistance gene *Pik-m* in rice. Int. Rice Res. Notes 21 (1996) 47.

147. K. Rybka, M. Miyamoto, I. Ando, A. Saito, S. Kawasaki, High resolution mapping of the indica-derived rice blast resistance genes. 2. *Pi-ta* and a consideration of their origin. Mol. Plant-Microbe Interact. 10 (1997) 517–524.

148. D. Chen, G.L. Wang, P.C. Ronald, Location of rice blast resistance locus *Pi-5(t)* in Moroberekan by AFLP bulk segregant analysis. Rice Genet. Newsl. 14 (1997) 95–97.

149. T. Imbe, S. Oba, M.J.T. Yanoria, H. Tsunematsu, A new gene for blast resistance in rice cultivar IR24. Rice Genet. Newsl. 14 (1997) 60–62.

150. D.H. Chen, M. dela Vina, T. Inukai, D.J. Mackill, P.C. Ronald, R.J. Nelson, Molecular mapping of the blast resistance gene, *Pi44(t)*, in a line derived from a durably resistant rice cultivar. Theor. Appl. Genet. 98 (1999) 1046–1053.

151. K. Fujii, Y. Hayano-Saito, N. Sugiura, N. Hayashi, N. Saka, T. Tooyama, T. Izawa, A. Shumiya, Gene analysis of panicle blast resistance in rice cultivars with rice stripe resistance. Breed. Res. 1 (1999) 203–210.

152. Z.H. Yu, D.J. Mackill, J.M. Bonman, S. McCouch, E. Guiderdoni, J.L. Notteghem, S.D. Tanksley, Molecular mapping of genes for resistance to rice blast (*Pyricularia grisea* Sacc.) Theor. Appl. Genet. 93 (1996) 859–863.

153. R.E. Tabien, Z. Li, A.H. Paterson, M.A. Marchetti, J.W. Stansel, S.R.M. Pinson, Mapping QTLs for field resistance to the rice blast pathogen in improved varieties. Theor. Appl. Genet. 105 (2002) 313–324.

154. G. Zhang, E.R. Anglis, M.L. Abenes, G.S. Khush, N. Huang, Molecular mapping of bacterial blight resistance gene on chromosome 8 in rice. Rice Genet. Newslett. 11 (1994) 142–144.

155. P.C. Ronald, B. Albano, R. Tabien, L. Abenes, K. Wu, S. McCouch, S.D. Tanksley, Genetic and physical analysis of the rice bacterial blight resistance locus, *Xa-21*. Mol. Gen. Genet. 236 (1992) 113–120.

156. S. Yoshimura, A. Yoshimura, R.J. Nelson, T.W. Mew, N. Iwata, Tagging *Xa-1*, the bacterial blight resistance gene in rice by using RAPD markers. Breed. Sci. 45 (1995) 81–85.

157. G. Zhang, E.R. Angeles, M.L.P. Abenes, G.S. Khush, N. Huang, RAPD and RFLP mapping of the bacterial blight resistance gene *xa-13* in rice. Theor. Appl. Genet. 93 (1996) 65–70.

158. X.H. Lin, D.P. Zhang, Y.F. Xie, H.P. Gao, Q.F. Zhang, Identifying and mapping a new gene for bacterial blight resistance in rice based on RFLP markers. Phytopath. 86 (1996) 1156–1159.

159. Y. Vikal, J.S. Sidhu, S. Singh, M. Sodhi, H.S. Dhaliwal, Tagging of bacterial blight resistance gene in rice using RAPD markers. Rice Genet. Newsl. 14 (1997) 92–94.

160. Q. Zhang, S.C. Lin, B.Y. Zhao, C.L. Wang, W.C. Yang, Y.I. Zhou, D.Y. Li, C.B. Chen, L.H. Zhu, Identification and tagging a new gene for resistance to bacterial blight (*Xanthomonas oryzae* pv. *oryzae*) from *O. rufipogon*. Rice Genet. Newsl. 15 (1998) 138–141.

161. A. Ghesquiere, L. Albar, M. Lorieux, N. Ahmadi, D. Fargette, N. Huang, S. McCouch, J.L. Notteghem, A major quantitative trait locus for rice yellow mottle virus resistance maps to a cluster of blast resistance genes on chromosome 12. Phytopath. 87 (1997) 1243–1249.

162. G. Pressoir, L. Albar, N. Ahmadi, I. Rimbault, M. Lorieux, D. Fargette, A. Ghesquiere, Genetic basis and mapping of the resistance to rice yellow mottle virus. II. Evidence of a complementary epistasis between two QTLs. Theor. Appl. Genet. 97 (1998) 1155–1161.

163. Y. Saito-Hayano, T. Teuji, K. Fujii, K. Saito, M. Iwasaki, A. Saito, Localization of the rice stripe disease resistance gene, *Stv-b^i*, by graphical genotyping and linkage analyses with molecular markers. Theor. Appl. Genet. 96 (1998) 1044–1049.

164. Sebastian, L.S., Ikeda, R., Huang, N., Imbe, T., Coffman, W.R., McCouch, S.R., 1996. Molecular mapping of resistance to rice tungro spherical virus and green leafhopper. Phytopath. 86: 25–30.

165. G.P. Yan, X.M. Chen, R.F. Line and C.R. Wellings, Resistance gene-analog polymorphism markers co-segregating with the *YR5* gene for resistance to wheat stripe rust. Theor. Appl. Genet. 106 (2003) 636–643.

166. K.P. Che, Q.C. Zhan, Q.H. Xing, Z.P. Wang, D.M. Jin, D.J. He, B. Wang, Tagging and mapping of rice sheath blight resistance gene. Theor. Appl. Genet. 106 (2003) 293–297.

167. Z.Q. Ma, M.E. Sorrells, S.D. Tanksley, RFLP markers linked to powdery mildew resistance genes *Pm1*, *Pm2*, *Pm3* and *Pm4* in wheat. Genome 37 (1994) 871–875.

168. L.H. Hartl, F.J. Weiss, U. Stephan, F.J. Zeller, A. Jahoor, Molecular identification of powdery mildew resistance genes in common wheat. Theor. Appl. Genet. 90 (1995) 601–606.

169. J.C.Nelson, M.E. Sorrells, A.E. van Deynze, Y.H. Lu, M. Atinkson, M. Bernard, P. Leroy, J.D. Faris, J.A. Anderson, Molecular mapping of wheat. Major genes and rearrangements in homoeologous groups 4, 5 and 7. Genetics 141 (1995a) 721–731.

170. J.C. Nelson, A.E. van Deynze, E. Antrique, M.E. Sorrells, Y.H. Lu, M. Merlino, M. Atinkson, P. Leroy, Molecular mapping of wheat. Homoeologous group 2. Genome 38 (1995b) 516–524.

171. V. Mohler, A. Jahoor, Allele specific amplification of polymorphic sites for detection of powdery mildew resistance loci in cereals. Theor. Appl. Genet. 93 (1996) 1078–1082.

172. A. Jahoor, Marker assisted breeding in cereal, specially with respect to synteny among loci for mildew resistance, in: P.K. Gupta (Ed), Genetics and Biotechnology in Crop Improvement, Rastogi Publ., 1998, pp. 237–254.

173. J. Jia, K.M. Devos, S. Chao, T.E. Miller, S.M. Reader, M.D. Gale, RFLP-based maps of the homoeologous group-6 chromosomes of wheat and their application in the tagging of *Pm12*, a powdery mildew resistance gene transferred from *Aegilops speltoides* to wheat. Theor. Appl. Genet. 92 (1996) 559–565.

174. L.L. Qi, M.S. Cao, P.D. Chen, W.L. Li, D.J. Lu, Identification, mapping and application of polymorphic DNA associated with resistance gene *Pm21* of wheat. Genome 39 (1996) 191–197.

175. J.K. Roug, E. Millet, J. Manisterski, M. Feldman, A powdery mildew resistance gene from wild emmer transferred into common wheat and tagged by molecular markers, in: A.E. Slinkard (Ed), Proc. 9th Int. Wheat Genet. Symp., Vol. 3, Iniv. Extension Press, Univ. of Saskatchewan, pp. 148–150.

176. L. Hartl, S. Mon, G. Schweiger, Identification of a diagonistic molecular marker for the powdery mildew resistance gene *Pm4b* based on fluorescently labelled AFLPs, in: A.E. Slinkard, (Ed), Proc. 9th Int. Wheat Genet. Symp. Vol. 3, Uinv. Extension Press, Univ. of Saskatchewan, Saskatoon, 1998, pp. 111–113.

177. D.J. Liu, J.Y. Liu, W.J. Tao, P.D. Chen, Molecular markers breeding wheat for powdery mildew resistance, in: A.E. Slinkard (Ed), Proc. 9th Int. Wheat Genet. Symp. Vol. 3, Uinv. Extension Press, Univ. of Saskatchewan, Saskatoon, 1998, pp. 128–131.

178. A. Cenci, D.R. Ovidio, O.A. Tanzarella, C. Ceoloni, E. Porceddu, Identification of molecular markers linked to *Pm 13*, an *Aegilops longissima* gene conferring resistance to powdery mildew in wheat. Theor. Appl. Genet. 98 (1999) 448–454.

179. C. Xie, Q. Sun, Z. Ni, T. Yang, E. Nevo, T. Fahima, Chromosomal location of a *Triticum dicoccoides*-derived powdery mildew resistance gene in common wheat by using microsatellite markers. Theor. Appl. Genet. 106 (2003) 341–345.

180. S. Liu, C.A. Griffey, M.A. Saghai Maroof, Preliminary report on molecular marker analysis of adult plant resistance to powdery mildew in winter wheat Massey, in: A.E. Slinkard (Ed), Proc. 9th Int. Wheat Genet. Symp., Vol. 3, Univ. Extension Press, Univ. of Saskatchewan, Saskatoon (1998), pp. 132–134.

181. T. Demeke, A. Laroche, D.A. Gaudet, A DNA marker for the *BT-10* common bunt resistance gene in wheat. Genome 39 (1996) 51–55.

182. L. Lintott, J. Davoren, D. Gaudet, B. Puchalski, A. Laroche, Development of molecular markers for resistance to common bunt in hexaploid wheats, in: A.E. Slinkard (Ed), Proc. 9th Int. Wheat Genet. Symp., Vol. 3, Univ. Extension Press, Univ. of Saskatchewan, Saskatoon (1998), pp. 126–127.

183. J.C. Nelson, J.E. Autrique, G. Fuentes-Davila, M.E. Sorrells, Chromosomal location of genes for resistance to karnal bunt in wheat. Crop Sci. 38 (1998) 231–236.

184. H.S. Bariana, S. Kailasapillai, G.N. Brown, P.J. Sharp, Marker-assisted identification of *Sr2* in The National Cereal Rust Control Prgramme in Australia, in: A.E. Slinkard (Ed), Proc. 9th Int. Wheat Genet. Symp., Vol. 3, Univ. Extension Press, Univ. of Saskatchewan, Saskatoon (1998), pp. 89–91.

185. S.J. Johnston, P.J. Sharp, R.A. McIntosh, Molecular markers for the *Sr2* stem rust resistance gene, in: A.E. Slinkard (Ed), Proc. 9th Int. Wheat Genet. Symp., Vol. 3, Univ. Extension Press, Univ. of Saskatchewan, Saskatoon, (1998), pp. 117–119.

186. J.G. Paull, M.A. Pallota, P. Langridge, 1994. RFLP markers associated with *Sr22* and recombination between chromosome 7A of bread wheat and the diploid species *Triticum boeoticum*. Theor. Appl. Genet. 89 (1994) 1039–1045.

187. G.M. Schachermayr, H. Siedler, M.D. Gale, H. Winzeler, M. Winzeler, B. Keller, Identification and localization of molecular markers linked to the *Lr9* rust gene of wheat. Theor. Appl. Genet. 88 (1994) 110–115.

188. M. Yamamori, An N-band marker for gene *Lr18* for resistance to leaf rust in wheat. Theor. Appl. Genet. 89 (1994) 643–646.

189. C. Peuillet, M. Messmer, G. Schachermayr, B. Keller, Genetic and physical characterization of the *Lr1* leaf rust locus in wheat (*Triticum aestivum* L.). Mol. Gen. Genet. 248 (1995) 553–562.

190. E. Autrique, R.P. Singh, S.D. Tanksley, M.E. Sorrells, Molecular markers for four leaf rust resistance genes introgressed into wheats from wild realtives. Genome 38 (1995) 75–83.

191. G.M. Schachermayr, M.M. Messmer, C. Feuillet, H. Winzeler, M. Winzeler, B. Keller, Identification of molecular markers linked to the *Agropyrum elongatum*-derived leaf rust resistance gene *Lr24* in wheat. Theor. Appl. Genet. 90 (1995) 982–990.

192. J.D. Procunier, T.F. Townley-Smith, S. Fox, S. Prashar, M. Gay, W.K. Kim, E. Czarnecki, P.L. Dyck, PCR based RAPD/DGGE markers linked to leaf rust resistance genes *Lr29* and *Lr25* in wheat (*Triticum aestivum* L.). J. Genet. Breed. 49 (1995) 87–92.

193. F. Dedryver, M.F. Jubier, J. Thouvenin, H. Goyeau, Molecular markers linked to the leaf rust resistance gene *Lr24* in different wheat cultivars. Genome 39 (1996) 568–573.

194. G.M. Schachermayr, C. Feuillet, B. Keller, Molecular markers linked for the detection of the wheat leaf rust resistance gene *Lr10* in diverse genetic backgrounds. Mol. Breed. 3 (1997) 65–74.

195. J.C. Nelson, R.P. Singh, J.E. Autrique, M.E. Sorrells, Mapping genes conferring and suspecting leaf rust resistance in wheat. Crop Sci. 37 (1997) 1928–1935.

196. H.M. William, D. Hoisington, R.P. Singh, D. Gonzalez-de-Leon, Detection of quantitative trait loci associated with leaf rust resistance in bread wheat. Genome 40 (1997) 253–260.

197. R. Seyfarth, C. Feuillet, B. Keller, Development and characterization of molecular markers for the adult leaf rust resistance genes *Lr13* and *Lr35* in wheat, in: A.E. Slinkard (Ed), Proc. 9th Int. Wheat Genet. Symp., Vol. 3, Univ. Extension Press, Univ. of Saskatchewan, Saskatoon, (1998), pp. 154–155.

198. F. Sacco, E.Y. Suarez, T. Naranjo, Mapping of the leaf rust resistance gene *Lr3* on chromosome 6B of Sinvalocho MA wheat. Genome 41 (1998) 686–690.

199. C.H. Danna, F. Sacco, L.R. Ingala, H.A. Saione, R.A. Ugalde, Cloning and mapping of genes involved in wheat-leaf rust interaction through gene-expression analysis using chromosome-deleted near-isogenic wheat lines. Theor. Appl. Genet. 105 (2002) 972–979.

200. G.L. Sun, T. Fahima, A.B. Korol, T. Turpeinen, A. Grama, Y.I. Ronin, E. Nevo, Identification of molecular markers linked to *Yr15* stripe resistance gene originated in wild emmer wheat, *Triticum dicoccoides*. Theor. Appl. Genet. 95 (1997) 622–628.

201. T. Fahima, V. Chague, G. Sun, A. Korol, Y. Ronin, M. Roder, A. Grama, E. Nevo, Identification and potential use of PCR markers flanking the *Triticum dicoccoides*-derived stripe rust resistance gene *Yr15* in wheat, in: 5th Int. Congr. Plant Mol Biol., Singapore (1997), pp. 21–27, Abstr. 249.

202. R.E. Knox, N.K. Howes, A monoclonal antibody chromosome marker analysis used to locate a loose smut resistance gene in wheat chromosome 6A. Theor. Appl. Genet. 89 (1994) 787–793.

203. J.D. Procunier, R.E. Knox, A.M. Bernier, M.A. Gray, N.K. Howes, DNA markers linked to a *T10* loose smut resistance gene in wheat (*Triticum aestivum* L.). Genome 40 (1997) 176–179.

204. W. Cao, G.R. Huges, H. Ma, Z. Dong, Development of DNA based markers for resistance to *Septoria nodorum* blotch in durum wheat, in: A.E. Slinkard (Ed), Proc. 9th Int. Wheat Genet. Symp., Vol. 3, Univ. Extension Press, Univ. of Saskatchewan, Saskatoon (1998), pp. 95–97.

205. S.B. Goodwin, X. Hu, G. Shaner, An AFLP marker linked to a gene for resistance to *Septoria tritici* blotch in wheat, in:c A.E. Slinkard (Ed), Proc. 9th Int. Wheat Genet. Symp., Vol. 3, Univ. *Extension Press*, Univ. of Saskatchewan, Saskatoon (1998), pp. 108–110.

206. J.A. Anderson, B.L. Waldron, B. Moreno-Sevilla, R.W. Stack, R.C. Frohberg, Detection of *Fusarium* head blight resistance QTL in wheat using AFLPs and RFLPs, in: A.E. Slinkard (Ed), Proc. 9th Int. Wheat Genet. Symp., Vol. 1, Univ. Extension Press, Univ. of Saskatchewan, Saskatoon (1998), pp. 135–137.

207. G.H. Bai, F.L. Kolb, G. Shaner, L.L. Domier (1998) Identification of AFLP markers linked to one major QTL controlling scab resistance in wheat, in: A.E. Slinkard (Ed), Proc. 9th Int. Wheat Genet. Symp. Vol. 3, Univ. Extension Press, Univ. of Saskatchewan, Saskatoon (1998), pp. 81–83.

208. P.H. Smith, R.M.D. Koebner, L.A. Boyd, The'development of a STS marker linked to a yellow rust resistance derived from the wheat cultivar Moro. Theor. Appl. Genet. 104 (2002) 1278–1282.

209. S. Singh, G.L. Guedira-Brown, T.S. Grewal, H.S. Dhaliwal, J.C. Nelson, H. Singh, B.S. Gill, Mapping, of a resistance gene effective against karnal bunt pathogen of wheat. Theor. Appl. Genet. 106 (2003) 287–292.

210. G.L. Sun, P.L. Bruckner, L.Y. Smith, R. Sears, T.J. Martin, Development of PCR markers linked to resistance to wheat streak mosaic virus in wheat. Theor. Appl. Genet. 93 (1996) 463–467.

211. S. Bentolila, C. Guitton, N. Bouvet, A. Sailland, S. Nykaza, G. Freyssinet, Identification of an RFLP marker tightly linked to the *Ht1* gene in maize. Theor. Appl. Genet. 82 (1991) 393–398.

212. M.D. McMullen, R. Louise, The linkage of molecular markers to a gene controlling the symptom response in maize to maize dwarf mosaic virus. Mol. Plant-Microbe Interact. 2 (1989) 309–314.

213. D.M. Bukeck, M.M. Goodman, W.D. Bearis, D. Grant, Quantitative trait loci controlling resistance to gray leaf spot in maize. Crop Sci. 33 (1993) 838–847.

214. H.G, Welz, A. Schechert, A. Pernet, K.V. Pixley, H.H. Geiger, A gene for resistance to the maize streak virus in the African CIMMYT maize inbred line CML202. Mol. Breed. 4 (1998) 147–154.

215. A. Perenet, D. Hoisington, J. Dintinger, D. Jewell, C. Jiang, M. Khairallah, P. Letourmy, J.L. Marchand, J.C. Glaszmann, D. Gonzalez de Leon, Genetic mapping of maize streak virus resistance in the tropical "Revolution" source 2. Resistance in line CIRAD 390 and stability across germplasm. Theor. Appl. Genet. 99 (1998a) 540–553.

216. A. Pernet, D. Hoisington, J. Franco, M. Isnard, D. Jewell, C. Jiang, M. Khairallah, J.L. Marchand, B. Reynaud, J.C. Glaszmannn, D. Gonzalez de Leon, Genetic mapping of maize streak virus resistance in the tropical "Revolution" source 1. Resistance in line D211 and stability against different virus clones. Theor. Appl. Genet. 99 (1998b) 524–539.

217. A.E. Melchinger, L. Kuntze, R.K. Gumber, T. Lubberstedt and E. Fuchs, Genetic basis of resistance to sugarcane mosaic virus in European maize germplasm. Theor. Appl. Genet. 96 (1998) 1151–1161.

218. M. Quint, R. Mihaljevic, C.M. Dussle, M.L. Xu, A.E. Melchinger and T. Lubberstedt, Development of RGA-CAPs markers and genetic mapping of candidate genes for sugarcane mosaic virus resistance in maize. Theor. Appl. Genet. 105 (2002) 355–363.

219. C.M. DuBle, M. Quint, A.E. Melchinger, M.L. Xu and T. Lubberstedt, Saturation of two chromosome regions conferring resistance to SCMV with SSR and AFLP markers by targeted BSA. Theor. Appl. Genet. 106 (2003) 485-493.

220. M. Heun, Mapping qualitative powdery mildew resistance in barley using RFLP map. Geneome 35 (1992) 1019–1025.

221. T. Lahaye, S. Hartmann, S. Topsch, A. Freialdenhoven, M. Yano and P. Lefert-Schulze, High-resolution genetic and physical mapping of the *Rar 1* locus in barley. Theor. Appl. Genet. 97 (1998) 526–534.

222. D. C. Abbot, E.S. Lagudah, A.D.H. Brown, Identification of RFLPs flanking a scald resistance gene on barley chromosome 6. J. Hered. 86 (1995) 152–156.

223. D.M.E. Poulsen, R.J. Henry, R.P. Johnston, J.A.G. Irwin and R.G. Rees, The use of bulk segregant analysis to identify a RAPD marker linked to leaf rust resistance in barley. Theor. Appl. Genet. 91 (1995) 270–273.

224. Qi, X., Niks, R. E., Stam, P., Lindhout, P., Identification of QTLs for partial resistance to leaf rust (*Puccinia hordei*) in barley. Theor. Appl. Genet. 96 (1998) 1205-1215.

225. S. Brunner, B. Keller and C. Feuillet, Molecular mapping of the *Rph7.g* leaf rust resistance gene in barley (*Hordeum vulgare* L.). Theor. Appl. Genet. 101 (2000) 783–788.

226. H. El Attari, P.M. Hayes, A. Rebai, G. Barrault, G. Guillaume-Dechamp and A. Sarrafi, Potential of doubled-haploid lines and localization of quantitative trait loci (QTL) for partial resistance to bacterial leaf streak (*Xanthomonas campestris* pv. *hordei*) in barley. Theor. Appl. Genet. 96 (1998) 95-100.

227. T. Toojinda, E. Baird, A. Booth, L. Broers, P. Hayes, W. Powell, W. Thomas, H. Vivar and G. Young, Introgression of quantitative trait loci (QTLs) determining stripe rust resistance in barley: an example of marker-assisted line development. Theor. Appl. Genet. 96 (1998) 123–131.

228. N.G. Paltridge, N.C. Collins, A. Bendahmane and R.H. Symons, Development of YLM, a codominant PCR marker closely linked to the *Yd2* gene for resistance to barley yellow dwarf virus. Theor. Appl. Genet. 96 (1998) 1170–1177.

229. A. Graner, S. Streng, A. Kellermann, A. Schiemann, E. Bauer, R. Waugh, B. Pellio and F. Ordon, Molecular mapping and genetic fine-structure of the *rym5* locus encoding resistance to different strains of the barley yellow mosaic virus complex. Theor. Appl. Genet. 98 (1999) 285–290.

230. S. Zhong and B.J. Steffenson, Identification and characterization of DNA markers associated with a locus conferring virulence on barley in the plant pathogenic fungus *Cochliobolus sativus*. Theor. Appl. Genet. 104 (2002) 1049–1054.

231. L. Arru, E. Francia and N. Pecchioni, Isolate-specific QTLs of resistance to leaf stripe (*Pyrenophora graminea*) in the 'Steptoe' × 'Morex' spring barley cross. Theor. Appl. Genet. 106 (2003) 668–675.

232. B.J. Oh, R.A. Frederiksen and C.W. Magill, Identification of molecular markers linked to head smut resistance gene (*shs*) in sorghum by RFLP and RAPD analysis. Phytopath. 84 (1994) 830–833.

233. J. Behare, H. Laterro, M. Sarfatti, D. Zamie, Restriction fragment length polymorphism mapping of the Stemphylium resistance gene in tomato. Mol. Plant Microbe Interact. 4 (1991) 489–492.

234.   J.G. VanderBeek, R. Verkerk, P. Zabel and P. Lindhout, Mapping strategy for resistance genes in tomato based on RFLPs between cultivars: *Cf9* (resistance to *Cladosporium fulvum*) on chromosome 1. Theor. Appl. Genet. 94 (1992) 106–112.

235.   M. Sarfatti, J. Katan, R. Fluhr and D. Zamir, An RFLP marker in tomato linked to the *Fusarium oxysporum* resistance gene I_2. Theor. Appl. Genet. 78 (1989) 755–759.

236.   T. Debener, H. Lehnackers, M. Arnold, J.L. Dang, Identification and molecular mapping of a single *Arabidopsis thaliana* locus determining resistance to a phytopathogenic *Psuedomonas syringae* isolate. Plant J. 1 (1991) 289–302.

237.   J. Chunwongse, T.B. Bunn, C. Crossman, J. Jiang, S.D. Tanksley, Chromosomal localization and molecular-marker tagging of the powdery mildew resistance gene (*Lv*) in tomato. Theor. Appl. Genet. 89 (1994) 76–79.

238.   L.M. Kawchum, D.R. Lynch, J. Hachey, P.S. Bains, F. Kulscar, Identification of a co-dominant amplified polymorphic DNA marker linked to the verticillium wilt resistance gene in tomato. Theor. Appl. Genet. 89 (1994) 661–664.

239.   N. Diwan, R. Fluhr, Y. Eshed, D. Zamir, S.D. Tanksley, Mapping of *Ve* in tomato: a gene conferring resistance to the broad-spectrum pathogen, *Verticillium dahliae* race 1. Theor. Appl. Genet. 98 (1999) 315–319.

240.   R. Messeguer, M. Ganel, M.C. de Vicente, N.D. Young, H. Bolkar, S.D. Tanksley, High resolution RFLP map around the root knot nematode resistance gene (*Mi*) in tomato. Theor. Appl. Genet. 82 (1991) 529–536.

241.   D. Danesh, S. Aarons, G.E. McGill, N.D. Young, Genetic dissection of oligo genic resistance to bacterial wilt in tomato. Mol. Plant Microbe Interact. 7 (1994) 464-471.

242.   C.C. Huang, Y.Y. Cui, C.R. Weng, P. Zabel, P. Lindhout, Development of diagnostic PCR marker is closely linked to the tomato powdery mildew resistance gene *Ol-1* on chromosome 6 of tomato. Theor. Appl. Genet. 101 (2000) 918–924.

243.   E. Ritter, T. Debener, A. Barone, F. Salamini, C. Gebhardt, RFLP mapping on potato chromosomes of two genes controlling extreme resistance to potato virus X (PVX). Mol. Gen. Genet. 227 (1991) 81–85.

244.   M.R. Marano, I. Malcuit, W. De Jong, D.C. Baulcombe, High-resolution genetic map of Nb, a gene that confers hypersensitive resistance to potato virus X in *Solanum tuberosum*. Theor. Appl. Genet. 105 (2002) 192–200.

245.   C. Leonards-Schippers, W. Gieffers, R. Schanfar-Pregl, E. Ritter, S.J. Krapp, Quantitative resistance to *Phytophthora infestans* in potato: a case study of QTL mapping in an allogamous plant spores. Genetics 137 (1994) 67–77.

246.   J.H. Hamalainen, V.A. Sorri, K.N. Watanabe, C. Gebhardt, J.P.T. Valkonen, Molecular examination of a chromosome region that controls resistance to potato Y and A potyviruses in potato. Theor. Appl. Genet. 96 (1998) 1036–1048.

247.   K.M. Polzin, L.L. Lorenzen, T.C. Oslon, R.C. Shoemaker, An unusual polymorphic locus useful for tagging *Rps1* resistance alleles in soyabean. Theor. Appl. Genet. 89 (1994) 226–232.

248.   Y.G. Yu, M.A. Saghai Maroof, G.R. Buss, P.J. Maughan, S.A. Toten, RFLP and microsatellite mapping of a gene for soybean mosaic virus resistance. Phytopath. 84 (1994) 60–64.

249.   T. Ashfield, J.R. Danzer, D. Held, K. Clayton, P. Keim, M.A. Maroof Saghai, D.M. Webb, R.W. Innes, *Rpg1*, a soybean gene effective against races of bacterial blight, maps to a cluster of previously identified disease resistance genes. Theor. Appl. Genet. 96 (1998) 1021–1031.

250.   E. Johnson, P.N. Miklas, J.R. Stavely, J.C. Martenez-Crusado, Coupling and repulsion-phase RAPDs for marker-assisted selection of Pp181996 rust resistance in common bean. Theor. Appl. Genet. 90 (1995) 659–664.

251.   S.O. Park, D.P. Coyne, J.M. Bokosi, J.R. Steadman, Molecular markers linked to genes for specific rust resistance and indeterminate growth habit in common bean. Euphytica 105 (1999) 133–141.

252.   S.D. Haley, L. Afanador, J.D. Kelly, Identification and application of a random amplified polymorphic DNA marker for the *I* gene (Polyvirus resistance) in common bean. Phytopath. 84 (1994) 157–160.

253.   R.O. Nodari, S.M. Tsai, R.L. Gilbertson, P. Gepts, Towards an integrated linkage map of common bean: mapping genetic factors controlling host-bacteria interactions. Genetics 134 (1993) 341–350.

254. R.A. Young, M. Melotto, R.O. Nodari, J.D. Kelley, Marker-assisted dissection of the oligogenic anthracnose resistance in the common bean cultivar, 'G2333'. Theor. Appl. Genet. 96 (1998) 87–94.

255. G.M. Timmerman, T.J. Frew, A.L. Miller, N.F. Weeden, W.A. Jermyn, Linkage mapping of *sbm-1*, a gene conferring resistance in *Pisum sativum*. Theor. Appl. Genet. 85 (1993) 609–615.

256. E. Dirlewanger, P.G. Issac, S. Ranade, M. Belajouza, R. Cousin, D. deVienne, Restriction fragment length polymorphism anaysis of loci associated with disease resistance genes and developmental traits in *Pisum sativum*. Theor. Appl. Genet. 88 (1994) 17–27.

257. D. Bai, R. Reeleder, J.E. Brandle, Identification of two RAPD markers tightly linked with the *Nicotiana delneyi* gene for resistance to black root rot of tobacco. Theor. Appl. Genet. 91 (1995) 1184–1189.

258. H. Yang, J. Kruger, Identification of an RAPD marker linked to the *Vf* gene for scab resistance in apples. Plant Breed. 112 (1994) 323–329.

259. M. Xu, E. Huaracha, S.S. Korban, Development of sequence characterized amplified regions (SCARs) from amplified fragment length polymorphism (AFLP) markers tightly linked to the *Vf* gene in apple. Genome 44 (2001) 63–70.

260. D.W. Wolff, J. Zhou, Potential utility of RAPD markers linked to *Fom2* gene in melon (*Cucumis melo* L.). Cucubit Genet. Coop. Rep. 19 (1996) 61–62.

261. N.D. Young, D. Danesh, D. Menancio-Hautea, L. Kumar, Mapping oligogenic resistance to powdery mildew in mungbean with RFLPs. Theor. Appl. Genet. 87 (1993) 243–249.

262. M.H. Falment, I. Kebe, D. Clement, I. Pieretti, A.M. Ristericci, J.A.K. N'Goran, C. Cilas, D. Despreaux, C. Lanaud, Genetic mapping of resistance factors to *Phytophthora palmivora* in cocoa. Genome 44 (2001) 79–85.

263. Oil palm genome mapping and QTL detection. Plant genomics at cirad. www.cirad.fr/presentation/programmes/biotrop/resultats/biositecirad/mapping/oilgm.htm.

264. D. Lespinasse, L. Grivet, V. Troispoux, M. Rodier-Goud, F. Pinard, M. Sequin, Identification of QTLs involved in the resistance to South American leaf blight (*Microcyclus ulei*) in the rubber tree. Theor. Appl. Genet. 100 (2000) 975–984.

265. V. Le Guen, M. Seguin, C.R.R. Mattos, Qualitative resistance of *Hevea* to *Phyllachora huberi* P. Henn. Euphytica 112 (2000) 211–217.

266. J.H. Daugrois, L. Grivet, D. Roques, J.Y. Hoarau, H. Lombard, J.C. Glaszamnn, A. D'Hont, A putative major gene for rust resistance linked with an RFLP marker in sugarcane cultivar R570. Theor. Appl. Genet. 92 (1996) 1059–1064.

267. M.L. Pilet, R. Delourme, N. Foisset, M. Renard, Identification of loci contributing to quantitative field resistance to blackleg disease, causal agent *Leptosphaeria maculans* (Desm.) Ces. et de Not., in winter rapeseed (*Brassica napus* L.). Theor. Appl. Genet. 96 (1998) 23–30.

268. M.J. Dauleux-Manzanares, R. Delourme, F. Baron, G. Thomas, Mapping of one major gene and of QTLs involved in resistance to clubroot in *Brassica napus*. Theor. Appl. Genet. 101 (2000) 885–891.

269. J. Zhao, J. Meng, Genetic analysis of loci associated with partial resistance to *Sclerotinia sclerotiorum* in rapeseed (*Brassica napus* L.). Theor. Appl. Genet. 106 (2003) 759–764.

270. M.B. Ratnaparkhe, D.K. Santra, A. Tullu and F.J. Muehlbauer, Inheritance of inter-simple-sequence-repeat polymorphisms in chickpea. Theor. Appl. Genet. 96 (1998) 348–353.

271. R.N. Morgan, J.P. Wilson, W.W. Hanna, P. Akins-Ozias, Molecular markers for rust and pyricularia leaf spot disease resistance in pearl millet. Theor. Appl. Genet. 96 (1998) 413–420.

272. B. Malek von, W.E. Weber, T. Debener, Identification of molecular markers linked to *Rdr1*, a gene conferring resistance to blackspot in roses. Theor. Appl. Genet. 101 (2000) 977–983.

273. J. Pauquet, A. Bouquet, P. This, A.F. Blondou-Adam, Establishment of a local map of AFLP markers around the powdery mildew resistance gene *Run1* in grapevine and assessment of their usefulness for marker-assisted selection. Theor. Appl. Genet. 103 (2001) 1201–1210.

274. M.S. Andres-Arnedo, R. Ortega-Gill, M. Arteaga-Luis, J.I. Hormaza, Development of RAPD and SCAR markers linked to the *Pvr4* locus for resistance to PVR in pepper (*Capsicum annuum* L.). Theor. Appl. Genet. 105 (2002) 1067–1074.

275. A.O. Akano, A.G.O. Dixon, C. Mba, E. Barrera, M. Fregene, Genetic mapping of a dominant gene conferring resistance to cassava mosaic disease. Theor. Appl. Genet. 105 (2002) 521–525.

276.   R. Mago, W. Spielmeyer, G.J. Lawrence, E.S. Lagudah, J.G. Ellis, A. Psyor, Identification and mapping of molecular markers linked to rust resistance genes located on chromosome 1RS of rye using wheat-rye translocation lines. Theor. Appl. Genet. 104 (2002) 1317–1324.

277.   F. Lheureux, F. Carreel, C. Jenny, B.E.L. Lockhart, M.L. Caruana-Iskra, Identification of molecular markers linked to banana streak disease expression in inter-specific *Musa* hybrids. Theor. Appl. Genet. 106 (2003) 594–598.

278.   Nishi, T., Tajima, T., Noguchi, S., Ajisaka, H., Negishi, H., 2003. Identification of DNA markers of tobacco linked to bacterial wilt resistance. Theor. Appl. Genet. 106: 765–770.

279.   A. Tullu, L. Buchwaldt, T. Warkentin, B. Taran, A. Vandenberg, Genetics of resistance to anthracnose and identification of AFLP and RAPD markers linked to the resistance gene in PI 320937 germplasm of lentil (*Lens culinaris* Medikus). Theor. Appl. Genet. 106 (2003) 428–434.

280.   C.M. Kreike, J.R.A. de Koming, J.A. Vinke, J.W. van Ooijen, C. Gebhardt, W.J. Saikema, Mapping of loci involved in quantitatively inherited resistance to the potato cyst nematode *Globodera restochiensis*. Theor. Appl. Genet. 87 (1993) 464–470.

281.   J. Voort van der Rouppe, W. Linderman, R. Folkertsma, R. Hutten, H. Overmars, E. Vosses van der, E. Jacobsen, J. Bakker, A QTL for broad-spectrum resistance to cyst nematode species (*Globodera* spp.) maps to a resistance gene cluster in potato. Theor. Appl. Genet. 96 (1998) 654–661.

282.   J.E. Bradshaw, C.A. Hackett, R.C. Meyer, D. Milbourne, J.W. NicolMc, M.S. Phillips, R. Waugh, Identification of AFLP and SSR markers associated with quantitative resistance to *Globodera pallida* (Stone) in tetraploid potato (*Solanum tuberosum*) with a view to marker-assisted selection. Theor. Appl. Genet. 97 (1998) 202–210.

283.   G.J. Bryan, K. McLean, J.E. Bradshaw, W.S. DeJong, M. Phillips, L. Castelli, R. Waugh, Mapping QTLs for resistance to the cyst nematode *Globodera pallida* derived from the wild potato species *Solanum vernei*. Theor. Appl. Genet. 105 (2002) 68–77.

284.   M.W. Ganal, R. Simon, S. Brommenschenkel, M. Arndt, M.S. Phillips, S.D. Tanksley, A. Kumar Genetic mapping of a wide spectrum nematode resistance gene (Hero) against *Globodera rostochiensis* in tomato. Mol. Plant-Microbe Interact. 8 (1995) 886–891.

285.   J.C. Veremis, A.W. Heusden van, P.A. Roberts, Mapping of novel heat-stable resistance to *Meloidogyne* in *Lycoperison peruvianum*. Theor. Appl. Genet. 98 (1999) 274–280.

286.   K. Boivin, M. Deu, J.F. Rami, G. Trouch, P. Hamont, Towards a saturated sorghum map using RFLP and AFLP markers. Theor. Appl. Genet. 98 (1999) 320–328.

287.   H.A. Agrama, G.E. Widle, J.C. Reese, L.R. Campbell, M.R. Tuinstra, Genetic mapping of QTLs associated with green bug resistance and tolerance in *Sorghum bicolor*. Theor. Appl. Genet. 104 (2002) 1373–1378.

288.   B.X. Qiu, P.R. Arelli, D.A. Sleoer, RFLP markers associated with soybean cyst nematode resistance and seed composition in a 'Peking' × 'Essex' population. Theor. Appl. Genet. 98 (1999) 356–364.

289.   A.A. Myburg, M. Cawood, B.D. Wingfield, A.M. Botha, Development of RAPD and SCAR markers linked to the Russian wheat aphid resistance gene *Dn2* in wheat. Theor. Appl. Genet. 96 (1998) 1162–1169.

290.   Z.Q. Ma, N.L.V. Lapitan, A comparison of amplified and restriction fragment length polymorphism in wheat. Cereal Res. Commun. 26 (1998) 7–13.

291.   X.M. Liu, C.M. Smith and B.S. Gill, Identification of microsatellite markers linked to Russian wheat aphid resistance genes *Dn4* and *Dn6*. Theor. Appl. Genet. 104 (2002) 1042–1048.

292.   Z.Q. Ma, B.S. Gill, M.E. Sorrells, S.D. Tanksley, RFLP markers linked to two Hessian fly resistance genes in wheat (*Triticum aestivum* L.) from *Triticum tauschii* (Coss.) Schal. Theor. Appl. Genet. 85 (1993) 750–754.

293.   I. Dweikat, H. Ohm, S. Mackenzie, F. Patterson, S. Cambron, R. Ratcliffe, Association of a DNA marker with the Hessian fly resistance gene *H9* in wheat. Theor. Appl. Genet. 89 (1994) 964–968.

294.   I. Dweikat, H. Ohm, F. Patterson, S. Cambron, Identification of RAPD markers for 11 Hessian fly-resistance genes in wheat. Theor. Appl. Genet. 94 (1997) 419–423.

295.   Y.W. Seo, J.W. Johnson, R.L. Jarret, A molecular marker associated with the *H21* Hessian fly-resistance gene in wheat. Mol. Breed. 3 (1997) 177–181.

296. I. Dweikat, W. Zhang, H. Ohm, Development of STS markers linked to Hessian fly-resistance gene *H6* in wheat. Theor. Appl. Genet. 105 (2002) 766–770.

297. K.J. Williams, S.P. Taylor, P. Bogacki, M. Pallotta, H.S. Bariana, H.Wallwork, Mapping of the root lesion nematode (*Pratylenchus neglectus*) resistance gene *Rlnn1* in wheat. Theor. Appl. Genet. 104 (2002) 874–879.

298. K.J. Williams, J.M. Fisher, P. Langridge, Identification of RFLP markers linked to the cereal cyst nematode resistance gene (*Cre*) in wheat. Theor. Appl. Genet. 83 (1994) 919–924.

299. R.F. Eastwood, E.S. Lagudah, R. Appels, A direct search for DNA sequences tightly linked to cereal cyst nematode resistance genes in *Triticum tauschii*. Genome 37 (1994) 311–319.

300. M. Bohn, B. Schulz, R. Kreps, D. Klein, A.E. Melchinger, QTL mapping of resistance against the European corn borer (*Ostrinia nubilalis* H.) in early maturing European dent germplasm. Theor. Appl. Genet. 101 (2000) 907–917.

301. N. Sardesai, A. Kumar, K.R. Rajyashri, S. Nair, M. Mohan, Identification and mapping of an AFLP marker linked to *Gm7*, a gall midge resistance gene and its conversion to a SCAR marker for its utility in marker aided selection in rice. Theor. Appl. Genet. 105 (2002) 691–698.

302. H. Hirabayashi, T. Ogawa, RFLP mapping of *Bph-1* (brown planthopper resistance gene) in rice. Breed. Sci. 45 (1995) 369–371.

303. T. Ishii, D.S. Brar, D.S. Multani, G.S. Khush, Molecular tagging of genes for brown plant hopper resistance and earliness introgressed from *Oryza australiensis* into cultivated rice, *O. sativa*. Genome 37 (1994) 217–221.

304. Fukuta, Y., Tamura, K., Hirae, M., Oya, S., Genetic analysis of resistance to green rice leafhopper (*Nephotettix cincticeps* UHLER) in rice parental line. Norin-PL6, using RFLP markers. Breed. Sci. 48 (1989) 243–249.

305. K. Tamura, Y. Fukuta, M. Hirae, S. Oya, I. Ashikawa, T. Yagi, Mapping of the *Grh1* locus for green rice leafhopper resistance in rice using RFLP markers. Breed. Sci. 49 (1999) 11–14.

306. M. Yamasaki, H. Tsunematsu, A. Yoshimura, N. Iwata, H. Yasui, Quantitative trait locus mapping of ovicidal response in rice (*Oryza sativa* L.) against whitebacked planthopper (*Sogatella furcifera* Horvath). Crop Sci. 39 (1999) 1178–1183.

307 P. Kadirvel, M. Maheswaran, K. Gunathilagaraj, Molecular mapping of quantitative trait loci (QTLs) associated with whitebacked planthopper in rice. Int. Rice. Res. Notes 24 (1999) 12–14.

308. V. Cevik and G.J. King, High-resolution genetic analysis of the *Sd-1* aphid resistance locus in *Malus* sp. Theor. Appl. Genet. 105 (2002) 346–354.

309 L. Bullrich, M.L. Appendino, G. Tranquilli, S. Lewis, J. Dubcovsky, Mapping of a thermo-sensitive earliness *per se* gene on *Triticum monococcum* chromosome 1A^m Theor. Appl. Genet. 105 (2002) 585–595.

310. M.C. Luo, J. Dvorak, Molecular mapping of an aluminium tolerance locus on chromosome 4D of Chinese spring wheat. Euphytica 91 (1996) 31–35.

311. C.R. Riede, J.A. Anderson, Linkage of RFLP markers to an aluminium tolerance gene in wheat. Crop Sci. 36 (1996) 905–909.

312. M.J. Gao, R.L. Travis, J. Dvorak, Mapping of protein polymorphisms associated with the expression of wheat *Kna1* locus under NaCl stress, in: A.E. Slinkard (Ed), Proc. 9th Int. Wheat Genet. Symp., Vol. 3, Univ. Extension Press, Univ. of Saskatchewan, Saskatoon (1998), pp. 105–107.

313. H. Raman, J.S. Moroni, K. Sato, B.J. Read, B.J. Scott, Identification of AFLP and microsatellite markers linked with an aluminium tolerance gene in barley (*Hordeum vulgare* L.). Theor. Appl. Genet. 105 (2002) 458–464.

314. Bentolila, S., Zethof, J., Gerats, T., Hanson, M. R., 1998. Locating the petunia *Rf* gene on a 650-kb DNA fragment. Theor. Appl. Genet. 96: 980–988.

315. Q. Zhang, B.Z. Shen, X.K. Dai, M.H. Mei, M.A. Saghai Maroof, Z.B. Li, Using bulked extremes and recessive class to map genes for photoperiod-sensitive genic male sterility in rice. Proc. Natl. Acad. Sci. USA 91 (1994) 8675–8679.

316. Mei, M. H., Chen, L., Zhang, Z. H., Li, Z. Y., Xu, C. G., Zhang, Q. F., *pms3* is the locus causing the

original photoperiod-sensitive male sterility mutation of 'Nongken 58S'. Sci. China Ser. C-Life Sci. (1999) 42: 316–322.

317. H.J. Koh, Y.H. Son, M.H. Heu, H.S. Lee, S.R. McCouch, Molecular mapping of a new genic male-sterility gene causing chalky endosperm in rice (Oryza sativa L.). Euphytica 106 (1999) 57–62.

318. H. Akagi, Y. Yokozeki, A. Inagaki, A. Nakamura, T. Fujimura, A codominant DNA marker closely linked to the rice nuclear restorer gene, Rf-1, identified with inter-SSR fingerprinting. Genome 39 (1996) 1205–1209.

319. X.L. Tan, A. Vanavichit, S. Amornsilpa, S. Trangoonrung, Genetic analysis of rice CMS-WA fertility restoration based on QTL mapping. Theor. Appl. Genet. 97 (1998) 994–999.

320. D.C. Yang, G.B. Magpantay, M. Mendoza, N. Huang, D.S., Brar, Construction of a contig for a fertility restorer gene, Rf2, in rice using BAC library and its sequence with Rf2 gene of maize. Rice Genet. Newsl. 14 (1997) 116–117.

321. G. Zhang, T.S. Bharaj, Y. Lu, S.S. Virmani, N. Huang, Mapping of the Rf-3 nuclear fertility-restoring gene for WA cytoplasmic male sterility in rice using RAPD and RFLP markers. Theor. Appl. Genet. 94 (1997) 27–33.

322. Y.W. Shen, Z.Q. Guan, J. Lu, J.Y. Zhuang, K.L. Zheng, M.W. Gao, X.M. Wang, Linkage analysis of a fertility restoring mutant generated from CMS rice. Theor. Appl. Genet. 97 (1998) 261–266.

323. F.Y. Yao, C.G. Xu, S.B. Yu, Li, J. X., Y.J. Gao, X.H. Li, Q.F. Zhang, Mapping and genetic analysis of two fertility restorer loci in the wild-abortive cytoplasmic male sterility system of rice (Oryza sativa L.). Euphytica 98 (1997) 183–187.

324. S. Fukuoka, H. Namai, K. Okuno, RFLP mapping of the genes controlling hybrid breakdown in rice (Oryza sativa L.). Theor. Appl. Genet. 97 (1998) 446-449.

325. S. Yanagihara, S.R. McCouch, K., Ishikawa, Y. Ogi, K. Maruyama, H. Ikehashi, Molecular analysis of the inheritance of the S⁻5 locus, conferring wide compatibility in indica/japonica hybrids of rice (O. sativa L.). Theor. Appl. Genet. 90 (1995) 182–188.

326. Z.Q. Ma, M. E. Sorrells, Genetic analysis of fertility restoration in wheat using restriction fragment length polymorphisms. Crop Sci. 35 (1995) 1137–1143.

327. A. Borner, V. Korzun, A. Polley, S. Malyshev, G. Melz, Genetics and molecular mapping of a male fertility restoration locus (Rfg1) in rye (Secale cereale L.). Theor. Appl. Genet. 97 (1998) 99–102.

328. A.V. Voylokov, V. Korzun, A. Borner, Mapping of three self-fertility mutations in rye (Secale cereale L.) using RFLP, isozyme and morphological markers. Theor. Appl. Genet. 97 (1998) 147–153.

329. M. Jean, G.G. Brown, B.S. Landry, Targeted mapping approaches to identify DNA markers linked to the Rfp1 restorer gene for the Polima CMS of canola (Brassica napus L.). Theor. Appl. Genet. 97 (1998) 431–438.

330. L. Wen, H.V. Tang, W. Chen, R. Chang, D. R., Pring, P.E. Klein, K.L. Childs, R.R. Klein, Development and mapping of AFLP markers linked to the sorghum fertility restorer gene rf4. Theor. Appl. Genet. 104 (2002) 577: 585.

331. R. Horn, B. Kusterer, E. Lazarescu, M. Priife, W. Friedt, Molecular mapping of the Rf1 gene restoring pollen fertility in PET1-based F₁ hybrids in sunflower (Helianthus annuus L.). Theor. Appl. Genet. 106 (2003) 599–606.

332. L. Liu, W. Huo, X. Zhu, T. Zhang, Inheritance and fine mapping of fertility restoration for cytoplasmic male sterility in Gossypium hirsutum L., Theor. Appl. Genet. 106 (2003) 461–469.

333. I. Coulibaly, J. Louarn, M. Lorieux, A. Charrier, S. Hamon, M. Noirot, Pollen viability restoration in a Coffea canephora and C. heterocalyx Stoffelen backcross. QTL identification for marker-assisted selection. Theor. Appl. Genet. 106 (2003) 311–316.

334. J.F. Rami, P. Dufour, G. Trouche, G. Fliedel, C. Mestres, F. Davrieux, P. Blanchard, P. Hamon, Quantitative trait loci for grain quality, productivity, morphological and agronomical traits in Sorghum (Sorghum bicolor L. Moench). Theor. Appl. Genet. 97 (1998) 605–616.

335. A.Y. Wang, W.P. Yu, R.H. Juang, J.W. Huang, H.Y. Sung, J.C. Su, Presence of three rice source synthase genes as revealed by cloning and sequencing of cDNA. Plant Mol. Biol. 18 (1992) 1191–1194.

336. G.D. Parker, K.J. Chalmers, A.J. Rathjen, P. Langridge, Mapping loci associated with flour colour in wheat (Triticum aestivum L.). Theor. Appl. Genet. 97 (1998) 238–245.

337. K. Kato, H. Miura, S. Sawada, Mapping QTLs controlling grain yield and its components on chromosome 5A of wheat. Theor. Appl. Genet. 101 (2002) 1114–1121.

338. M. Ahmad, Molecular marker-assisted selection of HMW glutenin alleles related to wheat bread quality by PCR-generated DNA markers. Theor. Appl. Genet. 101 (2000) 892–896.

339. M. Prasad, N. Kumar, P.L. Kulwal, M.S. Roder, H.S. Balyan, H.S. Dhaliwal, P.K. Gupta, QTL analysis for grain protein content using SSR markers and validation studies using NILs inbred wheat. Theor. Appl. Genet. 106 (2003) 659–667.

340. P. Sourdille, M.R. Perretant, G. Charmet, P. Leroy, M.F. Gautier, P. Jourdier, J.C. Nelson, M.E. Sorrells, M. Bernard, Linkage between RFLP markers and genes affecting kernel hardness in wheat. Theor. Appl. Genet. 93 (1996) 580–586.

341. D'Ovidio, O.D. Anderson, PCR analysis to distinguish between alleles of a member of a multigene family correlated with wheat bread-making quality. Theor. Appl. Genet. 88 (1994) 759–763.

342. E. Araki, H. Miura, S. Sawada, Identification of genetic loci affecting amylose content and agronomic traits on chromosome 4A of wheat. Theor. Appl. Genet. 98 (1999) 977–984.

343. L. Mokrani, L. Genetzbittel, F. Azanza, L. Fitamant, G.E. Chaarani-Al, A. Sarrafi, Mapping and analysis of quantitative trait loci for grain oil content and agronomic traits using AFLP and SSR in sunflower (*Helianthus annuus* L.). Theor. Appl. Genet. 106 (2002) 149–156.

344. J.M. Gion, P. Rech, J. Grima-Pettenati, D. Verhaegen, C. Plomion, Mapping candidate genes in *Eucalyptus* with emphasis on lignification genes. Mol. Breed. 6 (2000) 441–449.

345. G. Scovel, H. Meir-Ben, M. Ovasis, H. Itzhaki, A. Vainstein, RAPD and RFLP markers tightly linked to the locus controlling carnation (*Dianthus caryophyllus*) flower type. Theor. Appl. Genet. 96 (1998) 117–122.

346. M.A.R. Mian, R., Wells, T.E. Jr. Carter, D.A. Ashley, H.R. Boerma, RFLP tagging of QTLs conditioning specific leaf weight and leaf size in soybean. Theor. Appl. Genet. 96 (1998) 354–360.

347. T. Cadalen, P. Sourdille, G. Charmet, M.H. Tixier, G. Gay, C. Boeuf, S. Bernard, P. Leroy, M. Bernard, Molecular markers linked to genes affecting plant height in wheat using a doubled-haploid population. Theor. Appl. Genet. 96 (1998) 933–940.

348. P. Sourdille, Barloy, M. Bernard, Linkage between RFLP molecular markers and the dwarfing genes *Rht-B1* and *Rht-D1* in wheat. Hereditas 128 (1998) 41–46.

349. A.M. Torp, A.L. Hansen, I.B. Holme, S.B. Anderson, Genetic markers for haploid formation in wheat anther culture, in: A.E. Slinkard (Ed), Proc. 9th Int. Wheat Genet. Symp., Vol. 3, Univ. Extension Press, Univ. of Saskatchewan, Saskatoon (1998), pp. 159–161.

350. M.H. Ellis, W. Spielmeyer, K.R. Gale, G.J. Rebetzke, R.A. Richards, "Perfect" markers for the *Rht-B1b* and *Rht-D1b* dwarfing genes in wheat. Theor. Appl. Genet. 105 (2002) 1038–1042.

351. K. Kato, H. Miura, S. Sawada, QTL mapping of genes controlling ear emergence time and plant height on chromosome 5A of wheat. Theor. Appl. Genet. 98 (1999) 472–477.

352. J.A. Anderson, M.E. Sorrells, S.D. Tanksley, RFLP anlaysis of genomic regions associated to preharvest sprouting in wheat. Crop Sci. 33 (1993) 453–459.

353. G. Galiba, S.A. Quarrie, J. Sutka, A. Morgonuv, J.W. Snape, RFLP mapping of the vernalization (*Vrn1*) and frost resistance (*Fr1*) genes on chromosome 5A of wheat. Theor. Appl. Genet. 90 (1995) 1174–1179.

354. J. Dubcovsky, D. Lijavetzky, L. Appendino, G. Tranquilli, Comparative RFLP mapping of *Triticum monococcum* genes controlling vernalization requirement. Theor. Appl. Genet. 97 (1998) 968–975.

355. J.W. Snape, A. Semikhodskii, R. Srama, V. Korzun, L. Fish, S.A. Quarrie, B.S. Gill, T. Sasaki, G. Galiba, J. Sutka, Mapping vernalization loci in wheat and comparative mapping with other cereals, in: A.E. Slinkard (Ed), Proc. 9th Int. Wheat Genet. Symp., Vol. 3, Univ. Extension Press, Univ. of Saskatchewan, Saskatoon (1998), pp. 156–158.

356. G.A. Penner, J. Clark, L.Z. Bezte, D. Lisle, Identification of RAPD markers linked to a gene governing cadmium uptake in durum wheat. Genome 38 (1995) 543–547.

357. S.A. Quarrie, M. Gulli, C. Calestani, A. Steed, N. Marmiroli, Location of a gene regulating drought-induced abcisic acid production in the long arm of chromosome 5A of wheat. Theor. Appl. Genet. 89 (1994) 794–800.

358. G. Parker, C. Ken, R. Anthony, L. Peter, Identification of molecular markers linked to flour colour and miling yield in wheat, in: 5th Int. Congr. Plant Mol. Biol., Singapore, (1997), pp. 21–27, abstr. 246.

359. R.C. de la Pena, T.D. Murray, S.S. Jones, Linkage relations among eyespot resistance gene *Pch2*, endopeptidase EP-A1B, and RFLP marker *Xpsr121* on chromosome 7A of wheat. Plant Breed. 115 (1996) 273–275.

360. R.C. de la Pena, T.D. Murray, S.S. Jones, Identification of an RFLP interval containing *Pch2* on chromosome 7AL in wheat. Genome 40 (1997) 249–252.

361. J.D. Faris, J.A. Anderson, L.J. Francl, J.G. Jordahl, RFLP mapping of resistance to chlorosis induction by *Pyrenophora tritici-repentis* in wheat. Theor. Appl. Genet. 94 (1997) 98–103.

362. G.J. Allen, G.W. Jones, R.A. Leigh, Sodium transport measured in plasma membrane vesicles isolated from wheat genotypes with differing K^+/Na^+ discrimination traits. Plant Cell Enivron. 18 (1995) 105–115.

363. P.J. Conner, S.K. Brown, N.F. Weeden, Molecular marker analysis of quantitative traits for growth and development in juvenile apple trees. Theor. Appl. Genet. 96 (1998) 1027–1035.

364. G.C. Yencho, S.P. Kowalski, R.S. Kobayashi, S.L. Sinden, M.W. Bonierbale, K.L. Deahl, QTL mapping of foliar glycoalkaloid aglycones in *Solanum tuberosum* × *S. berthaultii* potato progenies: quantitative variation and plant secondary metabolism. Theor. Appl. Genet. 97 (1998) 563–574.

365. E. Dirlewanger, A. Moing, C. Rothan, L. Svanella, V. Pronier, A. Guye, C. Plomion, R. Monet, Mapping QTLs controlling fruit quality in peach (*Prunus persica* (L.) Batsch). Theor. Appl. Genet. 98 (1999) 18–31.

366. L. Crespel, M. Chirollet, C.E. Durel, D. Zhang, J. Meynet, S. Gudin, Mapping of qualitative and quantitative phenotypic traits in *Rosa* using AFLP markers. Theor. Appl. Genet. 105 (2002) 1207–1214.

367. A. Doligez, A. Bouquet, Y. Danglot, F. Lahogue, S. Riaz, C.P. Meredith, K.J. Edwards, P. This, Genetic mapping of grapevine (*Vitis vinifera* L.) applied to the detection of QTLs for seedlessness and berry weight. Theor. Appl. Genet. 105 (2002) 708–795.

368. T. Komatsuda, Y. Mano, Molecular mapping of the intermedium spike-C (*int-c*) and non-brittle rachis 1 (*btr 1*) loci in barley (*Hordeum vulgare* L.). Theor. Appl. Genet. 105 (2002) 85–90.

369. A. Schneider, S.A., Walker, M. Sagan, G. Duc, T.H.N. Ellis, J.A. Downie Mapping of the nodulation loci *sym 9* and *sym 10* of pea (*Pisum sativum* L.). Theor. Appl. Genet. 104 (2002) 1312–1316.

370. K. Schneider, R. Preg-Schafer, D.C. Borchardt, F. Salamini, Mapping QTLs for sucrose content, yield and quality in a sugarbeet population fingerprinted by EST-related markers. Theor. Appl. Genet. 104 (2002) 1107–1113.

371. D.J. Mackill, M.A. Salam, Z.Y. Wang, S.D. Tanksley, A major photoperiod-sensitivity gene tagged with RFLP and isozyme markers in rice. Theor. Appl. Genet. 85 (1993) 536–540.

372. C.Z. Laing, M.H. Gu, X.B. Pan, G.H. Liang, L.H. Zhu, RFLP tagging of a new semidwarfing gene in rice. Theor. Appl. Genet. 88 (1994) 898–900.

373. Y. Fukuta, T. Yagi, Mapping of a shattering resistance gene in a mutant line SR-5, induced from an indica rice variety, Nan-jing 11. Breed. Sci. 48 (1998) 345–348.

374. Sobrizal, K. Ikeda, P.L. Sanchez, A. Yoshimura, RFLP mapping of a seed shattering gene on chromosome 4 in rice. Rice Genet. Newsl. 16 (1999) 74–75.

375. S.Y. Lin, T. Sasaki, M. Yano, Mapping quantitative trait loci controlling seed dormancy and heading date in rice, *Oryza sativa* L., using backcross inbred lines. Theor. Appl. Genet. 96 (1998) 997–1003.

376. T. Yamamoto, Y. Kuboki, S.Y. Lin, T. Sasaki, M. Yano, Fine mapping of quantitative trait loci Hd-1, Hd-2 and Hd-3, controlling heading date of rice, as single Mendelian factors. Theor. Appl. Genet. 97 (1998) 37–44.

377. A. Kamoshita, L.J.Wade, M.L. Ali, M.S. Pathan, J. Zhang, Sarkarung, H.T. Nguyen, Mapping QTLs for root morphology of a rice population adapted to rainfed lowland conditions. Theor. Appl. Genet. 104 (2002) 880–893.

378. H.J. Ju, R. Bernardo, H.W. Ohm, Mapping QTL for popping explosion volume in popcorn with simple sequence repeat markers. Theor. Appl. Genet. 106 (2003) 423–427.

379. T. Zhang, Y. Yuan, J. Yu, W. Guo, R.J. Kohel, Molecular tagging of a major QTL for fiber strength in Upland cotton and its marker-assisted selection. Theor. Appl. Genet. 106 (2003) 262–268.

21. Studies on Male Meiosis in Cultivated and Wild *Vigna* Species

S. Rama Rao and S.N. Raina*

Cytogenetics Laboratory, Department of Botany, J.N. Vyas University, Jodhpur 342 005, Rajasthan, India

*Cellular and Molecular Cytogenetics Laboratory, Delhi University, Delhi-110 007, India

Abstract: Male meiosis was studied in 11 species and one sub-species which include both cultivated and wild ones to understand the cytogenetic mechanism underlying the speciation and evolution in the genus *Vigna*. The present observations together with earlier published data indicate that the most common gametic number is $n = 11$, $2n = 22$ and in all probability this gametic number ($n = 11$) is the basic number of the genus. Majority of the species studied presently showed normal eleven bivalents at diakinesis/metaphase I, and 0–2 univalents found in four taxa have been ascribed to early separation of precocious separation of rod bivalents. Complete bivalent formation in the teraploid species *V. galbrescens* indicates its allopolyploid origin. Apparently, non-random distribution of chiasmata in majority of the species investigated, seems to be an important cytogenetic phenomenon in the genus. B-chromosomes [1–2] were recorded in two species. The data collected on anphase I/II distribution of bivalents/chromosomes show that inversion heterozygosity in the genus is not occasional but might be at floating stage in the population(s). Further, the study of meiosis in these wild and cultivated species of *Vigna* clearly confirms that gene mutations and chromosomal repatterning played a significant role in speciation and evolution of the genus.

1. Introduction

The genus *Vigna* comprises about 120 species distributed widely in tropical and subtropical regions of both hemispheres. It is one of the most important genus of Fabaceae and some thirty species including *V. radiata* (mung bean), *V. mungo* (urd bean), *V. aconitifolia* (moth bean), *V. umbellata* (rice bean), *V. trilobata*, *V. vexillata*, *V. angularis* (adzuki bean), *V. lanceolata*, *V. mariana*, *V. ambacensis*, *V. fisheri*, *V. unguiculata* (cowpea), *V. reticulata* and *V. capensis* are cultivated extensively for their pulse (protein content 17–24%) crop, vegetable, fodder crop, cover crop, green manure and soil erosion control value [1–5]. The close morphological resemblance between *Vigna* and two closely related genera (*Phaseolus* and *Dolichos*) has made it difficult for the taxonomists to clearly delimit the species of the genus from that of *Phaseolus* and *Dolichos*, and it is the phylogenetic classifications [6–8] which has set at rest most of such confusion. The morphological features that distinguish *Vigna* from *Phaseolus* and *Dolichos* are curved, rather a coiled or twisted keel and lateral rather than a terminal stigma [6], respectively, and Verdcourt [7] has included yellow flowered species of *Phaseolus* in section certotropis piper in the subgenus *Certotropis* of the genus *Vigna*. Several species of *Phaseolus* including *P. aureus*, *P. radiatus*, *P. angularis*, *P. mungo*, *P. trilobatus*, *P. pubescens*, *P. calcaratus* have now been transferred to large and heterogenous genus *Vigna*. Much of our present cytological understanding of *Vigna* is restricted to mere chromosome numbers, a few inter-specific hybrids, nuclear DNA amounts and often conflicting accounts about the chromosome complements and associations both within and between the species [9–43]. The detailed studies on mitotic complements, male meiosis, colchitetraploidy, interspecific hybrids and various other cytogenetical parameters, an important prerequisite for providing evidence of past evolutionary events of theoretical and practical importance and for logical manipulations to the advantage of economically important taxa as in cereals, is very much limited in *Vigna*. The reasons could be factors like inherent difficulty in obtaining good analyzable cytological preparations, small chromosome size, overall stability of chromosome morphology and symmetry, and no success in raising cytogenetic stocks like translocation testers and/or aneuploids. Such factors have also proved an impediment in ascertaining precisely the genome relationships between species in a few successful interspecific hybrids. The information available from interspecific crosses between *V. radiata*, *V. angularis*, *V. mungo*, *V. umbellata*, *V. minima* and *V. trilobata* has confirmed that there exists a certain degree of homology between different genomes [21, 34–41, 44]. The failure of crosses between other species, especially those with wild species could not, however, be taken up due to lack of genomic homology between them. There is, for example, complete bivalent pairing in the F_1 hybrids between *V. umbellata* and *V. angularis*, raised by embryo rescue culture techniques [45].

Besides this, there is an overall stability in chromosome morphology and symmetry between the species of the genus *Vigna*. The species differentiation cannot be, in most cases, correlated with chromosome differentiation. In such case, meiosis could be yet another parameter for understanding cytogenetic system in the genus. The comprehensive study about chromosome associations, chiasma distribution and its frequency, and chromosome distribution during anaphase would also throw some light on the nature of cytogenetic mechanisms underlying evolution in the genus. In spite of several inherent disadvantages in the material a concerted attempt has been made to bring out details of male meiosis in eleven taxa comprising ten species and one subspecies as detailed below.

2. Material and Methods

The seeds of various species and subspecies of *Vigna* were kindly supplied by the United States Department of Agriculture (USDA), Maryland, USA and the National Bureau of Plant Genetic Resources, New Delhi, India. For meiotic analysis flower buds of appropriate size were collected from field grown plants and anthers were squashed in 1% aceto-carmine. On an average 25 cells were analyzed at diplotene/diakinesis and metaphase I for recording chromosome associations and recombinational frequencies through chiasma analysis. 15–20 cells were also analysed at AI/AII for distributional pattern of chromosomes. For percentage pollen stainability, the pollen grains were stained in 1:1 (glycerin : acetocarmine) mixture and on average 10 slides were scored for stainable pollen. Photomicrographs from temporary preparations were taken using Agfa-Copex Pan photonegative film (ASA–20).

3. Results

The meiotic data has been summarized in Tables 1 to 3.

Table 1. Average number, range of chiasmata, terminalization coefficient and pollen stainability in *Vigna* species

Species	DNA amount (×10⁻¹² g)	2n	No. of cells analysed	Chiasmata		Chiasmata terminalized	Unterminalized	Termina-lization coefficient	Percentage pollen stainability
				Mean	Range				
V. aconitifolia		22	26	20.92 ± 1.29	18-24	18.77 ± 1.77	2.15	0.89	87.09
V. aureus		22	20	20.85 ± 1.79	18-24	15.80 ± 2.05	4.95	0.76	90.00
V. luteola		22	20	18.80 ± 1.76	15-21	16.10 ± 2.26	2.70	0.85	88.39
V. mungo	2.83	22	25	21.84 ± 1.77	18-25	17.56 ± 1.29	4.28	0.80	89.80
V. radiata	2.67	22	25	20.60 ± 1.63	17-26	17.12 ± 1.94	3.48	0.83	94.00
V. repens	2.76	22	25	18.52 ± 1.80	15-24	14.24 ± 1.66	4.28	0.76	93.16
V. umbellata	2.84	22	25	18.00 ± 2.27	13-22	14.96 ± 1.96	3.04	0.83	82.30
V. unguiculata	3.03	22	26	19.80 ± 0.98	18-21	16.07 ± 1.62	3.73	0.81	99.10
V. unguiculata ssp. *sesquipedaceae*		22	20	19.90 ± 2.86	13-22	18.80 ± 2.70	1.10	0.94	80.34
V. sps. Tvnu-72		22	15	19.93 ± 2.37	13-22	16.40 ± 1.50	3.53	0.82	89.92
V. glabrescens	4.95	44	14	39.57 ± 2.50	32-42	39.57 ± 2.50		1.0	69.09

Table 2. Average number, range of associations at diakinesis/metaphase I

Species	2n	No. of cells analysed	Cells with 11/22 II		Rod bivalents			Ring bivalents			Total bivalents			Univalents		
			No.	Percentage	No.	Mean	Range	No.	Mean	Range	No.	Mean	Range	No.	Mean	Range
V. aconitifolia	22	26	26	100.0	52	2.04 ± 0.94	0–4	234	8.96 ± 1.09	6–11	286	11.0	9–11			
V. aureus	22	20	20	100.0	44	2.25 ± 1.00	1–4	174	8.76 ± 1.08	7–10	218	11.0	9–11			
V. luteola	22	20	20	100.0	72	3.60 ± 1.66	2–7	148	7.40 ± 1.66	4–9	220	11.0				
V. mungo	22	25	25	100.0	50	2.00 ± 1.00	0–4	225	9.00 ± 1.00	7–11	275	11.0				
V. radiata	22	25	25	100.0	54	2.16 ± 0.94	1–4	221	8.84 ± 1.20	6–10	275	11.0				
V. repens	22	25	24	96.0	98	3.92 ± 1.11	2–6	176	7.04 ± 1.17	5–9	274	10.96	10–11	2	0.08 ± 0.40	0–2
V. umbellata	22	25	23	92.0	108	4.32 ± 1.51	2–8	164	6.56 ± 1.58	3–9	272	10.88	9–11	6	0.24 ± 1.38	0–2
V. unguiculata	22	26	26	100.0	80	3.07 ± 0.74	2–5	206	7.92 ± 0.74	6–9	286	11.0				
V. unguiculata ssp. sesquipedaceae	22	20	18	90.0	31	1.50 ± 0.94	0–3	183	9.15 ± 1.50	4–11	214	10.65	10–11	4	0.60 ± 2.3	0–2
V. sps. Tvnu-72	22	15	13	86.7	33	2.20 ± 0.94	1–4	130	8.66 ± 1.40	5–10	163	10.86	8–11	4	1.57 ± 3.25	0–12
V. glabrescens	44	14	9	64.29	39	2.79 ± 1.63	0–5	258	18.42 ± 1.34	16–21	297	21.21	16–22	22	1.57 ± 3.25	0–12

Table 3. Anaphase I distribution (U = Univalents, B = Bivalents)

Species	2n	No. of cells analysed	Chromosome distribution	No. of cells	Percentage
V. aconitifolia	22	15	11:11	15	100.0
V. aureus	22	20	11:11	20	100.0
V. luteola	22	20	11:11	20	100.0
V. mungo	22	20	11:11	20	100.0
V. radiata	22	20	11:11	20	100.0
V. repens	22	20	11:11	16	80.0
			10:1U:11	4	20.0
V. umbellata	22	15	11:11	14	93.3
			10:1U:11	1	6.7
V. unguiculata	22	20	11:11	20	100.0
V. unguiculata ssp. sesquipidaceae	22	15	11:11	15	100.0
V. sps. Tvnu-72	22	15	11:11	15	100.0
V. glabrescens	44	25	22:22	18	72.0
			23:21	3	12.0
			21:1B:21	3	12.0
			22:2U:20	1	4.0

3.1 Associations

Diploids

V. aconitifolia (n = 11): All the cells analyzed had eleven bivalents (Figs. 1 to 3). On the average there were 8.92 ring and 2.0 rod bivalents. Same gametic number has been reported by Purseglove [15], Bhatnagar et al. [17], Sarbhoy [25] and Tschechow and Karataschowa [46].

V. aureus (n = 11): Eleven bivalents were observed in all the cells analyzed. The bivalents on the average resolved into 8.70 ring and 2.20 rod bivalents.

V. luteola (n = 11): Eleven bivalents, observed in all the cells, on the average resolved into 7.40 ring and 3.60 rod bivalents. The nucleolus in a few cells at diplotene/diakinesis was seen associated with as many as 2-4 bivalents. Same gametic number has been reported by Sen and Bhowal [48].

V. mungo (n = 11): Eleven bivalents were encountered in all the twentyfive cells analysed. On the average there were 9.0 ring and 2.0 rod bivalents. In this species also the nucleolus was observed to be associated with more than one bivalent. The same gametic number has been reported by several workers [9, 13, 14, 18, 19, 21, 22, 25, 28, 30, 34, 49, 50].

V. radiata (n = 11): All the cells analyzed had eleven bivalents at diakinesis/metaphase I. The average number of ring and rod bivalents was 8.80 and 2.16, respectively. As in *V. luteola* and *V. mungo* few cells in this species had 1–3 bivalents associated with the nucleolus. The same gametic number has been reported by various researchers [9–12, 14, 17, 19, 20–22, 26, 28, 30, 34, 49, 50]. Univalents ranging from 2 to 4 were observed in a few cells by Sarbhoy [25]. A variety of the species (*V. radiata* var. *glabra*), which was given the rank of separate species by later authors, had 22 II (2n = 44) instead of normal 11 II [16] prevalent in the species.

V. repens (n = 11): In 24 out of 25 cells eleven bivalents were observed (Figs. 5 and 6). In the

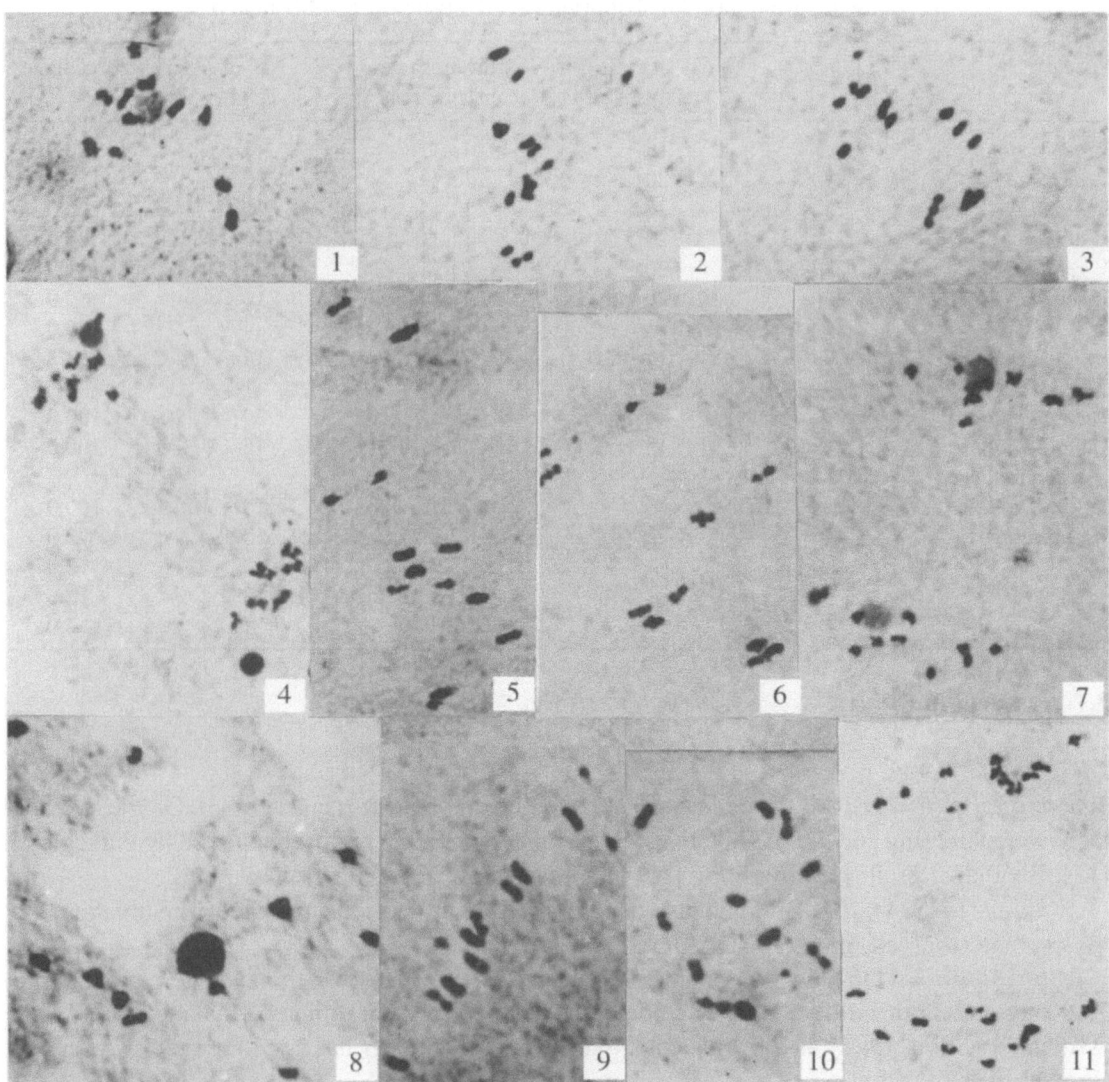

Figs. 1 to 11. Male meiosis in *Vigna*. Fig. 1. *V. aconitifolia*. Diakinesis, 11 II. Figs. 2 and 3. *V. aconitifolia*. Metaphase I, 11 II. Fig. 4. *V. aconitifolia* Anaphase I, 11: 11. Figs. 5 and 6. *V. repens*. Metaphase I, 11 II. Fig. 7. *V. repens*. Anaphase I 11:11 Fig. 8. *V. umbellata*. Diakinesis, 11 II. Figs. 9 and 10. *V. umbellata*. Metaphase I, 11 II. Fig. 11. *V. umbellata*. Anaphase I, 11: 11.

remaining one cell there were ten bivalents and two univalents. The average number of associations per cell was 10.96 II + 0.08 I. On the average 7.04 were ring and 3.92 rod bivalents.

V. umbellata ($n = 11$): Majority (92.0%) of the cells analyzed had eleven bivalents (Figs. 8 to 10). In the remaining cells besides bivalents, univalents ranging from 0 to 2 were also observed. The average number of associations per cell was 10.88 II + 0.24 I. On the average there were 6.56 ring and 4.32 rod bivalents. The same gametic number has been reported by a number of workers [21, 25, 26, 34, 51, 52].

V. unguiculata (*n* = 11): All the cells analyzed had eleven bivalents (Figs. 12 to 14). On the average there were 7.92 ring and 3.07 rod bivalents per cell. Same gametic number has been reported by Karpechenko [9], Mukherjee [54], while Kawakami [10], Rao [55], Floresca et al. [56], Miege [57], Faris [6], Frahm-Leliveld [12] and Yarnell [58] have all reported 12 as the gametic number.

V. unguiculata ssp. *sesquipedaceae* (*n* = 11): Majority (90%) of the cells analyzed had eleven bivalents (Figs. 16 to 18). In the remaining 10% cells univalents ranging from 0 to 2 were observed. The average number of associations per cell therefore was 10.65 II + 0.60 I. Ring bivalents (9.15) outnumbered rod bivalents (1.50).

V. sps. Tvnu 72 (*n* = 11): This species obtained from Dr. Barnhart, USA had eleven bivalents in 86.6% cells while in the remaining (13.3%) cells univalents ranging from 0 to 2 were observed. The average number of associations per cell therefore was 10.88 II + 0.26 I. Ring bivalents (8.66) were predominant over rod (2.20) bivalents.

Tetraploid
V. glabrescens (*n* = 22): The species, previously designated as *V. radiata* var. *glabra*, was the only one in the present investigation which had 2*n* = 44. Majority (64.3%) of the cells had normal 22 bivalents (Figs. 20 and 21). The other 35.7% cells had a mixture of bivalents and univalents ranging from 16 to 21 and 2 to 12, respectively. Multivalents were not observed in any of the cells analyzed. The average number of associations per cell was 21.21 II + 1.57 I. The mean frequency per cell of ring and rod bivalents was 18.42 and 2.79, respectively.

3.2 Chiasma Frequency
The average number of chiasmata per cell ranged from 18 to 24, 18 to 24, 15 to 21, 18 to 25, 17 to 26, 15 to 24, 13 to 22, 18 to 21, 13 to 22, 13 to 22 mean number being 20.92, 20.85, 18.80, 21.84, 20.60, 18.52, 18.00, 19.80, 19.90, 19.93 in *V. aconitifolia, V. aureus, V. luteola, V. mungo, V. radiata, V. repens, V. umbellata, V. unguiculata, V. unguiculata* ssp. *sesquipedaceae, V.* sps. Tvnu-72, respectively, out of which on the average 18.77, 15.90, 16.10, 17.56, 17.12, 14.24, 14.96, 16.07, 18.80, 16.40 were terminalized giving terminalization coefficient of 0.89, 0.76, 0.85, 0.80, 0.83, 0.76, 0.83, 0.81, 0.94 and 0.82, respectively. The corresponding values for *V. glabrescens* were 32–42, 39.57, 39.57 and 1.0, respectively. The maximum number of chiasmata observed for any bivalent was four. The most common observation was 1 chiasma in rod and 2–3 in ring bivalents. The average number of chiasmata per cell among the species with *n* = 11 was highest in *V. mungo* (21.84) and lowest in *V. umbellata* (18.0).

3.3 B-chromosomes
A single Feulgen positive B-chromosome was observed in two (*V. unguiculata, V. mungo*) out of the eleven species investigated presently. A large number of PMCs in these two species were, however, apparently without B-chromosome. They did not have any perceptible effect on the morphology, meiotic behaviour and pollen fertility.

3.4 Anaphase I, II
In seven out of the present eleven taxa anaphase I had equal distribution of chromosomes at the poles (Figs. 4, 7, 11, 15, 19 and 22; Table 3). Three species (*V. glabrescens, V. repens* and *V. umbellata*)

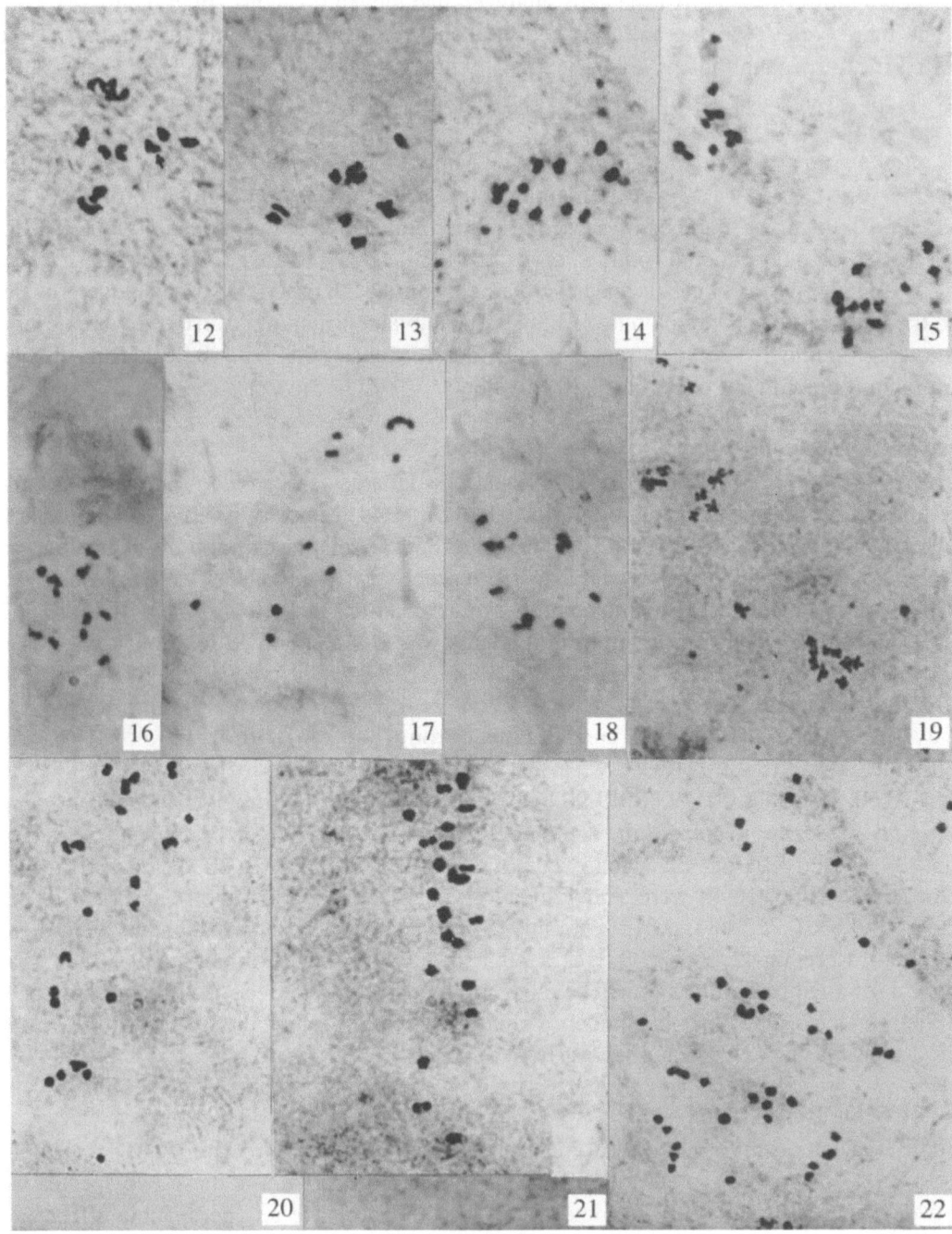

Figs. 12 to 22. Male meiosis in *Vigna*. Figs. 12 to 14. *V. unguiculata*. Metaphase I, 11 II. Note B-chromosome in Fig. 12. Fig. 15. *V. unguiculata*. Anaphase I, 11: 11. Figs. 16 to 18. *V. unguiculata* ssp. *sesquipedaceae*. Metaphase I, 11 II. Fig. 19. *V. unguiculata* ssp. *sesquipedaceae*. Anaphase I, 11: 11. Figs. 20 and 21. *V. glabrescens*. Metaphase I. 22 II. Fig. 22. *V. glabrescens*. Early anaphase I.

had also equal distribution of chromosomes in as many as 64.0, 80.0 and 93.3% cells, respectively. The remaining cells had unequal distribution and/or lagging univalents/bivalents (Table 3). In the same very species few cells analyzed at anaphase I and II had 1-2 bridge fragment configurations.

3.5 Pollen Stainability
All but one *V. glabrescens* had 82–99% stainable pollen. The pollen stainability in *V. glabrescens* was only 69.08%.

4. Discussion
Barring *V. glabrescens*, all the ten taxa investigated here had same gametic number ($n = 11$). Even in *V. glabrescens* the gametic number ($n = 22$) was multiple of $n = 11$. The previous data on male meiosis includes gametic numbers of 31 more species not presently investigated and it is evident from the combined data that as many as 36 species out of the total 51 have $n = 11$ followed by $n = 10$, $n = 9$, $n = 12$, $n = 22$ in 6, 4, 1, 1 species, respectively. The remaining three species are reported to have two gametic numbers ($n = 10, 11$; $n = 11, 12$) within the species.

In ascertaining the true basic number of the genus it is essential to have information regarding gametic and/or zygotic numbers of as many species as possible. From the review of literature, it is however clear that altogether 51 species out of total 150 species have only been evaluated for above aspects. Taking mitotic data also into consideration it becomes clear that there exists four basic numbers ($x = 9, 10, 11, 12$) in the genus and the most common among these is $x = 11$, met in about 78% of the species investigated so far. Two different patterns of origin of more than one basic number in a genus could be recognized. One that the genus might be polybasic in origin and second possibility would be that the genus had only one basic number and during the course of evolution one or more basic numbers originated from a relatively primitive basic number.

According to Frahm-Leliveld [12] and Goswami [27] the basic numbers of 10 and 12 in the genus are derivatives of $n = 11$. The reason for change in gametic number given are structural alterations including centric fusion. Similarly, Froni-Martinus [59] believes that $n = 9$ observed in *V. candida* might have arisen from $n = 11$ by structural alterations. Since all the species investigated here had either $n = 11$ or $n = 22$, the present author could not evaluate the comparative karyomorphology of the species with $n = 11$ to that of species with other basic numbers for making out alterations in morphology as a result of the change in basic number. The present author, however, supports the view of the other cytogeneticists referred above [12, 25, 27, 59], regarding 11 as the true basic number of the genus *Vigna* and all other basic numbers ($x = 9, 10, 12$) as derivative of $x = 11$ occurred during the course of evolution. The reasons for its validity are that this number is not only found in majority (78%) of the species analyzed cytologically so far, but is found in taxa that are not only morphologically distinct but are also widely distributed in tropics and subtropics of the world. Furthermore, the only polyploid species reported so far is built on this number. The occurrence of the two basic numbers within the species in *V. unguiculata* ($x = 11, 12$) and *V. vexillata* ($x = 10, 11$) is very interesting and hybridization between plants differing in basic numbers within species will be of importance in ascertaining the relationship of the two basic numbers within the species.

The plants representing *V. aconitifolia*, *V. aureus*, *V. luteola*, *V. mungo*, *V. radiata* and *V. unguiculata* were characterized by the presence of perfectly normal eleven bivalents at diakinesis/

metaphase I (Table 2). In comparison, a few cells in *V. repens, V. umbellata, V. unguiculata* ssp. *sesquipedaceae* and *V.* sps. Tvnu-72 had univalents, ranging from 2 in *V. repens* and *V. unguiculata* sp. *sesquipedaceae* to 6 in *V.* sps. Tvnu-72, in hardly 4.0, 2.0, 10.0 and 14.3% cells, respectively (Table 2). They in most cells behaved normally at anaphase I leading to organized and/or equal distribution to the respective poles. The occurrence of stray univalents might be attributed to early separation of synapsed homo/homeologues with or without formation of chiasmata or precocious separation of rod bivalents. Such behaviour will convert bivalents to pair of univalents which generally move to respective poles. The reason(s) for the prevalence of relatively high frequency (1.57) per cell of univalents (Table 2) in 35.21 per cent cells of the tetraploid (*V. glabrescens*) species is unknown.

The highest chiasma frequency in the diploid species was recorded in *V. mungo* (21.84) and lowest in *V. umbellata* (18.0) (Table 1). All other species had values between these two extremes. Majority of the species had in general 1 or 2 chiasmata in the bivalents. Due to very small size of bivalents, the exact location of chiasma even in early diplotene could not be made out clearly. They however seemed to be located at distal ends, and most of them got terminalized even at pro-metaphase I stage. Same situation existed in the lone tetraploid species (*V. glabrescens*) where average frequency of chiasma per cell was 39.57 and all of them were observed at the terminal region even at diplotene thus giving terminalization coefficient of one (Table 2).

The seemingly nonrandom distribution of chiasmata is corroborated by the fact that the frequency per cell in a particular species was not dependent on the size of chromosomes. In *V. umbellata* and *V. mungo*, for example, although there is a difference of 4.0 in chiasma frequency per cell, the 2C nuclear DNA amounts [42] are exactly the same. The occurrence of localization of chiasmata, which prevents some chromosome segments from recombining and thus keeping special gene combinations intact, apparently seems to be a strong feature at least in the species of *Vigna* investigated here. The presence of localized chiasma, proximal or distal, found in large number of plants [60, 61] has been attributed to the course of chromosome pairing, interference pattern or availability of only short segments for pairing of the chromosomes.

In eight out of eleven taxa presently investigated the distribution of chromosomes at anaphase I and presumably at anaphase II was normal and no abnormalities due to structural alteration and/or unequal distribution of chromosomes were observed (Table 3). However, in the remaining three species (*V. glabrescens*, $n = 22$, *V. repens*, $n = 11$, *V. umbellata*, $n = 11$) unequal distribution of chromosomes and lagging univalents/bivalents were encountered in a few cells. Similarly, bridge fragment configurations were also observed in these very same species, though only in stray cells. Occurrence of bridge fragment configuration has also been reported in some accessions of *V. aconitifolia* and *V. mungo* [23, 24, 28]. A dicentric bridge and an acentric fragment at anaphase I could result from a single crossover within the inverted region involving two nonsister chromatids or because of diagonal three strand double crossovers involving three chromatids within the inverted region. Dicentric bridge and an acentric fragment might also arise following chromosome breakage and reunion or by inverted crossing over during meiosis [62, 63]. The presence of two dicentric bridges and two acentric fragments at anaphase II observed in a single cell of *V. glabrescens* is the outcome of complementary four strand double crossovers involving all the chromatids within the inverted region and one crossover between centromere and the region outside the inverted loop. The low frequency of bridge fragment configuration in the cells investigated might be attributed to the failure of crossing over and/or nonpairing within and between the small inverted segments, respectively. From the present investigations together with earlier reports [23, 25, 28, 34], it is amply clear that inversion heterozygosity in the genus is not occasional as reported in large number of plants. They might be at a stage of floating inversions in the genus as in *Campanula* [64]. Inversion heterozygosity has been instrumental in establishing species relationships in *Drosophila, Lilium* and *Paeonia* [61, 65, 66]. How far they have got established in the genus *Vigna* could only be determined after a detailed study is conducted in as many species as possible.

The incidence of polyploidy in *Vigna* seems almost non-existent. So far, polyploidy has been reported in only one species (*V. glabrescens*) based on the basic number 11. The detailed meiosis of *V. glabrescens* conducted by Swindel et al. [16] and the present authors reveal that all the associations observed were in the form of bivalents. Other associations like quadrivalents were altogether absent, and therefore on the basis of chromosome associations it is evident that *V. glabrescens* having 2C DNA content of 4.95 pg is allotetraploid in nature. *V. glabrescens* has unmistakable morphological similarity with *V. radiata* and the fact remains that it was for a long time considered to be a variety of *V. radiata* (*V. radiata* var. *glabra*). One of the putative parent involved in its synthesis is therefore considered to be *V. radiata* [7] which has 2.67 pg of 2C nuclear DNA amount. The other species involved in its synthesis should have DNA content of the order 2.3 pg, an amount less than that of *V. ambacensis* (2.43), *V. oblongifolia* (2.55), *V. trilobata* (2.60), *V. angularis* (2.70), *V. repens* (2.76), *V. caracalla* (2.82), *V. mungo* (2.83), *V. umbellata* (2.84), *V. vexillata* (2.89), *V. parvifolia* (2.94), *V. unguiculata* (3.03) and more than *V. lancifolia* (2.13) for which the DNA amounts have been analysed so far. To determine the other species involved in the synthesis of *V. glabrescens*, therefore, one should find out the species having DNA amounts of the order of 2.13 pg and if they are more than one in number, the species should be involved in the synthesis of amphidiploids, *V. radiata* being common in all crosses, for the precise identification of other species involved in the synthesis of *V. glabrescens*.

The perusal of previous as well as present data on male meiosis and mitotic complements brings out clearly that due to inherent difficulty in obtaining good analyzable cytological preparations, very small size of chromosomes and overall stability of chromosome morphology and symmetry, the understanding of innate cytogenetic mechanisms underlying evolution, an important prerequisite for providing evidence of past evolutionary history of theoretical and practical importance and for logical manipulations to practical advantage of economically important taxa and wild relatives as in cereals is not forthcoming in the genus *Vigna*. The little information one could gather from the above studies so far is that (apparently) nonrandom distribution of chiasmata, cryptic structural hybridity and paracentric inversion might have played role in the evolutionary process of the genus. There has been no success in raising cytogenetic stocks like translocation testers and/or aneuploids. The limitations in conventional chromosome research has also provided an impediment in ascertaining precisely the genome relationships between species in a few successful interspecific hybrids including the species belonging Asiatic group. The little information one could gather from the crosses between *V. radiata, V. angularis, V. mungo, V. umbellata, V. minima* and *V. trilobata* is that there exists certain degree of homology between the genomes involved [13, 21, 32–41, 67, 68].

In an important genus like *Vigna* where cellular cytogenetics is of little consequence in determining phylogenetic relationships and genome architecture between species, the evidence of such differentiation could be obtained by molecular cytogenetics, such as relative quality and quantity of DNA change, longitudinal differentiation of chromosomes, overall sequence architecture of chromosomes, DNA/DNA hybridization, organelle DNA variation *in situ* hybridization of chromosomes such as FISH and McFISH etc. An attempt in this regard has already been made in the present laboratory and elsewhere. Microdensitometry measurements in 13 diploid species showed that the divergence and evolution of the species was accompanied by small but significant quantitative DNA variation ranging from 2.13 pg in *V. lancifolia* to 3.03 pg in *V. unguiculata* [42]. There was continuity in the distribution of DNA changes between the diploid complements of various species. The lone tetraploid species (*V. glabrescens*) had 4.95 pg of DNA. Significant differences between the species in areas of chromatin were also observed and compaction of DNA per unit area in interphase nuclei increases with increasing DNA amounts. Like *Vicia, Lathyrus, Festuca* and *Lolium* [69–71] the difference in DNA amount between *Vigna* species is equally distributed among all chromosomes within the complement irrespective of their size differences [42]. Molecular composition of DNA is available in only *V. radiata* and it has been reported that it consists of about 35% repetitive sequences [72, 73].

Acknowledgements

The authors are thankful to the United States Department of Agriculture (USDA), Maryland, USA and the National Bureau of Plant Genetic Resources, New Delhi for providing seed samples. SRR wish to thank Dr. Arun Kumar for his help in preparing the manuscript and illustrations.

References

1. K.P.S. Chandel, B.S. Joshi, K.C. Pant, Genetic resources of *Vigna* species in India, their distribution, diversity and utilization in crop improvement. Ann. Agric. Res. 3 (1982) 19–34.
2. Tropical Legumes, Resources for the future. National Academy of Sciences, Washington, 1979.
3. H.K. Jain, K.L. Mehra, Evolution, adaptation relationships and uses of the species of *Vigna* cultivated in India, in: R.J. Summerfield and A.H. Bunting (Eds). Advances in Legume Science. Proc. Int. Legume Congr. 1978, Royal Botanic Gardens, Kew 1 1980 pp. 459–468.
4. D.P. Chopra, T.A. Thomas, K.L. Mehra and R. Singh, Catalogue of moth bean germplasm. NBPGR Regional Station, Jodhpur, Rajasthan, 1981.
5. E. Kay, Food Legumes, Tropical Products Institute Crop and Product Digest 3, 1979.
6. D.G. Faris, The origin and evolution of the cultivated forms of *Vigna sinensis*, Can. J. Genet. Cytol. 7 (1965) 433–452.
7. B. Verdcourt, Studies in the Leguminosae-Papilionoideae for the 'Flora of tropical east Africa' IV, Kew Bull. 24 (1970) 507–569.
8. R. Marechal, J.M. Mascherpa, F. Stainier, Etude taxonomique d'un groupe complexe d'especes des genres *Phaseolus* et *Vigna* (Papilionaceae) sur la base de donnees morphologiques et pollinques traitees par l'analyse informatique, Boissiera 28 (1978) 1–273.
9. G.D. Karpechenko, On the chromosomes of the *Phasiolinae*, Bull. Appl. Bot. Plant Breed. 14 (1925) 143–148.
10. I. Kawakami, Chromosome numbers in Leguminosae, Bol. Mag. Tokyo 44 (1930) 319–329.
11. L.S.S. Kumar, A comparative study of autotetraploid and diploid types in mung bean (*Phaseolus radiatus* Linn.), Proc. Indian Acad. 21(B) (1945) 266–268.
12. J.A. Frahm-Leliveld, Cytological data on some wild tropical *Vigna* species and cultivars from cowpea and asparagus bean, Euphytica 14 (1965) 251–270.
13. D.N. De, R. Krishnan, Cytological studies of the hybrid. *Phaseolus aureus × P. mungo*, Genetica 37 (1966) 588–600.

14. R. Krishnan, D.N. De, Studies on pachytene and somatic chromosomes of *Phaseolus aureus*, The Nucleus 8 (1968) 7–16.

15. J.W. Purseglove, *Phaseolus aconitofolius* Jacq. Tropical crops: Dicotyledons. Longmann. Green and Co. Ltd., London, 1, 1968 pp. 286–287.

16. E.R. Swindell, E.E. Watt, M.G. Evans, A natural tetraploid mung bean of suspected amphidiploid origin, J. Hered. 64 (1973) 107.

17. C. P. Bhatnagar, R.P. Chandola, D.K. Saxena, S. Sethi, Cytotaxonomic studies on genus *Phaseolus*, Indian J. Genet. 34A (1974) 800–804.

18. B.D. Chaurasia, V.K. Sharma, Karyological studies in *Phaseolus mungo* Linn., Broteria Ser Trimest Cienc. Nat. 43 (1974) 33–34.

19. R.K. Chowdhury, J.B. Chowdhury, Induced amphidiploidy in *Phaseolus aureus* Roxb. × *P. mungo* L. hybrids. Crop. Improv. 1 (1974) 46–52.

20. S. Dana, N.D. Das, Natural amphidiploidy and *Phaseolus* hybrid. SABRAO J. 6 (1974) 219–222.

21. M.R. Biswas, S. Dana, Black gram × rice bean cross, Cytologia 40 (1975) 787–795.

22. R.K. Chowdhury, J.B. Chowdhury, V.P. Singh, An amphidiploid between *Vigna radiata* var. *radiata* and *Vigna mungo*. Crop. Improv. 4 (1977) 113–114.

23. R.K. Sarbhoy, Cytogenetical studies in the genus *Phaseolus* Linn. III. Evolution in the genus *Phaseolus*, Cytologia 42 (1977) 401–413.

24. R.K. Sarbhóy, Cytogenetical studies in genus *Phaseolus* Linn. I and II somatic and meiotic studies in fifteen species of *Phaseolus* (Part I), Cytologia 43 (1978) 161–170.

25. R.K. Sarbhoy, Cytogenetical studies in genus *Phaseolus* Linn. I and II somatic and meiotic studies in fifteen species of *Phaseolus* (Part 2), Cytologia 43 (1978) 171–180.

26. L.S. Joseph, J.C. Bouwkamp, Karyomorphology of several species of *Phaseolus* and *Vigna*, Cytologia 43 (1978) 595–600.

27. L.C. Goswami, Karyological studies of thirty-two varieties of black gram (*Phaseolus mungo* L.), Cytologia 44 (1979) 549–556.

28. S.S.N. Sinha, H. Roy, Cytological studies in the genus *Phaseolus* II. Meiotic analysis of sixteen species, Cytologia 44 (1979) 201–209.

29. J.A. Lackey, Chromosome numbers in the *Phaseoleae* (Fabaceae: Fabaideae) and their relation to taxonomy, Amer. J. Bot. 67 (1980) 595–602.

30. S. Sahai, R.S. Rana, Homology and differentiation in *Phaseolus*. Indian J. Genet. 40 (1980) 311–315.

31. U.C. Lavania, I.C. Lavania, Chromosome banding patterns in some Indian pulses, Ann. Bot. 49 (1983) 235–239.

32. N.K. Sen, A.K. Ghosh, Studies on the tetraploids of six varieties of green gram, Proc. Nat. Ins. Sci. India, Sec. B 26 (1960) 291–299.

33. M. Boiling, D.A. Sander, R.S. Matlock, Mung bean hybridization technique. Agron. J. 53 (1961) 54-55.

34. S. Dana, Cross between *Phaseolus aureus* Roxb. and *P. mungo* L., Genetica 37 (1966) 259–274.

35. S. Dana, Species cross between *Phaseolus aureus* Roxb. and *P. trilabus* Ait. Cytologia 31 (1966) 176–187.

36. S. Dana, Interspecific hybrid between *Phaseolus mungo* L. and *P. trilobus* Ait. J. Cytol. Genet. 1 (1966) 61–66.

37. A. Rabakoarihanta, D.W.S. Mok, M.C. Mok, Fertilization and early embryo development in reciprocal interspecific crosses of *Phaseolus,* Theor. Appl. Genet. 54 (1979) 47–53.

38. Y.P.S. Bajaj, S.S. Gosal, Induction of genetic variability in grain legumes through tissue culture, in: A.N. Rao (Ed) Tissue Culture of Economically Important Plants, National University, Singapore, 1982, pp. 25–41.

39. S.S. Gosal, Y.P.S. Bajaj, Interspecific hybridization between *Vigna mungo* and *V. radiata* through embryo culture. Euphytica 32 (1983) 129–137.

40. M.C. Gopinathan, C.R. Babu, K.R. Shivanna, Interspecific hybridization between rice bean (*Vigna umbellata*) and its wild relative (*Vigna minima*): fertility-sterility relationships, Euphytica 35 (1986) 1017–1022.

41. D.W.S. Mok, M.C. Mok, A. Rabakoarihanta, C.T. Shii, Phaseolus: Wide hybridization through embryo culture, in: Y.P.S. Bajaj (Ed) Biotechnology in Agriculture and Forestry, Springer, Berlin, 2, 1968 pp. 309–318.

42. A. Parida, S.N. Raina, R.K.J. Narayan, Quantitative DNA variation within and between chromosome complements of Vigna species (Fabaceae), Genetica 82 (1990) 125–133.

43. I. Galasso, L.S. Saponetti, D. Pignone, Cytotaxonomic studies in Vigna IV. Variation of the number of active and silent DNA sites in V. unguiculata populations, Caryologia 51 (1998) 95–104.

44. P.K. Gupta, J.R. Behl, Cytogenetics and origin of some pulse crops, In: M.S. Swaminathan, P.K. Gupta and U. Sinha (Eds) Cytogenetics of crop plants. Macmillan India Ltd., New Delhi, 1982, pp. 415–432.

45. C.S. Ahn, R.W. Hartman, Interspecific hybridization between mung bean (Vigna radiata (L.) Wilczek) and adzuki bean (V. angularis (Willd.) Ohwi & Ohashi), J. Amer. Soc. Hort. Sci. 103 (1978) 3–6.

46. W. Tschechow, N. Kartaschowa, Karyologisch-systematische Untersuchungen der Tribus Loteae and Phaseoleae, Uterfam, Papilionatae. Cytologia, 3 (1932) 221–249.

47. S,N. Raina, K. Yamamoto, M. Murakami, Intraspecific hybridization and its bearing on the chromosomal evolution in V. narbonensis, Plant Syst. Evol. 167 (1990) 201–207.

48. N.K. Sen, J.G. Bhowal, Cytotaxonomic studies on Vigna. Cytologia, 25 (1960) 195–207.

49. K. Shibata, Estudio citologicos de plantas columbianas silvestres. Y. cultivadas. J. Agric. Sci. Tokyo, Nogyo Daigaku, 8 (1962) 48–62.

50. S.S. Bir, S. Sidhu, Cytological observations on the north Indian members of family Leguminosae, The Nucleus 10 (1967) 47–63.

51. J.A. Frahm-Leliveld, Some chromosome numbers in tropical leguminous plants. Euphytica, 2 (1953) 46–48.

52. J.A. Frahm-Leliveld, Observations cytologiques sur quelques Legumineuses tropicales et subtropicales. Rev. Cytol. Biol. Veg., 18 (1957) 273–287.

53. A. Singh, R.P. Roy, Karyological studies in Trigonella, Indigofera and Phaseolus. The Nucleus, 13 (1970) 41–54.

54. P. Mukherjee, Pachytene analysis in Vigna. Chropmosome morphology in Vigna sinensis (cultivated), Sci. Cult. 34 (1968) 252–253.

55. N.S. Rau, Further contributions to the cytology of some crop plants of South India. J. Indian Bot. Soc. 8 (1929) 201–206.

56. E.T. Floresca, J.M. Capinpin, J.V. Pancho, A cytogenetic study of bush sitao and its parental types. Philippine Agriculturist 44 (1960) 290–298.

57. J. Miege, Quatrieme liste de numbers chromosomiques d'especes d' Afrique Occidentale. Rev. Cytol. Biol. Veg., 24 (1962) 149–164.

58. S.H. Yarnell, Cytogenetics of the vegetable crops. IV. Legumes. Bot. Rev., 31 (1965) 247–300.

59. E.R. Froni-Martinus, New chromosome number in the genus Vigna (Leguminosae-papilionoidae), Bull. Jardin Bot. Nat. Belgique 56 (1986) 129–133.

60. B. John, K.R. Lewis, The meiotic system, Protoplasmatologia, VI. Springer-Verlag, Wien. N.Y. 1965.

61. J. Sybenga, General Cytogenetics. North-Holland Publishing Co. Amsterdam, 1972.

62. H. Rees, J.B. Thompson, Localisation of chromosome breakage at meiosis, Heredity 9 (1955) 399–407.

63. H. Matsura, Chromosome studies on Trillium kamtschaticum Pall. and its allies XIX. Chromatid breakage and reunion at chiasmata, Cytologia 16 (1952) 48–57.

64. C.D. Darlington, A.E. Garidner, The variation system in Campanula persicifolia, J. Genet. 35 (1937) 97–128.

65. C.D. Darlington, The Evolution of Genetic Systems. Oliver and Boyd, Edinburgh, 1958.

66. R. Snow, Permanent translocation heterozygosity associated with an inversion system in Paeonia brownii, J. Hered. 60 (1969) 103–106.

67. H.J. Evans, Repair and recovery at chromosome and cellular levels: similarities and differences. Brookhaven Symp. Biol. 20 (1967) 111–131.

68. J. Smartt, Gene pools in Phaseolus and Vigna cultigens, Euphytica 30 (1981) 445–449.

69. R.K.J. Narayan, A. Durrant, DNA distribution in chromosomes of Lathyrus species, Genetica 61 (1983) 47–53.

70. S.N. Raina, H. Rees, DNA variation between and within chromosome complements of *Vicia* species, Heredity 51 (1983) 335–346.
71. S.N. Raina, H. Rees, Variation in chromosomal DNA associated with the evolution of *Vicia* species, Kew Chromosome Conference, Allen and Unwin, London, 2 (1983) 360.
72. W.F. Thompson, M.G. Murray, Sequence organization in pea and mung bean DNA and a model for genome evolution, in: D.R. Davies, D. Hopwood (Eds), Plant Genome, John Inter. Charity, England 1980.
73. M.G. Murray, J.D. Palmer, R.E. Cuellar, W.F. Thompson, DNA sequence in the pea genome, Biochemistry 17 (1979) 5781–5790.

Plant Biotechnology and Molecular Markers
P.S. Srivastava, Alka Narula and Sheela Srivastava (Editors)
Copyright © 2004 Anamaya Publishers, New Delhi, India

22. Transgenic Crops for Abiotic Stress Tolerance

Deepti Tayal[1,2], P.S. Srivastava[2] and K.C. Bansal[1]*

[1]National Research Centre on Plant Biotechnology, Indian Agricultural Research Institute,
New Delhi 110 012, India *e-mail: kailashbansal@hotmail.com

[2]Centre for Biotechnology, Faculty of Science, Jamia Hamdard, New Delhi 110 062, India

Abstract: Crop cultivars that are high yielding and also possess tolerance to abiotic stresses mainly drought and salinity are always on the shopping list of Indian farmers. However, in the past it had been technically difficult to produce such improved cultivars through conventional breeding due to the complex nature of the traits involved. Currently, the modern transgenic approach is being utilized to develop cultivars that are tolerant to abotic stresses. Efforts have been made and are currently in progress worldwide to understand the genetic and molecular basis of this complex trait. Substantial progress has been made in the identification of genes involved in abiotic stress tolerance, and their transfer to model plant species as well as crop species of economic importance for increased stress tolerance. In this chapter, we have reviewed results of the economically important transgenic crop plants that have been developed with altered expression of the genes implicated in stress tolerance. Most of the transgenic crops have shown enhanced tolerance to abiotic stress factors in pot culture experiments. However, data on their field performance under real stress situation are lacking. Nevertheless, analysis of the transgenic plants has proved useful in providing basic understanding of the function of the stress-induced genes in stress tolerance. Further progress is anticipated by the rapid discovery of more novel genes on a genome wide basis followed by determination of their precise physiological role in stress tolerance through functional genomics, and subsequent generation of transgenic crops with enhanced tolerance to multiple abiotic stresses and improved yields.

1. Introduction

Abiotic stresses represent the most limiting environmental factors affecting agricultural productivity. To overcome these limitations and to improve production, to feed the ever-increasing population, it is imperative to develop crop cultivars that are stress tolerant. When crop plants are subjected to environmental stress conditions, they fail to express their full genetic potential for production. The effect of stress depends on the developmental stage, genotype of plant species as well as duration and intensity of the stress. Generally, plants respond to these stresses under low or moderate levels, but when the stress levels exceed a certain critical level (which varies from crop to crop), the physiological mechanisms imparting tolerance to plants start breaking down causing ultimately plant death. Consequently, the abiotic stress factors cause a massive loss to the productivity of crop plants. According to the ICRISAT Report [1], biotic and abiotic stress factors lead to a loss of US$ 15.74 billion in five most important crops of semi arid tropics— sorghum, pear millet, pigeonpea, chick pea and groundnut. These crops are the main food source for poor people of the developing countries. Amongst the various stresses affecting crop plants, loss due to abiotic stresses is much more significant as compared to the losses that occur due to insect/pests, weeds and diseases (Fig. 1).

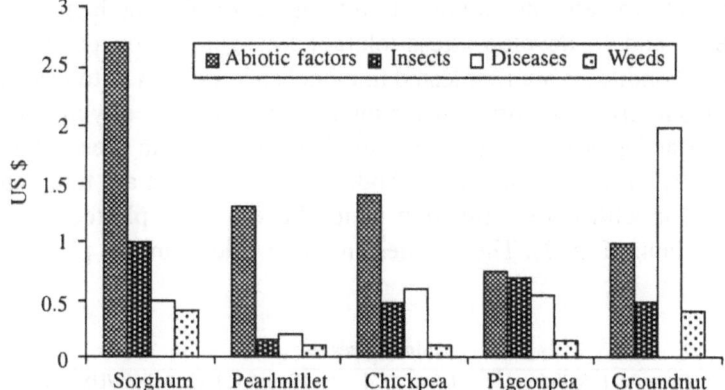

Fig. 1   Loss due to abiotic factors, insects, diseases and weeds
(*Source:* ICRISAT[1])

Classical plant breeding methods involving inter-specific or inter-generic hybridization and *in vitro* induced variation have been applied to improve the abiotic stress tolerance of various crop plants but without much success. The conventional breeding strategies are limited by the complexity of stress tolerance traits, low genetic variance of yield components under stress condition and lack of efficient selection criteria. It is important, therefore, to look for alternative strategies to develop stress tolerant crops. Recently, marker assisted selection of specific traits that are linked to yield, e.g. osmotic adjustment, membrane stability or physiological tolerance indices, has been recommended. However, QTL that are linked to tolerance at one stage in plant development can differ from those linked to tolerance at other stages. Furthermore, desirable QTLs can require extensive breeding to restore suitable traits along with the introgressed tolerance trait. The best alternative, therefore, is the direct introduction of genes by genetic engineering to incorporate tolerance traits in target crops.

Research over the past two decades has provided a better understanding of the molecular biology of stress responses in plants. Many genes and gene products have been identified which get induced upon exposure of plants to various abiotic stresses—drought, salinity, low and high temperature stress, etc. Consequently, biotechnological tools have been applied to transfer some of these useful genes implicated in stress tolerance to plants. In addition to these stress-induced proteins, genes encoding enzymes of the biosynthetic pathways of different osmolytes such as proline, glycine betaine, trehalosc, sorbitol, pinitol, etc. have been cloned and exploited in improving abiotic stress-tolerance in plants through genetic engineering.

In this chapter we have made an attempt to summarize the progress made towards understanding the role of different genes implicated in stress tolerance and the genetic engineering efforts towards developing stress tolerant transgenics in crop plants of economic importance.

2.   Genes involved in Abiotic Stress Tolerance

Under different abiotic stress conditions, a large number of genes show elevated transcript levels in plants (Table 1). Up-regulation of these genes does not always confirm their role in stress tolerance. Changes in gene expression may be due to disruption of physiological and metabolic processes of the cell. However, precise physiological function of any such gene can be studied by its altered expression (overexpression or suppression) in transgenic plants. Indeed, transgenic approach has emerged as a valuable tool in determining or confirming the precise function of the stress-induced genes and to develop stress-tolerant transgenic crop plants. Normally, genes isolated under stress conditions are first tested in model species such as tobacco and *Arabidopsis* for their role in stress tolerance before transferring them to economically important crop species. Stress-induced genes and gene products that accumulate under abiotic stresses have been reviewed by Shinozaki et al. [2], Grover et al. [3] and Abdin et al. [4]. There are four categories of stress induced genes/proteins with known function which have been exploited for generating stress tolerant transgenic plants (Fig. 2). These genes and their role in model plant species have been described below.

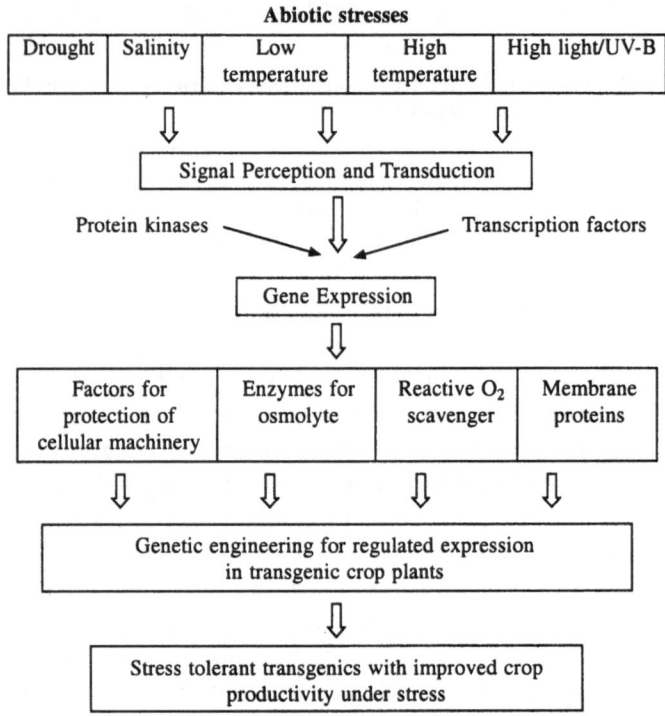

Fig. 2   Schematic representation of stress perception and transduction, stress-induced gene expression, and genetic engineering utilizing the candidate genes for developing stress tolerant transgenic crop plants.

2.1   Genes Involved in Osmolyte Biosynthesis

Osmolytes are highly soluble compatible solutes that are neutral at physiological pH. Being neutral at physiological pH, increased concentration of compatible osmolytes does not interfere with macromolecular conformation. Enhanced accumulation of osmolytes under stress conditions lead to the lowering of osmotic potential of cells which inturn results in uptake of water and maintenance of cell turgor. Various kinds of osmolytes are known to accumulate in plants under stress such as proline, mannitol, glycine betaine, trehalose, etc. [5].

Proline, an important osmolyte is synthesized from glutamate by the catalytic action of enzyme Δ^1-pyrroline 5 carboxylate synthetase (P5CS). Overexpression of this enzyme in transgenic tobacco showed enhanced biomass production, better plant growth and flower development

Table 1. **Some examples of osmotic stress induced genes and gene products**

Category	Proteins	Genes	Reference
Genes involved in osmolyte biosynthesis	*Enzymes for synthesis of:*		
	Proline	*P5CS*	[6]
	Polyols-mannitol, ononitol	*MtlD, IMT*	[8,9]
	Fructans, Trehalose	*SacB, TPSI*	[10,11]
	Polyamine-putrescine	*ADC, ODC*	[12,13]
	Quarternaryamine-glycine betaine	*codA, CDH, CMO*	[15-17]
	Osmotin-induced proline	*Osmotin*	[7]
Genes encoding factors for protection of cellular machinery	*Antifreeze proteins*	*AFP1, AFP2*	[18, 19]
	LEA proteins	*LEAI, Dehydrins, HVA1, LEAIV*	[20-23]
	LEA-like proteins	*COR14, COR15*	[24, 25]
	Osmotin	*Osmotin*	[7]
Genes encoding membrane proteins	*Water channel proteins*		
	Aquaporins	*γ-TIP, PM28A, AthH$_2$*	[26-28]
	Transport proteins		
	H$^+$ ATPase	*AHA3, PMA2*	[29,30]
	Ca^{2+} ATPase	*LCA1*	[31]
	K$^+$ transporters	*HKt1, Hak1*	[32,33]
	K$^+$ channels	*AKt1, AKt2*	[34,35]
	Na$^+$/H$^+$ antiporter	*AtNHX1*	[36]
Genes for reactive oxygen scavenger proteins	Superoxidase dismutase	*Cu/Zn SOD, MnSOD, FeSOD*	[37-39]
	Ascorbate peroxidase	*Apx*	[40]
	Glutathione synthetase	*GS*	[41]
	Glutathione reductase	*GR*	[42]
	Catalase	*Catalase*	[43]
	Glyoxalase	*GlyI, GlyII*	[44, 45]
Genes encoding transcription factors	Ethylene responsive element binding factors (ERF)	*CBF1, Tsi, DREBF1*	[46-48]
	Basic domain Leucine Zipper (BZIP)	*ABF3, ABF4*	[50]
	Myb and Myc like protein	*Atmyb2, rd22BP1*	[51]
Genes encoding protein kinases	*Mitogen activated protein kinases*		
	MAPK	*At MPK 3/6, At MPK4*	[52]
	MAPKK	*At MPKK 4/5, At MKK 1/2*	
	MAPKKK	*At MEKK1, At ANP1*	
	Ca^{2+} dependent protein kinases	*AtCDPK1, AtCDPK2*	[53]

under drought and salinity stress conditions [6]. We recently reported accumulation of free proline in transgenic tobacco plants over-expressing osmotin gene [7]. The transgenic plant showed tolerance to osmotic stress caused by drought and salinity; however, the precise role of osmotin in imparting tolerance could not be ascertained.

Among polyols, mannitol overproduction through *E. coli* mannitol 1-phosphate dehydrogenase (*MtlD*) gene expression in transgenic tobacco provided enhanced tolerance against salinity stress [8]. Overaccumulation of D-mannitol upto a remarkable limit of 600 mM in cytosol provided osmotic tolerance to transgenic tobacco [9]. Similarly, fructan synthase (*Sac B*) gene responsible for fructan biosynthesis, showed tolerance to freezing and PEG-mediated water stress in transgenic tobacco [10]. Engineering of trehalose metabolism by transferring trehalose 6-phosphate synthase (*TPS1*) gene in tobacco plants showed improved drought tolerance. However, the transgenic plants exhibited stunted growth, reduced sucrose content and lancet-shaped leaves [11]. Experiments are in progress to circumvent these growth related problems.

Polyamines are also known to have positive effects on plants exposed to abiotic stresses. Spermine and spermidine are two major polyamines synthesized from putrescine. Ornithine decarboxylase (ODC) and arginine decarboxylase (ADC) are the key enzymes involved in putrescine biosynthesis. Although the role of putrescine in stress tolerance remains to be elucidated, the biosythesis is stimulated in the presence of osmotic stress [12-14].

A quaternary amine, glycine betaine is another important osmolyte whose enhanced accumulation was observed in halophytes and bacterium under drought and salinity stress. Choline oxidase (COD) from *Arthrobacter globiformis*, or choline dehydrogenase (CDH) and choline monoxygenase (CMO) in plants are the key enzymes involved in glycine betaine biosynthesis. Transgenic tobacco and *Arabidopsis* plants producing COD, CDH and CMO have shown enhanced tolerance against salinity stress [15, 16]. However, it has been reported that availability of choline is a limiting factor in glycine betaine producing transgenic plants. This problem can be overcome to some extent by exogenous choline supply [17]. Engineering osmolyte biosynthesis is emerging as a viable approach in producing transgenics for enhanced tolerance to osmotic stresses in plants. However, the major focus is on enhanced biosynthesis of trehalose and glycine betaine through genetic engineering.

2.2   Genes Encoding Factors for Protection of Cellular Machinery

Protection factors such as antifreeze proteins (AFPs) bring about lowering of freezing point by inhibiting binding of additional water molecules to ice crystals. Larger ice crystals have more harmful effects on tissues as compared to small crystals. When a synthetic fusion protein (based on type I AFP) was expressed in yeast [18], inhibition of recrystallization was observed as a result of AFP expression. Transformed yeast cells also showed a two-fold increase in survival after rapid freezing. In another study, no effect on freezing tolerance in transgenic tobacco plants expressing the type II antifreeze protein was observed [19]. Further studies are required for defining the exact role of AFPs in abiotic stress tolerance.

Another kind of protection factors are late embryogenesis (LEA) proteins that are highly hydrophilic. They accumulate in seeds during desiccation. Group II LEA proteins, also known as dehydrins help in maintaining folded form of proteins and thereby function as chaperons [20]. These proteins are generally induced during cold acclimatization and dehydration. Overexpression of group I LEA proteins from wheat in yeast cells showed attenuation of growth inhibition in high osmolarity media. However, their role in freezing tolerance remains unknown [21]. Group III Lea proteins have been suggested to function against desiccation tolerance by sequestration of ions [22]. But there is no report available on transfer of HVA1 gene encoding LEA III proteins in model species. Overexpression of group IV LEA protein in yeast has shown tolerance against low temperature and salinity stress [23].

Another group of LEA-like proteins are hydrophilic COR proteins having repeated amino acid sequence motifs forming amphipathic α-helix. Based on this property, these proteins have been suggested to increase freezing and dehydration tolerance by stabilizing proteins and membranes. Constitutive expression of cold regulated *COR15a* gene in transgenic *Arabidopsis* plants showed an increase in both chloroplast and protoplast freezing tolerance [24]. However, effect of low temperature was not measured in transgenic *Arabidopsis* plants over-expressing the barley gene *Cor14b* [25].

2.3 Genes Encoding Membrane Proteins

The membrane proteins involved in osmotic stress tolerance include water channel and transport proteins. Water channel proteins control cellular water transport in response to drought and salt stress. The recently identified aquaporins are complex family of water channel proteins having control over water flux in and out of the cell. Aquaporins also maintain proton gradient for osmotic balance by preventing the ion flow through water channel. Phosphorylation of aquaporins through membrane bound protein kinase has been suggested as an essential factor for regulation of activity of water channel proteins [26,27). Transgenic *Arabidopsis* plants with antisense construct of plasma membrane aquaporin have revealed the role of aquaporins in maintaining cytosolic osmoregulation [28].

In high saline environments, plants take up excessive amounts of Na^+ and Cl^- at the cost of K^+ and Ca^{2+}. K^+ is required as a cofactor for many enzymes and Ca^{2+} is essential in signal transduction. A number of transport proteins play an important role in maintaining ion homeostasis under stress condition. In salt tolerant plants, H^+ ATPase maintains H^+ ion-flux across the plasma membrane [29, 30]. Ca^+ homeostasis for reducing toxic effects of NaCl is maintained by Ca^+ ATPase [31]. K^+ transporters and K^+ channels maintain K^+ and Na^+ uptake for mediating ion homeostasis [32–35]. Na^+/H^+ antiporters use electrochemical proton gradient for transporting Na^+ into vacuole. This gradient is provided by vacuolar H^+ translocating enzymes. Compartmentation of Na^+ into vacuole helps in accumulating water into the cell and thus in maintaining osmotic balance. Consequently, overexpression of the gene encoding vacuolar Na^+/H^+ antiporter showed tolerance to 200 mM NaCl in *Arabidopsis thaliana* [36].

2.4 Genes for Reactive Oxygen Species Scavenger Proteins

Under stress conditions, plants produce various active oxygen species (AOS) such as superoxide $O_2^{\cdot}$, hydrogen peroxide H_2O_2, and hydroxyradical $OH^{\cdot}$. Plants generally respond to these active oxygen species by inducing antioxidant system involving superoxide dismutase (SOD), ascorbate peroxidase (APx), glutathione synthetase (GS), glutathione reductase (GR) and catalase enzymes. For detailed analysis of contribution of these antioxidant enzymes to stress tolerance, a large number of experiments have been conducted with transgenic model plants overproducing the antioxidant enzymes [37–43]. Most of these transgenics provided tolerance against oxidative stress, and photooxidative and ozone damage.

Glyoxalate system is known for being involved in protection against cytotoxicity. The first evidence investigating the role of glyoxalase I enzyme in imparting tolerance to plants under salinity stress came through the studies of Veena et al. [44]. The same group has now overexpressed glyoxalase II in tobacco either independently or in concert with *gly* I . Transgenic plants inheriting both the genes showed many fold increase in salinity tolerance over the single gene transgenics depicting a synergistic effect of the *gly* I and *gly* II genes [45].

2.5 Genes Encoding Transcription Factors

Transcription factors play an important role in controlling the expression of stress-responsive genes. Few important families of transcription factors are:

2.5.1 Ethylene Responsive Element Binding Factors (ERF)

All ERFs are suggested to have a conserved 58-59 amino acid domain that can bind to C-repeat/dehydration responsive element (DRE). DRE motifs are involved in regulation of ABA independent gene expression under drought, salinity and cold stress. Therefore, overexpression of single *ERF* gene may help in improving tolerance to a range of abiotic stresses. The role of *ERF* gene in freezing tolerance was confirmed, through overexpression of CRT/DRE binding factor CBF1 in *Arabidopsis thaliana* [46]. The *ERF* gene imparted tolerance to multiple stress factors such as drought, salinity and cold stresses imposed together [47]. Overexpression of tobacco stress induced gene (*Tsi*) in transgenic tobacco further confirmed the role of *ERF* gene in conferring tolerance to osmotic stress [48].

2.5.2 bZIP Transcription Factor

bZIPs belong to a large family of transcription factor genes and possess a basic domain adjacent to leucine-zipper motif. A number of bZIP proteins are found to be involved in stress signaling [49]. The first genetic evidence of importance of bZIP proteins in stress tolerance was provided by overexpressed ABRE binding factor/ABA responsive element binding protein of *bZIP* family in transgenic *Arabidopsis thaliana* [50].

2.5.3 Myb and Myc Binding Proteins

Myb-like proteins contain helix turn helix related motif and Myc-like proteins have basic helix loop helix domain for DNA binding. Expression of this class of transcription factors is induced by ABA. In *Arabidopsis,* application of exogenous ABA induces a dehydration responsive gene *rd22*. Expression of this gene requires protein synthesis as revealed by the use of cycloheximide, an inhibitor of protein synthesis. The promoter of *rd22* contains a 67bp DNA sequence, which is sufficient for the expression of the gene. Abe et al. [51], identified the presence of MYB and MYC recognition sites in the 67bp region by transforming tobacco plants with this region. cDNA encoding MYB related DNA binding protein was termed as *At MYB2* and gene encoding MYC related protein was given the name *rd22 BP1*.

2.6 Protein Kinases

2.6.1 Mitogen Activated Protein Kinases (MAPKs)

MAPKs are serine/threonine protein kinases which phosphorylate a number of substrates involved in various cellular responses including gene expression. They play essential role in plant signal transduction pathways. MAPK cascade is regulated by MAPK kinases (MAPKK) and MAPKK kinases (MAPKKK). In this cascade, signal is sensed by MAPKKK first that phosphorylates the MAPKK, which in turn phosphorylates the MAPK. A number of abiotic stress factors such as wounding, low temperature, high osmolarity, high salinity and reactive oxygen species act as a signal in activating MAPK cascade. To our knowledge there seems to be no report as yet on transgenic with overexpression of MAPK cascade genes. Studies are underway on cloning these genes on the basis of sequence homology and specific antibody recognition. A detailed analysis of mitogen activated protein kinase signaling cascade has been presented by Guillaume et al. [52].

2.6.2   Calcium Dependent Protein Kinases

A number of abiotic stress factors such as cold, salt and drought elevate Ca^{2+} levels in cells for achieving control over various cellular mechanisms. Ca^{2+} influx mediates this control by phosphorylation/dephosphorylation of various proteins through Ca^{2+} dependent protein kinases (CDPKs). These kinases contain a calmodulin like regulatory domain and a Ca^+ binding site at C terminal. Around 40 different CDPKs have been investigated in *Arabidopsis thaliana*. Sheen [53] introduced eight CDPK isoforms of *Arabidopsis* into maize protoplasts, and found that only two isoforms, AtCDPK1 and AtCDPK2 induced the expression of specific stress genes thereby suggesting the presence of specific CDPK isoforms for different stress signaling pathways.

3.   Development of Stress-Tolerant Transgenic Crops

3.1   Wheat

Wheat is an important cereal crop.There are only few reports on transgenics for abiotic stress tolerance in this economically important crop. For instance, improved biomass productivity and water use efficiency was observed when wheat cultivar Hi-Line was transformed with *HVA1* gene encoding LEAIII protein. These transgenic lines were shown to have higher dry mass, root fresh and dry weight and shoot dry weight as compared to control plants [54]. For elucidating the role of *HKT1*, transformed wheat with sense and antisense construct of *HKT1* were raised [55]. The transgenic plants exhibited better growth and reduced Na^+/K^+ ratios as compared to control plants under saline conditions.

The role of mannitol accumulation in imparting stress tolerance is known in model transgenic plants. Based on this fact, transgenic plants were generated with *mtlD* gene of *E.coli* in sense and antisense orientation [56]. Wheat plants do not synthesize mannitol by their own metabolism. The transgenic wheat plants with *mtlD* gene accumulated very low level of mannitol which was not sufficient for osmotic adjustment. However, the transgenic plants showed improved growth under water stress and salinity conditions probably due to protein stabilizing effect of osmolytes under stress [56].

3.2   Rice

Rice is a highly drought and salt sensitive crop. A number of studies have been performed for its improvement through genetic engineering. In 1998, Sakamoto et al. [57] developed transgenic rice by introducing *codA* gene from *Arthrobacter globiformis* for glycinebetaine synthesis. These transgenic plants could not show tolerance against salinity stress but their stress recovery rate was high. They showed that transgenic plants with CodA enzyme targeted to chloroplasts were more efficient in protecting photosynthetic machinery against stress than transgenic plants with *codA* expression in cytosol. Further, the role of *codA* gene in improving salt stress tolerance was confirmed by Mohanty et al. [58], in transgenic lines of Indica rice. In a recovery period after exposure to 0.15 M NaCl for one week, the transgenic plants survived well whereas control plants failed to recover and died. For elucidating the role of photorespiration in protection against salt stress, transgenic rice plants over expressing chloroplast glutamine synthetase (GS2)

Table 2. Some examples of transgenic crop plants tolerant to abiotic stresses

Transgenic crop	Gene introduced	Source of the gene	Performance of transgenics under stress	Reference
Wheat	HVA1	Barley	Improved biomass under water deficit conditions	[54]
	HKT1	Wheat	Enhanced growth under salinity	[55]
	MtlD	E. coli	Improved growth under water stress and salinity	[56]
Rice	CodA	A. globiformis	Early recovery from salt induced damage	[57]
	CodA	A. globiformis	Tolerance against salt stress	[58]
	GS2	Rice	Enhanced tolerance to salt stress and cold stress	[59]
	P5CS	Mothbean	Increased biomass under salt and water stress	[60]
	OSCDPK7	Rice	Improved tolerance against cold and salt/ drought	[61]
	HVA1	Barley	Significantly increased tolerance to water deficit and salt stress	[62]
	HVA1	Barley	Improvement in drought and salt tolerance	[63]
	PMA80 & PMA1959	Wheat	Enhanced dehydration and salt stress tolerance	[64]
	ADC	Oat	Enhanced tolerance to drought and salinity	[65]
	SAMDC	Tritordeum	Enhanced NaCl stress tolerance	[66]
	Catalase	Wheat	Improved tolerance against low temperature stress	[67]
	OtsA +OtsB	E.coli	High tolerance against drought, salinity and low temperature stress	[68]
	TPS + TPP	E.coli	Enhanced tolerance to drought, salinity and cold stress	[69]
Mustard	Glutathione synthetase	E.coli	Enhanced cadmium accumulation and tolerance	[70]
	Glutathione reductase	E.coli	Targeted expression in chloroplast showed cadmium tolerance	[71]
	CodA	A. globiformis	Tolerance against salt and water stress	[72, 73]
	AtNHX1	A. thaliana	Enhanced salt tolerance	[74]
	Osmotin	Tobacco	Enhanced drought and salt tolerance	[75]
Soybean	Antisense P5CR	A. thaliana	Tolerance to drought stress as compared to control ones	[76]
Alfalfa	MnSOD	Tobacco	Enhanced freezing stress tolerance	[77]
	MnSOD	Tobacco	Enhanced water deficit tolerance	[78]
	MnSOD	Tobacco	Enhanced winter survival	[79]
	FeSOD	A. thaliana	Increased winter survival, no change in oxidative stress tolerance	[80]
	MitMnSOD + ChlMnSOD	Tobacco	Improved biomass, stress tolerance not detected	[81]
	Alfin	Alfalfa	Improvement against salinity tolerance	[82]

Crop	Gene	Source	Effect	Reference
Cotton	*MnSOD*	–	Enhanced tolerance to photooxidative and low temperature effect	[83]
	MnSOD	–	No tolerance conferred to low temperature and high light	[84]
	Apx	Pea	Protection to photosynthesis against moderate chilling and high photon flux density	[85]
	GR	*A. thaliana*		
Potato	Cu, ZnSOD	Tomato	Enhanced tolerance to oxidative stress	[86]
	Osmotin like protein	Potato	No appreciable role in freezing tolerance but showed increased tolerance to late-blight	[88]
	Osmotin	Tobacco	Enhanced tolerance to drought and salt stress	[89]
	AFP (synthetic antifreeze protein)	Synthesized based on Winter flounder	Enhanced tolerance to freezing stress	[90]
	OtsA	*E.coli*	No trehalose accumulation, abiotic stress tolerance not determined	[91]
	OtsB			
	TPS1	*S. cerevisiae*	Improved drought tolerance	[92]
	Glyceraldehyde -3 phosphate dehydrogenase	*Oyster mushroom*	Improved salt tolerance	[93]
	CDSP32	Potato	Enhanced tolerance to oxidative damage	[94]
Tomato	*HAL1*	Yeast	Improved salt tolerance	[95]
	ATNHX1	*A. thaliana*	Improved fruit yield and K^+/Na^+ selectivity	[96]
	CBF1	*A. thaliana*	Tolerance against 200mM NaCl stress, enhanced tolerance to water deficit stress, catalase activity increased and H_2O_2 decreased	[97]
	CBF3	Tobacco	Elevated tolerance to chilling and oxidative stresses, catalase activity induced	[98]
Oat	HVA1	Barley	Higher osmotic tolerance	[99]
Carrot	*ODC*	Mouse	Response against abiotic stress not studied	[100]

were generated [59]. One of the transgenic lines retained more than 90% PSII activity whereas control plants lost it completely after two weeks of stress. The same transgenic line also exhibited resistance to cold stress as observed in a preliminary experiment. Zhu et al. (60) overexpressed full length cDNA of Δ^1-pyrroline 5 carboxylate synthetase (P5CS) in rice under ABA-inducible promoter complex (AIPC) [60]. The transgenic plants showed better fresh root weight as compared to controls under salt stress (100 mM NaCl). The transgenics showed higher growth rate as compared to the control plants under water stress as well.

A full-length cDNA encoding CDPK was cloned from rice and overexpressed in rice under the control of CaMV 35S promoter for detecting its physiological function [61]. The transgenic plants showed tolerance against cold and salt/drought stresses. Overexpression of the *OsCDPK* in rice induced the expression of many other genes such as *rab16A, SalT* and *Wsi18* under salt/drought but not under cold stress. This suggested the presence of two distinct ABA-induced pathways using a single CDPK, one that is induced by salt/drought stress, whereas, the other induced by cold stress.

For engineering Lea group of genes in rice, suspension culture of rice, *Oryza sativa* L. (cv. Nipponbare) were transformed with *HVA1* gene encoding the LEA III group of proteins [62]. Later, the *HVA1* gene was overexpressed for the improvement of abiotic stress tolerance in Basmati rice [63]. The transgenic plants maintained growth rates higher than control plants under water deficit and salt stress conditions. However, the use of stress-inducible promoter gave better results in term of stress tolerance than the constitutive promoter under stress conditions. Further, transgenic plants harboring *PMA80* gene (encoding LEA II group protein) and *PMA 1959* gene (encoding LEA group I protein) were developed separately [64] and the role of these proteins against dehydration and salt tolerance was studied. The tolerance level of transgenic plants with *PMA80* gene was higher than the plants with *PMA 1959* gene.

In an attempt to decipher the role of polyamines in stress tolerance, Malabika Roy and Ray Wu [65, 66] generated transgenic plants with *adc* gene and *samdc* gene, respectively. Both the types of transgenic plants which accumulated polyamines to a significant level exhibited increased tolerance to environmental stress to almost an equal extent.

The evidence confirming tolerance against low temperature stress in rice came through overexpression of catalase gene. The transgenic rice plants overexpressing the catalase gene displayed less damage as compared to control plants against a treatment of 5°C for 8 days. This enhanced tolerance to cold stress in transgenics was attributed to higher detoxification of H_2O_2 by enhanced catalase activity [67].

Tolerance to multiple abiotic stresses was introduced by the overexpression of trehalose biosynthetic genes (*OtsA* and *OtsB*) as fusion gene in rice plants [68]. The transgenic plants showed better growth, less photo-oxidative damage and more favourable mineral balance than that of the non-transgenic controls under drought, salinity and low temperature stresses. More recently, Jhang et al. [69] reported tolerance against drought, salinity and low temperature by introducing gene encoding a bifunctional fusion protein trehalose 6-phosphate synthase and trehalose 6-phosphate phoshatase, in transgenic rice plants. High level of trehalose accumulation resulted in multiple stress tolerance that was attributed to enzymatic activities of both the enzymes.

3.3 Mustard

Mustard is one of the important oilseed crops grown all over the world. *Brassica juncea*, the

Indian mustard is the second most important oilseed crop in India. A number of transgenics have been developed in *Brassica* species with improved abiotic stress tolerance. Among them, overexpression of glutathione synthetase showed enhanced accumulation and tolerance to cadmium [70]. In another study, overexpression of glulathione reducatase (GR) targeted to cytoplasm did not show any cadmium tolerance, whereas targeted expression in chloroplasts did show higher cadmium tolerance [71].

Transgenic mustard showing tolerance to salinity stress have been developed [72]. Glycine betaine biosynthesis pathway gene *codA* encoding choline oxidase was introduced into *B. juncea*. The transgenic plants showed significantly improved performance as compared to control plants in terms of chlorophyll loss, photosystem II activity and shoot growth under stress conditions [73]. A very interesting example of salt tolerance came through overexpression of *AtNHX1* in transgenic *B. napus*. These plants grew well in the presence of 200 mM NaCl, flowered and set seeds. An increase in proline content was observed attributing to osmotic adjustment [74].

Our recent studies have shown that overexpression of osmotin gene in transgenic Indian mustard enabled plants to tolerate drought and salinity stresses. The transgenic plants exhibited increased level of water retention by excised leaves at the laboratory bench as compared to the wild type plants of cultivar Pusa Jaikisan. In addition, loss of chlorophyll in the presence of salt stress (100-200 mM NaCl) was retarded in transgenic leaf discs [75].

3.4 Soybean

Soybean is an important source of nutrition to human beings. Antisense soybean transgenic plants with L-Δ^1-pyrroline 5 carboxylate reductase (*P5CR*) gene under the control of an inducible heat shock promoter (IHSP) confirmed the potential role of proline in stress tolerance. Investigation of antisense plants under stress conditions provides a means to understand plant metabolic pathways. The IHSP was fully activated at 32 and 42°C along with mannitol stress. The antisense expression of *P5CR* gene resulted in significant decrease in proline accumulation in transgenic plants. In contrast, control plants showed higher proline content, thus exhibiting better growth under similar stress conditions [76]. This clearly suggests that overexpression of genes encoding enzymes required for proline biosynthesis will lead to increased stress tolerance in transgenic crops including soybean.

3.5 Alfalfa

Medicago sativa is an important perennial forage legume all over the world. First transgenic alfalfa plant with improved tolerance to abiotic stress was developed by McKersie et al. [77] by overexpressing MnSOD cDNA under CaMV35S promoter. The transgenic plants showed more rapid growth recovery after exposure to freezing stress than that of control plants. After 3 years of field trials in 1996, data suggested that these plants also showed improved survival to water deficit stress, as determined by chlorophyll fluorescence and electrolyte leakage. During these experiments only few transgenic alfalfa plants were obtained [78]. Moreover, variety RA3 used for transgenic development was popular at the time of experiment but showed poor agronomic performances. Consequently, two different clones of alfalfa, N4 and S4 were transformed with *MitSOD* and *ChlSOD* genes [79]. Results confirmed the hypothesis that MnSOD overexpression improves survival of transgenic alfalfa against abiotic stresses. Further McKersie et al. [80] overexpressed *FeSOD* in transgenic alfalfa for investigating its role in stress tolerance as compared to the *MnSOD*. The transgenics showed increased *FeSOD* activity along with increased winter survival. However, this improvement in winter survival was not due to improvement in oxidative stress tolerance associated with photosynthesis.

For testing the synergy between SOD transgenes and stress tolerance, gene-pyramiding studies were conducted. Samisk et al. [81], crossed a hemizygous Mit-MnSOD plant and hemizygous *Chl-MnSOD* transgenic plants. F1 progeny containing joint expression of the two genes (*Mit-MnSOD + Chl-MnSOD*) had lower shoot and storage organ biomass compared to either of the parent, whereas, the progeny containing either of the transgene had significantly higher shoot and storage organ biomass.

In another experiment, overexpression of transcription factor Alfin1 in transgenic alfalfa improved growth properties of plants exposed to 128 mM NaCl stress for 17 days [82]. *Alfin1*, cDNA encodes zinc finger family of transcription factor which binds to promoter fragment of root-specific *MsPRP2* gene. The *MsPRP2* gene is also induced by NaCl stress. The transgenic plants with *Alfin1* overexpression showed MsPRP2 accumulation, thereby confirming the role of *Alfin1* in imparting enhanced NaCl stress tolerance to alfalfa plants.

3.6   Cotton

As photosynthesis in cotton is highly sensitive to low temperature and high light stress, focus has, therefore, been on developing transgenic to protect the photosynthetic machinery under abiotic stresses. Overproduction of MnSOD in chloroplasts of cotton conferred a substantially enhanced tolerance to photo-oxidative stress (high light) and low temperature [83]. However, Payton et al. [84] failed to achieve photosynthetic stability against low temperature stress in transgenic cotton overexpressing chloroplast *MnSOD* [84]. Attempts were also made to improve cotton for tolerance against abiotic stresses by overproducing chloroplast targeted glutathione reducatase (GR) and ascorbate peroxidase (APx) [85]. Elevated levels of GR or Apx activity improved photosynthetic capacity after chilling treatment at 10°C and high photon flux exposure.

3.7   Potato

Potato is highly sensitive to abiotic stresses. Perl et al. [86] developed transgenic plants by transforming potato tubers with Cu/Zn superoxide dismutase. These transgenic lines showed elevated tolerance to superoxide generating herbicide paraquat (methyl viologen). Induction of osmotin-like protein by low temperature stress in potato was shown by Zhu et al. [87]. However, transgenic potato expressing sense and antisense genes for osmotin-like proteins showed no statistical difference among sense/antisense transgenics and control plants against low temperature stress measured as electrolyte leakage. This ruled out the possibility of role of osmotin-like proteins as a major freezing tolerance determinant [88]. Our unpublished results with osmotin overexpressed transgenic potato have confirmed the role of osmotin protein in imparting tolerance to osmotic stresses caused by drought and salinity [89].

The first evidence of potato transgenics tolerating freezing stress came through the expression of a synthetic *AFP-PHA* (antifreeze protein gene fused to phytohemagglutinin) gene construct. Phytohemagglutinin acted as signal peptide directing the antifreeze protein molecule to extra cytoplasmic space where ice crystallization occurs. Transgenic plants showing maximum level of *AFP* expression showed the highest degree of tolerance against freezing stress [90] as evidenced by significantly reduced electrolyte leakage in transgenics as compared to the wild type plants.

Goddijin [91] tried to develop stress tolerant transgenic potato by engineering trehalose biosynthesis. However, surprisingly no trehalose accumulation was observed in transgenics, which was attributed to trehalase activity. Later, the role of trehalose as an osmoprotectant was confirmed by expressing *TPS1* (Trehalose 6-phosphate synthase) gene in potato plants [92]. Although, the transgenic potato plants showed abnormal morphological characteristics, such as dwarfism, yellowish lancet shaped leaves and aberrant root development, drought resistance capacity of these plants was significantly increased.

Overexpression of glyceraldehyde 3-phosphate dehydrogenase in transgenic potato showed increased tolerance to salt stress [93]. Overexpression of chloroplastic drought-induced stress protein (CDSP32) conferred protection to transgenic potato against photooxidative stress induced by incubation with either methyl viologen or t-butyl hydroperoxide or by exposure to low temperature. On the contrary, plants without CDSP32 expression showed enhanced damage to photosynthetic membrane [94].

3.8 Tomato

Tomato is a widely grown vegetable crop. Higher ability to withstand salt tolerance was observed in transgenic lines of tomato expressing *HAL* gene [95]. Further elucidation of these transgenics for long term salinity effects showed many improved characteristics as compared to control. On exposure to 35 mM NaCl concentration, 58% reduction in fruit yield was observed in control plants whereas in transgenic plants expressing *AtNHX1* the loss was 30%. Similarly, loss of leaf water content was higher in controls than the transgenics under 100 mM NaCl. These plants also maintained higher K^+/Na^+ selectivity values [96]. A remarkable example of tolerance to 200mM NaCl and preserving fruit quality at such a higher concentration was shown in transgenic tomato plants overexpressing Na^+/H^+ antiporter gene [97]. Accumulation of salts was observed in leaves without affecting the fruit quality.

Overexpression of a transcription factor gene encoding *CBF1* in transgenic tomato conferred enhanced tolerance to water deficit stress. Lack of water for 4 weeks showed 80% survival in transgenic plants as compared to less than 6% survival of control plants. Water content of transgenics was relatively high during stress treatment. However, these plants showed retardation in growth resulting into reduction in number and fresh weight of fruits [98].

3.9 Oat

Oat, a cereal crop serves as an important component of human and animal diets. This crop requires sufficient water for growth and grain production. Overexpression of *HVA1* in transgenic plants showed higher osmotic tolerance than non-transgenics. Under NaCl and mannitol mediated stresses, there were significant differences in wilting, death of old leaves and necrosis of young leaves between the transgenic and non-transgenic plants [99].

3.10 Carrot

To our knowledge, there seems be no report on transgenic carrot development tolerant to drought, salinity or low temperature stress. Efforts have been made in this direction to increase polyamine levels by expressing ornithine decarboxylase (ODC) in transgenic cell lines of carrot. Detailed metabolic studies of transgenic lines revealed higher rate of putrescine anabolism as well as catabolism producing spermine and spermidine as compared to non-transgenic cell lines [100]. However, the effect of abiotic stress was not tested on these transgenic cell lines.

4   Conclusions and Future Perspectives

Development of crop cultivars tolerant to abiotic stresses is an important goal of national and international institutions engaged in plant research. Both traditional plant breeding methods and transgenic technology are being employed to achieve the above objective. Since conventional breeding approaches were not found sufficient, scientists are now trying to explore the advantages of the transgenic technology to develop transgenic crops tolerant to abiotic stresses viz. drought, salinity, cold and high temperature, etc. Although numerous studies have demonstrated the feasibility of developing such transgenics in an array of crop species, substantial data are lacking on the response of these transgenics subjected to field stress conditions.

Abiotic stress tolerance is a complex trait that is controlled by multiple genes. Studies in early 1990s demonstrated that a battery of genes get up-regulated in plants that are exposed to drought or salinity stress. However, function of majority of these stress-induced genes/gene products remained largely unknown. With the advent of high throughput sequencing of genes (genomics) and proteomics, more and more ESTs/cDNA/genes or proteins are being added to the list by the global effort with little information on elucidation of their function or the mechanism of stress tolerance in plants. Genome wide approaches coupled with reverse genetics approach will surely allow deciphering the role of specific gene / gene combinations in stress tolerance. Undoubtedly, studies on stress signal perception and transduction have identified genes that play a significant role in controlling the expression of stress-induced genes. As a result, transgenic development with genes encoding transcription factors and/or protein kinases have provided tolerance to multiple stresses to significantly high levels, and has increased the hope of generating transgenic crops cultivars with improved stress tolerance.

Although there are numerous examples of transgenics over-expressing genes encoding enzymes for increased osmolyte biosynthesis, stress-induced proteins, ROS scavengers and membrane proteins, their field performance is awaited. To make the transgenic route more effective with respect to tolerance against abiotic stresses at the field level, it will be important that crop species-specific research programmes are undertaken for developing transgenics considering the crop phenology, water requirement, type of stress experienced by the crop, growth stages sensitive to stress, and the existing response of the crop to a given stress. For each crop, it is necessary to understand the basis of effective engineering strategies leading to greater stress tolerance. Information on genetic regulation and complex interaction of genes with environment is scanty and further studies in this direction will help understanding the molecular mechanism of stress tolerance in plants. Difference in molecular response of plants subjected to drought and salinity stress also need to be addressed, as salt-specific effects are different than drought [101]. For making the transgenic technology more effective, it is required that transgenes are expressed in a specific tissue in a developmental or stress-inducible manner. Moreover, integration of cellular and whole plant response is required for combining increased stress tolerance with high yields.

Acknowledgement

DT thanks Council of Scientific and Industrial Research (CSIR), New Delhi for financial assistance in the form of SRF.

References

1. The medium term plan, International Crop Research Institute for semi arid tropics (ICRISAT), Patancheru, India 1 (1992) 7–9.

2. K.Y. Shinozaki, M. Kasuga, Q. Liu, K. Nakashima, Y. Sakuma, H. Abe, Z.K. Shinwari, M. Seki, K. Shinozaki, Biological mechanisms of drought stress response, JIRCAS working report (2002) 1–8.

3. A. Grover, A. Kapoor, O.S. Lakshmi, S. Agarwal, C. Shai, S.K. Agarwal, M. Agarwal, H. Dubey, Understanding molecular alphabets of the plant abiotic stress responses, Curr. Sci. 80 (2001) 206–216.

4. M.Z. Abdin, R.U. Rehman, M. Isras, P.S. Srivastava, K.C. Bansal, Abiotic stress related genes and their role in conferring resistance in plants, Indian J. Biotech. 1 (2002) 225–244.

5. R. Serraj, T.R. Sinclair, Osmolyte accumulation: can it really help increase crop yield under drought conditions, Plant Cell Environment 25 (2002) 333–341.

6. P.B. Kavi Kishor, Z. Hong, G.H. Miao, C.A.A. Hu, D.P.S. Verma, Overexpression of Δ^1 pyrroline 5-carboxylate synthetase increase proline production and confers osmotolerance in transgenic plants, Plant Physiol. 108 (1995) 13 87–1394.

7. S. Barthakur, V. Babu, K.C Bansal, Overexpression of osmotin induces proline accumulation and confers tolerance to osmotic stress in transgenic tobacco. J. Plant Biochem. Biotech. 10 (2001) 31–37.

8. M.C. Tarczynski, R.G. Jensen, H.J. Bohnert, Stress protection of transgenic tobacco by production of osmolyte mannitol, Science 259 (1993) 508–510.

9. Sheveleva, W. Chmara, H.J. Bohnert, R.G. Jensen, Increased salt and drought tolerance by D-mannitol production in transgenic *Nicotiana tabacum* L, Plant Physiol. 115 (1997) 1211–1219.

10. C. Romero, J.M. Belles, J.L. Vaya, R. Serrano, F.A. Culianez-Macia, Expression of the yeast trehalose 6-phosphate synthase in transgenic tobacco plants: Pleiotropic phenotypes include drought tolerance, Planta 201 (1997) 293–297.

11. E.A.H. Pilon-Smits, M.J.M. Ebskamp, M.J. Paul, M.J.W. Jeuken, P.J. Weisbeek, S.C.M. Smeekens, Improved performance of transgenic fructan accumulating tobacco under drought stress, Physiol. Plant 107 (1995) 125–130.

12. M.B. Watson, R.L. Malmberg, Regulation of *Arabidopsis thaliana* (L) Heynh. arginine decarboxylase by potassium deficiency stress. Plant Physiol. 111 (1996) 1077–1083.

13. J. Michael, J.M. Furze, R.J.C. Rhodes, D. Burtin, Molecular cloning and functional identification of a plant ornithine decarboxylase cDNA, Biochemical J. 314 (1996) 241–248.

14. A.W. Galston, R. Kaur-Sawhney, T. Altabella, A.F. Tiburcio, Plant polyamine in reproductive activity and response to abiotic stress, Bot. Acta 110 (1997) 197–207.

15. H. Hayashi, A. Murtardy, P.Z. Deshnium, M. Ida, N. Murata, Transformation of *Arabidopsis thaliana* with *codA* gene for choline oxidase: accumulation of glycine betaine and enhanced tolerance to salt and cold stress, Plant J. 12 (1997) 133–142.

16. G. Lilius, N. Holmberg, L. Bulow, Enhanced NaCl stress tolerance in transgenic tobacco expressing bacterial choline dehydrogenase, Biotechnology 14 (1996) 177–180.

17. M.L. Nuccio, B.L. Russel, K.D. Nolte, B. Rathinasabapathi, D.A. Gage, A.D. Hanson, The endogenous choline supply limits glycine betaine synthesis in transgenic tobacco expressing choline monooxygenase, Plant J. 16 (1998) 487–496.

18. R.L. McKown, G.J. Warren, Enhanced survival of yeast expressing an antifreeze gene analogue after freezing, Cryobiology 28 (1991) 474–482.

19. K. D. Kendwarad, J. Brandle, J. McPherson. P. L. Dawes, Type II fish antifreeze protein accumulation in transgenic tobacco does not confer frost resistance, Transgene Res. 8 (1999) 105–117.

20. T.J. Close, Dehydrins : Emergence of a biochemical role of a family of plant dehydration proteins, Plant Physiol. 97 (1996) 795–803.

21. A. Swire Clark, W.R. Jr Marcotte, The wheat LEA protein Em functions as an osmoprotective molecule in *Saccharomyces cerevisiae,* Plant Mol. Biol. 39 (1999) 117–128.

22. N.K. Singh, D.E. Nelson, David Kuhn, P.M. Hasegawa, R.A. Bressan, Molecular cloning of osmotin and regulation of its expression by ABA and adaptation to low water potential, Plant Physiol. 90 (1989) 1096–1101.

23.   R. Imai, L. Chang, A. Ohta, E.A. Bray, M. Tyagi, A-LEA class gene of tomato confers salt and freezing tolerance when overexpressed in *Saccharomyces cerevisiae*, Gene 170 (1996) 243 –248.

24.   N.N. Artus, M. Uemura, P.L. Streponkus, S.J. Gulmour, C. Lin, M.F. Thomashow, Constitutive expression of the cold regulated *Arabidopsis thaliana COR 15a* gene affects both chloroplast and protoplast freezing tolerance, Proc. Natl. Acad. Sci. USA 93 (1996) 13404–13409.

25.   C. Crosatti, P. Polveriono de Laureto, R. Bassi, L. Cattivelli, The interaction between cold and light controls the expression of the cold regulated barley gene *cor14b* and the accumulation of the corresponding protein, Plant Physiol. 119 (1999) 671–680.

26.   C. Maurel, R.T. Kado, J. Guern, M J. Chrispeels, Phosphorylation regulates the water channel activity of the seed specific aquaporin γ-TIP, EMBO J. 14 (1995) 3028–3035.

27.   I. Johansson, M. Karlesson, V.K. Shukla, M.J. Chrispeels, C. Larsson, P. Kjellbom, Water transport activity of the plasma membrane aquaporin PM28A is regulated by phosphorylation, Plant Cell 10 (1998) 451–459.

28.   R. Kaldenhoff, A. Kalling, J. Meyers. U. Karmann, G. Rupple, G. Ritcher, The blue light – responsive AthH2 gene of *Arabidopsis thaliana* is primarily expressed in expanding as well as differentiating cells and encodes a putative channel protein of plasmalemma, Plant J. 7 (1995) 87–95.

29.   J.M. Pardo, R. Serrano, Structure of a plasma membrane $H^+/ATPase$ gene from the plant *Arabidopsis thaliana*, J. Biol. Chem. 264 (1989) 8557–8562.

30.   M. Boutry, B. Michelet, A. Goffeau, Molecular cloning of a family of plant genes encoding a protein homologous to plasma membrane H^+ translocating ATPases, Biochem. Biophys. Res. Commun. 162 (1989) 567–574.

31.   L.E. Wimmers, N.N. Ewing, A.B. Bennett, Higher plant Ca^{2+} ATPase: Primary structure and regulation of mRNA abundance by salt, Proc. Natl. Acad. Sci. USA 89 (1992) 9205–9209.

32.   P. Schachtman, J.J. Schreoder, Structure and transport mechanism of a high affinity potassium uptake transporter from higher plants, Nature 370 (1994) 655–658.

33.   G.E. Santa Maria, F. Rubio, J. Dubcovsky, A. R.Navarro, The *HAK1* gene of barley is a member of a large family and encodes a high affinity potassium transporter, Plant Cell 9 (1997) 2281–2289.

34.   J.A Anderson, S.S. Huprikar, L.V. Kochain, W.J. Lucas, R.F. Gaber, Functional expression of a probable *Arabidopsis thaliana* potassium channel in *Saccharomyces cerevisie*, Proc. Natl. Acad. Sci. USA 89 (1992) 3736–3740.

35.   H. Sentenac, N. Bonneaud, M. Minet, F. Lacroute, J.M. Salmon, Cloning and expression in yeast of a plant potassium ion transport system, Science 256 (1992) 663–665.

36.   M.P. Apse, G.S. Aharon, W.A. Snedden, E. Blumwald, Salt tolerance conferred by overexpression of a vacuolar Na^+/H^+ antiport in *Arabidopsis*, Science 285 (1999) 1256–1258.

37.   J.M. Tepperman, P. Dunsmuir, Transformed plants with elevated levels of chloroplastic SOD are not more resistant to superoxide toxicity, Plant Mol. Bio. 14 (1990) 501–511.

38.   C. Bowler, L. Slooten, S. Vandenbranden, R. De Rycke, J. Botterman, C. Sybesma , M. Van Montagu, D. Inze, Manganese superoxide dismutase can reduce cellular damage mediated by oxygen radicals in transgenic plants, EMBO J. 10 (1991)1723–1732.

39.   W. Van Camp, H. Willeken, C. Bowler, M. Van Montagu, D. Inze, P. Reupold-Popp, H. Jr. Sandermann C. Langebartels, Elevated levels of superoxide dismutase protect transgenic plants against ozone damage, Biotechnology 12 (1994) 165–168.

40.   J. Wang, H. Zhang, R.D. Allen, Overexpression of an *Arabidopsis* peroxisomal ascorbate preoxidase gene in tobacco increases protection against oxidative stress, Plant Cell Physiol. 40 (1999) 725–32.

41.   M. Strohm, L. Jouanin, K.J. Kunert, C. Pruvost, A. Polle, C.H. Foyer, H. Rennenberg, Regulation of glutathione synthesis in leaves of transgenic popular (*Populus tremula + P. alba*) over expressing glutathione synthetase, Plant J. 7 (1995) 141–145.

42.   M. Aono, A. Kubo, H. Saji, K. Tanaka, N. Kondo, Enhanced tolerance to photooxidative stress of transgenic *Nicotiana tabacum* with high chloroplastic glutathione reductase activity, Plant Cell Physiol. 34 (1993) 129–135.47.

43.   Y. Miyagawa, M. Tamoi, S. Shigeoka, Evaluation of the defense system in chloroplasts to photooxidative

stress cased by paraquat using transgenic tobacco plants expressing catalase from *Escherichia coli*, Plant Cell Physiol. 41 (2000) 311–20.

44. Veena, V.S. Reddy, S.K. Sopory, Glyoxalase 1 from *Brassica juncea*: molecular cloning, regulation and its overexpression confer tolerance in transgenic tobacco under stress, Plant J. 17 (1999) 385–395.

45. S.L. Singla-Pareek, M.K. Reddy, S.K. Sopory, Manipulation of glyoxalase pathway leads to salinity tolerance in transgenic tobacco, 2nd International Congress of Plant Physiology on Sustainable Plant Productivity Under Changing Environment. Jan 8–12 (2003) pp 454.

46. K.R. Jaglo-ottosen, S.J. Gilmour, D.G. Zarka, O. Schabenberger, M.F Thamshow, *Arabidopsis* CBF1 overexpression induces *cor* genes and enhances freezing tolerance, Science 280 (1998) 104–106.

47. M. Kasuga, Q. Liu, S. Miura, K. Yamaguchi-Shinozaki, K. Shinozaki, Improving plant drought, salt and freezing tolerance by gene transfer of a single stress inducible transcription factor, Nat. Biotechnol. 17 (1999) 287–291.

48. J.M. Park, C.J. Park, S.B. Lee, B.K. Ham, R. Shin, K.H. Peak, Overexpression of the tobacco *Tsi* gene encoding an EREBP/AP2 type transcription factor enhances resistance against pathogen attack and osmotic stress in tobacco, Plant Cell 13 (2001) 1035–1046.

49. M. Jakoby, B. Weisshaar, W. Droge-Lasar, J. Vicente-Carbajosa, J. Tiedemann T. Kroj, F. Parcy, bZIP transcription factors in *Arabidopsis*, Trends Plant Sci. 7 (2002) 106–111.

50. J. Kang, H. Choi, M. Im, S.K. Kim, *Arabidopsis* basic leucine zipper proteins that mediate stress responsive abscissic acid signaling, Plant Cell 14 (2002) 343–357.

51. H. Abe, K. Yamaguchi-Shinozaki, T. Urao, T. Iwasaki, D. Hosikawa, K. Shinozaki, Role of *Arabidopsis* Myc and Myb homologs in drought and abscisic acid regulated gene expression, Plant Cell 9 (1997)1859–1868.

52. T. Guillaume, A. Tsuneaki, L.C. Wan, S. Jen, Plant mitogen activated protein kinase signaling cascade, Curr. Opin. Plant Biol. 4 (2001) 392–400.

53. J. Sheen, Ca^{+2} dependent protein kinases and stress signal transduction in plants, Science 274 (1996) 1900–1902.

54. E. Sivamani, A. Bahieldin, J.M. Wraith, T. Al-Niemi, W.E. Dyer, T.D. Ho, R. Qu, Improved biomass productivity and water use efficiency under water deficit conditions in transgenic wheat constitutively expressing the barley *HVA1* gene, Plant Sci. 155 (2000) 1–9.

55. S. Laurie, K.A. Feeney, F.J. Mathuis, P.J. Heard, S.J. Brown, R.A. Leigh. A role for HKT1 in sodium uptake by wheat roots, Plant J. 32 (2002) 139–49.

56. T. Abebe, A.C. Guenzi, B. Martin, J.C. Cushman, Tolerance of mannitol-accumulating transgenic wheat to water stress and salinity, Physiol. Plant 131 (2003)1748–1755.

57. A. Sakamoto, Alia, N. Mutata, Metabolic engineering of rice leading to biosynthesis of glycine betaine and tolerance to salt and cold, Plant Mol. Biol. 38 (1998) 1011–1019.

58. A. Mohanty, H. Kathuria, A. Ferjani, A. Sakamoto, P. Mohanty, N. Murata, A.K. Tyagi, Transgenics of an elite indica rice variety *Pusa basmati* I harbouring the *codA* gene are highly tolerant to salt stress, Theor. Appl. Genet. 106 (2002) 51–57.

59. H. Hoshida, Y. Tanaka, T. Hibino, Y. Hayashi, A. Tanaka, T. Takabe, Enhanced tolerance to salt stress in transgenic rice that over expresses chloroplast glutamine synthetase, Plant Mol. Biol. 43 (2000) 103–111.

60. B. Zhu, J. Su, M. Chang, D.P.S. Verma, Y.L. Fan, R. Wu, Overexpression of a Δ^1-pyrroline-5carboxylate synthetase gene and analysis of tolerance to water and salt-stress in transgenic rice, Plant Sci. 139 (1998) 1–48.

61. Y. Saijo, S.Hata, J. Kyozuka, K. Shimamoto, K. Izui. Overexpression of a single Ca^{2+}-dependent protein kinase confers both cold and salt/drought tolerance on rice plants, Plant J. 23 (2000) 19–327.

62. D. Xu, X. Duang, B. Wang, B. Hong, T.H.D. Ho, R. Wu, Expression of a late embrogenesis abundant protein gene, HVA1, from barley confers tolerance to water deficit and salt stress in transgenic rice, Plant Physiol. 110 (1996) 249–257.

63. J.S. Rohila, R.K. Jain, R. Wu, Genetic improvement of basmati rice for salt and drought tolerance by regulated expression of a barley HVA1 cDNA, Plant Sci. 163 (2002) 525–532.

64. Z. Cheng, J.P. Targolli, X. Huang, R. Wu, Wheat *LEA* genes, *PMA80* and *PMA1959*, enhance dehydration tolerance of transgenic rice (*Oryza sativa* L.), Molecular Breeding 10 (2002) 71–82.

65. M. Roy, R. Wu, Arginine decarboxylase transgene expression and analysis of environmental stress tolerance in transgenic rice, Plant Sci. 160 (2001) 869–875.

66. M. Roy, R. Wu, Overexpression of *S-adenosylmethionine decarboxylase* gene in rice increases polyamine levels and enhances sodium chloride stress tolerance, Plant Sci. 163 (2002) 987–992.

67. T. Matsumura, N. Tabayashi, Y. Kamagata, C. Souma, H. Saruyama, Wheat catalase expressed in transgenic rice can improve tolerance against low temperature stress, Plant Physiol.116 (2002) 317–327.

68. A.K. Garg, J.K. Kim, T.G. Owens, A.P. Ranwala, Y.D. Choi, L.V. Kochian, R.J. Wu, Trehalose accumulation in rice plants confers high tolerance levels to different abiotic stresses, Proc. Natl. Acad. Sci. USA 99 (2002) 15898–15903.

69. C. Jhang, S.J. Oh, J.S. Seo, W.B. Choi, S.I. Song, C.H. Kim, Y.S. Kim, H.S. Seo, Y.D. Choi, H.B. Nahm, J.K. Kim, Expression of a bifunctional fusion of the *Esherichia coli* genes for trehalose-6-phosphatase synthase and trehalose–6–phosphate phosphatase in transgenic rice plants increases trehalose accumulation and abiotic stress tolerance without stunting growth, Plant Physiol. 131 (2003) 516–524.

70. Y.L. Zhu, E.A.H Pilon-Smits, L. Jouanin, N. Terry, Overexpression of Glutathione synthetase in Indian mustard enhances cadmium accumulation and tolerance, Plant Physiol. 119 (1999) (1) 73–80.

71. E.A.H. Pilon-Smits, Y.L. Zhu, T. Sears, N. Terry, Overexpression of glutathione reductase in *Brassica juncea*: Effects on cadmium accumulation and tolerance, Plant Physiol. 110 (2000) 455–460.

72. K.V.S.K. Prasad, P. Sharmila, P.A. Kumar P.P. Saradhi, Enhanced tolerance of transgenic *Brassica juncea* to choline confirm successful expression of bacterial *codA* gene, Plant Sci. 159 (2000) 233–242.

73. K.V.S.K. Prasad, P. Sharmila, P.A. Kumar, P.P. Saradhi, Transformation of *Brassica juncea* (L.) with bacterial *codA* gene enhances its tolerance to salt stress, Molecular Breeding 6 (2000) 489–499.

74. X. Zhang, J.N. Hodson, J.P. Williams, E. Blumwald, Engineering salt tolerant *Brassica* plants: Characterization of yield and seeds oil quality in transgenic plants with increased vacuolar sodium accumulation, Proc. Natl. Acad. Sci. USA 98 (2001) 12832–12836.

75. D. Tayal, P.S. Srivastava, K.C. Bansal, Development of transgenic Indian mustard for abiotic stress tolerance, 2nd International Congress of Plant Physiology on Sustainable Plant Productivity Under Changing Environment, Jan 8–12 (2003) pp 479.

76. J.A. Ronde, M.H. Spreeth, W.A. Cress, Effect of antisense L-Δ^1-pyrroline-5-carboxylate reducatase transgenic soybean plants subjected to osmotic and drought stress, Plant Growth Regulation 32 (2000) 13–26.

77. B.D. Mckersie, Y. Chen, M.D. Beus, S.R. Bowley, C. Bowler, D. Inze, K.D. Halluin, J. Botterman, Superoxide dismutase enhances tolerance of freezing stress in transgenic alfalfa (*Medicago sativa* L.). Plant Physiol. 103 (1993) 1155–1163.

78. B.D. Mckersie, S.R. Bowley, E. Harjanto, O. Leprince. Water-deficit tolerance and field performance of transgenic alfalfa overexpressing superoxide dismutase, Plant Physiol. 111 (1996) 1177–1181.

79. B.D. Mckersie, S.R. Bowley, K.S. Jones, Winter survival of transgenic alfalafa overexpressing superoxide dismutase, Plant Physiol. 119 (1999) 839–847.

80. B.D. Mckersie, J. Murnaghan, K.S. Jones, S.R. Bowley Iron-superoxide dismutase expression in transgenic alfalfa increases winter survival without a detectable increase in photosynthetic oxidative stress tolerance, Plant Physiol. 122 (2000) 1427–1437.

81. K. Samisk, S. Bowley, B. Mckersie, Pyramiding Mn superoxide dismutatse transgenes to improve persistence and biomass production in alfalfa, J. Exp. Bot. 53 (2002) 1343–1350.

82. I. Winicov, D.R. Bastola, Transgenic overexpression of the transcription factor *Alfin1* enhances expression of the endogenous *MsPPR2* gene in alfalfa and improves salinity tolerance of the plants, Plant Physiol. 120 (1999) 473–480.

83. R.D. Allen, Dissection of oxidative stresses tolerance using transgenic plants. Plant Physiol. 107 (1995) 1049–4054.

84. P. Payton, R.D. Allen, N. Trolinder, A.S. Holaday, Overexpression of chloroplast targeted Mn superoxide

dismutase in cotton does not alter the reduction of photosynthesis after short exposure to low temperature and high light intensity, Photosynthesis Res. 52 (1997) 233–244.

85. P. Payton, R. Webb, D. Kornyeyev, R. Allen, A.S. Holaday, Protecting cotton photosynthesis during moderate chilling at high light intensity by increasing chloroplastic antioxidant enzyme activity, J. Exp. Bot. 52 (2001) 2345–2354.

86. R.P. Perl, Treves, Galilli, D. Aviv, E. Shalgi, S. Malkin, E. Galun, Enhanced oxidative stress defense in transgenic potato expressing tomato Cu, Zn, superoxide dismutases, Theo. Appl. Genet. 85 (1993) 568–576.

87. B. Zhu, T.H.H. Chen, P.H. Li, Activation of two osmotin-like protein genes by abiotic stimuli and fungal pathogen in transgenic potato plants, Plant Physiol. 108 (1995) 929–937.

88. B. Zhu, T.H.H Chen, P.H. Li, Analysis of late–blight disease resistance and freezing tolerance in transgenic potato plants expressing sense and antisense genes for an osmotin-like protein, Planta 198 (1996) 70–77.

89. V. Babu, K.C. Bansal, Osmotin overexpression in transgenic potato plants provide protection against osmotic stress, 5th International Symposium on Molecular Biology of Potato, Bogensee, Germany, (1998) Aug 2–6.

90. J.G. Wallis, H. Wang, D.J. Guerra, Expression of a synthetic antifreeze protein in potato reduces electrolyte release at freezing temperatures, Plant Mol. Biol. 35 (1997) 323–330.

91. O.J. Goddijin, T.C. Verwoerd, E. Voogd, W.H.H. Krutwagen, R.W. de Graff, J. Poels, K. Van Dun, A.S. Ponstein, B. Damm, J. Pen, Inhibition of trehalase activity enhances trehalose accumulation in transgenic plants, Plant Physiol. 113 (1997) 181–190.

92. E.T. Yeo, H.B. Kwon, S.E. Han, J.T. Lee, J.C. Ryu, M.O. Byu, Genetic engineering of drought resistant potato plants by introduction of *trehalose-6-phosphatase synthase* (TPSI) gene from *Saccharomyces cerevisiae*, Mol. Cells 10 (2000) 263–268.

93. M.J. Jeong, S.C. Park, M.O. Byun, Improvement of salt tolerance in transgenic potato plants by glyceraldehydes-3 phosphate dehydrogenase gene transfer, Mol Cells 12 (2001) 185–189.

94. M. Broin, S. Cuine, F. Eymery, P. Rey, The plastidic 2–cysteine peroxiredoxin is a target for thioredoxin involved in the protection of photosynthetic apparatus against oxidative damage, The Plant Cell 14 (2002) 1417–1432.

95. C. Gisbert, A.M. Rus, M.C. Bolarin, J.M. Lopez-Coronado, I. Arrillaga, C. Montesinos, M. Caro, R. Serrano, V. Moreno. The yeast *HAL1* gene improves salt tolerance of transgenic tomato, Plant Physiol. 123 (2000) 393–402.

96. A.M. Rus, M.T. Estan, C. Gisbert, B. Garcia, R. Serrano, M. Caro, V. Moreno, M. C. Bolarin, Expressing the yeast *HAL1* gene in tomato increases fruit yield and enhances K^+/Na^+ selectivity under salt stress, Plant Cell Environ. 24 (2001) 875–880.

97. H.X. Zhang, E. Blumwald, Transgenic salt-tolerant tomato plants accumulate salt in foliage but not in fruit, Nature Biotechnol. 19 (2001) 765–768.

98. T.H. Hsieh, J.T. Lee, Y.Y. Charng, M.T. Chan, Tomato plants ectopically expressing *Arabidopsis* CBF1 show enhanced resistance to water deficit stress, Plant Physiol. 130 (2002) 618–626.

99. S.B. Maqbool, H. Zhong, Y. El-Maghraby, A. Ahmad, B. Chai, W. Wang, Sabzikar, M.B. Sticklen, Competence of oat (*Avena sativa* L.) shoot apical meristems for integrative transformation, inherited expression and osmotic tolerance of transgenic lines containing HVA1, Theor. Appl. Genet. 105 (2002) 201–208.

100. S.E. Andersen, D.R. Bastola, S.C Minocha, Metabolism of polyamines in transgenic cells of carrot expressing a mouse ornithine decarboxylase cDNA, Plant Physiol. 116 (1998) 299–307.

101. R. Munns, Comparative physiology of salt and water stress, Plant Cell and Environment 25 (2002) 239–250.

Plant Biotechnology and Molecular Markers
P.S. Srivastava, Alka Narula and Sheela Srivastava (Editors)

23. Cell Differentiation in Shoot Meristem: A Molecular Perspective

Jitendra P. Khurana*, Lokeshpati Tripathi, Dibyendu Kumar, Jitendra K. Thakur and Meghna R. Malik

Department of Plant Molecular Biology, University of Delhi South Campus, New Delhi 110021, India

Abstract: The basic body plan of higher plants is laid down during embryogenesis, however, the entire adult plant develops post-embryonically through the activity of two meristems (shoot and root apical meristems) established originally at the opposite ends of the embryo. This article focuses on the shoot apical meristem (SAM), which is primarily responsible for the formation of leaves and stems in the vegetative phase and converts into reproductive meristem at a specific stage of development. The SAM comprises a central zone harboring a reservoir of pluripotent stem cells and a peripheral zone, which gives rise to primordia for organs such as leaves and flowers. Studies in the past decade have unravelled some of the molecular pathways that determine stem cell fate in the central portion of the SAM as well as regulate organ formation from peripheral zone of SAM. These studies are providing insight into the information flow between various zones and cell layers of the SAM that helps in stabilizing the size of the stem cell population, so vital for cellular proliferation and regulation of plant growth and development.

1. Introduction

A vascular plant begins its life cycle as a simple unicellular zygote. The zygote develops into the embryo and eventually into the mature sporophyte. These developmental events involve cell division, enlargement and differentiation, and organization of cells into tissues or organs. After a certain period of vegetative growth, the plant enters the reproductive stage with the development of specialized structures. The most striking feature of plant development is that plants continue to develop new organs after embryogenesis. Thus, portions of embryonic tissue persist in the adult and juvenile tissues in plants throughout their life cycle. These tissues are a specialized group of stem cells, called meristems, primarily concerned with the formation of new cells. Stem cell populations act as self-renewing populations of cells that give rise to one or more differentiated cell types. Angiosperm shoot and root apical meristems consist of such a population of cells located at their tips. The primary growth initiated in the apical meristems expands the plant body, increases its surface, leading eventually to reproductive development. In addition, many plants possess additional extensive meristems, the cambia, which aid in increasing the volume of the conducting system and forming supporting and protective cells.

During vegetative growth, shoot apical meristems (SAMs) give rise to repeating units of shoot called phytomers. Depending on their location within the shoot, the successive phytomers differ in internode length, leaf size and shape, axillary bud etc. A typical phytomer consists of a node to which leaf is attached, a subtending internode and an axillary bud at the base of the leaf (Fig. 1). The axillary bud has a meristem similar to SAM and can give rise to indeterminate

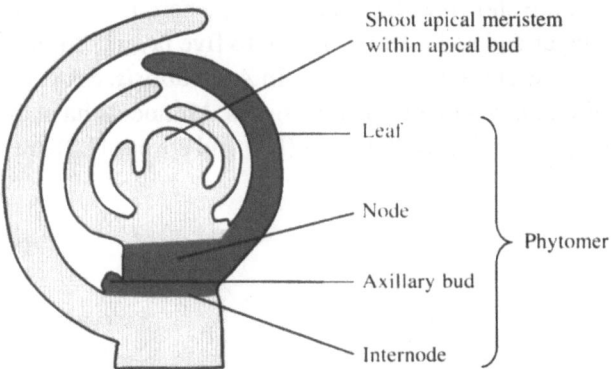

Shoot apical meristem
within apical bud

Leaf

Node

Axillary bud

Internode

Phytomer

Fig. 1   Schematic representation of the plant shoot apex (adapted from Langdale, 1994)

structures. SAMs are formed at distinct times and locations in the embryo, in the axils of the leaves and in modified form as floral meristems. The plant architecture is essentially determined by the pattern of meristem formation.

SAMs perform two main functions—maintaining a self-renewing population of initial cells and producing the fundamental parts of shoot system—leaves and stem. These two functions must be spatially coordinated. Leaves develop on the flanks of SAM in a defined arrangement, whereas initial cells are present in the center of the SAM. Unravelling the communication pathways that guide the SAM cells to continuously coordinate these two processes is critical for understanding and ultimately manipulating the plant form. Mutants defective in SAM formation and/or function have been identified in *Arabidopsis* and some other plant species. These studies have not only helped in understanding the functional domains of SAM but also indicated that the localized signaling between different cells and regions of SAM is an important component of SAM functioning.

2.   Structure and Organization of the Shoot Apical Meristem

Vegetative shoot apices vary in size, shape, cytological zonation and meristematic activity. The shape and size of the apex change during the development of a plant from embryo to reproduction, and also in response to seasonal variations [1]. The average diameter of SAM is about 100 to 200 microns, although SAMs as small as 50 microns and as large as 900 microns have been reported from diverse species. The size and shape of SAM also vary within a plant during plastochron (time interval between production of successive leaf primordia). Toward the end of the plastochron, just before the emergence of a distinct leaf primordium, the SAM is largest, whereas at the beginning of the plastochron, just after the leaf primordia has been produced, the SAM is at its smallest [2]. The changes in the morphology of the shoot apex occurring during one plastochron may be referred to as plastochronic changes.

In dicotyledonous plants like *Arabidopsis*, the SAM, covered and protected by leaves in a bud, consists of a small dome of cells, organized into regions with different functions and fates. The outermost region of meristem consists of one or more sheets of cells that divide anticlinally and together comprise the tunica. The tunica lies above corpus, which comprises of a mass of cells that divide obliquely and periclinally.

Cell layers within SAM are shown as L1, L2, L3 etc., where L1 refers to the outermost layer, L2 to the next subjacent layer and so on (Fig. 2). One to five layers of tunica have been reported for dicots, with two layers present in most species. In *Arabidopsis*, SAM consists of three layers — L1 and L2 comprise the tunica while L3 is the corpus. Monocots have one to four layers with one and two predominating. In maize, both tunica (L1) and corpus (L2) are single layered [1, 3].

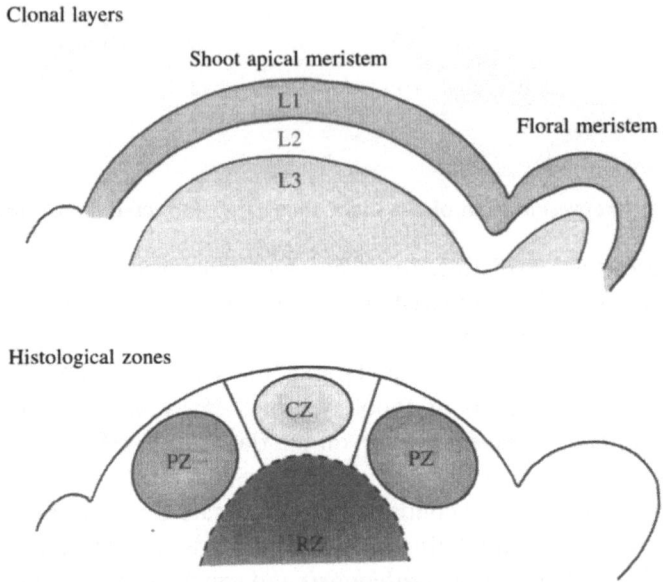

Fig. 2   **Diagrammatic sketch of *Arabidopsis* SAM showing the clonal layers (L1-L3) and the histological zonation (adapted from Fletcher and Meyerowitz, 2000). PZ, peripheral zone; CZ, central zone; RZ, rib meristem zone.**

The outermost layer L1 gives rise to a cell layer that covers all organs while the cells in L2 layer give rise to sub-epidermal tissue, the procambium and the part of the ground meristem. The cells in the L3 layer give rise to the rest of the ground meristem and pith. However, studies using genetic mosaics have shown that the position of the cell and not its clonal origin determines its fate [4]. All three layers contribute to organ formation and the growth of the stem, indicating that coordination of cell proliferation and cell fate specification is required during development [5]. One of the mechanisms responsible for communication of developmental signals could involve the transfer of signalling molecules, proteins and RNA, through plasmodesmata that establish a cyclic continuity between neighboring cells [6].

SAMs are also radially organized into zones. This pattern is termed cytohistological zonation, which is reflective of different rates of mitotic activity of cells in different regions of the meristem. The SAM consists of two different zones: (i) a central zone made up of stem cells (or central mother cells) that are large and exhibit low mitotic index and (ii) a peripheral zone of apical initial cells, which are undifferentiated and exhibit higher mitotic index, which surrounds the central zone and is the site of leaf formation. Both the central and peripheral zones overlap the tunica and corpus. At the flanks of the meristem is the morphogenic zone where organ primordia are formed [7]. Initials in the peripheral zone lying below the central zone constitute the rib meristem, which gives rise to the pith of shoot axis and is responsible for shoot elongation. In addition to their cell division rates, these zones differ in terms of cytology, expression pattern of marker genes and membrane potentials [5].

The capacity of the SAM for self-perpetuation and cell-type specification implies that SAM harbors an extensive communication system. In apical meristems, cells are interconnected via plasmodesmata, which are probably used to exchange molecules. The shoot meristem is compartmentalized into a central and peripheral symplastic field, as revealed by use of tracer dyes, that could restrict the diffusion of molecules to the cells within their boundaries [8]. Microinjection techniques have revealed that cells in the central zone are cytoplasmically coupled in two symplasmic fields: one within the tunica and other within the corpus, which harbor distinct signal networks. In addition, cells in tunica peripheral zone appear to be coupled into symplasmic ring, surrounding the central field. These observations substantiate the tunica-corpus concept as well as the apical zonation model, and indicate that tunica-corpus concept is physiologically relevant!

3. Origin of Shoot Meristem During Embryogenesis

In dicots, the SAM develops between the two cotyledon primordia, in the central portion of the embryo. In contrast, SAM develops laterally on the embryo in monocots, at the base of the single scutellum (Fig. 3). In maize, the first five leaves are produced during embryogenesis whereas in *Arabidopsis*, two small leaf primordia are detectable at the end of embryogenesis [2,3]. The origin and development of SAM during embryogenesis has been equivocal with respect to whether cotyledons are formed from the SAM or if the SAM and cotyledons arise independently. According to one school of thought, the cotyledons and SAM respond to positional signals and establish their respective cell fates independently in the apical region of the globular embryo [2]. Alternatively, specification of either one may require prior specification of the other.

The vegetative meristem with characteristic tunica-corpus structure is not evident until the torpedo stage of embryogenesis in *Arabidopsis* [9]. After cotyledons and provasculature are clearly distinguishable, the apical histological zonation is visible and this is considered to be an indicator for the activity of SAM [10]. It has thus been presumed that either the apical portion of the globular embryo forms the SAM or the SAM is formed only after the tunica-corpus structure is evident at the early torpedo stage of embryogenesis.

It has been shown by histological studies and clonal analysis in different species, such as *Arabidopsis* [9] and cotton [11], that the cotyledons and SAM develop from distinct regions of the globular embryo. The existence of mutations that adversely affect SAM formation but not cotyledons (e.g. *STM* mutations described below) suggest that cotyledons are specified independent of SAM specification. At the same time, mutations leading to loss of both the cotyledons and the SAM exist. Such mutations could prevent the specification of cotyledons and thereby SAM formation or affect the gene products involved in signal transduction to both SAM and cotyledon primordia [2,9]. Alternatively, there are evidences to support the view that cotyledons are homologous to leaves and that they are the first products of embryonic SAM. Thus, in bipolar embryos, the SAM originates in the globular stage before the cotyledons arise, at the same time when root apical meristem is also defined [10].

Recent molecular genetic analysis of mutants defective in embryonic shoot meristem development has improved our understanding greatly. These mutants were identified during systematic screens

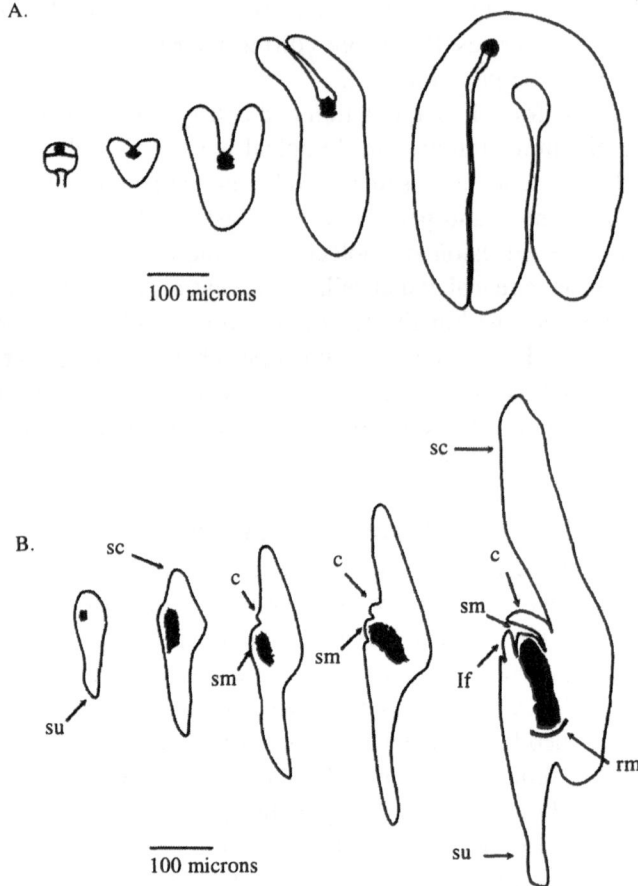

Fig. 3   SAM formation during embryogenesis. A. *Arabidopsis*: Shaded cells indicate pattern of *STM* expression. The diagrams (from left to right) represent different stages of embryo development from globular to mature stage. B. Maize: Shaded areas show pattern of *kn1* expression. The maize SAM develops on the side of the embryo; sm, shoot apical meristem; c, coleoptile; sc, scutellum; If, first leaf; su, suspensor; rm, root apical meristem (adapted from Evans and Barton, 1997).

for mutants affected in SAM formation or embryo development. Some of these are described below in some detail.

The lesions in the *GURKE* gene cause characteristic apical defects and the cotyledons are either reduced, absent or reduced to knob-like structures [12-13]; roots are also short and hypocotyls malformed. It could be ascribed to abnormal cell divisions occurring within the apical region leading to establishment of no or only rudimentary cotyledonary primordia during early heart-stage of embryogenesis. Post-embryonically also, *gurke* seedlings give rise to abnormal leaves and stem-like structures, suggesting that *GURKE* gene, although involved primarily in the organization of the apical region in the embryo, may play a role in post-embryonic development [13]. The *gk* mutations are allelic with *emb22* mutations, which do not have apical deletion phenotype but are generally defective in morphogenesis and cellular differentiation and form abnormal, thick tube-shaped leaves in culture. This suggests *gk* alleles to be weak alleles of *emb22* [2,12].

Mutations in *MONOPTEROS* gene affect both root and shoot apical meristem formation. Mutant alleles of the *mp* gene eliminate both the hypocotyls and the root and many mutant seedlings have two fused cotyledons or single cotyledon and also lack a shoot meristem [12]. *MP* gene encodes a protein with features of transcriptional regulator harboring a DNA binding domain that possibly binds upstream regulatory elements of auxin-inducible promoters and thus modulate gene activities in response to auxin signals [14]. However, the role of *MONOPTEROS* in embryonic SAM formation is still unclear.

The *Arabidopsis* mutants defective in *TOPLESS (TPL)* gene fail to form cotyledons and SAM during embryogenesis. Severely affected *tpl* mutants consist of only root and hypocotyls with no SAM or cotyledons, while the less severely affected seedlings may have both cotyledons and SAM depending on the severity of phenotype. The *TPL* gene may be required for specifying shoot fate in a general way or some aspect of embryonic pattern formation, but not for post-embryonic SAM function [2].

The mutant analysis has also shown *WUSCHEL* gene to be necessary for maintaining structural and functional integrity of the shoot meristem. The *wus* mutant apices have aberrant organization and form a flat structure with abnormal cells in comparison to wild-type shoot meristem. The SAM stem cells are miss-specified and undergo differentiation without becoming incorporated into organ primordia, terminating into flat apices. Subsequently, new shoot meristems are initiated, which again terminate after making a few leaves. The *wus* plants also initiate numerous adventitious meristems, which form only a few organs and terminate prematurely. This stop/start phenotype suggests separate regulatory mechanisms for shoot meristem initiation and maintenance. In *wus* embryos, the shoot meristem appears to be initiated but displays defective organization, suggesting *WUS* gene product is required for central zone function [15]. *WUS* encodes a novel sub-type of the homeodomain protein family [16]. Homeodomain proteins regulate transcription in many diverse species and are generally involved in developmental or cell type specification. The characteristic homeodomain consists of about 60 amino acids that bind DNA in a sequence-specific manner. The helix-loop-helix-turn-helix structure and twelve highly conserved amino acid residues are also essential features [17]. The WUS homeodomain conforms to these features and is about 30% identical and 45%-50% similar to homeodomain sequences from diverse organisms. *WUS* expression is initiated in the four sub-epidermal cells of the apical region of embryo at the 16-cell stage and becomes gradually confined to sub-epidermal cells in the center of the embryonic SAM through several asymmetric divisions. These *WUS*-expressing cells specify the overlying cells to maintain their specification as stem cells [16].

The maize homeobox gene *KNOTTED 1 (KN1)* is a useful molecular marker for the SAM [18-19]. The onset of *KN1* expression during embryogenesis coincides with the first histological recognition of SAM in maize embryos [19]. *KN1* is not expressed in determinate products of meristems such as leaves and floral organs and, even within the vegetative meristem, the *KN1* mRNA declines tremendously in cells destined to form the next leaf [20]. *KN1* was first defined by dominant gain-of-function mutations affecting leaf development [21]. In both maize and *Arabidopsis*, *KN1*-like genes are represented as a multigene family. The *KN1* homeobox (*KNOX*) genes characterized in maize share a high degree of similarity within and outside the homeodomain and show overlapping expression patterns [22-23]. *KN1* mutant analysis indicates that maize *KN1* gene product serves to maintain the cells in a meristematic or undifferentiated state and thus *KN1* may be involved in maintenance of the morphogenetic zone of the SAM [20–21].

In *Arabidopsis, SHOOTMERISTEMLESS (STM)* displays a pattern of expression very similar to *KN1* [24]. *STM* expression is first detected in a few cells in the center of the apical half of the mid-globular stage embryos, when the embryo consists of about 32-64 cells. In late-globular stage embryos, *STM* mRNA appears in a stripe across the top half of the globular embryo and later becomes restricted to the notch between the cotyledons in the heart, torpedo and mature stage embryos [24]. Histological analysis indicates that *STM* mRNA accumulates in the cells predicted to give rise to SAM [9,24]. The expression of *STM* mRNA persists in seedling and adult plants and is detectable in vegetative, axillary, inflorescence and floral meristem. However, *STM* mRNA is absent from leaves and leaf primordia, and even within the meristem, *STM* expression is down-regulated at the site of primordia formation. The analysis of *Arabidopsis* seedlings carrying mutations at the *STM* locus has indicated that *STM* gene affects embryonic SAM initiation and spacing of the cotyledons [9,25]. The *stm* mutant shoot meristems, however, initiate primordia in aberrant phyllotaxis with phenotypic defects and the cells in the center appear to be incorporated into the ectopic primordia and undergo differentiation, suggesting *STM* is required to maintain undifferentiated population of cells within the SAM. Alternatively, it is proposed that *STM* is required for the initiation of the shoot meristem after appearance of cotyledons, as in *stm* mutants cells in presumptive SAM fail to organize into a tunica-corpus structure [9]. On the basis of these observations, it has been proposed that in the peripheral region, *STM* is required to inhibit organ outgrowth and differentiation, while in the central region, *STM* is required to inhibit differentiation and to inhibit SAM-specific program of development [26].

The *STM* encodes a KN1-type of homeodomain protein of 382 amino acid residues and is, therefore, likely to act as transcriptional regulator [24]. A possible target for *STM* is *UNUSUAL FLORAL ORGANS (UFO)* gene, as *UFO* accumulates in early heart-shaped embryo in *STM*-dependent manner, although *ufo* mutants are unaffected in SAM development. However, *UFO* negatively regulates growth of inflorescence meristem and floral meristems, but promotes the expression of floral organ identity gene *APETALA3* [27].

Using transposon-mediated activation tagging, a mutant of *Arabidopsis, drn-D (Dornroschen)*, has been identified where shoot meristem activity is arrested prematurely, with the formation of radicalized lateral organs [28]. The expression of the homeobox gene STM is downregulated during development of *drn-D* mutant. Strikingly, the expression of *CLV3* and *WUS*, which act antagonistically to regulate stem cell fate in meristems is increased in *drn-D* mutants. The cloning of DRN gene, revealed that it encodes an AP2/ERF-type transcription factor that is probably involved in regulation of gene expression patterns, and eventually the cell fate in developing meristems. Mutations in *CUC1* and *CUC2 (CUP-SHAPED COTYLEDON)* cause defects in separation of cotyledons, sepals and stamens as well as in the formation of SAM in *Arabidopsis*. The *CUC1* and *CUC2* genes are functionally redundant and thus the defects described above are most apparent in *cuc1cuc2* double mutants. These two genes encode members of NAC family of proteins that share a highly conserved N-terminal domain termed the NAC domain [29–30]. Petunia *NO APICAL MERISTEM (NAM)* gene is another member of this family [31]. Petunia embryos in *nam* mutation background fail to develop a SAM and cotyledons are partially fused. The *NAM* transcripts are first detected at late heart stage of embryogenesis and accumulate in cells at the boundaries of meristems and primordia, indicating that *NAM* helps in determining positions of meristem and primordia [31]. However, *cuc1cuc2* and *nam* mutants give rise to adventitious shoots showing normal vegetative and reproductive development, suggesting that *CUC1*, *CUC2* and *NAM* genes are not essential for SAM maintenance during later development [31–32]. Another gene encoding NAC-domain protein, *CUC3*, with high similarity to CUC1 and CUC2 has been identified [33]. The *CUC3* mutant analysis indicates that *CUC3* functions in establishing the boundary between the cotyledons and in SAM formation. Expression analysis in the overexpressor lines and in loss-of-function mutants suggests that *CUC1*, *CUC2* and *CUC3* act upstream of *STM* and are redundantly required for *STM* expression through yet other unidentified factors [30, 32].

In the developing SAM, the *ZWILLE/PINHEAD* (*ZLL/PNH*) gene is also required for maintaining stem cells in an undifferentiated state. In *zll/pnh* mutants, the cells in the SAM primordium do not maintain *STM* expression and differentiate, and the defective SAMs that are formed terminate shortly after germination. The adventitious meristems developing on these mutants resemble the wild-type, although occasionally *zll/pnh* mutants do not initiate meristems in the axils of the cauline leaves. However, the defect is limited to secondary inflorescences and may be secondary to fasciation. Thus, *ZLL/PNH* seems to be required specifically for SAM formation in the embryo [34]. According to another study, there is no correlation between fasciation and failure to form axillary meristems and that *ZLL/PNH* is required for the efficient formation of axillary meristems during post-embryonic development. This is consistent with the persistent expression of *ZLL/PNH* mRNA in the meristem and the adaxial leaf domain [35]. The expression of *ZLL/PNH* is first detected at four-celled stage. As the development proceeds, the expression becomes progressively confined to presumptive SAM region and the provascular tissue.

The *ZLL/PNH* gene encodes a member of novel family of proteins found in many eukaryotes and that includes the product of *ARGONAUTE1* (*AGO1*) gene involved in leaf development and meristem cell maintenance [36]. The rabbit translation initiation factor eIF2C is another family member, suggesting a role for *ZLL/PNH* and *AGO1* in translational control of development [37]. As mutations in *ZLL/PNH* result in specific defects, the gene could be involved in tissue- and/or stage-specific translational control. *ZLL/PNH* function is necessary for regulating spatial *STM* expression at later stages of embryogenesis and, thus, it is proposed that *ZLL/PNH* relays positional information required to maintain stem cells of developing shoot meristem in an undifferentiated state and is required for partitioning of the embryo apex by regulating spatial STM expression [34–35]. Embryos doubly mutant for *ZLL/PNH* and *AGO1* fail to progress to bilateral symmetry and do not accumulate *STM* transcripts, suggesting these genes could encode partially redundant functions in regulating growth and gene expression patterns during embryogenesis [35]. Recent work on *Drosophila* homologue of *ZLL/PNH* and *AGO1*, called *PIWI*, suggests that these genes have an ancestral function in stem-cell maintenance [38]. Thus, signaling from differentiated cells to stem cells may represent a basic mechanism for stem cell maintenance among diverse eukaryotes [37].

3.1 Axillary Shoot Meristem Formation

Shoot apical meristems form throughout the lifecycle of most higher plants. Three types of SAMs form post-embryonically: vegetative SAMs in lateral positions along the main shoot axis, inflorescence lateral SAMs, and floral meristems. Molecular genetic analysis indicates that axillary SAMs share many common genetic determinants with the primary SAM formed in the embryo [39]. However, the formation of the axillary shoot is inhibited by the shoot apical meristem and depends on the subtending leaf primordium. Axillary meristem formation in *Arabidopsis* occurs in two waves: an *acropetal wave* forms during vegetative development, and a *basipetal wave* forms during reproductive development [40]. Axillary meristems are believed to originate from the SAM as detached meristematic cells, in the axils of leaf primordia [40]. However, in some cases as *Heracleum*, axillary meristems appear to arise from leaf primordium rather than from separate meristematic cells [7, 41].

Numerous mutants affected in axillary meristem initiation have been identified, for example *auxin resistant* mutants in maize and *decreased apical dominance* mutants in petunia [42]. In tomato, *lateral suppressor* (*ls*) mutant prevents the initiation of the vegetative lateral SAMs but not of floral or sympodial mersitems. In this mutant, the SAM was shown to be smaller than normal, which may restrict the initiation of axillary bud primordia. The *ls* mutant has no effect when present in L1 layer, whereas chimeras with the mutation in L2 and L3 have *ls* phenotype, suggesting that internal cell layers are important for regulating lateral SAM formation. Another mutant in tomato, *torosa-2* (*to-2*), reduces the number of vegetative axillary buds. Both *ls and to-2* mutants have reduced cytokinin levels suggesting a possible cause of the mutant phenotype. However, increasing cytokinin levels in the mutants failed to increase lateral SAM formation although dormancy of existing buds was eliminated in *to-2* [2, 42].

In *Arabidopsis*, two mutants identified with defects in axillary bud formation are in fact defective in primary SAM. For example, *pinhead* mutations affect embryonic SAM where a leaf or pin-like organ terminates the growth of the meristem [43]. In addition, the *pinhead* mutations reduce the inflorescence SAMs in the axils of cauline and rosette leaves and are defective in floral meristem formation. In comparison, the *rev* (*revoluta*) mutants unusually develop large leaves, stems and floral organs but reduced numbers of vegetative and floral axillary shoots. These mutants are defective in apical meristem activity resulting in premature termination of shoot apex and in formation of abnormal or incomplete structures in place of axillary shoots, indicating *REVOLUTA* plays a role in meristem maintenance [41]. The *REV* gene has been identified and found to encode a predicted homeodomain leucine zipper transcription factor that also contains START sterol-lipid binding domain [39]. *REV* gene is expressed at the earliest stages of lateral shoot meristem and floral meristem formation. Within the inflorescence shoot meristem, *REV* expression appears to mark the next 3–5 flower primordia forming on the flanks of the shoot meristem, and its expression matches that of *WUS* and *STM* at stage 1 and stage 2, respectively. These observations suggest that *REV* acts to establish meristem identity or activates other meristem regulators. Within the organs, *REV* is expressed largely on the adaxial portion of the organ and may be involved in establishing adaxial fate in cotyledons, leaves and floral organs. *REV* is also expressed in developing vasculature and may be required for proper differentiation of vascular elements within the SAM. The different activities of *REV* appear to be functionally distinct [39]. Recently, another gene involved in axillary meristem formation has been identified by the analysis of an *Arabidopsis* mutant, *las* (*lateral suppressor*), which are unable to form lateral shoots during vegetative development [44]. This study also examined the transcript accumulation of LAS and STM and showed that *LAS* gene works upstream to other regulators of shoot branching like *REV*. Members of the MADS box gene family are highly conserved transcription factors that play diverse roles in regulating plant development. Potato *MADS box 1* (*POTM1*) expresses most abundantly in vegetative meristems of potato (more specifically in tunica and corpus layers), the procambium, the lamina of newly formed leaves and developing axillary meristems [45]. The transgenic suppression lines of POTM1 exhibited decreased apical dominance and enhanced axillary bud growth, coupled with 2-3 fold increase in cytokine levels. This implies that *POTM1* regulates cell growth in vegetative meristems and indirectly regulates axillary bud development.

3.2 Adventitious Shoot Meristem Formation

Meristems also develop at locations other than those formed during embryogenesis, in the axils of leaves and during reproductive development. These meristems form normally on many different organs of adventitious shoots in a variety of plant species, such as root bearing shoots of *Convolvulus arvensis*, and shoots initiating from cambial tissue of tree stumps. Epiphylly is another example wherein organs or shoots develop upon a leaf as observed in plants such as *Bryophyllum*, in which plantlets form along the margin of the blade. An epiphyllous shoot may represent fusion or displacement of a normal axillary meristem or may be a true adventitious shoot. Displacement of normal axillary buds into organ surfaces results in meristem formation followed by change in meristem position relative to other plant parts due to differential growth, e.g., *Coryphantha* [42]. However, in *Bryophyllum*, vegetative adventitious shoots are rather initiated by remeristemization of differentiated tissues of mature leaves. In *Bryophyllum*, epiphyllous shoots arise from cells of the leaf margin that stop dividing and remain blocked in G1 phase of the cell cycle. In mature leaf, these cells become reactivated to form an undifferentiated meristem that acquires zonation and forms a small shoot [42].

Adventitious SAMs have also been produced in transgenic plants, overexpressing *KN*-like genes and cytokinin biosynthetic pathway genes. Tobacco and *Arabidopsis* transgenic plants overexpressing maize *KN1* gene show lack of apical dominance and are severely dwarfed [23, 46]. Leaves are thickened and lobed and, in severe cases, small shoots originate from these diminutive leaves [46–47]. However, the ectopic expression of *KN*-like genes does not lead to adventitious shoot production in species such as tomato. The analysis of transgenics overexpressing a related gene *KNAT1* in *Arabidopsis* showed that simple leaves are transformed into lobed leaves with stipules in the sinus, the region at the base of the two lobes. Ectopic meristems also arise in the sinus region close to the veins. The shoot-like characteristics of these leaves suggest that *KN1*-related genes may have an important role in the regulation of leaf diversity [47].

The overexpression of *isopentenyltransferase* (*ipt*), bacterial gene involved in cytokinin production, in transgenic tobacco plants leads to adventitious shoot meristem formation at the site of *Agrobacterium* infection as well as the shooty phenotype of the transformed cells in culture. The phenotype of the tobacco plants expressing either *ipt* or *KN1*-like genes are quite similar. It has been observed that *ipt* transgenics have higher steady-state mRNA levels of *KNAT1* and *STM*, similar to cytokinin overproducing shoot meristem mutant *amp1* [48], suggesting that cytokinins possibly act upstream of *KNAT1* and *STM* in the same pathway. This provides a link between the hormone and developmental genes and indicates for a probable role for cytokinins in the SAM formation [48–49].

Another study has reported that ectopic expression of TBP-2, the TATA box binding protein, induces apical shoot proliferation in *Arabidopsis* [50]. The ectopic meristem-like structures arose essentially from highly undifferentiated leaf primordium or from young ectopic shoots during more advanced vegetative growth phases. The expression of some shoot meristem regulatory genes such as *STM*, *KNAT1*, and *CLV1* is altered in *Arabidopsis* apical shoots. This suggests that

TBP-2 protein might be needed for apical meristem function and that high TBP-2 levels prevent cell differentiation in undifferentiated cells, possibly by increasing transcription initiation and, thus, play a role in controlling shoot production [50].

4. Coordination of SAM Proliferation

The SAMs remain relatively constant in size throughout the life cycle of most higher plants, but still continue to produce lateral organs. This indicates that cells within the SAM somehow continually assess their relative positions and subsequently decide to divide, differentiate, maintain *status quo* or restrict cell proliferation in the SAM. Several mutants accumulating rather too many cells in the SAM have been identified in *Arabidopsis*. The first class is represented by the recessive *clavata* mutants (*clv1, clv2, clv3*). *clavata* is derived from the latin word *clava* and means 'club'. The *clavata* mutants harbor excess cells in the central zone of the SAM and floral meristems, resulting in fasciation of the shoot and generation of flowers with extra floral organs and distortion of siliques, giving them a club-shaped appearance. Although the central zones of both the shoot and floral *clv*-meristems appear enlarged, the organs initiating from the peripheral zone grow normally [51–53]. The comparative analysis of cell division patterns suggests that increase in meristem size in *clv* SAMs is due to reduction in drafting of these cells into the peripheral zone, rather than increase in the cell division rates in the central zone [54]. The second class of mutants include the *mgoun* mutants (*mgo1* and *mgo2*), which display reduction in the number of leaves and floral organs, larger meristems and fasciation of the inflorescence stem. The molecular genetic analysis of these mutants indicates that *MGO* genes play a role in organ primordia initiation and determining their number [55].

The epistasis analysis revealed that *CLV1* and *CLV3* act in the same pathway to control meristem-cell proliferation [52]. The shoot and floral meristems of *clv2*, third mutant in this group, are phenotypically similar to those of weak *clv1* and *clv3* mutants, and *clv1/clv3* are epistatic to *clv2* with regard to size of shoot meristem and the number of flowers initiated by the floral meristem. Thus *CLV2*, like *CLV1* and *CLV3*, helps in preventing the accumulation of undifferentiated cells at the shoot and flower meristems [53]. However, *clv2* mutants display other organ defects (e.g. reduced anthers/stamens) as well, suggesting that while *CLV2* is required to regulate meristem development, it also functions independent of *CLV1* and *CLV3* to regulate organ development [53].

The cloning and molecular analysis of CLV genes has provided new insights into the mechanism by which they regulate meristem cell proliferation. Both *CLV1* and *CLV2* genes encode transmembrane leucine-rich repeat (LRR) transmembrane proteins [56–57]. The CLV2 receptor-like protein (RLP) carries only a short C-terminal domain, whereas the CLV1 protein harbors a C-terminal serine/threonine kinase domain. There is evidence to suggest that CLV1 and CLV2 most likely form a heterodimeric receptor molecule localized in the plasma membrane [57]. LRRs are a common motif of protein-binding domains both in plants and animals suggesting that the CLV receptor interacts with an extracellular protein ligand. LRR-receptor kinases have been implicated in signal transduction cascades and more than 50 genes have already been identified in diverse plant species in many cases. For example, the *Arabidopsis ERECTA* (*ER*) gene encodes an LRR-receptor kinase. It is expressed in SAMs and flowers and is thought to mediate cell-cell communication to accelerate cell division and elongation [58–59]. The *BRASSINOSTEROID INSENSITIVE 1* (*BRI1*) gene also encodes an LRR-receptor kinase that most likely acts as a receptor for plant steroid hormone brassinolide [60–61]. Mutations in *BRI1* gene cause dwarfism and plants grown in dark display light-grown phenotype. However, analysis of CLV1/BRI1 chimeric receptors in *clv1* mutant background provides evidence that CLV1 and BRI1 kinase domains are not interchangeable [62]. The results of this study also indicate that CLV1 can act outside the meristem to regulate the pedicel length in *erecta* mutant background. In addition, several plant disease resistance genes encode LRR-receptors or receptor kinases that enable plants to sense and respond appropriately to specific bacterial and fungal pathogens [63].

Interestingly, *CLV3* encodes a secretary protein of 96 a.a. and carries an 18 a.a. N-terminal signal peptide, suggesting CLV3 protein may be extracellular ligand, which may interact with the extracellular domains of CLV1 and CLV2 receptors [37, 64]. *CLV3* mRNA is detected mainly in L1 and L2 layer of the central zone and probably demarcates the stem cells in these layers; it is however not detected in the flanks of the meristem. In contrast, *CLV1* is expressed mostly in an underlying domain in the L3 layer and is not detected in the L1 layer. The *CLV2* mRNA is detected in all the shoot tissues of the plant. This suggests that *CLV3* may signal in a non-cell autonomous manner from overlying to the underlying regions of *Arabidopsis* SAM [56–57, 64]. However, until recently the experimental evidence that CLV3 acts as an extracellular signaling molecule was lacking. Employing genetic and immunological assays, CLV3 has been shown to localize to the apoplast [66]. Apoplastic localization permits CLV3 to signal from the stem cell population to the organizing centre in the underlying cells, activating the CLV1/CLV2 receptor complex. Essentially a similar conclusion was drawn in a parallel study [65] whereby it was shown that CLV3 functions as a mobile intercellular signal but its spread is regulated by its receptor CLV1. This enables the shoot meristem to permit the peripheral cell differentiation and yet maintains a stable niche for the stem cells in the middle. Biochemical and genetic analysis shows that CLV1 function depends on the presence of functional CLV2 and CLV3, and that CLV3 acts as a ligand for CLV1 as a part of multimeric complex [57, 67]. *In vivo* CLV1 forms an inactive complex of approximately 185 kDa, which is thought to consist of a CLV1 disulfide linked to CLV2, and an active complex of approximately 450 kDa, containing 185 kDa complex, and a type-2C kinase-associated protein-phosphatase (KAPP), which has been shown to act as a negative regulator of CLV1 signal transduction pathway [68–69].

The role of *WUS* gene product in CLV signal transduction pathway has been implicated (Fig. 4). As mentioned earlier, the *WUS* gene promotes stem cell fate and encodes a homeodomain transcription factor that is expressed in the L3 layer of the shoot and flower meristems throughout development. Mutations in *WUS* or the *CLV* genes have opposite phenotypes, indicating that these genes promote and restrict stem cell formation in the central zone, respectively. *wus* mutants are largely epistatic to *clv* mutants, indicating *WUS* functions downstream of and could be a target gene for repression by *CLV* genes [15–16]. Recent studies have shown that *WUS* mRNA is not confined to its normal expression domain in *clv* mutants, but expands both apically and laterally, indicating *CLV* signaling restricts the boundary of *WUS* expression [70–71]. The enlarged size of the shoot meristem in *clv* mutants may be a consequence of deregulation of *WUS* activity, as a result of which more stem cells would be specified, causing expansion of the central zone and eventual fasciation of the meristem. Indeed, constitutive expression of *CLV3* in transgenic *Arabidopsis* plants caused severe reduction in the levels of the *WUS* transcripts and phenotypically these plants resembled *wus* mutants [70]. Transgenic plants overexpressing *WUS* under the control of *CLV* promoter resembled *clv* mutants with large and fasciated meristems

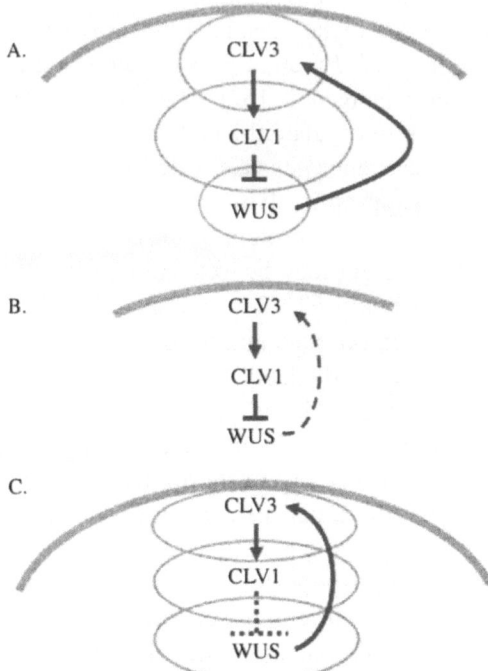

Fig. 4   Control of stem cell maintenance. A. Expression of *CLV3*, *CLV1* and *WUS* share common regions. Expression of *WUS* is restricted to very thin area by CLV3 signaling involving CLV1 receptor complex. B. Expression of *WUS* is decreased due to constitutive CLV3 signaling. C. When CLV pathway is broken, restriction over limited area of *WUS* expression is relieved, resulting into excessive stem cell accumulation and expansion of the meristem. The arrow indicates positive regulation and blunt-ended line indicates negative regulation (adapted from Fletcher, 2002).

[71]. The expression of *WUS* under the control of *ANT* promoter that confers expression in organ primordia and developing organs, leads to the termination of leaf formation and a large bulge of cells similar to meristem cells is formed. In addition, these cells express the stem cell marker *CLV3*, suggesting that *WUS* is sufficient to induce *CLV3* expression at the correct position and thus specify stem cell identity [71]. In a recent study, it has been found that expression of *CLV3* depends on *WUS* function only in the embryonic shoot meristem. At later stages of development, *WUS* stimulates *CLV3* expression together with *STM* gene [70].

It is possible that there may be additional factors that function together with *WUS* and are regulated by *CLV* to promote stem cell pathway. Mutations in the *POLTERGEIST* (*POL*) gene have been identified as partial suppressors of meristem defects in *clv mutants. When CLV* genes are functional, *pol* mutants are nearly indistinguishable from wild type plants. Like *WUS*, *POL* may either encode for a transcription factor or a protein that regulates one [72]. It is proposed that *WUS* together with other genes like *POL* acts to promote stem cell fate and *CLV3* expression in the overlying cells in a non-cell-autonomous manner. *CLV3* signaling would act to repress the activity of these regulatory factors. This mutual regulation, involving positive and negative interactions, provides a feedback system maintaining the meristem size [70].

The role of *CLV* and *STM* genes in regulating SAM cell proliferation is rather antagonistic. Unlike *wus* and *clv* mutants, *clvstm* double mutants display an additive phenotype, suggesting that they act in separate pathways [73]. SAM development is defective in both *stm* and *wus* mutants as both *stm* and *wus* mutants lack stem cells, however, terminal phenotypes are different in two cases. Genetic analysis indicates that *STM* acts upstream of *WUS* and, although *WUS* and *STM* expressions are initiated independently of each other in different meristem domains, expression of one cannot be maintained in absence of the other. Therefore, while *WUS* (and possibly *POL*) may be required to specify stem cells, *STM* activity allows their progeny to proliferate before incorporation into lateral organ primordia [16, 25]. The stem cell promoting activity of *STM* and *WUS* genes is restricted by *CLV* genes. It has been observed that mRNAs of various components of the signaling network accumulate in non-overlapping domains of the meristem, suggesting the involvement of all the cell layers in regulating meristem structure. This indicates the existence of an intercellular communication network through symplasmic domains and exchange of cytoplasmic components [74–75].

5. Regulation of SAM by the Lateral Organ Primordia

The *STM* gene is required for SAM development in *Arabidopsis* and is expressed throughout the meristem but is absent in organ primordia, indicating that *STM* expression is repressed in cells that give rise to organ primordia [24]. In maize, expression of *KNOX* genes is regulated by *ROUGH SHEATH 2* (*RS2*), a MYB protein found only in lateral organ primordia and their initials [76–77]. In *rs2* mutants, *KNOX* genes are ectopically expressed in developing leaves indicating that *RS2* represses *KNOX* gene expression in lateral organs. *PHANTASTICA* (*PHAN*), an *RS-2* related gene in *Antirrhinum*, is expressed in organ founder cells and required to inhibit expression of an *STM* orthologue *AmSTM1* in lateral organs [78]. *asymmetric leaves 1* (*as1*) is a mutation in *Arabidopsis* that disrupts development of cotyledons, leaves and floral organs. Loss of *AS1* activity leads to misexpression of *KNOX* genes, *KNAT1* and *KNAT2*, closely related to *STM*, while *STM* expression itself is unchanged, and *as1stm* double mutants have *as1* phenotype except failure to form flowers, suggesting that *STM* is a negative regulator of *AS1* [79]. In stem cells, *STM* negatively regulates *AS1*, while in the organ founder cells *STM* is downregulated permitting *AS1* expression, which in turn downregulates *KNAT1* and *KNAT2* genes. Thus, *stm* mutants fail to develop a meristem due to misexpression of *AS1* in stem cells causing them to differentiate. Lack of *AS1* in *as1stm* double mutant allows meristem function possibly by derepression of *KNAT1* and *KNAT2*. The *clv1*, *clv3* and *wus* mutants show additive interactions with *as1*, suggesting *AS1* along with *STM* acts independently of these genes in stem cell specification pathway [79]. Recently, another gene *ASYMMETRIC LEAVES2* has been identified, which along with *AS1* is involved in establishing the entire vein system and repression of class I *knox* genes in the leaves [80].

6. Meristem Patterning and Floral Determination

Floral meristems arise post-embryonically from the shoot apical meristem. Floral meristems are modified shoot meristems since they have a similar structure and function. Since the two meristems are functionally very similar, it has long been speculated that same genes may be involved in regulating their function. In *Arabidopsis*, *LEAFY* and *APETALA* genes are involved in floral meristem specification and mutations in these genes convert floral meristems into shoot meristems

[81–82]. Conversely, mutations in *TERMINAL FLOWER* gene convert shoot meristems into floral meristems [83]. An important difference between the two is that floral meristems are determinate structures, producing flowers with fixed number of parts namely stamens, petals, sepals and carpels, while shoot meristems produce indefinite number of leaves.

Two groups have independently shown that *AGAMOUS* (*AG*) gene, which has a role in meristem termination and floral organ patterning, interacts with *WUS* [84–85]. *AG* is a MADS box gene expressed in third and fourth whorls of developing flower and specifies stamens and carpels. The *ag* mutants develop indeterminate flowers containing only sepals and petals [86]. In contrast, *wus* mutants display premature termination of SAM and floral meristems after formation of a few organs, and the flowers formed lack carpels and most stamens. Thus, it is possible that the two genes interact to control floral development.

In *ag* mutants, *WUS* expression persists in flowers whereas in wild type it is switched off when carpel primordia are established. Also, in plants with reduced *WUS* expression in floral meristems, *AG* expression domains are smaller, indicating *WUS* is required for *AG* activation. Moreover, over expression of *WUS* using *LFY* and *APETALA* 3 (*AP* 3) promoters can cause ectopic formation of stamens and carpels. It has been observed that *wus* mutants are epistatic over *ag* mutants, suggesting that *AG* functions as a negative regulator of *WUS* [84–85]. *In vitro*, WUS protein binds to consensus homeodomain target sites within the *AG* regulatory region and these sites are necessary for expression of *AG* reporter gene *in planta* [85]. However, activation of *AG* expression by WUS is restricted to floral meristems, suggesting requirement of additional flower specific factors. It has been shown that one of these factors could be floral meristem identity gene *LEAFY* [84-85]. Endogenous *WUS* is unable to activate *AG* expression in *lfy* mutants [84] however, overexpression of *WUS* can activate *AG* promoter in absence of LFY suggesting LFY requirement is not absolute [85]. LFY protein directly binds to sites in the *AG* regulatory region and binding of both WUS and LFY is essential for the expression of *AG* since activation of *AG* reporter gene in yeast occurs only when both *LFY* and *WUS* are coexpressed. However, the two proteins bind their recognition sites independently and binding is not cooperative [85].

These data suggest that stem cell termination in floral meristems requires a autoregulatory mechanism involving WUS and AG. WUS is responsible for activation of *AG* expression in the center of the floral meristem. AG once established, represses WUS either alone or in combination with other factors, which results in the termination of the stem cell maintenance. This mechanism is restricted to floral meristem since *AG* activation by WUS requires LFY [84–85]. This loop is analogous to negative feed back loop between WUS and CLV3, which regulates stem cell population in shoot apical meristem.

7. Conclusions and Perspectives

The development of higher plants depends on the activity of the shoot meristem, a dynamic structure consisting of self-renewing stem cells. The shoot meristem arises early during embryogenesis and subsequently forms the basic subunits of the shoot, leaf and stem, in repeated patterns. SAM maintenance depends on two antagonistic processes: stem cell self-renewal and organ initiation. In order to achieve this, the cells need to be placed in proper fields of positional information. Recent molecular and genetic studies have identified many components of the intercellular pathways that play important roles in regulating meristem function. As a result of these studies, the view that emerges is that SAM is a dynamic structure in which cell fate is not predetermined; rather cell fate is in accordance with their positions relative to each other. These studies also bring into limelight the significance of interactions between the SAM and the developing leaf being necessary for axis specification. Thus, cells in the SAM continuously modify their gene expression patterns in accordance with their environment. In this regard, a number of receptors, protein ligands and putative transcription factors have been identified, which are essential for meristem regulation. Sequencing of the *Arabidopsis* genome has helped identify many possible components of the meristem signaling machinery such as target genes and receptor molecules, however, their potential roles in SAM development need to be identified. The signaling pathways known are just the tip of the iceberg and other pathways involving plant hormones and other morphogens may also exist. The emphasis therefore must lie on understanding the coordination between the events occurring during development and the intercellular communication between the cells in the meristem. These areas offer a great deal of challenges and promises for the future.

References

1. K. Esau, in: Plant Anatomy (2nd ed.). John Wiley & Sons, Inc., New York. (1965).
2. M.M.S. Evans, M.K. Barton, Genetics of angiosperm shoot apical meristem development, Annu. Rev. Plant Physiol. Plant Mol. Biol. 48 (1997) 673–701.
3. J.I. Medford, F.J. Behringer, J.D. Callos, K.A. Feldmann, Normal and abnormal development in the *Arabidopsis* vegetative shoot apex, Plant Cell 4 (1992) 631–643.
4. S. Poethig, Genetic mosaics and cell lineage analysis in plants, Trends Genet. 5 (1989) 273–277.
5. U. Brand, M. Hobe, R. Simon, Functional domains in plant shoot meristems, Bioessays 23 (2001) 134–141.
6. W.J. Lucas, S. Bouche-Pillon, D.P. Jackson, L. Nguyen, L. Baker, B. Ding, S. Hake, Selective trafficking of KNOTTED1 homeodomain protein and its mRNA through plasmodesmata, Science 275 (1995) 1980–1983.
7. T.A. Steves, I.M. Sussex, The structure of the shoot apex, In: Patterns in Plant Development, edn. 2, Cambridge University Press, Cambridge, (1989) 46–61.
8. P.L. Rinne, C. van der Schoot, Sympalsmic fields in the tunica of the shoot apical meristems coordinate morphogenetic events, Development 125 (1998) 1477–1485.
9. M.K. Barton, R.S. Poethig, Formation of the shoot apical meristem in *Arabidopsis thaliana*: an analysis of the development in the wild type and in the *shootmeristemless* mutant, Development 119 (1993) 823–831.
10. D.R. Kaplan, T.J. Cooke, Fundamental concepts in the embryogenesis of dicotyledons: a morphological interpretation of embryo mutants, Plant Cell 9 (1997) 1903–1919.
11. M.L. Christianson, Fate map of the organizing shoot apex in *Gossypium*, Am. J. Bot. 73 (1986) 907–916.
12. U. Mayer, R.A. Torres-Ruiz, T. Berleth, S. Misera, G. Jürgens, Mutations affecting body organization in the *Arabidopsis* embryo, Nature 353 (1991) 402–407.
13. R.A. Torres-Ruiz, A. Lohner, G. Jürgens, The *GURKE* gene is required for normal organization of the apical region in the *Arabidopsis* embryo, Plant J. 10 (1996) 1005–1016.
14. C.S. Hardtke, T. Berleth, The *Arabidopsis* gene *MONOPTEROS* encodes a transcription factor mediating embryo axis formation and vascular development, EMBO J. 17 (1998) 1405–1411.
15. T. Laux, K.F.X. Mayer, J. Berger, G. Jürgens, The *WUSCHEL* gene is required for shoot and floral meristem integrity in *Arabidopsis*, Development 122 (1996) 87–96.

16. K.F.X. Mayer, H. Schoof, A. Haecker, M. Lenhard, G. Jürgens, T. Laux, The role of *WUSCHEL* in regulating stem cell fate in the *Arabidopsis* shoot meristem, Cell 95 (1998) 805–815.

17. W.J. Gehring, M. Affolter, T. Bürglin, Homeodomain proteins, Annu. Rev. Biochem. 63 (1994) 487–526.

18. D. Jackson, B. Veit, S. Hake, Expression of maize *KNOTTED1* related genes in the shoot apical meristem predicts the patterns of morphogenesis in the vegetative shoot, Development 120 (1994) 405–413.

19. L.G. Smith, D. Jackson, S. Hake, Expression of *KNOTTED1* marks the shoot meristem formation during maize embryogenesis, Dev. Genet. 16 (1995) 344–348.

20. E. Vollbrecht, L. Reiser, S. Hake, Shoot meristem size is dependent on inbred background and presence of the maize homeobox gene, *KNOTTED1*, Development 127 (2000) 3161–3172.

21. E.Vollbrecht, B. Veit, N. Sinha, S. Hake, The developmental gene *KNOTTED1* is a member of a maize homeobox gene family, Nature 350 (1991) 241–243.

22. R.A. Kerstetter, E. Vollbrecht, B. Lowe, B. Veit, J. Yamaguchi, S. Hake, Sequence analysis and expression patterns divide the maize *KNOTTED*–like genes into two classes, Plant Cell.6 (1994) 1877–1887.

23. C. Lincoln, J. Long, J. Yamaguchi, K. Serikawa, S. Hake, A *KNOTTED1*-like homeobox gene in *Arabidopsis* is expressed in the vegetative meristem and dramatically alters leaf morphology when overexpressed in transgenic plants, Plant Cell 6 (1994) 1859–1876.

24. J.A. Long, E.I. Moan, J.I. Medford, M.K. Barton, A member of the *KNOTTED* class of homeodomain proteins encoded by the *STM* gene of *Arabidopsis*, Nature 379 (1996) 66–69.

25. K. Endrizzi, B. Moussian, A. Haecker, J.Z. Levin, T. Laux, The *SHOOTMERISTEMLESS* gene is required for maintenance of undifferentiated cells in *Arabidopsis* shoot and floral meristems and acts at a different regulatory level than the meristem genes *WUSCHEL* and *ZWILLE*, Plant J. 10 (1996) 967–979.

26. J.A. Long, M.K. Barton, The development of apical embryonic pattern in *Arabidopsis*, Development 125 (1998) 3027–3035.

27. A. Samach, J.E. Klenz, S.E. Kohlami, E. Risseeuw, G.W. Haughn, W.L. Crosby, W.L. The *UNUSUAL FLORAL ORGANS* gene of *Arabidopsis thaliana* is an F-box protein required for normal patterning and growth in the floral meristem, Plant J. 20 (1999) 433–445.

28. T. Kirch, R. Simon, M. Grunewald, W. Werr, The *DORNROSCHEN/ENHANCER OF SHOOT REGENERATION 1* gene of *Arabidopsis* acts in the control of meristem cell fate and lateral organ development, Plant Cell 15 (2003) 694–705.

29. M. Aida, T. Ishida, H. Fukaki, H. Fujisawa, M. Tasaka, Genes involved in organ separation in *Arabidopsis*: An analysis of the *cup-shaped cotyledon* mutant, Plant Cell 9 (1997) 841–857.

30. S. Takada, K. Hibara, T. Ishida, M. Tasaka, The *CUP-SHAPED COTYLEDON1* gene of *Arabidopsis* regulates shoot apical meristem formation, Development 128 (2001) 1127–1135.

31. E. Souer, A. van Houwelingen, D. Kloos, J. Mol, R. Koes, The *NO APICAL MERISTEM* gene of petunia is required for pattern formation in embryos and flowers and is expressed at meristem and primordia boundaries, Cell 85 (1996) 159–170.

32. M. Aida, T. Ishida, M. Tsaka, Shoot apical meristem and cotyledon formation during embryogenesis: interaction between the *CUP-SHAPED COTYLEDON* and *SHOOT MERISTEMLESS* genes, Development 123 (1999) 1563–1570.

33. C.W. Vroemen, A.P. Mordhorst, C. Albrecht, M.A.C.J. Kwaaitaal, S.C. de Vries, The *CUP-SHAPED COTYLEDON3* gene is required for boundary and shoot meristem formation in *Arabidopsis*, Plant Cell 15 (2003) 1563–1577.

34. B. Moussian, H. Schoof, A. Haecker, G. Jürgens, T. Laux, Role of *ZWILLE* gene in the regulation of central shoot meristem cell fate during *Arabidopsis* development, EMBO J. 17 (1998) 1799–1809.

35. K. Lynn, A. Fernandez, M. Aida, J. Sedbrook, M. Tsaka, P. Masson, M.K. Barton, The *PINHEAD/ZWILLE* gene acts pleiotropically in *Arabidopsis* development and has overlapping functions with *ARGONAUTE1* gene, Development 126 (1999) 469–481.

36. K. Bohmert, I. Camus, C. Bellini, D. Bouchez, M. Caboche, C. Benning, *AGO1* defines a novel locus of *Arabidopsis* controlling leaf development, EMBO J. 17 (1998) 170–180.

37. J.C. Fletcher, E.M. Meyerowitz, Cell signaling within the shoot meristem, Curr. Opin. Plant Biol. 3 (2000) 23–30.

38. D.N. Cox, A. Chao, J. Baker, L. Chang, D. Qiao, H. Lin, A novel class of evolutionarily conserved genes defined by *PIWI* are essential for stem cell self-renewal, Genes Dev. 12 (1998) 3715–3727.

39. D. Ostuga, B. DeGuzman, M.J. Prigge, G.N. Drews, S.E. Clark, *REVOLUTA* regulates meristem initiation at lateral positions, Plant J. 25 (2001) 223–236.

40. V. Grbi_, A.B. Bleecker, Axillary meristem development in *Arabidopsis thaliana*, Plant J. 21 (2000) 215–223.

41. P.B. Talbert, H.T. Adler, D.W. Parks, L. Comai, The *REVOLUTA* (*REV*) gene is necessary for apical meristem development and for limiting cell divisions in the leaves and stems of *Arabidopsis thaliana*, Development 121 (1995) 2723–2735.

42. R.A. Kerstetter, D. Laudencia-Chingcuanco, L.G. Smith, S. Hake, Loss-of-function mutants in the maize homeobox gene *KNOTTED1* are defective in shoot meristem maintenance, Development 124 (1997) 3045–3054.

43. J. McConnel, M.K. Barton, Effect of mutations in the *PINHEAD* gene of *Arabidopsis* on the formation of the shoot apical meristems, Dev. Genet. 16 (1995) 358–366.

44. T. Greb, O. Clarenz, E. Schafer, D. Muller, R. Herrero, G. Schmitz, K. Theres, Molecular analysis of the *LATERAL SUPPRESSOR* gene in *Arabidopsis* reveals a conserved control mechanism for axillary meristem formation, Genes Dev. 17 (2003) 1175–1187.

45. F.M. Rosin, J.K. Hart, H. Van Onckelen, D.J. Hannapel, Suppression of a vegetative MADS box gene of potato activates axillary meristem development, Plant Physiol. 131 (2003) 1613–1622.

46. N.R. Sinha, R.E. Williams, S. Hake, Overexpression of the maize homeobox gene *KNOTTED-1* causes a switch from determinate to indeterminate cell fates, Genes Dev. 7 (1993) 787–795.

47. G. Chuck, C. Lincoln, S. Hake, *KNAT1* induces lobed leaves with ectopic meristems when overexpressed in *Arabidopsis*, Plant Cell 8 (1996) 1277–1289.

48. H. Rupp, M. Frank, T. Werner, M. Strnad, T. Schmülling, Increased steady state mRNA levels of the *STM* and *KNAT1* homeobox genes in cytokinin overproducing *Arabidopsis thaliana* indicate a role for cytokinins in the shoot apical meristem, Plant J. 18 (1999) 557–563.

49. G. Frugis, D. Giannino, G. Mele, C. Nicoldi, A.M. Innocenti, A. Chiapetta, M.B. Bitonti, W. Dewitte, H.W. Oncklen, D. Marioth, Are homeobox knotted-like genes and cytokinins the leaf architects? Plant Physiol. 119 (1999) 371–373.

50. Y. Li, F. Dubois, D. Zhou, Ectopic expression of TATA-box binding protein induces shoot proliferation in *Arabidopsis*, FEBS Lett. 489 (2001) 187–191.

51. S.E. Clark, M.P. Running, E.M. Meyerowitz, *CLAVATA1*, a regulator of meristem and flower development in *Arabidopsis*, Development 119 (1993) 397–418.

52. S.E. Clark, M.P. Running, E.M. Meyerowitz, *CLAVATA3* is a specific regulator of shoot and floral meristem development affecting the same processes as *CLAVATA1*, Development 121 (1995) 2057–2067.

53. J.M. Kayes, S.E. Clark, *CLAVATA2*, regulator of meristem and organ development in *Arabidopsis*, Development 125 (1998) 3843–3851.

54. P. Laufs, O. Grandjean, C.Jonak, K. Kieu, J. Trass, Cellular patterns of the shoot apical meristem in *Arabidopsis*. Plant Cell 10 (1998) 1375–1389.

55. P. Laufs, J. Dockx, J. Kronenberger, J. Trass, *MGOUN1* and *MGOUN2*: two genes required for primordium initiation at the shoot apical and floral meristems in *Arabidopsis thaliana*, Development 125 (1998) 1253–1260.

56. S.E. Clark, R.E. Williams, E.M. Meyerowitz, The *CLAVATA1* gene encodes a putative receptor kinase that controls shoot and floral meristem size in *Arabidopsis*, Cell 89 (1997) 575–585.

57. S. Jeong, A.E. Trotochaud, S.E. Clark, The *Arabidopsis CLAVATA2* gene encodes a receptor-like protein required for the stability of the *CLAVATA1* receptor-like kinase, Plant Cell 11 (1999) 1925–1933.

58. K.U. Torii, N. Mitsukawa, T. Oasumi, Y. Matsura, R. Yokoyama, R.F. Whittier, Y. Komeda, The

Arabidopsis ERECTA (*ER*) gene encodes a putative receptor protein kinase with extracellular leucine-rich-repeats, Plant Cell 8 (1996) 735–746.

59. R. Yokoyama, I. Takahashi, A. Kato, K.U. Torii, Y. Komeda, The *Arabidopsis ERECTA* (*ER*) gene is expressed in the shoot apical meristem and organ primordia, Plant J. 15 (1999) 301–310.

60. J. Li, J. Chory, A putative leucine-rich receptor kinase involved in brassinosteroid signal transduction, Cell 90 (1997) 929–938.

61. K. Schumacher, J. Chory, Brassinosteroid signal transduction: still casting the actors, Curr. Opin. Plant Biol. 3 (2000) 79–84.

62. A. Dievart, M. Dalal, F.E. Tax, A.D. Lacey, A. Huttly, J. Li, S.E. Clark, *CLAVATA1* dominant-negative alleles reveal functional overlap between multiple receptor kinases that regulate meristem and organ development, Plant Cell 15 (2003) 1198–1211.

63. B.J. Staskawicz, F.M. Ausubel, B.J. Baker, J.G. Ellis, J.D. Jones, Molecular basis of plant disease resistance, Science 268 (1995) 661–667.

64. J.C. Fletcher, U. Brand, M.P. Running, R. Simon, E.M. Meyerowitz, Signaling of cell fate decisions by *CLAVATA3* in *Arabidopsis* shoot meristems, Science 283 (1999) 1911–1914.

65. M. Lenhard, T. Laux, Stem cell homeostasis in the *Arabidopsis* shoot meristem is regulated by intracellular movement of *CLAVATA3* and its sequestration by *CLAVATA1*, Development 130 (2003) 3163–3173.

66. E. Rojo, V.K. Sharma, V. Kovaleva, N.V. Raikhel, J.C. Fletcher, CLV3 is localized to the extracellular space, where it activates the *Arabidopsis* CLAVATA stem cell signaling pathway, Plant Cell 14 (2002) 969–977.

67. A.E. Trotochaud, S. Jeong, S.E. Clark, CLAVATA3 a multimeric ligand for CLAVATA1 receptor kinase, Science 289 (2000) 613–617.

68. J.M. Stone, A.E. Trotochaud, J.C. Walter, S.E. Clark, Control of meristem development by *CLAVATA1* receptor kinase and kinase-associated-protein-phosphatase interactions, Plant Physiol. 117 (1998) 1217–1225.

69. A.E. Trotochaud, T. Hao, G. Wu, Z. Yang, S.E. Clark, The CLAVATA1 receptor like kinase requires CLAVATA3 for its assembly into a signaling complex that includes KAPP and a Rho-related protein, Plant Cell 11 (1999) 393–405.

70. U. Brand, J.C. Fletcher, M. Hobe, E.M. Meyerowitz, R. Simon, Dependence of stem cell fate in *Arabidopsis* on a feedback loop regulated by *CLV3* activity, Science 289 (2000) 617–619.

71. H. Schoof, M. Lenhard, A. Haecker, K.F.X. Mayer, G. Jürgens, The stem cell population of *Arabidopsis* shoot meristems is maintained by a regulatory loop between *CLAVATA* and *WUSCHEL* genes, Cell 100 (2000) 633–644.

72. L.P. Yu, E.J. Simon, A.E. Trotochaud, S.E. Clark, *POLTERGEIST* functions to regulate meristem development downstream of *CLAVATA* loci, Development 127 (2000) 1661–1670.

73. S.E. Clark, S.E. Jacobson, J.Z. Levin, E.M. Meyerowitz, The *CLAVATA* and *SHOOTMERISTEMLESS* loci competitively regulate meristem activity in *Arabidopsis*, Development 122 (1996) 1567–1575.

74. P. Doerner, Plant stem cells: The only constant thing is change, Curr. Biol. 10 (2000) R826–R829.

75. S.E. Clark, Cell signaling at the shoot meristem, Nat. Rev. Mol. Cell Biol. 2 (2001) 276–284.

76. M.C. Timmermans, A. Hudson, P.W. Becraft, T. Nelson, ROUGH SHEATH 2, a Myb protein that represses *KNOX* homeobox genes in maize lateral organ primordial, Science 284 (1999) 154–156.

77. M. Tsiantis, R. Schneeberger, J.F. Golz, M. Freeling, J.A. Langdale, The maize *ROUGH SHEATH 2* gene and leaf development programs in monocot and dicot plants, Science 284 (1999) 154–156.

78. R. Waites, H.R. Selvadurai, I.R. Oliver, A. Hudson, The *PHANTASTICA* gene encodes a MYB transcription factor involved in growth and dorsoventrality of lateral organs in *Antirrhinum*, Cell 93 (1998) 779–789.

79. M.E. Byrne, R. Barley, M. Curtis, J. Arroyo, M. Dunham, A. Hudson, R.A. Matiensseon, *ASSYMETRIC LEAVES 1* mediates leaf patterning and stem cell function in *Arabidopsis*, Nature 408 (2000) 967–971.

80. E. Semiarti, Y. Ueno, H. Tsukaya, H. Iwakawa, C. Machida, Y. Machida, Y. The *ASYMMETRIC LEAVES2* gene of *Arabidopsis thaliana* regulates formation of a symmetric lamina, establishment of venation and repression of meristem-related homeobox genes in leaves, Development 128 (2001) 1771–1783

81. J.L. Bowman, J. Alvarez, D. Weigel, E.M. Meyerowitz, D.R. Smyth, Control of flower development in *Arabidopsis thaliana* by *APETALA1* and interacting genes, Development 119 (1993) 721–743.

82. D. Weigel, J. Alvarez, D.R. Smyth, M.F. Yanofsky, E.M. Meyerowitz, *LEAFY* controls floral meristem identity in *Arabidopsis,* Cell 69 (1992) 843–859.

83. J. Alvarez, C.L. Guli, X.-H. Yu, D.R. Smyth, *terminal flower*: a gene affecting inflorescence development in *Arabidopsis thaliana*. Plant J. 2 (1992) 103–116.

84. M. Lenhard, A. Bohnert, G. Jürgens, T. Laux, Termination of stem cell maintenance in *Arabidopsis* floral meristems by interactions between *WUSCHEL* and *AGAMOUS*, Cell 105 (2001) 805–814.

85. J.U. Lohmann, R.L. Hong, M. Hobe, M. Busch, F. Parcy, R. Simon, D. Weigel, A molecular link between stem cell regulation and floral patterning in *Arabidopsis*, Cell 105 (2001) 793–803.

86. M.F. Yanofsky, H. Ma, J.L. Bowman, G.N. Drews, K.A. Feldman, E.M. Meyerowitz, The protein encoded by the *Arabidopsis* homeotic gene *agamous* resembles transcription factors, Nature 346 (1990) 35–39.

Index

Abiotic stress, 70, 94, 159-61, 174, 196, 206, 296, 346, 347, 348, 350, 352, 353, 355-60
Abortifacient, 88-90
Abrus precatorius, 88
Acacia, 171
ACC synthase, 61
Acentric fragment, 340
Acetobacter, 65
Achillea ospenifolia, 100
Aconitum heterophyllum, 51, 52
Acropetal wave, 374
Actigard, 159
Actin, 35-37
Actinomycetes, 274
Actinomycin D, 10
Adapter molecules, 290
Adaptive evolution, 43
Additive phenotype, 379
Adenosine diphosphate, 133
Adrenaline, 133
Aeginetia indica, 53-55
Aegle marmelos, 197
Aequorea victoria, 34
AFLP, 80, 82, 99,100, 173, 174, 207, 290-92, 294, 296, 298, 299, 301-04
AFP-PHA, 358
ag mutants, 380
AGAMOUS (AG) gene, 174, 380
Agave amaniensis, 92, 94
Aglycones, 299
Agmatine, 62
AGO1, 373
Agrobacterium, 21, 67, 78, 81, 99, 101, 103, 175, 176, 234, 235, 375
Agrobacterium rhizogenes, 66, 67, 99-102, 139, 176, 212, 247, 285
Agrobacterium tumefaciens, 99, 101, 103, 104, 175, 176, 212, 234
Agronet, 198
Agroperlite, 208
Agropyron repens, 54
Ajmalicine, 92, 95, 120, 123
Ajmaline, 103
Albizzia, 171
Albuginacae, 157

Albugo candida, 156-59, 162, 301
Alfalfa, 61, 97, 354, 357, 358
Alfin, 358
Alginate, 81, 138, 172, 210, 211, 280, 282
Allele mining, 311
Allium cepa (onion), 133, 263
Allium sativum, 100, 133
Allium wallichii, 88
Allometry, 53
Allopolyploid, 18, 331, 341
Aloe vera, 267
Alternaria alternata, 310
Alternaria blight, 156
Alternaria brassicae, 156
Alternaria brassicola, 20, 156
Alternaria solani, 294, 304
Amaranthaceae, 158
Amaranthus caudatus, 94
Amber mutations, 63
American chestnut, 171
γ-aminobutyric acid (GABA), 61
Aminobutylcanavalmine, 61
Aminobutylhomo-SPD, 60
Aminocyclopropylcarboxylic acid, 62
Aminopropyl pyrroline, 61
Aminopropylcanavalmine, 61
Ammi majus, 88, 92, 94, 101
Amomum subulatum, 268
Amp1, 375
Amphipathic, 350
Amylose, 298
Anabasine, 103
Anacardiaceae, 244, 245, 248
Anacardium occidentale, 244, 245, 253, 257, 258
Anchusa officinalis, 123
Ancymidol, 170
Androgenesis, 1-3, 5-7, 10-13, 21, 25, 233
Androgenic, 2, 3, 5-9, 12, 13, 104, 228, 233
Anethum graveolens, 270
Anise, 270
Anisodus acutangulus, 98, 282, 286
Annona, 229, 230
Anogeissus, 187-90, 197
ANOVA, 219
ANT promoter, 378

Anthraquinone, 82, 94, 96
Anthuriums, 271
Antiasthamatic, 88
Anticarcinogenic, 90, 91, 133
Anticholesterol, 132
Anticholinergic, 89
Antidiabetic, 89, 90
Antifertility, 89
Antifreeze proteins (AFPs), 349
Antihepatotoxic, 90
Antiinflammatory, 88, 89, 91, 133
Antileukaemic, 88
Antimalarial, anti-HIV, 88
Antioxidant, 132, 133, 197, 351
Antirrhinum, 379
Antisenescence, 63
Antisense, 34, 39, 66-69, 237
Antispasmodic, 88
Antistress, 91
Antithrombosis, 133
Antitumor, 89, 91
Antiulcer, 133
Antixenosis, 310
APETALA genes, 372, 379, 380
Aphanomyces euteiches, 303
Aphanomyces root rot, 303
Apium graveolens, 269
Apocynaceae, 217
Apomixis, apomictic, 130
Apoplast, 377
Apple, 226, 262, 266, 295, 296, 299, 309, 310
Apple scab, 311
Apricot, 262
Aquaporins, 349, 351
Arabidopsis, 34, 35, 38, 48, 49 63-65,174, 195, 302, 347, 348, 350-55, 367-72, 374-77, 379, 381
Aralia cordata, 123
Arginine decarboxylase (ADC), 61-70, 104, 348, 356
ARGONAUTE1 (AGO1) gene, 373
Armoracia rusticana, 158
Arnesyl diphosphate synthase, 104
Arnica montana, 88
Aromatic plants, 261, 267
Arteannuin B, 101
Artemisia, 83, 88, 92, 100, 101,104
Artemisinin, 92, 101, 104
Artemisnic acid, 101
Arthrobacter globiformis, 348, 353, 354
Arthritis, 129, 132
Artocarpus heterophyllus, 230

Ascochyta, 295, 304
Ascochyta blight, 304
Ascorbate peroxidase (Apx), 175, 349, 351, 358
Asparagus, 97, 276
Asterad-, Asteraceae-type, 47, 49, 50, 54
'Asymmetric leaves' 1, 2 *(as1, 2)*, 379
Atherosclerosis, 132, 133
Atropa, 286
Atropa acuminata, 88
Atropa baetica, 100
Atropa belladonna, 97, 98, 101, 103, 282
Atropine, 101, 103
Azadirachta indica, 99, 100, 171
Azotobacter, 274

Bacillus thuringiensis (Bt), 176, 177, 273, 310
Bacopa monniera, 84, 88, 94
Bacterial wilt, 69
Balanites aegyptiaca, 195, 197, 200, 201
Balsamodendron mukul, 129, 133
Bambusa arundinacea (bamboo), 171, 172, 188, 190, 191
Banana, 210, 229, 234, 262, 264-66, 273, 295
Barley, 12, 21, 33, 36, 97, 294, 299, 309, 310, 354, 355
Barley gene *Cor, 14b*, 351
Barley yellow dwarf, 294
Barley yellow mosaic, 294
ß-carotene, 150
B-chromosome, 331, 337
ß-element, 131
Berberine, 95, 96, 123, 124
Beta vulgaris, 92, 93, 285
Betacyanins, 94
Beta-glucan, 95
Betaine synthesis, 174
Betalains, 92
Betelvine, 268
Betula pendula, 171
Bioprocesses, 79, 117
Bioreactor, 117-125, 137, 139
Bio-safety, 169, 170, 177
Biotic stress, 19, 95, 156, 159,160,162, 191, 212, 346
Black pepper, 268
Blackleg, 156
Blackmold resistance, 310
Black-plum, 197, 202
Bombyx mori, 206
Boraginaceae, 158
Boswellia, 130, 133

Botrytis, 96
Bradyrhizobium, 274
Brassica, 7, 8, 11, 18, 21, 37, 144-146, 148, 151, 156-62, 295, 297, 356
Brassica aegyptiaca, 198-202
Brassica campestris, 18, 19, 21-24, 98,145, 146, 157-161, 302
Brassica campestris-oleracea, 302
Brassica carinata, 18-20, 22, 24, 157-160
Brassica carteri, 133
Brassica juncea, 18, 19, 22-24, 144-48, 151, 157-62, 301, 302, 356, 357
Brassica napus (rape seed), 2, 3, 5, 6, 8, 11, 18-25, 145-150, 157-162, 301, 302, 357
Brassica nigra, 18, 19, 21, 157, 158
Brassica oleracea, 18, 19, 157-160, 302
Brassica rapa (turnip rape), 301, 302
Brassica serrata, 133
Brassica spinenscens, 160
Brassica tournefortti, 162
Brassinolide, 377
BRASSINOSTEROID INSENSITIVE 1 (BRI1) gene, 376, 377
BRI1 kinase, 377
Bridge fragment, 339, 340
Broccoli, 276
Bronowski gene, 146
Bryophyllum, 50, 55, 375
Bunium persicum, 270
Burseraceae, 129, 130, 133
Butanedioic acid mono (2, 2-diemthyl-hydrazide) diaminozide, 172
t-butyl hydroperoxide, 359

3-carene, 131
Citrus sinensis × Poncirus trifoliata, 230
Cabbage, 263
Cadaverine, 63, 103
Caesin kinase, 63
Caffeic acid, 175
Calamus, 173
Calathea, 271
California bay plant, 150
Calmodulin, 34, 353
Calystegia sepium, 103
Cameleon, calcium indicator, 35, 38
Campanula, 341
Camphor, 269
Camphorene, 131
CaMV 35S promoter, 65-67, 356, 357
Canavalmine, 61

Candida albicans, 95
Canola, 144, 147, 150
Capparis decidua, 197
Capric acid, 149,151
CAPS, 290, 291, 296, 301, 304, 308, 311
Capsaicin, 95
Capsella bursa-pastoris, 49, 54, 158
Capsicum frutescens, 95
N-carbomyl putrescine, 62
Cardamom, 268
Cardiac glycosides, 89
Cardiac lipidosis, 145
Cardioprotective, 133
Cardiovascular, 90
Carica papaya, 230, 232, 235
Carica papaya × C. cauliflora, 232
Carnation, 298
Carotenoids, 150, 285, 306
Carrot, 12, 66, 97, 263, 359
Caryophyllaceae-type, 47
Caryophyllene, 131
Cashew nut, 244-46, 248, 250-53, 255-59
Cassava, 210, 295
Castanea dentata, 171
Casuarina, 171, 175
Catalase, 349, 355, 356
Catharanthene, 92, 96, 123
Catharanthus, 65, 82, 85, 88, 92, 94-96, 101, 103, 104, 120-25, 267, 282, 286
Cauliflower, 263
Cauligenic/caulogenesis, 43, 229
Cauline, 374
CDPK, CDPK isoforms, 353, 356
Celastrus paniculatus, 197, 202
Celery, 97, 269, 276
Central cell, 31-34, 36, 38
Ceratonia siliqua, 171
Ceratophyllaceae, 53
Ceratophyllum demersum, 53, 54
Cercospora zeamaydis, 294
Cereal cyst nematode, 296
Chalara elegans, 295
Chaperon, 350
Chenopodiaceae, 158
Chenopodiate/Chenopodiaceae-type, 47, 49
Chenopodium, 85
Chinese cabbage, 276
Chinese cassia, 269
Chitinases, 174
Chitosan, 95
Chlamydomonas, 61

Chlorella, 61

2-chloroethyltrimethyl ammonium chloride (CCC), 172

Chlorophytum borolivilianum, 267

Cholestane, 131

Cholesterol, 92, 124, 125, 129, 132-34, 157

Choline, 348

Choline dehydrogenase (CDH), 348

Choline monoxygenase (CMO), 348

Choline oxidase (COD), 348, 357

Chrysanthemum, 272

Cicer (chick pea), 295, 346, 304

Cichorium/chicory, 97, 99, 100, 282

Cinchona, 92, 94, 95, 101, 103, 282

Cinchonidine/cinchonine, 101

Cinnamomum, 269

Cinnamoyl, 123

Citrus, 94, 97, 202, 228, 229, 231, 233, 234, 261, 262, 266

Citrus acida, 230

Citrus aurantifolia, 230, 232

Citrus aurantium, 230, 232

Citrus berryi, 129

Citrus clementina, 232

Citrus grandis, 230, 232

Citrus halimii, 230

Citrus incisa, 133

Citrus jambhiri, 230, 232, 235

Citrus ledgeriana, 101

Citrus limetoides, 230

Citrus limon, 195, 197-202, 230, 232

Citrus madurensis, 230

Citrus microcarpa, 232

Citrus nobilis, 232

Citrus paradisi, 230, 232

Citrus reticulata, 230, 232, 235

Citrus sinensis, 230, 232, 235

Citrus unshiu, 232

Cladosporium fulvum, 294

Clavata mutants, 376

Clerodendrum, 89

Clitoria ternatea, 97

Clove, 270

Club root, 156

CoA ligase, 175

Cochliobolus sativus, 294

Cocoa, 295

Coconut/ coconut water, 18, 149-151, 209, 231, 232

CodA, 353

Codonopsis pilosula, 100

Coffee, 245, 297

Coix, 33

Colchicine, 2-4, 6, 24, 25, 332

Cold stress/tolerance, 234, 300, 353

Coleus, 89, 95, 282, 285

Colletotrichum, 295

Combretaceae, 197

Commiphora, 129-33, 135-139, 142, 267

Conessine, 123, 125

Coniferyl alcohol, 96

Constitutive CLV3 signalling, 378

Convolvulaceae, 158

Convolvulus arvensis, 375

Coptis japoinca, 123, 124

Copulus, 187

Cordyline, 271

Coriandrum sativum, 151

Corky root rot, 304

Corn earworm, 303, 310

Corpus, 367-69, 374

Corylus avellana, 95

Coryphantha, 375

Cosmid, 304

Cotton, 175, 297, 299, 358

p-Coumaric acid, 175

Crassula multicava, 51

CRD analysis, 219

Crepenylic acid, 151

Crepis, 151

Crocus sativus, 85, 270

Cronartium ribicola, 306

Cruciferae-type, 47

cry, 176

Cryopreservation, 78-81, 97, 98, 170, 171, 206, 210, 211, 233, 278-286

Cucumber mosaic virus, 304

Cudrania tricuspidata, 121, 122

Culms, 172

Cuphea, 151

Curcuma, 268

Curdlan, 95

Cyanobacterium, 172

Cyclin genes, 33

Cycloheximide, 352

Cysteine, 9, 11

Cytokine, 375

Dactylorhiza maculata, 54

Dahlia, 272

Dalbergia, 171, 172

DAMD primers, 207

Datura, 65, 92, 100, 103, 286

Datura candida hybrid, 101

Datura ferox, 101

Datura innoxia, 1, 5, 21, 89, 95, 97, 98, 101, 104, 282

Datura metel, 9

Datura quercifolia, 102

Datura stramonium, 96, 101, 102, 104, 282

Datura wrightii, 101

Daucus carota, 52, 92

DDRT-PCR, 293

Decanol, 125

Decarboxylases, 65

Dehydrins, 350

Dehydrogenase, 61

Dendrocalamus, 171, 172, 188, 191

Dephosphorylation, 63, 353

de-sanguinarine, 93

Desaturase, 149, 150

Diamineoxidase (DAO), 61, 65, 66, 68, 69

Dibutylphthalate, 125

Dicamba, 231, 232

Dieffenbachia, 271

α-difluoromethylarginine (DFMA), 64

α-difluoromethyllysine (DFML), 64

α-difluoromethylornithine (DFMO), 64, 69

Digenic, 158, 162

Digitalis, 97

Digitalis lanata, 92, 121, 122, 282

Digitalis obscura, 100

Digitalis thapsi, 282, 285

Digitoxin, 92

Dimethylsulfoxide (DMSO), 81, 125, 280, 281, 285

Dioscorea, 82, 85, 281, 286

Dioscorea alata, 97, 98

Dioscorea balanica, 98, 282

Dioscorea bulbifera, 86, 89, 94, 95, 98, 100, 282, 284

Dioscorea caucasia, 282

Dioscorea cayenensis, 100

Dioscorea deltoidea, 92, 95, 97, 281-83

Dioscorea floribunda, 97, 98, 281, 282

Dioscorea rotundata, 100

Diosgenin, 82, 92, 95, 96, 100, 197, 202, 282

Diplocarpon, 295

Diplotaxis, 159, 162

Diterpenoids, 131

Diuraphis noxia, 296

Diuretic, 90

DL-*β*-phenyllactic, 95

DNA chips, 174

DNA finger printing/ marker, 169, 193, 207, 212, 301, 306, 307

Dolichos, 331

Dominant marker, 290

Double haploids, 18-22, 25, 147-149, 300

Double low, 144, 147

Downy mildew, 157, 159, 162

d-pseudoephedrine, 92

drn-D (Dornroschen) gene, 372

Drosophila, 341, 373

DTT, 208

Duboisia, 100

Duplicate gene, 158, 162

Durable stem rust, 293

Dysaphis devecta, 296

Ectopic, ectopic primordia, 372, 375, 379

Eicosenoic, 145

EIF2c, 373

Electrofusion, 32, 209

Electroporation, 81, 235

Elettaria cardamomum, 268

Elicitors, Elicitation, 93, 95, 96, 281

Elite genotypes, 98

Embryogenesis, 4, 6, 9-13, 22-25, 31, 34, 36, 43, 44, 46, 49, 51-53, 55, 56, 148, 258, 366, 369-71, 373, 375, 380

Embryogenic, 4-13, 22, 23

Embryoid, 3, 4, 11, 21, 43, 44, 49, 50-52, 55, 56

Embryoidogeny, embryoidogenesis, 43, 44, 49, 51-53, 55

Embryonal suspensor mass (ESM), 148, 173

Encapsulated, Encapsulation, 81, 98, 210, 211, 228, 234, 237, 280-84

Endoparasitic, 69

Endophytes, 170

Endosperm, 31, 33, 34, 36-38

1-ephedrine, 92

Ephedra, 92

Epicotyl, 80, 171

Epigenetic, 161, 231

Epiphylly/epiphyllous, 375

Epiphysis, 44, 46, 49-51, 53, 55, 56

Epistasis, 376, 377, 380

ERECTA (ER) gene, 376, 377

Eriobotrya japonica, 232

Erodium cicutarium, 46

Errera's law, 44

Eruca sativa, 159, 162

Erucic acid, 20, 21, 144, 145, 147-50, 157, 302

Erysiphe cruciferarum, 156
Erysiphe graminis, 292, 294
Erysiphe polygoni, 295
Escherichia coli, 63, 65, 353-55
Eschscholtzia, 95, 96
Ethephon, 134, 135
Ethylene, 67, 174, 306
Ethylene glycol, 281
Eucalyptus, 98, 99, 171, 175, 184, 188, 191, 192, 282, 298, 306
Eucalyptus camaldulensis, 188, 192
Eucalyptus citriodora, 188, 192
Eucalyptus grandis, 306
Eucalyptus tereticornis, 188, 192
Eucalyptus urophylla, 306
Eugenia spp., 232
Euonimus macroptera, 50
Eupatorium cannavulgaris, 65
Euphoria longan, 232

Faba bean, 35
Fabaceae, 331
Fatty acids, 144, 145, 157
Feeder cells, 33
Feijoa, 228, 232, 233
Fertility restoration, 297
Festuca, 342
Ficus, 171, 271
Flavonoids, 80, 131, 157
Flooding tolerance, 300
Floral meristems, 379
Fluridone, 11
Foeniculum vulgare, 270
Forage legume, 357
Frankia, 175
Fraxinus angustifolia, 226
Freeze preservation, 97, 281, 352, 355, 357
Frost-hardy, 156, 279
Fructan synthase (Sac B), 348
Fructans, 349
Fucus serratus, 34
Furanocoumarins, 80
Fusarium, 69, 294, 295, 303, 304
Fusarium head blight, 293
Fusiform rust disease, 174
Fusogen, 209

Galangal, 269
Gall midge biotypes, 296, 300, 308
Gallium mollugo, 82
Garcinia mangostana, 229, 230

Gelatin, 81
Gelrite, 81
Gemmorhizogenesis, 43, 49
KNOTTED gene, 371
Gene bank, 210, 212, 290
Gene pyramiding, 289, 303, 307, 358
Gene silencing, 237
Gene tagging, 312
Generative cell, 3-6, 23
Genetic diseases, 291
Genetic fidelity, 78, 99, 201
Genetic markers, 207, 289
Genetic mosaics, 368
Genomics, 311, 360
Genotypic apomixis, 311
Gentiana scabra, 282
Gerbera, 271
Geum urbanum, 47, 48
Gingenoside, 122
Ginger, 268
Ginseng saponin, 122, 123
Ginsenoside, 102, 103, 285
Globodera, 296
Glucobrassicanapin, 146
Gluconapin, 146
Glucosinolates, 144, 146-48
Glutamine, 2, 135, 209, 231
Glutamine synthetase, 353
Glutathione reductase (GR) (gor), 175, 349, 351, 354, 356, 358
Glutathione synthetase (GS) (gshII), 349, 351, 354, 356
gly I, 351
Glyceraldehyde 3-phosphate dehydrogenase, 359
Glycine betaine, 347-349, 353, 357
Glycine max, 98
Glycoalkaloids, 93, 299
Glycoprotein, 11, 146
Glycosides, 80, 102
Glyoxalase/ glyoxalate, 349, 351
Gmelina, 171
Gourds, 263
Graminad, 46, 47, 49, 51, 54
Grapes, 261, 262, 295, 299
Green beans, 263
Green fluorescent protein (GFP), 34
Groundnut, 18, 25, 156, 346
Guanylhydrazone, 64
Guargum, 81
Guava, 226, 228, 229, 233, 234, 262, 266, 273
Guazuma crinita, 97

Guggul, 129, 130, 132-34
Guggulipid, 132
Guggulster, 138
Guggulsterol-I, II, III, IV, V, VI, 130, 131, 138
Guggulsterone, 129-33, 137, 139
Gum-resin, 129-134, 138
GURKE gene, 370
Gus, GUS, 4, 35, 36, 67, 212, 234-36
Gymnadenia conopsea, 49
Gymnema sylvestris, 267
Gynogenesis, gynogenic, 21, 209, 228

5-hydroxyferulic acid o-methyltransferase, 175
6-ß-hydroxy hyoscyamine, 103
8-hydroxyquinoline, 249
H_2O_2, 351, 356
Hairy roots, 95, 100, 101, 103, 212, 284, 304
Haploids, 1, 5, 6, 11, 19, 21, 209, 228
Heat-shock proteins (HSPs), 7, 174
Helianthus, 61, 257
Helicoverpa zea, 296, 303
Heliminthosporium turcicum, 294
Hepatitis B, 89
Heracleum, 374
Herbicide resistance, 176, 206, 234
Heterodera glycines, 296, 302
Heterophasic, 43, 44, 49, 50, 55, 56,
Hevea brasiliensis, 104
Hingota, 195, 197
Holarrhena, 89, 117, 123, 125
Holoparasitic, 55
Holostemma annulare, 97, 282
Homeodomain/protein, 371, 372, 377, 380
Homophasic, 43, 44, 49, 50, 55, 56
Homo-SPD synthase, 65
hpt, 235
Hydrophilic COR proteins, 350
Hydroxycinnamic acids, 60
Hydroxyphenols, 257
Hyoscyamine, 92, 95, 100-02, 104
Hyoscyamus albus, 96, 100, 102, 103
Hyoscyamus desertorum, 102
Hyoscyamus muticus, 97, 102, 103
Hyoscyamus niger, 5, 9, 10, 65
Hyoscyamus × gyorffyi, 102
Hyper-cholesterolemia, 132
Hyperlipidemia, 132, 133
Hyperplasia/Hypertrophy, 157
Hypocholesterolemic, 132
Hypoglycemic, 133
Hypolipidemic, 130-33

Hypophysis, 44, 46, 47, 49-51, 53, 55, 56

IAA-oxidase, 9
Idioblast, 146
IHSP, 357
Imidazoline, 20
Immobilization, 97, 138, 139
Indian bdellium, 129
Indian long pepper, 268
Indole alkaloids, 85, 92, 124
Introgression/introgressed, 160, 162, 304, 310, 347
Inula racemosa, 133
inversion heterozygosity, 331, 341
Ipomea batatas, 98
ipt, 375
Ischemic, 132
Isonicotinic acid, 159
Isopentenyltransferase (ipt), 375
Isoplexis canariensis, 89
Isoproterenol, 133
Isothiocyanates, 146
Isozymes, 99, 206, 207, 304
ISSR-ISSR primers, 207, 295, 301; 303

Jaccard's coefficient, 82
Jaceosidin, 95
Jack pine, 174
Jamun, 195, 197
Jasmonate/Jasmonic acid, 95, 96
Java long pepper, 268
Jojoba, 149
Juglans, 257

Kaempferia, 269
Kala zira, 270
Kale, 276
Kasturi turmeric, 268
Kentucky bluegrass, 311
Kinase-associated protein-phosphatase (KAPP), 377
Kinases, 349, 360
Kinks, 189, 193
Klebsiella, 274
KN1 gene, 370-72, 375
KNAT1, 2, 375, 379
KNOX genes, 371, 379

Larch, 175
Large cardamom, 268
las (lateral suppressor), 374

Late blight of potato, 306
Late embryogenesis (LEA) proteins, 349, 350, 356
Lathyrus, 342
Lauric acid, 149-51
Lauryo-ACP thioesterase, 150
Lavendula angustifolia, 269
Laxative, 89
LDC (lysine decarboxylase), 103
L-DOPA, 93, 94
Leek lettuce, 276
Leishmania, 64, 65
Lentil, 65, 295
Lepidine, 92, 94
Lepidium, 92, 94
Leprosy, 90
Leptosphaeria maculans, 156, 160, 161, 295, 302
Leucaena hybrids, 188
Leucine zipper, 349
Leucine-zipper motif, 352
Leucoderma, 88
Leveillula tourica, 294
Lignan, 96
Lignin, 62, 175, 176
Lilium, 341
Limonene, 131, 132
Linoleic/linolenic acid, 21, 145, 148, 149, 302
Linseed, 149
Lipid peroxides, 132, 133
Lipoproteins, 132, 133
Liposome, 81, 234
Liquidambar styraciflua, 175
Litchi, 230, 231, 261, 266
Lithospermum erythrorhizon, 82, 87, 94, 96, 103, 121-23
Loblolly pine, 171, 174, 176, 306
Loganin, 95, 124
Lolium, 342
Loose smut, 293
LRR-receptor/kinases, 376, 377
Lubimin, 96
Lupinus polyphyllus, 96
Luteovirus, 294
Lycopene, 305
Lycopersicon, 304, 310
Lysine, 103, 145
Lysine decarboxylase, 64

MADS box gene, 374, 380
Magnaportha grisea, 300, 308
Maize viruses, 294

Maize *(Zea mays)*, 7, 12, 21, 31-38, 65, 68, 294, 296, 299, 311, 353, 368, 370, 371, 374, 379
Mangifera indica (mango), 228, 230-32, 234, 236, 262, 273
Mangosteen, 229
Mannitol 1-phosphate dehydrogenase (MtlD), 348
Map-based cloning, 304
MAPK kinases, 352
Marjoram, 269
Marjorana hortensis, 269
Marker assisted selection/breeding, 289, 300, 302, 303, 307-09
Maytenus emarginata, 197
McFISH, 342
Medicago sativa, 98, 357
Meloidogyne, 294, 296, 305
Melon, 295
Memory vitalizer, 88
Mentha, 97, 98, 269
Metallothionein, 11
Methionine, 61, 62, 67, 145
Methyl jasmonate (MeJa), 95, 96, 103
Methyl viologen, 175, 359
mgo genes, 376
Microcyclus ulei, 295
Micrografting, 244, 248
Microinjection, 35, 36, 81, 234, 369
Microprojectile, 36, 234-36
Microsatellite markers, 290, 300, 301
Microshoots, 171, 246-48, 258
Microspore embryogenesis, 7, 8
Microtubers, 266
Mildew, 311
Milkwhey, 82
Minimata disease, 176
Misexpression, 379
Mitotic complements, 332
Molecular markers, 78, 80, 81, 98-100, 161, 206, 289, 291, 300-03, 308-12
Monoclonal antibody, 8
Monofluoromethylarginine (MFMA), 64
Monofluoromethylornithine (MFMO), 64
Monogenic, 158, 162, 300, 302
Monoterpenes, 131, 132
Monozygotic, 43, 49
Morina kokanica, 48
Morinda citrifolia, 94, 96
Moringa oliefera, 100
Morphactin, 134
Morus (mulberry), 206-12
Morus alba, 208-10, 212

Morus australis, 208
Morus bombycis, 208, 211
Morus cathayana, 208
Morus indica, 208, 212
Morus laevigata, 207, 208
Morus lhou, 208
Morus multicaulis, 208, 211
Morus nigra, 210
Morus serrata, 208
mtlD, 353
Mucuna, 94
Mukulol, 131
Multigene family, 371
Multimeric complex, 377
Musa, 228-30, 232, 233, 236
Mushroom, 355
Mustard, 97, 147, 356, 357
Mutagenesis, 147-49, 151, 159
Myb and Myc like protein, 349, 352, 379
Mycorrhizae, 186, 274
Myocardial, 132, 133, 145, 157
Myrcene, 131, 132
Myrciaria cauliflora, 232
Myricyl alcohol, 131
Myristic acid, 149, 151
Myristica fragrans, 151
Myrosinase, 146
Myrtaceae, 195, 197

Naphthoquinone, 97
Napin, 11
Napoleiferin, 146
Nelumbo, 48, 53, 54
Nelumbonaceae, 53
Neomycin phosphotransferase (npt II), 175, 234-36
Neurospora, 65
Nicotiana, 82
Nicotiana plumbaginifolia, 282
Nicotiana rustica, 8, 104
Nicotiana sylvestris, 65, 282
Nicotiana tabacum, 3-5, 8, 9, 11, 94, 97, 98, 103, 104, 123, 124, 282
Nicotine, 62, 66, 104
Nifedipine, 133
Nitella, 61
Nitriles, 146
Nitrogen-fixing bacteria, 175
NO APICAL MERISTEM (NAM) gene, 372
nopaline, 234
NorSPD/NorSPM, 60, 61
Norway spruce, 176

Nothapodytes foetida, 89, 94
Nucellar/embryos, 43, 44, 50, 51, 231, 248, 249

Obesity, 129, 132
Oenothera biennis, 46
Oidium lycopersicum, 294
Oil palm, 295, 298
Oilseed, 145, 149, 151, 156, 301, 356
Okazaki fragments, 63
Okra, 263
Olea europe, 282
Oleanane triterpenes, 96
Oleanolic acid, 132
Oleic, 148, 149
Oleic acid, 20, 21, 145, 148-50, 302
Oleiferous, 25, 144, 156
Oleogum-resin, 129, 130, 133
Oligogalacturonide, 96
Oligonucleotide fingerprinting, 99, 100
Omega-3-desaturase, 302
Onagrad, 49-52, 54, 55
Ononitol, 349
Orange, 97
Orchidaceae, 53
Orchids, 271
Oregano, 269
Organoleptic, 304
Origanum vulgare, 269
Ornithine, 61, 62, 67
Ornithine decarboxylase *(odc)*, 61-70, 104, 348, 359
Orobanchaceae, 53, 55
Orseolia oryzae, 296, 308
Orthologue, 379
Oryza sativa (Rice), 36, 65-69, 97, 211, 245, 291, 296, 297, 299, 300, 307, 308, 353, 354, 356
Osmolytes, 60, 347, 348, 353, 360
Osmoprotectant, 359
Osmoregulation, 351
Osmotic stress/osmotic tolerance, 233, 348, 351, 352, 355
Osmotin, 348, 349, 357, 358
Osteoarthritis, 133
Ostrinia nubilalis, 296
Oxidation, 244, 350
Oxidative stress, 174, 175, 355, 357
Oxygenation, 119, 120
Ozone, 351

Paclitaxel, 124
Paeonad, 46, 47
Paeonia, 49-51, 341

Palm, 18, 149, 150

Palmitic, 21, 149

Panax ginseng, 97, 98, 100, 102, 103, 122, 124, 282, 284-86

Panax notoginseng, 121, 123

Panax quiniquefolium, 100, 282

Papaver, 93, 96, 282, 284

Papaya, 228, 229, 231, 133, 234, 262, 273

Papaya ring spot virus, 235

Parsley, 269, 276

Particle bombardment, 81, 175, 235

Patchouli, 267

Paulownia fortunei, 188, 192

PCR/markers, 81, 284, 285, 290, 291, 304, 308, 310, 312

Pea, 65, 66, 68, 295, 299, 303, 304, 355

Peach, 273, 299, 300

Pear, 273

Pearl millet, 295, 346

Pecan, 99

Pedigree selection, 147

Peganum harmala, 85, 89, 103

Pepper, 295

Peppermint, 269

Perilla frutescens, 123

Peronospora parasitica, 156, 157

Peroxidase, 173, 207

Petroselenic acid, 151

Petroselinum crispum, 269

Petunia, 297, 300, 372, 374

PgEMB22, 27 and 29, 174

Pgq (Panax hybrid), 103

PHANTASTICA (PHAN), 379

Pharbitis nil, 65

Phaseolus, 95, 331

Phenetic dendrogram, 82

Phenolics, 9, 93, 157, 158, 225, 244, 246, 250, 256, 257

Phenylalanine ammonialyase (PAL), 175

Phenylpropanoids, 63

Philodendron, 271

Phloroglucinol, 217, 218, 222, 226

Phoma lingam, 20

Phosphorylation, 6, 8, 63, 351, 353,

Photooxidative, 351, 355, 356, 359

Phyllanthus, 89, 267

Phyllochora herberi, 295

Phyllostachys, 171

Phytoalexins, 95, 159

Phytohemagglutinin, 358

Phytomers, 366

Phytophthora, 96, 294, 295, 306

Picea, 174, 258

Picrorhiza kurroa, 97

Pigeonpea, 346

Pimpinella anisum, 270

Pine blister rust, 306

Pineapple, 262, 266, 273

ß-pinene, 131, 132

PINHEAD (ZLL/PNH), 373

Pinitol, 347

Pinoresinol, 96

Pinus, 299

Pinus banksiana, 174

Pinus elliottii, 306

Pinus lambertiana, 306

Pinus palustris, 306

Pinus taeda, 174, 306

Piper barberi, 268

Piper, 268

Piperad-type, 47

Plagiotrophy, 196, 201, 225

Plantago major, 100

Plantago ovata, 89

Plasmodiophora brassicae, 156, 160, 295, 302

Plastochronic, 367

Platantera bifolia, 49, 48

Plectonema boryanum, 172

Pluchea lanceolata, 87

Plum, 273

Plumbago, 90

Pluronic PE 6100, 120

Poa pratensis, 311

Poaceae, 53, 55, 190

Podophyllotoxin, 123, 124

Podophyllum hexandrum, 121, 123, 124

Polemonium, 48, 49

POLTERGEIST (POL) gene, 378

Polyamine oxidase (PAO), 61

Polyamines, PAs, 60-70, 100, 231, 259, 348, 356

Polyethylene glycol (PEG), 32, 174, 235, 348

Polygonum, 46, 282

Polypropylene glycol 1025 and 2025, 120

Polyvinylpyrrolidone (PVP), 9, 249, 250, 255

Pome, 229

Pomegranate, 266

Poncirus trifoliata, 230, 235

Populus/poplar, 95, 171, 175, 188, 191,193

Populus alba, 175

Populus canescens, 175

Populus deltoides, 188, 189, 193

Populus euphratica, 188, 193

Populus tremula, 175
Portulacca grandiflora, 94
Portulaceae, 158
Potato, 66, 67, 210, 263, 264, 266, 267, 275, 294, 296, 299, 355, 358, 359, 374
Potato virus, 294, 304
Potexvirus, 294
Potyvirus, 294
Powdery mildew, 156, 293
Pratylenchus neglectus, 296
Precursor, 81, 95, 122, 124, 125
Pregnane, 131, 133
Pregnenolone, 134
Primula obconica, 98
Progoitrin, 146
Proline, 81, 249, 280, 347-49, 357
Prosopis cineraria, 171, 197, 226
Protein kinase, 351, 352
Protocorm, 53, 97
Protoderm, 44, 49, 50, 53, 55
Prunus, 266
PRV cp gene, 234
Pseudomonas, 294, 295
Psidium guajava, 230
Psoralea corylifolia, 90
Psoriasis, 90
Puccinia helianthi, 309, 310
Puccinia hordei, 294
Puccinia melanocephala, 295
Puccinia recondite, 293
Puccinia striiformis, 292, 294
Puccinia substriata, 295
Pulse treatment, 200
Pumpkins, 263
Puromycin, 10
Purseglove, 335
PUT-methyl transferase (PMT), 62
PUT-N-methyltransferase., 65
Putrescine, PUT, 60-63, 66-70, 123, 249, 349, 359
PVS$_2$, 280, 281, 284, 285
Pyramid genes, 307-10
Pyrenophora graminea, 294
Pyricularia grisea, 292
Pyrrolidine alkaloids, 104
D1-pyrroline-5-carboxylate synthetase (P5CS), 348
Pyrroline/pyrrolidine ring, 61, 62

QTA, 306
QTL, 173, 212, 289, 300-03, 306, 310, 311, 347
Quaking aspen, 175
Quercetin, 131

Quercus, 225, 226, 257
Quinidine, 101
Quinine, 89, 92, 95, 101

Rab16A, 356
Radiata pine, 170, 173
Ralstonia solanacearum, 69, 295
Ramosus genes, 304
Ranunculaceae, 51, 52
Ranunculus sceleratus, 50, 51
RAPD, 80, 82, 99, 100, 173, 174, 207, 282, 284, 285, 290-94, 296-304, 306, 309
Rapeseed, 18, 144, 146, 150, 151, 157
Raphanus sativus, 158
Raspberry, 266
Rauwolfia, 90, 93, 97, 103
Red fluorescent protein, 37
Rehmannia sp., 100
Remeristemization, 375
Reporter gene, 31, 35
Reserpine, 95
Resin/ducts, 129, 130, 133, 137
rev (revoluta) gene, 374
RFLP, 12, 82, 100, 161, 290-92, 294, 296-303, 306, 308, 310, 311
RGA-CAPs, 294, 300
Rhamnus, 85
Rheumatoid, 132
Rhizobium, 60, 274, 186, 299
Rhizoctonia solani, 292
Rhododendron, 257
Rhodotorula rubra, 96
Rhus typhina, 248
Rib meristem, 369
Rice stripe, 292
Rice tungro, 300
Rice yellow mottle virus, 292
Richinolic acid, 151
Ring spot virus, 234
Robinia pseudoacacia, 170
Rorrippa islandica, 158
ROS scavengers, 360
Rose, 274, 295, 299
Rosmaric acid, 285
Rosmarinic acid, 95, 123
Rough lemon, 235
Rubber, 295
Rutaceae, 195, 197
Rynchosporium secalis, 294

'Syn' seeds (artificial seeds), 78, 81, 97, 139, 172, 210, 228, 234, 245

NaCl/salt-stress/tolerance, 172, 233, 296, 297, 353, 355-59
Sabinene, 131
Saccharomyces cerevisiae, 63, 95, 355
Sacred basil, 269
Saffron, 270
Sage, 269
Salai guggul, 130
Salvia officinalis, 269
SAM, 367, 369, 371-75, 379, 380
SAM decarboxlase, 62-69, 356
SAMdc (S-adenosylmethionine), 61
Samdc/spd syn, 70
Samdc-odc, 66
SAM-S (S-Adenosyl Methionine Synthetase), 174
Sandalwood, 61, 171, 172, 174
Sanguinaria canadensis, 96
Sanguinarine, 96
Sap protein, 207
Sapogenin, 92, 94, 197
Saponin, 102
Saturated linkage maps, 291, 300
SCAR marker, 290, 291, 293, 294, 296-99, 302, 309, 311, 312
Scavenger proteins, 351
Schizaphids graminum, 296
Sclerotinia sclerotiorum, 96, 156, 295
Scopolamine, 92, 100-03
Scopolia, 103
Secale cereale (rye), 32, 95, 297
Secologanin, 95, 124
Secondary embryogenesis, 255
Semidwarf gene, 298, 299, 308
Senecio vernalis, 65
Septoria, 293
Sequoia sempervirens, 173
Serotonin, 103, 133
Serpentine, 92, 103, 120, 123
Sesquipedaceae, 333-35
Sesquiterpenes, 131-133
Shattering-resistance, 299
Shikonin, 82, 87, 93-95, 97, 103, 118, 121, 123
Shoot apical meristem (SAM), 366
SHOOTMERSITEMLESS (STM), 372
Shorter rotation, 98
Signal transduction, 34, 38, 63, 351, 352, 376, 377
Silver birch, 171
Silver staining, 291
Silybin, 93
Silybum marianum, 90, 93

Sinapis alba, 158-60
Sinigrin, 146
Sisymbrium officinale, 158
ß-sitosterol, 131
Small heat-shock protein (smHSP), 2
Sn-2 acyl transferase gene, 151
SNP, 82, 291, 311
Solanaceae, 47
Solanaceous, 66
Solanum, 93
Solanum aviculare, 93, 104
Solanum bulbocastanum, 306
Solanum eleagnifolium, 103
Solanum khasianum, 90
Solanum nigrum, 95
Solanum tuberosum, 10
Solanum xanthocarpum, 95
Solasodine, 93, 95, 103
Somaclonal, 99, 159, 161, 228, 233, 244, 265, 285
Somatic embryogenesis, 55, 61, 67, 82, 94, 99, 124, 136, 170-73, 201, 209, 228, 229, 231-33, 237, 248, 252, 253, 257-59
Somatic hybridization, 159, 160, 206, 209
Sorghum, 33, 257, 294, 296, 297
Soybean, 18, 65, 95, 175, 226, 295, 296, 298, 302, 303, 309, 310, 357
Soybean death syndrome (SDS), 303
Spathiphyllum, 271
Spearmint, 269
Spermidine (SPD), 60, 62, 359
Spermine (SPM), 60, 62-70, 348, 359
Spike disease, 174
Spinach, 65, 159
Sporisorium reilianum, 294
Spruce, 174, 175
Squash, 263
SSR markers, 173, 290, 292, 294, 296, 298, 299, 301-03, 310
Stearic acid, 149, 150, 303
Stearoyl ACP desaturase gene, 150
Stemphylium vesicarum, 294
Sterculia foetida, 90
Steroid/steroidal drugs/alkaloids/sterols, 80, 90, 95, 104, 130, 131, 133, 197
Stevia rebaudiana, 90
Stizolobium, 93, 94
Strawberry, 264-66
Strictocidine synthase, 104
Stripe rust, 293, 301
Suckers, 164, 228, 265
Sugar apple, 228, 233

Sugar beet, 299, 309
Sugarcane, 264, 267, 295, 306
Sugarcane mosaic virus, 294
Summer turnip rape, 145
Sunflower, 18, 297-99, 309
Superoxidase dismutase (SOD), 132, 133, 349, 351, 358
Sweet gum, 175
Syngonium, 271
Syzygium, 195, 197-02, 230, 270

Tabernaemontana divaricata, 95
TATA box, 375
Taxane, 123
Taxol, 123
Taxus, 121-25
TBP-2 protein, 376
Teak, 171
Tecomella undulata, 197, 226
Tectona grandis, 171, 225
TERMINAL FLOWER, 380
Terminalia arjuna, 133
Terminalization coefficient, 337, 340
Terpenoid indole alkaloid (TIA), 104
Tet-inducible, 67, 68
Tetraamine SPM, 61
Thalictrum rugosum, 85, 96
Thalidomide, 176
Thermophilic red algae, 60
Thermosensitive, 297
Thermo-SPM, 60
Thevetia neriifolia, 87
Thiocyanates, 146
Thuja occidentalis, 85
α-thujene, 131
Thymus vulgaris, 269
Thyroid, 133
Thyrsostachys siamensis, 172
Tilletia indica, 294
Tobacco (see also *Nicotiana*), 7, 9, 10, 35, 63-70, 104, 175, 212, 286, 294-96, 304, 306, 309-11, 348, 352, 354, 355, 359, 374, 375,
Tomatoes, 263
TOPLESS (TPL) gene, 371
Tracer dyes, 369
Tranquillizer, 90
Transferase, 66
Transformation, 81, 99, 103, 104, 169-71, 173, 175, 219, 234, 237, 258, 286
Transmembrane leucine-rich repeat (LRR), 376
Transporters, 351

Trehalase, 348, 349, 356, 358, 359
Trehalose 6-phosphate phoshatase, 356
Trehalose 6-phosphate synthase, 348, 356, 359
Triacyl glycerol, 151, 159
Triamine SPD, 61
2,4,5-trichlorophenoxy acetic acid, 136
Trichoderma virideae, 96
Trifolium repens, 98, 257, 282
Triglycerides, 132, 133
Triiodothyronine, 133
Tripterygium wilfordii, 96
Triterpenes, 132
Triticum, 51, 55
Triticum aestivum, 3, 48, 49, 51, 54
Triticum dicoccoides, 301
Tritordeum, 65, 354
Tropane alkaloids, 62, 100
Tropinone, 95
Trypanosoma, 64, 65
Tryptamine, 103, 123, 124
Trytophane decarboxylase, 103
Tuberculosis, 88
Tungro, 292
Tunica/tunica-carpus, 367-69, 372, 374
Turmeric, 268
Tylophora indica, 90, 100
Typhonium, 90

Ubiquinone, 10, 82
Ubiquitin, 8
Ubiquitin promoter, 35, 68
uidA, 176
Umbelluria californica, 150
UNUSUAL FLORAL ORGANS (UFO), 372
UPGMA, 82
Uromyces appendiculatus, 295

Valeriana jatamansi, 90
Vanilla fragrans (vanilla), 267, 269
Venturia inaequalis, 295
Vernolic acid, 151
Verticillium dahliae, 69, 96, 294
Vesicular arbuscular mycorrhiza (VAM), 173, 274
vid A, 236
Vigna, 331-33, 336, 341
Vigna aconitifolia (moth bean), 331, 333-37, 339, 340, 354
Vigna ambacensis, 331, 341
Vigna angularis (adzuki bean), 331, 332, 341
Vigna aureus, 333-35, 337, 339
Vigna candida, 339

Vigna capensis, 331
Vigna caracalla, 341
Vigna radiata (mung bean), 295, 331-35, 337, 339, 342
Vigna fisheri, 331
Vigna galbrescens, 331, 333-35, 337-42
Vigna lanceolata, 331
Vigna lancifolia, 342
Vigna luteola, 333-35, 337, 339
Vigna mariana, 331
Vigna minima, 332, 341
Vigna mungo (urd bean), 331-35, 337, 339-41
Vigna oblongifolia, 341
Vigna parvifolia, 341
Vigna repens, 333-37, 340, 341
Vigna reticulata, 331
Vigna sps, Tvnu-72, 333, 337, 340
Vigna trilobata, 331, 332, 341
Vigna umbellata (rice bean), 331-337, 340, 341
Vigna unguiculata (cowpea), 331, 333-35, 337-42
Vigna vexillata, 331, 339, 341
Vinblastine, 92
Vinca rosea, 122
Vindoline, 92
Visnagin, 101
Vitamin A, 150, 211
Vitex negundo, 91
Vitis, 257
Vitrification, 98, 211, 279-81

Water stress, 356
Wax esters, 149

Wheat, 11, 33, 35, 36, 51, 97, 292, 296- 98, 301, 309, 353, 354
Wheat streak mosaic virus, 294
White blisters, 157
White guggul, 130
White rust, 147, 156-62
White spruce, 174
Withaferin A, 93
Withania, 91, 93, 94, 99, 103, 267
Withanolide D, 93, 94, 103
Wound induced promoters, 176
Wrightia tinctoria, 226
Wrightia tomentosa, 217-26
Wsi18, 356
wus mutant, 371, 372, 374, 377-80
WUSCHEL gene, 371

Xanthan, 95
Xanthine oxidase, 132, 133
Xanthomonas, 292, 294, 295, 300, 307
Xanthotoxin, 92, 94

Yeast odc, 104
Yellow rust, 293

Zantedeshia, 271
Zinc finger family, 358
Zingiber officinale, 268
zip, 232
Zizyphus, 197, 266
Zucchini, 62